Sustainability

An Environmental Science Perspective

Sustainability

An Environmental Science Perspective

John C. Ayers

CRC Press
Taylor & Francis Group
Boca Raton London New York

CRC Press is an imprint of the
Taylor & Francis Group, an **informa** business

CRC Press
Taylor & Francis Group
6000 Broken Sound Parkway NW, Suite 300
Boca Raton, FL 33487-2742

Printed on acid-free paper

International Standard Book Number-13: 978-1-4987-5265-7 (Hardback)

Visit the Taylor & Francis Web site at
http://www.taylorandfrancis.com

and the CRC Press Web site at
http://www.crcpress.com

Contents

List of Figures

List of Tables

Abbreviations

A	Affluence, usually expressed as per capita GDP
AMD	Acid mine drainage
B	Biocapacity (hectares)
BAU	Business as usual
C	Consumption or resource use intensity
CAFO	Concentrated (or confined) animal feeding operation
CCS	Carbon capture and storage (or sequestration)
CFL	Compact fluorescent light
CO_2	Carbon dioxide
CO_2e	Carbon dioxide equivalent
DOE	U.S. Department of Energy
EF	Ecological footprint (hectares)
EIA	U.S. Energy Information Administration (a division of the DOE)
EPA	U.S. Environmental Protection Agency
EROEI	Energy return on energy invested
EU	European Union
FCA	Full cost accounting
GCC	Global climate change
GDP	Gross domestic product
GMO	Genetically modified organism
HDI	Human development index
HVAC	Heating, ventilating, and air conditioning
HVDC	High voltage direct current
I	Environmental impact
IPCC	UN Intergovernmental Panel on Climate Change
LCA	Life cycle assessment
LED	Light emitting diode
NGO	Nongovernmental organization
OPEC	Organization of Petroleum Exporting Countries
P	Population
PETM	Paleocene-Eocene Thermal Maximum
PPP	Purchasing power parity
PV	Photovoltaic
RPUD	Raw material, production, use, and disposal
SES	Socioecological system
SNF	Spent nuclear fuel
SRM	Solar radiation management
T	Technological inefficiency
TPES	Total primary energy supply
TVA	Tennessee Valley Authority
UK	United Kingdom
UN	United Nations
U.S.	United States

Acknowledgments

Thanks to the following Vanderbilt colleagues for their help, especially Jim Clarke, George Hornberger, and David Hess for reviewing most of the chapters, and Larisa DeSantis (Chapter 15), Mike Vandenbergh (Chapter 8), Dalia Abbas (Chapter 14), Andrea George (Chapter 1). Students in my "Sustainability Systems Science" class (Spring 2010, 2013, and 2015) and my freshman writing seminar "Sustainability: An Environmental Science Perspective" (Fall 2009, Spring 2012) provided feedback on early chapter drafts, as did Leslie Labruto and Jonathan Gilligan. Vanderbilt undergraduate Jacob Graham also provided feedback. This book is dedicated to my wife Mary, who patiently endured my agonizing over the subject matter and the writing of this book; and to my children Alicia and Austin, whose future I kept thinking about as I wrote.

About the Author

Dr. John C. Ayers is a geochemist who received his PhD from Rensselaer Polytechnic Institute in 1991. He is currently a professor at Vanderbilt University in the Department of Earth and Environmental Sciences and the Department of Civil and Environmental Engineering. Ayers has been an author of 36 peer-reviewed journal articles, 3 book chapters, and 2 encyclopedia entries. He has been PI or co-PI on eight National Science Foundation Grants and one grant from the Office of Naval Research. He served as associate editor for *The American Mineralogist* and for the *Geochemical Transactions of the American Chemical Society*, and is a fellow of the Mineralogical Society of America. He has been a registered professional geologist and a GIS consultant, and is a board certified environmental scientist. At Vanderbilt he has taught 16 different courses covering aspects of geochemistry, environmental science, and sustainability, and has served as Director of Graduate Studies and Department Chair.

1 Introduction

There is always an easy solution to every human problem—neat, plausible, and wrong.

H.L. Mencken
American journalist and essayist

Note: New terms highlighted in **bold** can be explored in greater detail online. Simply type the word or phrase into the address bar of your web browser.

In 2010, the Deepwater Horizon drilling rig exploded and sank, causing a massive oil spill that affected a large swath of coastline and threatened the viability of marine fisheries and the thriving tourism industry in Louisiana. A four-year drought helped trigger a Civil War in Syria in 2011. Wildfires consumed hundreds of thousands of acres of forests in the western United States in 2014–2015. On a global scale, marine fisheries have collapsed, causing rising prices and the expansion of environmentally destructive fish farming. Rising sea surface temperatures fuel more intense hurricanes, and rising sea levels combined with destruction of protective coastal wetlands exacerbate the damage from hurricanes. Global terrorism, funded by payments to Middle Eastern fundamentalist regimes for oil, is expanding in response to widening social inequalities. The ever-increasing human population is straining resource supply systems and causing biodiversity loss and local ecosystem collapses. Loss of social, environmental, and economic capital is causing societal collapse in failed states. What do these recent news stories have in common? They reveal the consequences of unsustainable policies and practices.

But there are signs of hope. Globally, the percentage of people living in poverty or dying from violence is decreasing. Adult illiteracy and child mortality are also decreasing, while life expectancy has increased. Urbanization is decreasing fertility and per capita environmental impact and increasing educational opportunities. The Internet revolution has broken down barriers between social groups and countries and increased the pace of economic and technological development. Adoption of communication tools such as smartphones is making it easier for developing countries to provide education and other services and to speed the growth of their economies with lower upfront costs and environmental impacts. These developments suggest another less pessimistic view: that scientific and social progress combined with economic development driven by globalization and the free market will lift the developing world out of poverty and increase human well-being, and that the free market will find solutions to environmental problems and substitutes for scarce natural resources.

Which of these two views of the relationship between humans and their environment is closer to the truth? Is human civilization headed toward collapse, or toward ever greater levels of human well-being? Will environmental degradation and unchecked human population growth limit or even undermine human development? Or will society reverse the current negative trends and find a way to further improve human well-being, while maintaining earth's ecosystem services that support it? What approaches would be most effective at making human society more sustainable? And what role does science play in the process?

Clearly humans have made mistakes during the rapid expansion and development of society that has occurred since the Industrial Revolution. Our global civilization is now experiencing growing pains, and we need to make informed decisions so that we and our descendants can avoid societal collapse and maintain a high quality of life. In his book *Collapse*, Jared Diamond (2005) lists eight environmental problems that contributed to the collapse of past civilizations: deforestation and habitat destruction, soil problems (erosion, salinization, and soil fertility losses), water management problems, overhunting, overfishing, effects of introduced species on native species, human population growth, and increased per capita impact of people. For modern civilization, Diamond adds four new global problems: human-caused climate change, buildup of toxic chemicals in the environment, energy shortages, and full human utilization of the earth's photosynthetic capacity. In this book we will explore these problems by asking: Are the risks associated with these problems small or large? Can we rely on new technologies to solve these problems? When we run short of a natural resource, can we expect to find a reliable substitute? We will examine potential solutions to these environmental problems, focusing on a small number that are effective and, in most cases, easy to implement.

The good news is that we already have solutions to most of today's sustainability crises. We don't have to give up everything to live sustainably, we just have to live smarter. We don't have to make radical changes; even minor adjustments can greatly decrease our environmental footprints. We have to work with corporations, since no real advances in environmental protection have come without the support of the corporate sector. And we have to consider the impact of every dollar that we spend, so that we don't support activities that are harmful to us.

This book aims to give readers the information and the tools they need to understand the causes of sustainability crises and to achieve sustainability while maintaining a high quality of life. The overriding theme is that "Earth is still our only home." Once citizens know what sustainability is and why it is important, they can elect political candidates who advocate for workable solutions to sustainability challenges. Currently in the United States, ideological thinking has somehow convinced some ultraconservatives that taking care of our home is a bad thing. Conversely, some ultraliberals or radical environmentalists tend to focus only on saving the home and not its inhabitants. To achieve sustainability, members of society must work together toward this common goal, avoiding the ideologies of the political extremes.

A home represents a good metaphor for the relationship between humanity and Earth: conscientious homeowners must continuously monitor the health of their house and spend time and money to keep it in repair so it can continue to provide services such as shelter, sanitation, and provision of drinking water. Homeowners must also take care of the inhabitants, providing them with healthy food, and high quality education and healthcare. These are all essential services that keep the home functioning properly. On the larger scale, both society and the Earth must continue to provide these services to even the least fortunate in order for human civilization to be sustainable.

This book examines the sustainability challenges our society currently faces or will face in the coming decades. It reflects the current scientific consensus stated in the 2006 Millennium Ecosystem Assessment Reports,* the IPCC Fourth Assessment Report: Climate Change 2007,† and the 2007 UNEP Global Environment Outlook 4 report (UNEP 2007). We will use an objective, evidence-based approach, not anecdotes, to identify solutions that will improve human well-being. This is the basis for the new discipline called "sustainability science."

1.1 WHAT ARE SUSTAINABILITY AND SUSTAINABILITY SCIENCE?

Here we briefly introduce the concepts of **sustainability** and **sustainability science** before examining them in greater detail in later chapters. The Brundtland Commission first defined **sustainable development** as development that "meets the needs of the present without compromising the ability of future generations to meet their needs" (WCED 1987). Sustainability promotes human well-being now and in the future through economic and social development and environmental protection. Sustainability therefore requires the balancing of economic prosperity, social fairness, and environmental responsibility.

---

* http://www.millenniumassessment.org/en/index.aspx.

† http://www.ipcc.ch/publications_and_data/publications_and_data_reports.htm.

Sustainability science uses rigorous scientific methods to better understand the relationships between humans and the environment with the purpose of promoting a sustainable future.* The National Academy of Sciences states that sustainability science addresses "the interactions between natural and social systems and how those interactions affect the challenge of sustainability: meeting the needs of present and future generations while substantially reducing poverty and conserving the planet's life support systems.†" As an emerging field, sustainability science is evolving.

Like the previous Industrial and Agricultural Revolutions, the current Sustainability Revolution will have a major, lasting impact on societies worldwide. Sustainability science can develop technologies and provide information to policy makers that will help society achieve sustainability goals. But it must also identify the physical limits to growth of the human population and the global economy; identify the limits of resilience and sources of vulnerability in socio-ecological systems; and find ways to incentivize sustainable practices. These terms and concepts will be developed in detail in coming chapters.

This is a time of rapid advances in our understanding of the relationship between humans and the environment, and the public usually only catches glimpses of these exciting developments. We will approach these issues holistically, pointing out, but not dwelling on, the problems, and focusing on the possible solutions. The goal is to convey to the reader an understanding of how science works, and how it interacts with other areas of human endeavor (politics, ethics, psychology, economics, etc.); this is particularly useful when trying to understand the development of a public consensus on controversial and complex issues. Topics like global climate change and societies' responses (or lack of responses) to it are fascinating, and the rapid growth of our knowledge in this area is truly exhilarating.

1.2 PHILOSOPHY

The premises of this book, which will be explored in detail in succeeding chapters, are that the size of a sustainable human population is limited by constraints on available resources; that fossil fuels temporarily relieved these limits by allowing us to produce more food, leading to a rapid increase in population; that human environmental impacts have exceeded biophysical limits (planetary boundaries) for some sustainability indicators, leading to climate change and the sixth great mass extinction; and that we can return to sustainability by transitioning to renewable energy sources and adopting a sustainable lifestyle to preserve essential ecosystem services and maintain a high level of human well-being. The objective is to convince readers of the need for change, and to empower them to make the necessary changes.

This book uses a utilitarian approach to developing sustainable practices. The societal objective is long-term prosperity for all, regardless of religion, gender, skin color, or country of origin; all people should have the same rights and be entitled to the same opportunities in life. Since there are physical limits to material wealth (but not human well-being), this requires the fair, but not necessarily equal, distribution of resources. Over time, developed countries must decrease their material consumption in order to accommodate increasing material consumption in developing countries while remaining within global planetary boundaries, but this can be accomplished without decreasing quality of life. Economic and social development is favored over economic growth alone.

As a society, we have an ethical obligation to make decisions that do not harm our fellow human beings and future generations. Currently, over seven billion people share the planet, and everyone will be better off if we remain aware of our responsibilities toward our neighbors and act accordingly. We should use resources wisely and share them, not squander them. Thus, an essential

* Andersson et al. The Ruffolo Curriculum on Sustainability Science: 2008 Edition. Harvard University, Dec. 2008 (www.hks.harvard.edu/centers/cid/publications/research-fellow-graduate-student-working-papers/cid-graduate-student-and-postdoctoral-fellow-working-paper-no.-32).

† PNAS Sustainability Science website (www.pnas.org/site/misc/sustainability.shtml).

ingredient for a sustainable society is protection of the commons. Not long ago bodies of freshwater, fish, and game were regarded as shared resources. A new threat to sustainability is the privatization and commoditization of the commons. In a sustainable society, access to clean air and water is a right, not a privilege. Yet today many oppose the protection of the most essential public good, the air we breathe. A few industries in a few countries emit most air pollutants, including greenhouse gases that are causing global climate change. Although air pollution has enormous economic costs, the costs are not paid by the polluters, but by society at large. Since there is no economic incentive for these industries to reduce pollution, government regulations are needed to protect the commons.

Many environmental topics are controversial. As a result, there are many conflicting claims made in the media, some of which are not supported by the data. To counter this rising tide of misinformation, some of which is promulgated by special interest groups, the book presents data from the most reliable sources available and summarizes the data in charts and tables scattered throughout the book. In most cases, the data are available for free on the Internet, and some of the data sources are updated annually.

This book emphasizes the use of data for making policy decisions. People can disagree on opinions and even interpretations, but there is no room for disagreement on facts. Education of the public on environmental issues, increased transparency, and reliance on facts should lead to consensus building and adoption of policies and lifestyles that promote the greater good. Too often in the past, decisions have been based on uninformed opinions and "gut feelings," but now nearly everyone can quickly obtain credible, well-researched information published on the Internet by reliable sources (government agencies, academic publications, NGO reports, etc.) and use it to make informed decisions. The democratization of information by the Internet has revolutionized the decision-making process and now enables citizens to question effectively the assertions of politicians and business leaders. This empowerment can help citizens, armed with accurate information, to make the best possible decisions for the future, but only if they are able to accurately assess the reliability of information.

To become sustainable, we must be open to new ideas, experiment with new approaches, find the approaches that are effective, and then implement those approaches on a large scale. This empirical approach to sustainable living is evidence-based; instead of rejecting approaches up-front based on philosophical or ideological grounds, we need to test all of the approaches and then continue to use "whatever works." As noted by Nobel-prize winner Elinor Ostrom, "recommendations of reform may be based on naive ideas about which kinds of institutions are 'good' or 'bad' and not on an analysis of performance" (Ostrom 1990). We have to approach these problems with open minds and be willing to abandon our preconceived notions.

This book takes a novel approach to sustainability by characterizing it as *a strategy for reducing risk to current and future generations.* Individuals and communities must identify essential resources (food, water, energy, education, and medical care) and examine the risks associated with their current delivery mechanisms. Impending risks to individuals and communities include climate change, resource shortages, pollution, and increases in likelihood and severity of natural disasters like earthquakes and floods. Conservation, efficient resource use, and addition of redundancy to resource supply systems can greatly reduce these risks. We will see that by abandoning the ABCs of unsustainability (automobiles and airplanes, beef, and coal) and replacing them with sustainable substitutes (bicycles and mass transit, locally grown organic vegetables, and clean, renewable energy sources), a community can become sustainable. Widespread adoption of these strategies can save many lives and increase the average level of human well-being.

1.3 OUTLINE

Although sustainability is an integrated discipline, the chapters of this book are organized by subject. However, as the book progresses, the problems covered become more complicated and interdisciplinary; each book chapter builds on what has been presented in earlier chapters.

Following this brief introduction in Chapter 1, Chapter 2 defines sustainability and human well-being, introduces some sustainability measures, and then introduces the concepts of capital stocks and flows, renewable and nonrenewable resources, and the laws of sustainability. In Chapter 3, the ImPACT identity is introduced along with an overview of its components: Population, Affluence, Consumption, and Technology. Chapter 3 also shows that the best measure of environmental sustainability, the ecological footprint, is a special case of the ImPACT identity. Chapter 4 introduces the concepts of risk and resilience. It explores the system dynamics approach of modeling resource stocks and flows in complex systems with feedback loops. It also examines how a system can become unstable and collapse, and then shows an example of societal collapse, the failed state of Haiti. Chapter 5 explores sustainable development, the properties of sustainable societies, and some of the social problems that sustainable development can help solve, including poverty and the negative impacts of globalization and urbanization.

The focus then shifts to energy and climate. Chapter 6 discusses limits to the production of natural resources, using oil as an example. This leads to a discussion in Chapter 7 of global climate change and the problems it may cause. Chapter 8 examines possible solutions to these problems and the social challenges presented by global climate change. Nonrenewable energy resources other than oil are covered in Chapter 9 and renewable energy sources in Chapter 10. Sustainable solutions to the energy-climate problem are reviewed in Chapter 11. Besides energy, human societies need water (Chapter 12) and food (Chapter 13). Energy availability and climate change both affect the ability of society to provide these resources. Further challenges to sustainability include waste and pollution (Chapter 14) and biodiversity loss (Chapter 15). We conclude with an evaluation of where society is headed (Chapter 16). The emphasis is on using the concepts of sustainability to steer civilization toward a desirable future.

1.4 AN OPTIMISTIC APPROACH

This book aims to give the reader a firm scientific grounding in the ways in which our society has developed unsustainably, and to offer practical guidance for developing solutions for the future. It examines approaches to restoration, preservation, and management of environmental resources including forests, soil, water, food, and biodiversity and gives examples of how readers can measure and reduce their environmental impact.

Our odds of success will be improved if we view each environmental problem as an opportunity. Sustainable living is a "win-win" for present and future generations. Ample business opportunities exist for innovative thinkers to supply creative solutions to environmental problems. We need people who can invent or design:

- Renewable, nonpolluting energy systems that do not consume all of our land and water
- Inexpensive, effective water treatment systems for the developing world
- Accounting systems that accurately measure the useful amounts of economic, social, and environmental capital
- Integrated resource management systems that use holistic approaches and accurate data to supply necessary resources to the greatest number of people without harming the environment
- Information systems to supply the knowledge necessary to make decisions that promote sustainability
- Agricultural systems that produce the maximum amount of healthy food at the lowest possible economic, social, and environmental cost
- Effective integrated waste management systems that minimize total waste and pollution
- Building systems that require less material, energy, and water
- Property-right systems that ensure built-in sustainability incentives
- Health systems that keep people healthy and productive at low economic, social, and environmental cost
- Monitoring systems that can provide global environmental data essential for effective decision-making

We also need politicians who understand environmental problems and the structural problems that have caused them, and have the courage to defy vested interests, end perverse subsidies that aid unsustainable industries, and make the necessary changes in the legal system to help the innovations listed above flourish.

Civilization is now at a crossroads. It may be that our society will take serious action on these environmental problems only when they reach the crisis stage—when it is too late. Alternatively, humanity could muster its courage and strength and alter our current course. As Jared Diamond notes, there is reason for optimism: "For the first time in history, we face the risk of a global decline. But we also are the first to enjoy the opportunity of learning quickly from developments in societies anywhere else in the world today, and from what has unfolded in societies at any time in the past" (Diamond 2005).

What will the world be like if we are successful? As stated by the United Nations Environment Programme, "Imagine a world in which human well-being for all is secure. Every individual has access to clean air and water, ensuring improvements in global health. Global warming has been addressed, through reductions in energy use, and investment in clean technology. Assistance is offered to vulnerable communities. Species flourish as ecosystem integrity is assured. Transforming these images into reality is possible, and it is this generation's responsibility to start doing so" (UNEP 2007).

It is in our power to solve the problems discussed in this book. First, however, we must acknowledge and identify those problems, and then discuss possible solutions to them. Let's get started!

2 Sustainability and Human Well-Being

Sustainability is receiving increasing levels of public interest and media coverage, yet most people are unclear on how it is defined, why it is desirable, or how it can be achieved. In this chapter we will define sustainability, analyze its components, and begin to explain how it can be measured. Only when sustainability is clearly defined and measurable can we as a society accurately know our objectives and create policies that will help us meet those objectives.

The concept of **sustainability** was ill defined until 1987, when the UN World Commission on Environment published "Our Common Future." The report is often referred to as the "Brundtland Report" because the Norwegian prime minister, Gro Harlem Brundtland, chaired the Commission (WCED 1987). The authors defined **sustainable development** as development that "meets the needs of the present without compromising the ability of future generations to meet their needs." It therefore emphasized both intragenerational (current) and intergenerational (future) equity (see Chapter 5). A system or process is considered sustainable if it can be maintained indefinitely; it is unsustainable if it is likely to fail sometime in the future. Sustainable development is the process of building a sustainable system (community, city, country, etc.) that meets human needs indefinitely.

While the Brundtland Report defined sustainability in terms of meeting human *needs*, sustainability now usually refers to maintaining human *well-being* now and in the future (UNEP 2007; GFN 2009). The Stiglitz report (2009) argues that we need to develop accurate measures of human well-being so that we can set well-being targets and gauge progress toward meeting those targets. Measures of human well-being should include material living standards such as income, consumption, and wealth; health; education; personal activities including work; political voice and governance; social connections and relationships; environment (present and future conditions); and insecurity, of an economic as well as a physical nature. The objective of sustainable development is therefore to increase quality of life by improving all of the components of human well-being. While all these components affect human well-being, many are not included in existing well-being indicators.

The most widely used measure of well-being in a country is the **human development index** (HDI), developed by the United Nations Development Program.* It takes into account progress in life expectancy, education, and standard of living as measured by the **gross domestic product** (GDP). The **human well-being index** is an average of indices of health and population, wealth, knowledge, community, and equity. We will tend to avoid using indices because they are not measured in physically meaningful units such as dollars, hectares, or years (Stiglitz, Sen, and Fitoussi 2009).

2.1 SUSTAINABILITY COMPONENTS

There is enough in the world for everyone's need, but not enough for everyone's greed.

Frank Buchman
Founder of the Oxford Group

As shown in Figure 2.1a, the three components of sustainability are social, economic, and environmental. Mnemonics to help remember these components are the three Ps (people, prosperity, and the planet) or ECO^3 (Eco cubed: ecovillages, economies, and ecosystems). To have a high level of

* UN Human Development Report 2009, http://hdr.undp.org/en.

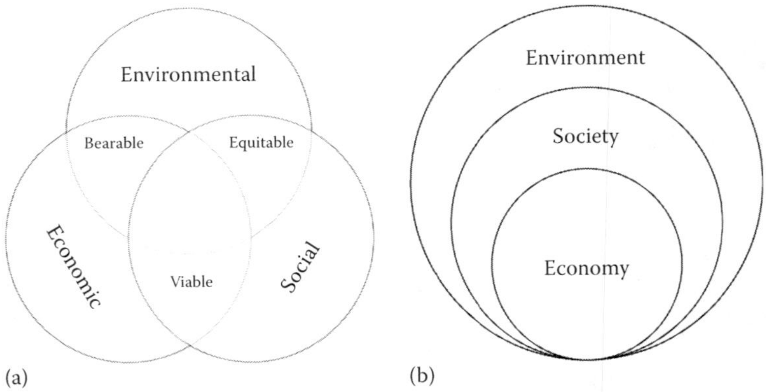

FIGURE 2.1 (a) The three spheres of sustainability. (b) The nested spheres of sustainability.

well-being we need healthy ecovillages, economies, and ecosystems. A more accurate illustration makes clear that in fact the economy is a subset of society, which in turn is a subset of the environment (Figure 2.1b). Throughout this book we will find that we cannot have a healthy, sustainable economy without a sustainable society, which in turn requires a sustainable environment. Thus, the goal of sustainability is to maximize our economic, social, and environmental well-being.

One measure of well-being is the accumulated amounts of **economic capital**, **social capital**, and **natural capital** (Goodland and Daly 1996). All three types of capital provide for our well-being and help meet our needs, and therefore are critical components of sustainability. Therefore, to meet present needs without compromising the ability of future generations to meet their own needs, we must preserve economic, social, and natural capital for future generations.

Preserving capital requires maintaining or increasing economic, environmental, and social security, which reduces risk and increases **resilience**. As we'll see in Chapter 4, resilience is the capacity of an individual or a **system** to survive disturbances unchanged (Mazur 2013). Think of what happens during a natural disaster. Some individuals will be able to cope well and be resilient, while others will be devastated and may not even survive. Those who prepare in advance by choosing a safe place to live (natural capital), maintaining a social support system of family and friends who can provide assistance (social capital), and keeping a stash of money and emergency supplies (economic capital) are more likely to survive a natural disaster unscathed.

A system is a set of interacting or interdependent components forming an integrated whole. Examples of systems include national and global economies, social communities and civilizations, and ecosystems. We need to understand how a system works, how it responds to perturbations such as natural disasters, and what the risks are, before we can devise reliable strategies for making the system resilient so that capital stocks can be preserved. For example, a coastal community can increase its resilience by evaluating current and future risks from hurricanes and then implementing policies that ensure future development does not increase risks. They can also develop emergency plans, build hurricane shelters, expand the capacity of evacuation routes, purchase hazard insurance for government assets, and set up an emergency fund to cover economic costs. All of these approaches would increase capital that could be used to avert disaster when a hurricane strikes.

To quantify sustainability and measure progress toward it, we need to estimate the amounts of economic, social, and natural capital over time. We can easily do this for economic capital because the market efficiently determines the value of each component of economic capital, but it is more difficult to accurately estimate the value of social and natural capital over time. **Full cost accounting** (FCA), also known as triple bottom line accounting, is used to keep track of all three forms of capital and to identify sustainable purchasing and policy options (Schaltegger and Burritt 2000). For now, we will assume that we can accurately measure social and natural capital.

Keeping the rate of capital consumption (demand) less than or equal to the rate of capital production (supply) grows or preserves capital. For example, your checking account balance is a measure of your economic capital. The rate at which you withdraw money should not exceed the rate at which you deposit it. You save (accumulate) capital when your deposit rate exceeds your withdrawal rate. The more economic capital you have saved in a safe place such as a bank savings account, the more resilient you will be, as the money you save can be used to bail you out of all sorts of unanticipated calamities (health problems, home repairs, stock market crash, etc.). In this example, the amount of accumulated capital (the "stock") changes over time if the deposit rate (inflow) is different from the withdrawal rate (outflow). **Stock and flow** diagrams are used to illustrate how capital stocks change over time (more on this below).

Sustainable strategies maintain economic, social, and natural capital through effective use, protection, and diversification of assets. For example, we can increase social capital by building effective social institutions such as schools and hospitals. These institutions provide jobs, an educated and healthy workforce, and space for social interactions that build healthy communities. Economic sustainability means a **steady-state economy** that equitably provides the resources required for all citizens to enjoy healthy, productive lives. Environmental sustainability means preserving biodiversity and therefore **ecosystem services**, the benefits that ecosystems provide to humanity, including purification of air and water.

Can we be sustainable by growing or preserving only one or two of the three types of capital? Here we must recognize and define two types of sustainability: weak and strong (Goodland and Daly 1996). **Weak sustainability** assumes that substituting one form of capital for another is possible, that is, we can compensate for a reduction in one type of capital (typically natural capital) with an increase in another type of capital (typically e.g., economic capital). This approach measures sustainability using the total value of the aggregate capital stock, that is, the combined values of economic, social, and natural capital. In contrast, **strong sustainability** argues that substitution is not always possible between different types of capital, and therefore that we must maintain economic, social, and natural capital stocks independently (Ott 2003). For example, economic capital cannot necessarily substitute for social or natural capital; some ecosystem services cannot be replaced. Also, loss of social and natural capital may be irreversible. Once a species goes extinct, it is lost forever.

As an example of building economic capital without social capital, consider the true story of a Nashville, Tennessee, firefighter who saved all of his money but was not on speaking terms with any of his family. When he died, he left $800,000 in a trust for an orphanage, but because he had no one who was close to him to administer the trust, he appointed an acquaintance as the executor. That person turned out to be a con man who squandered the money, leaving the children's home almost nothing.* While the firefighter successfully built economic capital, he lacked social capital. To be sustainable, individuals and society must build or preserve capital in all three areas of sustainability. Building only one or two legs of a three-legged stool results in an unstable stool.

Another argument against substitution is that natural resources often serve many purposes, and we have to find substitutes for each one. For example, trees provide many ecosystem services, even in urban areas. Trees hold moisture, and they moderate temperature. Trees effectively store the **greenhouse gas** carbon dioxide for long periods of time, helping to moderate climate change. In cities that lack trees, temperature swings are greater than in forested areas, and during hot summer days the "heat island" effect can make temperatures unbearably high. Furthermore, many studies have shown that people feel happier and less stressed when their environment includes abundant plants and trees rather than asphalt roads and concrete buildings (Wilson 1998). It would be impossible for us to find substitutes for all of the services trees provide. There are many other examples of resources for which we lack adequate substitutes. Thus, in this book we advocate strong sustainability, which requires that we balance accounts independently for all three components of sustainability.

* http://www.wsmv.com/story/14807859/judge-orders-childrens-home-money-returned-9-24-2010, retrieved 6/1/2016.

2.2 CAPITAL SUPPLY AND DEMAND

A sustainable state of a system is one that we can sustain indefinitely by preserving economic, social, and environmental capital. Capital stocks include resources that we can use to attain a high level of well-being. A recurring theme in this book is that our society will experience global shortages of some natural resources in the near future because we have been using them unsustainably. Some resources will become scarce within the lifetimes of the **baby boomers** (people born between 1946 and 1964). Others will become scarce during the lifetimes of their children (**Generation X** born between 1965 and 1982, and **Millennials** born between 1983 and 1999).

To understand the problem of resource shortages, we must introduce some terms and concepts. **Renewable resources** can be continuously replenished by natural processes. **Nonrenewable resources** are fixed in quantity on a human timescale, so the faster we use them, the sooner they disappear. Fossil fuels are considered nonrenewable resources because nature requires millions of years to produce oil and coal from plants, so the current rate of consumption greatly exceeds the rate of production. Overall, we are at greater risk of running out of nonrenewable resources than renewable resources. However, a renewable resource can become scarce if used faster than it is renewed, or if its renewal capacity is undermined by overharvesting. For example, some marine fisheries have collapsed because overfishing has reduced populations of certain fish species to critically low levels. If left alone to breed, they may regenerate their populations to preexisting levels—but that could take many decades. You may have noticed that certain species of fish such as cod have become more expensive or harder to find in grocery stores; this is because the commercial catch of cod has declined rapidly in recent years. In the year 2000, the World Wide Fund for Nature added cod to its list of endangered species, stating that the worldwide cod catch had decreased 70% in the preceding 30 years due to decreasing cod abundance.* Thus, the cod-fishing industry has been unsustainable, and environmentally minded consumers should avoid eating cod. Table 2.1 compares the situations where the rate of renewable resource use is less than, equal to, or greater than the rate of replenishment, leading to resource stocks increasing, remaining constant, or declining, respectively.

In the case of renewable resources, capital stocks consist of a nonrenewable portion called **principal** and a renewable portion called **interest**. To protect economic, social, and natural capital, individuals and society must spend only the interest and not the principal in each category. Fiscal conservatives use this approach to achieve economic sustainability.

Since the human population is growing at an exponential rate (Section 3.4), our rate of capital use is also increasing at an exponential rate. This is sustainable only until growth causes the rate of renewable resource consumption to exceed the rate of production, that is, the rate of replenishment. The sustainability level of a growing population changes over time. For example, for a renewable resource, society usually begins in a state where the rate of consumption is less than the rate of replenishment (Table 2.1, Category 3). Eventually, the population becomes so large that humans consume the resource at the same rate that nature produces it (Category 2). This optimal state allows the greatest number of people to live sustainably. If population and the resource consumption rate continue to increase so that the replenishment rate cannot keep up with increasing consumption, the system will become unsustainable, leading to environmental degradation, loss of natural capital, and resource shortages (Category 1). We will show in Chapter 3 that humanity began using many renewable resources faster than they could be replenished beginning in the 1970s, and the imbalance has continued to grow.

Let's look at the renewable natural resource water from a lake as an example (Figure 2.2). The stock of water is the amount of water stored in the lake. Inflow is a measure of supply, that is, the rate of water replenishment. Outflow is a measure of demand, that is, the rate of water withdrawal or consumption. The stock of water reflected by the level of water in the lake will fall if inflow is less than outflow, rise if inflow is greater than outflow, and remain steady if inflow equals outflow.

---

* http://www.panda.org/about_our_earth/all_publications/?12982/The-Barents-Sea-Cod-the-last-of-the-large-cod-stocks.

TABLE 2.1
Sustainable Use of Renewable Natural Resources

Consumption of Renewable Resources	State of Environment	Sustainability
1. More than nature's ability to replenish	Environmental degradation	Not sustainable
2. Equal to nature's ability to replenish	Environmental equilibrium	Steady state
3. Less than nature's ability to replenish	Environmental renewal	Environmentally sustainable

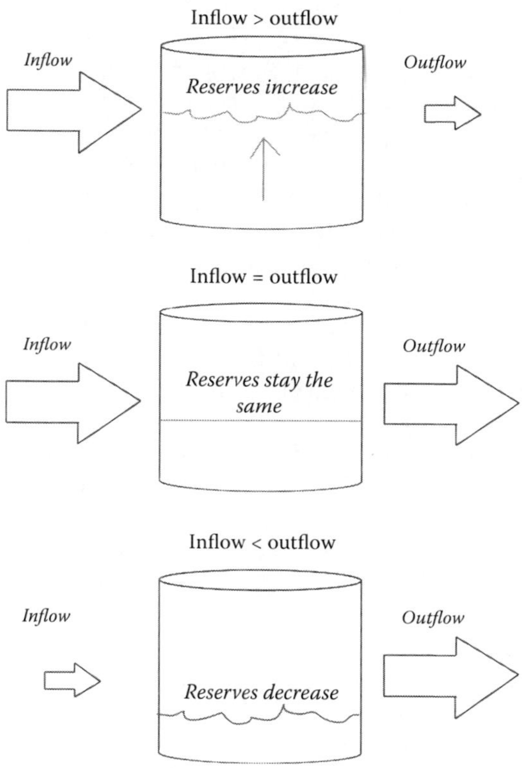

FIGURE 2.2 Illustration of the concept of steady state for a lake containing the renewable resource water. Arrow size is proportional to the water flow rate (flux). In a steady state, the amount of water in the reservoir does not change because inflow equals outflow.

We can temporarily use a renewable resource such as lake water faster than it is replenished. When we do, the lake water level drops. If we continue to withdraw water faster than it is replenished, the lake will eventually dry up.

Resource use is sustainable if the rate of withdrawal is less than or equal to the rate of replenishment. When inflow equals outflow, the stock (water level) remains constant and the system is in a **steady state**. Steady state systems are stable because they don't change with time. An equivalent phrase is **net zero**: When the amount of resources we consume equals the amount we produce, we are "net zero." For example, a household is sustainable if it consumes no more resources than it produces, that is, it is net zero for energy, water, and food. Sustainable use of renewable resources is desirable because it guarantees long-term availability of those resources.

We can see from these examples that *renewable resources are flow or rate limited*, while *nonrenewable resources are stock limited* (Gleick and Palaniappan 2010). We essentially have an infinite supply of renewable resources such as water and solar energy because they are constantly replenished, but the rate at which we use them is limited by the flow rate. Dynamic systems such as our hypothetical lake can be modeled by using stocks and flows. Modeling allows us to evaluate future scenarios in which changes in the environment or in human resource use affect the supply of renewable resources such as water over time. For example, if the inflow rate decreases during a drought, we can model the rate of water depletion and estimate when the lake will go dry.

For critical renewable natural resources such as water, it is wise to establish a safety margin by maintaining a demand (rate of withdrawal) that is less than the average rate of replenishment. For example, during a drought the rate of replenishment drops. To maintain sustainability, the demand must decrease at the same rate as supply. During droughts, communities usually accomplish this through water conservation efforts or by temporarily drawing on nonrenewable supplies such as **fossil water**.

Food serves as another example of a renewable resource that should be sustainably managed. We can estimate the maximum number of people that can be fed in a specific community, given the limitations of fertile land area, annual precipitation, and so on. Yet what if a drought or flood wipes out half of the crop? To increase resilience, the community could limit the population to half the maximum number that they could feed in a good year. In good years, the community could store the surplus food to use in lean years. When the community has the maximum number of people that can be fed in a good year, it operates at the physical limit of the food system because it cannot afford to save a surplus for lean years. During a lean year, the food supply system cannot keep pace with demand. The **overshoot** of the system is the amount that demand exceeds supply. As the degree of overshoot (the gap between supply and demand) increases, the risk of starvation increases. Thus, we can increase **food security** by using the food supply system sustainably, that is, by maintaining a demand on renewable systems that is less than the maximum rate of replenishment. The wisest approach is to reduce demand to the lowest anticipated level of food production.

How can consumption of a nonrenewable resource be sustainable, since by definition its replenishment rate is zero? For our purposes, it can be sustainable only if it is replaced with a renewable resource at a rate high enough to maintain the stock. Imagine a steady state economy in which the only energy options are nonrenewable coal and renewable solar. To be sustainable it must maintain a constant supply of energy indefinitely by increasing solar energy production at the same rate that coal consumption is decreased.

The **stock and flow model** can be applied not only to resource consumption but also to waste production. To achieve sustainability, we must not only decrease our consumption of renewable resources to a sustainable level, but also decrease the waste produced so as not to exceed the waste absorption capacity of the environment. In this example, society is the source of pollutants and the environment is called the sink. Pollution results when the rate of waste production (waste outflow from the economic system and inflow to the environment) exceeds the rate of waste absorption or elimination so that wastes accumulate in the environment. Ideally, we should eliminate waste by producing no more than we can consume. We can consume waste by treating it as a resource and finding productive ways to use it (e.g., recycled paper, use of **graywater** in landscaping, etc.).

These concepts are summarized in the Laws of Sustainability. According to ecological economist Herman Daly (1990), the sustainable rate of use can be no greater than:

1. The rate of regeneration of a **renewable resource**
2. The rate at which a renewable resource, used sustainably, can be substituted for a **nonrenewable resource**
3. The rate at which a **pollutant** can be recycled, absorbed, or rendered harmless in its sink

Now that we have introduced the concepts of weak and strong sustainability and the Laws of Sustainability, we can understand a more sophisticated definition of sustainability (GFN 2009): “Sustainability is a commitment to human well-being—well-being that lasts. Maintaining sufficient amounts of social, economic and natural capital in order to make well-being last necessitates the following: either (a) using each type of capital no faster than it regenerates (strong sustainability), or (b) if these assets are used faster, that these assets are not depleted faster than other human-made processes are able to compensate for the lost capital (weak sustainability)."

2.3 THE VALUE OF ECOSYSTEM SERVICES AND COSTS OF ENVIRONMENTAL DEGRADATION

Our focus is primarily on the sustainability of human societies, especially at the small scale of families and communities. That is what we are most familiar with, and it is where we have the most influence and can take personal action. However, we cannot maintain human society without also preserving our supporting ecosystems and their biological diversity because we depend on ecosystem services. Most people agree that "the quality of the environment is important both to their own well-being and to the common good" (Adams 2006). Ecosystems provide many life-support functions that we take for granted, such as purification of air and water (Millennium Ecosystem Assessment 2005b). Without these ecosystem services, we would likely perish. Preserving them will help society maintain public health and safety. Changing our lifestyles in order to preserve ecosystems and the services they provide is key to achieving sustainability.

In the field of ecological economics, ecosystems are considered capital assets because they provide services such as water purification. Many examples illustrate that relying on nature to provide ecosystem services is cost-effective. According to the MEA (Millennium Ecosystem Assessment 2005b), a four-year research effort by 1,360 of the world's leading scientists commissioned to measure the actual value of natural resources to humans and the world, ecosystems provide provisioning, regulating, and cultural services. Our focus will be on provisioning and regulating services because we can scientifically quantify them. Among other things, provisioning services involve the production of renewable resources such as wood, food, and fresh water. Regulating services keeps our environment hospitable by reducing environmental change by, for example, controlling pests and diseases and regulating climate (Cardinale et al. 2012). Altogether, ecosystem services have been valued at a minimum of $33 trillion annually in 1994 dollars (Costanza et al. 1997).

Humans are an integral part of ecosystems. Because the human population is increasing at an exponential rate, our impact on the environment is increasing at an exponential rate (Chapter 3). Environmental degradation results in the loss of ecosystem services. Many societies in the past collapsed because of environmental degradation and loss of ecosystem services (Section 4.7).

The poor are the most reliant on ecosystem services, since wealthy people can purchase substitutes for natural resources such as bottled water or pay others to provide food. As a result, degradation of ecosystem services disproportionately harms poor people. For example, in southern Honduras, the removal of soil-anchoring vegetation for agricultural purposes diminished soil absorbency and resulted in both desertification (a negative effect) and the reduction of malaria-bearing mosquito populations (which initially seemed like a positive effect). Over time, the land became uncultivable. When Hurricane Mitch struck in 1998, much of the soil was washed away in mudslides that displaced half the population and reduced agricultural production by 95% (Bright 2000). Meanwhile, communities in the area developed a lower immunity to malaria. When the low-immunity population moved north to rainforests that could support them, they encountered high concentrations of malaria-bearing mosquitoes, which led to a rapid rise in the number of malaria cases. This chain of unpredictable events illustrates the complexity of our environment and its sensitivity to human-driven change. To reduce the potential for negative environmental surprises, we should practice the **precautionary principle**, a risk management approach which states that we should not change a system unless we are reasonably confident that those changes will not cause harm to the public or the environment. As an example of ecosystem services and the effects of environmental degradation, we will look at forest ecosystems and the problem of deforestation.

2.3.1 Forest Ecosystems and Deforestation

Forests have a high density of trees and host a high diversity of wildlife. Primeval forests covered most of Europe and North America 10,000 years ago, but now only small pockets of primary (old growth) forests exist. Globally forests cover 30% of land area, compared to prehuman coverage of ~50%. The importance of forests led the United Nations to declare 2011 the International Year of Forests.

Forests provide a wide range of ecosystem services. They host roughly 80% of the earth's species, and in many regions produce abundant organic matter that accumulates to form thick, rich soils. Forests play an essential role in the global cycling of nutrients such as carbon and nitrogen (Raven, Berg, and Johnson 1995). Forests act as both sources and sinks. For example, forest trees are a sink for atmospheric carbon dioxide (CO_2), which they store through the process of photosynthesis:

$$CO_2 + H_2O = CH_2O + O_2 \tag{2.1}$$

Carbon dioxide can be stored in the form of woody matter represented by the organic molecule CH_2O for hundreds to thousands of years. This chemical reaction shows that trees are also a source for atmospheric oxygen (O_2). In fact, trees are the primary source of the oxygen we breathe. Forest trees exert a strong control on local climate by providing shade and in some cases reducing evaporation from soil (Siriri et al. 2013), keeping an area cool and humid even during hot, dry spells. Finally, forests provide wood, which has many unique qualities that make it a versatile and irreplaceable resource used for fuel, construction, and the arts. The timber industry is an important part of the global economy, accounting for roughly 1% of the world GDP, about U.S. $200 billion annually.* Roughly 1.6 billion people depend on forests for their livelihoods, and about 300 million people live in forests.*

Trees are therefore a valuable and renewable resource. However, trees are treated as a nonrenewable resource in most parts of the world because they are being harvested faster than they regrow—an unsustainable practice (Meadows, Randers et al. 2004). Unfortunately, because trees grow slowly, it is easy to harvest them faster than they replenish, which results in deforestation. Deforestation and associated soil problems contributed to the collapse of all ancient societies examined in Jared Diamond's book *Collapse* (2005). This suggests that each country should make forest preservation, **reforestation**, and **afforestation** (establishment of new forests) high priorities.

Deforestation is primarily caused by the expansion of logging and large-scale industrial agriculture. Today much deforestation is caused not by rural farmers, but by corporations, clearing trees for palm oil and cattle production. In the 1980s and 1990s, deforestation of tropical forests was the primary source of new agricultural land (Gibbs et al. 2010). The consequences of deforestation include loss of timber and other forest building materials, soil erosion and increased sediment loads in rivers, loss of soil fertility, loss of watershed protection and potential hydroelectric power, decreased rainfall, decreased biodiversity, and desertification (Diamond 2005). Particularly troublesome is slash and burn farming, where trees are simply chopped down and burned, amounting to a complete waste of a valuable resource. Trees store the greenhouse gas CO_2 through photosynthesis (Equation 2.1). Deforestation eliminates this sink for atmospheric CO_2, and burning the wood releases the CO_2 stored in the tree (the reverse of Equation 2.1), making slash and burn farming a double whammy for climate change. Deforestation is thus a leading cause of global warming (Brown 2009). Most deforestation today occurs in tropical forests, and current harvesting rates could result in unprotected tropical forests disappearing as early as 2054 (Meadows, Randers et al. 2004).

The harvesting of hardwood presents an excellent example of how society adapts to a resource shortage. In the early years of the United States, hardwood was abundant. However, as the nation

---

* http://wwf.panda.org/about_our_earth/about_forests/importance/economicforest/, retrieved January 20, 2011.

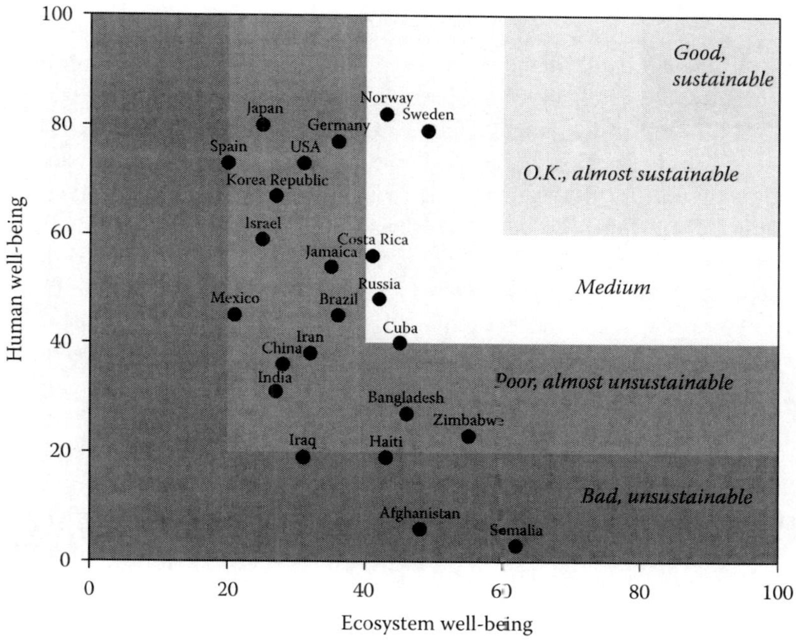

FIGURE 2.3 IUCN sustainability diagram. (Data from Prescott-Allen, Robert. 2001. *The Well-being of Nations: A Country-by-Country Index of Quality of Life and the Environment*. Island Press.)

expanded and population grew, Americans began to harvest hardwoods unsustainably. Today almost no old-growth forest remains in the United States, and the price for hardwood has skyrocketed. Even expensive furniture is often made from wood laminates or from softwoods like pine rather than solid hardwoods; cheap furniture is made from pressed particle board, which is made of small pieces of wood and sawdust bonded together by adhesives. As a society we have accepted this change. But will we always be able to find substitutes when we run short of natural resources?

So how can society make forest use sustainable? We in the United States can help grow the sustainable forestry industry by choosing lumber certified with the stamp of the Forest Stewardship Council (FSC), which promotes the harvesting only of mature trees and is opposed to forest clearcutting. Research and development and regulations could boost the efficiency of wood-burning stoves and lumber and paper mills (Meadows, Randers et al. 2004). We must also pressure politicians to eliminate logging subsidies and to set aside forests for protection. In 2010, the protected forest area increased to 460 million hectares, 12.5% of the total forest area.* Climate change mitigation efforts are now focusing on afforestation in the tropics because trees in the tropics grow faster and remove CO_2 from the atmosphere at a much higher rate than trees in temperate regions (~50 vs. 13 kg CO_2/year) (Brown 2011). Some developing countries have been able to reverse the trend of deforestation, and by promoting sound land management policies, they have achieved net increases in forest land cover while at the same time increasing agricultural production (Lambin and Meyfroidt 2011).

What can you do as an individual? To reduce demand for wood and slow the rate of deforestation, try to decrease the amount of wood you consume by reducing paper use and recycling paper. Reduce your use of disposable paper products, including paper napkins (use linen), tissues (use handkerchiefs), plates, and newspapers. Read articles on your computer or on an e-reader. Besides saving paper, this saves a great deal of office space. Demand for wood products should be decreased to the maximum sustainable harvesting rate where inflows (tree growth) equal outflows (tree harvesting).

* UN Food and Agriculture Organization, Global Forest Resources Assessment 2010 (Rome: 2010), p. 60.

2.3.2 Measures of Environmental Degradation

To measure progress toward sustainability we need quantitative measures. Environmental degradation is a loss of natural capital, so we need a measure of natural capital. One useful measure is the **Ecosystem Well-being Index**, which is an average of indices of land, water, air, species and genes, and resource use (Prescott-Allen 2001). Figure 2.3 shows composite Human Well-being and Ecosystem Well-being Indices for the United States and a sample of other countries for comparison. Note that the United States plots in the "almost unsustainable" field because its Ecosystem Well-being Index is low. No countries plot in the "sustainable" or "almost sustainable" fields. Sweden is the closest to being sustainable, plotting squarely in the "medium" field, while Canada, Norway, Costa Rica, and Russia also plot in the medium field. Clearly, our society has a long way to go before we achieve sustainability. In Chapter 3 we will quantify human impacts on the environment.

WEB RESOURCES

- Earth 2100 (video): https://www.youtube.com/watch?v=LUWyDWEXH8U
- Sustainability easily explained: https://www.youtube.com/watch?v=_5r4loXPyx8
- Worldwatch Institute: Vision for a Sustainable World: http://www.worldwatch.org/
- Yale Environment 360: http://www.e360.yale.edu/

HOMEWORK PROBLEMS

1. Use the Internet to find a measure of environmental sustainability other than the Ecosystem Well-being index. Which index do you think is preferable, and why?
2. Choose one of the countries plotted in Figure 2.3. Are its human well-being and ecosystem well-being index scores high or low? Speculate on the reasons why.
3. Imagine you are an alien approaching Earth from space. Your planet has scarce resources, so your culture has developed taboos against wasting them. You see a view similar to one seen from the International Space Station. What strikes you as unsustainable human activity on the global scale? You then land your spaceship and explore communities. What do you think stands out as the least sustainable common practice of humans on a local scale?

3 The Environmental ImPACT

3.1 INTRODUCTION

Sustainability is achieved when the supply of social, economic, and environmental capital exceeds the demand. As a global society, our current lifestyle is unsustainable because for many types of capital, demand exceeds supply. The human population of seven billion in 2015 combined with high per capita consumption rates has caused humanity to exceed Earth's ability to support us.

This book presents data and references to support the following statements (also see Schmidt-Bleek 2007). By the year 2100, Earth's population will increase to roughly 11 billion (Section 3.4), yet we are already pushing the limits of our resource use. Humans use over half the accessible water (Chapter 12) and about 40% of the energy from the sun captured by plants through photosynthesis. An estimated 50% of the global land area has been impacted due to direct human influence; more than half of global wetlands, which contribute to the hydrologic cycle and to biological diversity, have been lost, and the quality and productivity of 23% of the usable land area has decreased. Forest areas have been diminished from 6 billion to 3.9 billion hectares over the course of human history (Section 2.3.1). We have created a large hole in the ozone layer (Section 8.3.4) and increased the concentration of carbon dioxide in the atmosphere by more than a third in the last 175 years, resulting in warming of the atmosphere by ~0.7°C in the last 100 years (Chapter 7). The rate of species extinction both on land and in the ocean is growing rapidly, placing us in the midst of Earth's sixth mass extinction event, one primarily caused by humans (Chapter 15). The overuse of numerous marine fish stocks is putting the ecological balance of the oceans and coastal ecosystems at risk; more than one-quarter of all fish stocks are currently depleted or threatened by depletion, and a further 50% are being fished at the biological limit (Section 13.2.3). These changes demonstrate that humans are now capable of changing the environment on a global scale.

Critical shortages of resources, extreme pollution, plummeting biodiversity, and global climate change all point to a need for reduced resource consumption and a halt to population growth. While these negative trends are cause for great concern, some can be reversed, but only a major effort will restore the Earth's ability to sustain us (Brown 2009). The sooner we take action and the more drastic the positive changes we make, the less risk we, and our offspring, will face in the future.

This chapter introduces a simple approach to estimating the impact of human activities on environmental resources. This is an important first step toward sustainability: we must be able to identify and quantify the most unsustainable activities so that we can prioritize future actions aimed at achieving sustainability.

3.2 THE ImPACT IDENTITY

Each of us has an impact on the environment. Adding up the impacts of all individuals gives us the total environmental impact of society. Because Earth can provide a limited amount of resources, reducing the average environmental impact of individuals increases the number of people that can meet their needs. This is measured by the **carrying capacity**, defined as "the maximal population size of a given species that an area can support without reducing its ability to support the same species in the future" (Daily and Ehrlich 1992). But how can we accurately estimate and reduce our environmental impact? A simple formula called the "ImPACT identity" can give an accurate estimate of the environmental impact of individuals and social groups and can help us identify strategies for reducing it.

The ImPACT identity was preceded by the **IPAT** identity, introduced by Ehrlich and Holdren (1972) to quantify the environmental impact of humans. The equation is

$$I = P \times A \times T \tag{3.1}$$

where P = population, A = affluence or per capita consumption, and T = technology measured as the impact per unit of production or consumption.* T is the quantity we usually know the least about, so it is often solved for using the equation $T = I/P \times A$, where T is the impact per unit of economic activity (York, Rosa, and Dietz 2003). In effect, T lumps together all of the unknown driving forces of environmental impact.

In the IPAT identity, the term T lumps the effects of conservation and efficiency together, causing confusion. **Conservation** means reducing resource use (e.g., using fewer gallons of gasoline by driving fewer kilometers), while **efficiency** means increasing the work or yield per unit of resource (e.g., getting more kilometers per liter of gasoline). Conservation and efficiency are the two main tools we have for making our resource use more sustainable. Splitting the T term in IPAT into two we obtain the **ImPACT** identity for calculating the environmental impact I (Waggoner and Ausubel 2002):

$$\begin{aligned} \text{ImPACT } I &= \text{Population} \times \text{Affluence} \times \text{Consumption per unit of affluence} \times \text{Technology inefficiency} \\ &= P \times A \times C \times T \end{aligned} \tag{3.2}$$

The new term C is consumption per unit of affluence, sometimes referred to as "intensity of resource consumption." The product $A \times C$ is the amount of consumption, which can be decreased through conservation. The term T measures resource use inefficiency, which can be reduced through technological innovation (Table 3.1). We can use **ecotechnology** and **ecodesign** to increase efficiency, thereby reducing T and the environmental impact. In our example of driving an automobile, the environmental ImPACT in terms of amount of gas consumed is calculated as:

$$\begin{aligned} I_{\text{driving}} \text{ (liters gas)} &= P \times A \times C \times T \\ &= \text{persons} \times \$ \text{ spent on gas/person} \times \text{kilometers/\$ of GDP} \times \text{liters gas/kilometer} \end{aligned} \tag{3.3}$$

For this example, we find using **dimensional analysis** that the unit of I is liters of gasoline, which is equivalent to the amount of greenhouse gas emitted by auto use. We can see that the units of the product $A \times C$ is kilometers/person, a measure of the resource consumption level that can be reduced through conservation. The term T represents the inefficiency of the auto, which would be measured as gallons per mile in the United States and which could be reduced by making the car smaller and more efficient. This type of analysis can also be used to estimate the amount of carbon dioxide emitted by burning fossil fuels for energy. This application of the ImPACT identity is especially important because of its implications for global climate change, so it is named the **Kaya Identity** after the person who first stated it (Kaya and Yokoboi 1993, more in Section 11.2.1).

The ImPACT identity gives us guidance on how to become sustainable. As seen in Equation 3.2, the environmental ImPACT of humanity I is the product of four components: population P, affluence A, intensity of resource consumption C, and inefficiency of resource use determined by technology T. As noted by Waggoner and Ausubel (2002), the four *f*actors that can reduce the impact of each component in the ImPACT identity are parents P, workers A, consumers C, and producers T

* It is helpful to use dimensional analysis to check the units of each variable and make sure the units of the desired quantity on the left-hand side are correct:
Environmental Impact = # of persons × (amount consumed/# of persons) × (environmental impact/amount consumed).

TABLE 3.1
Variables in the ImPACT Identity

Variable	Related Measures/Examples	Strategies for Reducing
Impact *I*	Ecological footprint, Kaya identity	Reduce *P*, *A*, *C*, and *T*
Population *P*		Family planning
Affluence *A*	Per capita consumption	
Intensity of resource consumption *C*	Consumption per unit of affluence	Conservation: reduce consumption
Inefficiency of technology *T*	Impact per unit of consumption	Technological innovation can increase efficiency

Source: Derived from text in York, R. "$I = P \times A \times T$ Equation." In *Berkshire Encyclopedia of Sustainability*, edited by I. Spellerberg, D.S. Fogel, S.E. Fredericks, and L.M. Butler Harrington, pp. 194–197. Vol. 6, Measurements, Indicators, and Research Methods for Sustainability. Great Barrington, MA: Berkshire, 2012.

(Figure 3.1). Parents can choose to reduce population by choosing to have fewer children; workers can work fewer hours or choose jobs that bring them greater satisfaction than income; consumers can reduce their intensity of use through conservation, a process called **dematerialization**; and producers can decrease their environmental impact by increasing efficiency.

We can use the ImPACT identity to estimate the environmental impact of using resources, known as **resource footprints**. For example, the most serious environmental impact of burning oil is the release of the greenhouse gas carbon dioxide (CO_2). To estimate the environmental impact of oil use we calculate the **carbon footprint** of oil consumption as the amount of CO_2 released as follows:

$$\begin{aligned} CO_2 \text{ footprint oil(kg)} &= P \times A \times C \times T = \text{persons} \times \frac{\text{dollars}}{\text{person}} \times \frac{\text{unit enery}}{\text{dollar GDP}} \times \frac{\text{mass } CO_2 \text{ produced}}{\text{unit energy}} \\ &= \text{population} \times \text{dollars spent on oil per person} \\ &\quad \times \text{oil use intensity} \times \text{emission intensity of oil} \end{aligned} \tag{3.4}$$

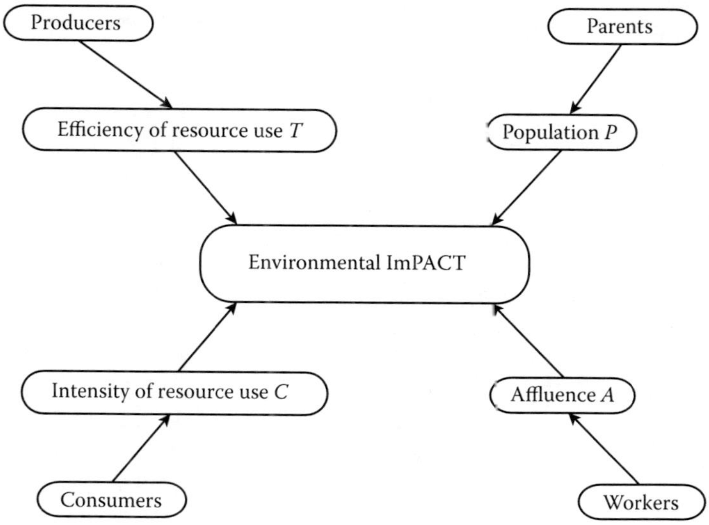

FIGURE 3.1 The environmental ImPACT of humanity *I* is the product of four factors: population *P*, affluence *A*, intensity of resource use *C*, and efficiency of resource use determined by technology *T*.

Let's calculate the carbon footprint of oil use in the United States in 2012. In that year, the U.S. population was roughly 310 million people. Using data from the EIA (2014) and the EPA (2004), we find that per capita consumption *A* was \$2,194 per person on oil. The energy intensity *C* has units of energy/unit GDP, and represents how much economic growth was generated per unit of energy, which is the same for all energy types. In 2012, the U.S. energy intensity was 7,000 BTU/\$ of GDP, where BTU is British Thermal Units, a measure of energy content (EIA 2014, Table A20). The energy inefficiency *T* is the amount of CO_2 released per unit of energy (also known as the emission intensity), which for oil is 0.063 g CO_2/BTU. Multiplying these factors to obtain the environmental impact of burning oil in the United States in 2012:

$$I = P \times A \times C \times T = 310 \times 10^6 \text{(persons)} \times 2{,}194\left(\frac{\$}{\text{person}}\right) \times 7{,}000\left(\frac{\text{BTU}}{\$}\right) \times 0.063\left(\frac{\text{g}\,\text{CO}_2}{\text{BTU}}\right)$$

$$= 3.0 \times 10^{14}\,\text{g}\,\text{CO}_2 = 330\,\text{million short tons}\,\text{CO}_2 \quad (3.5)$$

So the United States alone emitted 330 million tons of CO_2 in 2012 just from burning oil. In general, the higher the per capita income, the more energy people use, so the environmental ImPACT measured as CO_2 emissions correlates strongly with income (see the chart at www.bit.ly/1Ph7OHi). One of the goals of sustainable development is to decouple environmental impacts from economic growth. For example, while Singapore has a high per capita GDP, its per capita CO_2 emissions are relatively low. As we'll see, this is primarily because the population density of Singapore is very high, which leads to more efficient resource use.

The IPAT and ImPACT identities have some shortcomings. One is that they assume that impacts are directly proportional to each of the parameters *P*, *A*, *C*, and *T*. This may not always be true. For example, the **environmental Kuznets curve** suggests that environmental impact has a nonlinear dependence on affluence; as *A* increases, *I* initially increases, levels off, and then begins to decrease (Figure 3.2). This relationship has been found to hold for certain types of pollutants. As countries develop, they first build up their economy, and then when they have accumulated enough economic capital they invest it in pollution abatement in order to increase their environmental capital. The environmental Kuznets curve shows that environmental impact does not always increase linearly as per capita income increases.

The IPAT and ImPACT identities are useful for conceptualizing the driving forces of environmental impacts and as accounting tools, as long as we recognize that the true functional forms may be different, that is, *I* does not always exactly equal $P \times A \times C \times T$ (York, Rosa, and Dietz 2003).

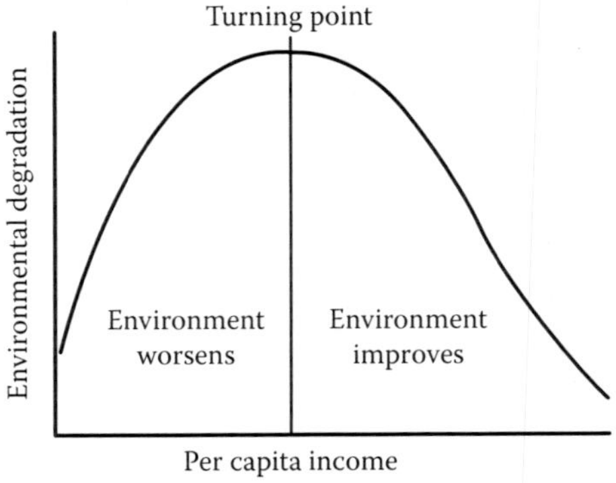

FIGURE 3.2 The environmental Kuznets curve.

Also, the factors in the ImPACT identity are not strictly independent. For example, increasing *A* has been shown to decrease growth in *P*, a phenomenon known as the **demographic transition** (Section 3.4.2). According to **Jevons Paradox**, decreasing *T* by improving energy efficiency can cause an increase in energy consumption $A \times C$ (Section 3.6.1). So we must be cautious and not assume that, for example, a decline in one factor in PACT will necessarily cause Impact *I* to decline (York 2012). The IPAT and ImPACT identities can be used for qualitative or semi-quantitative analysis of environmental impacts.*

In later chapters we will use the ImPACT identity to calculate carbon footprints and water footprints. In the next section we introduce another application of the ImPACT identity, one that aggregates many different environmental impacts to estimate the total environmental impact of individuals and society: the **ecological footprint**.

3.3 BIOCAPACITY AND THE ECOLOGICAL FOOTPRINT

> The economic and technological triumphs of the past few years have not solved as many problems as we thought they would, and, in fact, have brought us new problems we did not foresee.
>
> **Henry Ford II**

> Nature cannot be fooled.
>
> **Richard Feynman**
> *American physicist*

Many people think that the Earth has a limitless capacity to meet our resource needs and absorb the waste that we produce, but there is a fixed amount of productive land available on Earth. The Earth has roughly 100 million km² of biologically productive land and sea (Wann 2007). Dividing that up

* For quantitative analysis of environmental impacts, scientists use a more general form of the IPAT equation that can account for both linear and nonlinear relationships:

$$I = P^p A^a T^t \quad (1)$$

If *I* has a linear dependence on a parameter, the exponent for that parameter will have a value of one. To assess the relative importance of the terms in the IPAT identity, we take the logarithm of Equation 1:

$$\log I = \log EF = p \times \log P + a \times \log A + t \times \log T \quad (2)$$

Scientists use approaches like STIRPAT (Stochastic Impacts by Regression on Population, Affluence, and Technology) to accurately quantify the influences of *P*, *A*, and *T* on the environmental impact. However, as mentioned above, the value of *T* is usually unknown and therefore is lumped into the intercept, which is $\log T + \log e$, where *e* is the error (York, Rosa, and Dietz 2003). In 2010, *I* used the latest estimates of each parameter for each country and performed multiple linear regressions to estimate the coefficients, with log(EF) as the dependent variable and log(GDP) and log(Pop_millions) as the independent variables. *I* obtained the best-fit equation:

$$\log(EF) = -0.830 + (0.438 \times \log(GDP)) + (0.933 \times \log(Pop\_millions)) \quad (3)$$

The fit parameters were correlation coefficient $r = 0.97$ and number of countries = 145. Note that the coefficient for the population term is close to a value of one, indicating that EF is directly proportional to population. The percentage of the variation of EF explained by affluence (per capita GDP at PPP) is 16% and population 79%, which sum to 95%, meaning that technology *T* can explain less than 5%. We conclude that population is the most important parameter in the IPAT identity, followed by affluence and then technology.

In a study that used a similar approach but that used a wider range of independent variables, York et al. (2003) found that the variation in ln(EF) values across 142 countries using data from 1996 could be explained using independent variables expressed in natural log form that accounted for population, nondependent population, land area per capita, latitude, GDP per capita, and percentage urban (their model 4, the Human Ecology Model). Their findings show that EF increases with increasing population, increasing percentage of population between the ages of 19 and 65, increasing latitude (more resources are required to live in colder climates), increasing land area per capita (more resources per capita or lower population density), increasing GDP, and increasing urbanization (all else being constant, a measure of modernization). Like our example, they found that the coefficient of ln(population) is not significantly different from one. They interpreted the residual (the difference between observed and predicted) of the regression fit as a measure of *T*, the eco-efficiency of each country. The residual values ranged from a high value of 2.76 for the UAE to a low value of 0.52 for Iceland, meaning that Iceland is 2.76/0.52 = 5.3 times more efficient than the UAE. Although this gives us hope that some countries can dramatically improve efficiency, the maximum possible increase seems to be less than some environmentalists suggest (York, Rosa, and Dietz 2003).

among the 7.0 billion humans on Earth in 2012 leaves about 0.2 km^2 (~49 acres) per person. Is that enough to maintain a high level of well-being?

The **ecological footprint** (EF) is an estimate of the amount of land required to provide adequate resources and absorb wastes. People with high consumption and waste production levels require large amounts of land to meet their needs, and therefore have high environmental impacts. The per capita EF can be multiplied by population to estimate the total human impact on the environment, expressed as the sum of all effects of resource extraction, pollution emission, energy use, biodiversity destruction, urbanization, and the other consequences of physical growth (Meadows, Randers, and Meadows 2004). The form of the equation is

$$\text{EF}(\text{km}^2) = P \times A \times C \times T = \text{persons} \times (\text{dollars/person}) \times (\text{mass resources consumed/dollar}) \times (\text{land area/mass resources consumed}) \tag{3.6}$$

where km is kilometer, and km^2 is square kilometers, the SI unit for area.* Many EF calculators give results in global hectares gha, with 1 gha = 0.01 km^2. Notice that the EF equation has the same form as the ImPACT equation, so the EF is equivalent to the environmental impact *I*, and we use the terms footprint and impact synonymously. The EF has units of area because it is calculated by summing the land areas required to provide each of the ecological services required to maintain our current quality of life. It is a useful measure of sustainability because it has meaningful units, unlike other measures of sustainability such as the Human Well-being Index and other indices (Stiglitz, Sen, and Fitoussi 2009).

While the EF measures resource demand, **biocapacity** (*B*) measures resource supply, expressed as the area of available productive land. Our use of ecosystem services and resources is sustainable when supply is greater than or equal to demand, that is, when $B - \text{EF} \geq 0$. A system where $B - \text{EF}$ is < 0 is in **ecological overshoot** and unsustainable. Unfortunately, the global EF of 2×10^8 km^2 exceeds the biocapacity of 1.2×10^8 km^2, so the ecological overshoot is -0.8×10^8 km^2 (Figure 3.3), meaning we have exceeded the earth's capacity to provide resources and absorb our waste (Hails 2008).

You can estimate your EF by using an online calculator such as the Ecological Footprint Quiz by Redefining Progress (myfootprint.org). Different EF calculators do the calculation in different ways, so only one calculator should be used when making footprint comparisons. Remember that the EF represents the land area in km^2 required to maintain the current quality of life. However, it is hard to remember land areas in km^2, so the EF is often divided by the biocapacity in km^2 to obtain a unitless number. When this is done using the global average biocapacity, the resulting EF is expressed as the number of Earths required to maintain the current standard of living. From the global numbers given above, the normalized EF equals $\text{EF}/B = 2 \times 10^8\ \text{km}^2/1.2 \times 10^8\ \text{km}^2 = 1.7$ Earths. The Earth's carrying capacity, the population that Earth's biocapacity can support sustainably, would be the current population divided by the normalized EF = 7 billion/1.7 Earths = 4.2 billion people/1 Earth.

Figure 3.4 plots the global ecological footprint and biocapacity from 1961 to 2009. The Earth first went into ecological overshoot about 1970, and since that time the overshoot has continuously increased, a trend that is unsustainable. In 2011, the EF was 1.5 Earths, meaning that humanity used ecological services, including renewable resources and natural waste absorption, 1.5 times faster than Earth could renew them. To live sustainably, humanity must reduce its total ecological

* SI stands for **International System of Units**. One m^2 equals 10^{-4} ha (hectares) = 2.47×10^{-4} acres.

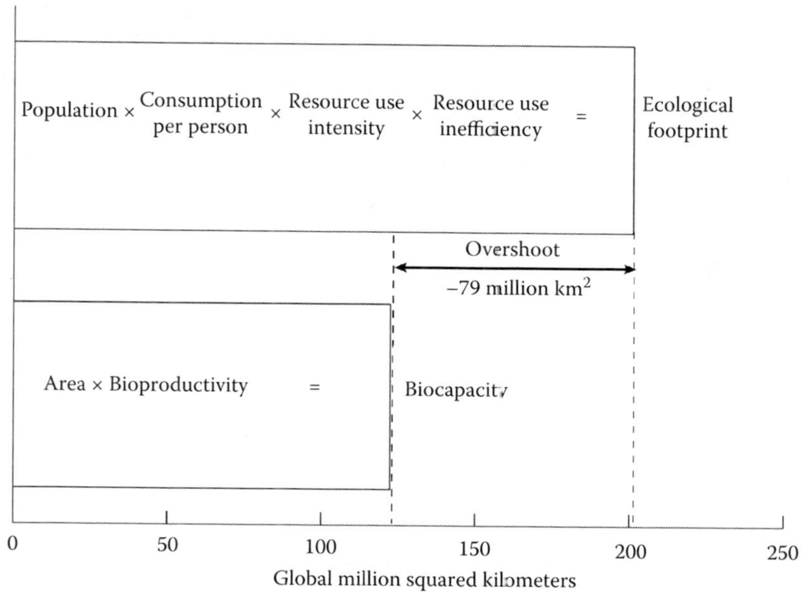

FIGURE 3.3 Global biocapacity and ecological footprint values in 10^6 km^2 in the year 2008. (Data from Hails, Chris. 2008. "Living Planet Report." World Wildlife Federation, Zoological Society of London, and the Global Footprint Network. http://www.footprintnetwork.org/download.php?id=505.)

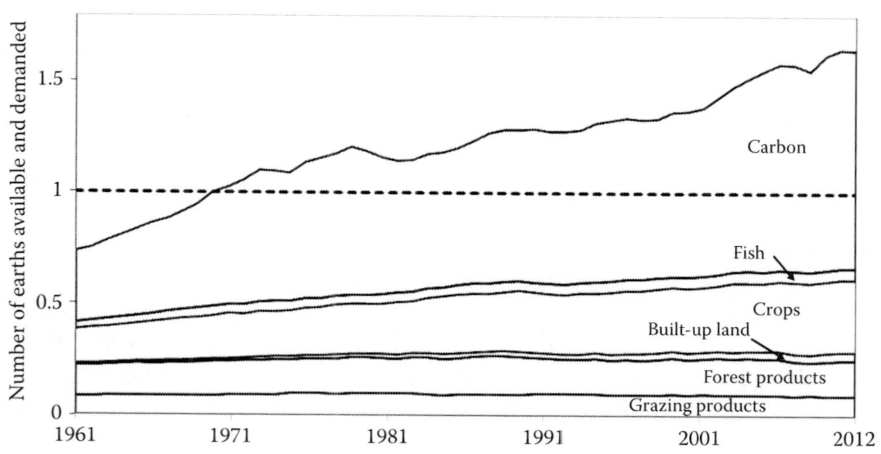

FIGURE 3.4 Global biocapacity (dashed horizontal line) and components of the global ecological footprint from 1961 to 2012, all divided by the biocapacity, which yields the number of earths. (©Global Footprint Network 2016. National Footprint Accounts, 2016 Edition. Licensed and provided solely for noncommercial informational purposes. Contact Global Footprint Network at www.footprintnetwork.org to obtain more information or obtain rights to use this and/or other data.)

footprint to one Earth by reducing one or more of the factors in $I = P \times A \times C \times T$. In 2011, the largest component of the global ecological footprint was the **carbon footprint**, followed by cropland (Figure 3.4). This makes clear that society must focus on decreasing the carbon footprint, which as we'll see is best accomplished by reducing the use of fossil fuels.

Given the current population, how much do we have to decrease the average EF to reach a sustainable level? If everyone on Earth shared the global biocapacity equally, then Americans would

TABLE 3.2
Comparison of Resource Consumption Levels for Groups with Low, Average, and High Ecological Footprints

Consumption Measures	Fair Earth-Share: 1 Planet	World Average: 1.5 Planets	High Consumption: 3 Planets
Daily calorie supply	2,424	2,809	3,383
Meat consumption (kg/year)	20	40	100
Living space (m^2)	8	10	34
People per household	5	4	3
Home energy use (kWh/year)	2,300	3,500	9,300
Motor vehicle travel (km/year)	582	2,600	6,600
Air travel (km/year)	125	564	2,943
CO_2 emissions (tons/year)	2	4	14

Source: Moore and Rees (2013). From *State of the World 2013: Is Sustainability Possible* by The Worldwatch Institute. Copyright ©2013 Worldwatch Institute. Reproduced by permission of Island Press, Washington, DC.

have to reduce their EF by 79%.* Realistically, each country will have to keep their EF less than their biocapacity. For the United States in 2012, U.S. biocapacity was 3.8 gha and EF 8.2 gha, so the average American would have to reduce his or her EF by almost 50% (Global Footprint Network 2016). We can achieve this reduction without sacrificing our quality of life. Much of this book outlines how we can achieve this goal, but for now, suffice it to say that a person can most effectively reduce his or her EF by reducing his or her use of the ABCs of unsustainability: automobiles and airplanes, beef, and coal.

Table 3.2 compares countries that use their fair share of one planet, the world average of 1.5 planets, and high-consumption countries of 3 planets (the average American in 2011 consumed 3.9 planets). People in high-consumption countries travel more and live in larger homes housing fewer people. They also consume more calories, especially meat, and cause emissions of much larger quantities of CO_2.

Because EF = $P \times A \times C \times T$, the average EF or environmental impact per person should increase as average income A increases, which is what we observe when comparing countries (Figure 3.5). Countries plotting above the solid line in Figure 3.5 have ecological footprints that exceed global per capita biocapacity. One of the primary goals of sustainability is to decouple environmental impact and economic prosperity. For example, although Brunei plots above the solid line that defines the global biocapacity (the maximum sustainable EF),† it manages to have a relatively low EF while having an average income level similar to the United States. Norway has roughly double the per capita GDP of the United States but still has a significantly lower EF. The Dominican Republic is an example of a country with EF = 1.5 gha less than the global biocapacity of 1.7 gha, but that still has moderate values of GDP ($6,000) and HDI (0.71). By living sustainably, we can move toward the solid line that defines sustainability without sacrificing our prosperity.

* From the Global Footprint Network 2016 Public Data Package (www.footprintnetwork.org/atlas), the average American EF = 8.2 gha. The global biocapacity is 1.7 gha. The number of Earths required for everyone to live like an American = 8.2/1.7 = 4.7 Earths. To reduce the consumption level to one Earth would require cutting consumption to 1/4.7 = 0.21 or 21%, meaning Americans must reduce their consumption by 100 – 21 = 79%. Put another way, if we decreased the American EF to 0.21 × 8.2 = 1.7 it equals the global biocapacity.

† Of course each country has a different biocapacity. In the absence of international trade, each country should aim to be self-sustaining by maintaining its EF below its biocapacity.

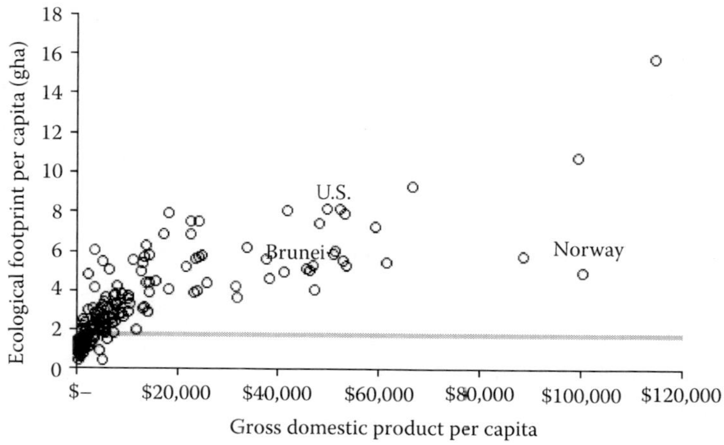

FIGURE 3.5 Gross domestic product at purchasing power parity (international dollars per person) versus ecological footprint (global hectares per person) using 2012 data. The horizontal line indicates global per capita biocapacity of 1.7 gha. (GDP data are 2012 values from the International Monetary Fund World Economic Outlook Database, Published 2015. EF data for 2012 ©Global Footprint Network 2016. National Footprint Accounts, 2016 Edition. Licensed and provided solely for noncommercial informational purposes. Contact Global Footprint Network at www.footprintnetwork.org to obtain more information or obtain rights to use this and/or other data.)

A map with symbols proportional to "# of Earths" for each country shows large disparities in average national per capita ecological footprints (Figure 3.6). The higher the # of Earths for a country, the more unsustainable are its citizens, and the more they exceed their fair shares of global resources. In general, countries that are highly industrialized have high per capita ecological footprints. Examples include many EU countries, the United States, and Australia. Interestingly, most countries with high ecological footprints fall within a relatively narrow range of latitudes in the northern hemisphere.

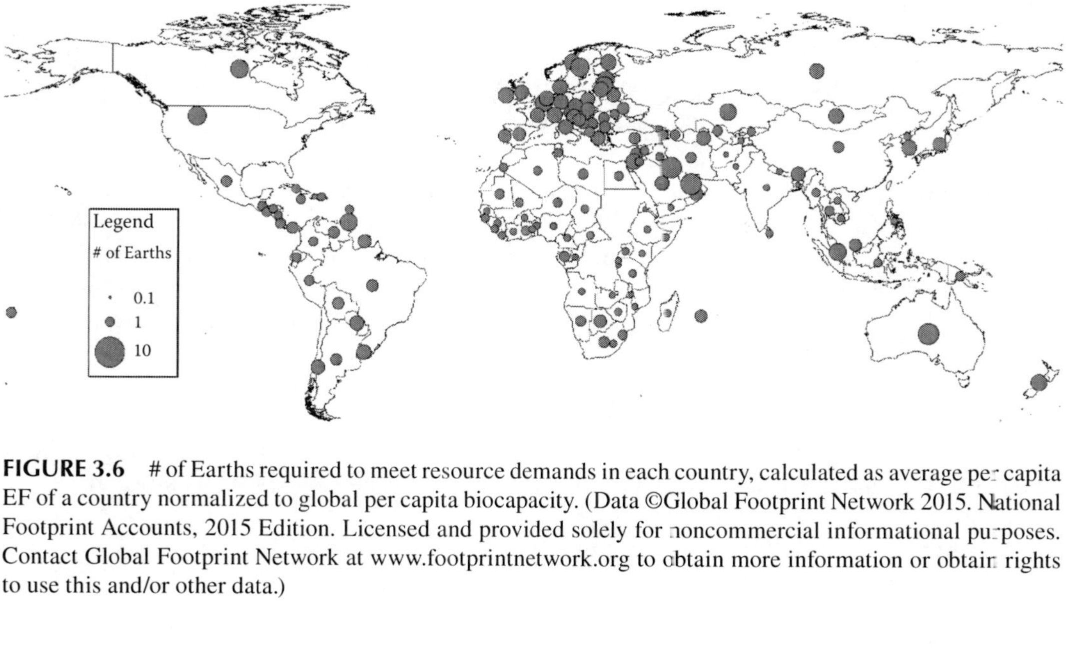

FIGURE 3.6 # of Earths required to meet resource demands in each country, calculated as average per capita EF of a country normalized to global per capita biocapacity. (Data ©Global Footprint Network 2015. National Footprint Accounts, 2015 Edition. Licensed and provided solely for noncommercial informational purposes. Contact Global Footprint Network at www.footprintnetwork.org to obtain more information or obtain rights to use this and/or other data.)

Because EF and therefore ImPACT have been increasing for decades (Figure 3.4), reductions in C and T are not keeping pace with increases in P and A. Sustainability can be achieved only by decreasing the environmental impact as measured by the EF until it is less than the biocapacity. The ImPACT identity tells us that society can decrease its ecological footprint and become more sustainable by decreasing population P, consumption $A \times C$, and resource use inefficiency T. Let's look at each of these factors in more detail, focusing on ways to decrease each factor and reduce the overall environmental ImPACT.

3.4 POPULATION

3.4.1 Problem: Exponential Growth of Population

> The raging monster upon the land is population growth. In its presence, sustainability is but a fragile theoretical construct.
>
> **E.O. Wilson**
> *The Diversity of Life, 1999*

When human numbers were small, the atmosphere and oceans could adequately absorb our wastes. Now with over seven billion people and population still increasing at an exponential rate, we have overwhelmed the capacity of the environment to absorb our wastes, so they have begun to accumulate on a global scale. For example, fossil fuel burning has increased the concentration of carbon dioxide in the atmosphere, which is causing global climate change and ocean acidification (Chapter 7). Currently we have a state of **overpopulation** because we have more people than the Earth can support over the long term, that is, the global ecological footprint or environmental impact is greater than the global biocapacity, and we are in ecological overshoot.

In 1798 Thomas Malthus published a paper titled "An Essay on the Principle of Population." Malthus argued that agricultural production would limit human population growth. Unchecked population growth could lead to a **Malthusian catastrophe** in which widespread starvation would reduce the population to a level that could be supported at a subsistence level. This bleak view of the future was later supported by studies of animal populations. For example, at times when food is abundant, deer populations increase at an exponential rate, but when food is scarce for extended periods of time (due to drought, extended winters, etc.) deer will die of starvation in large numbers, and the deer population plummets. This cycle can be repeated indefinitely. **Malthusianism** has more recently been extended from food to all resources that are essential for maintaining our current lifestyle, including metals and oil, and people who subscribe to this view are labeled "Malthusians." Malthusians subscribe to the philosophy of Malthus that resources are finite and are threatened by human population growth, and that unrestrained growth eventually leads to collapse.

In contrast, a **Cornucopian** contends that Malthusian limits do not apply to human populations because our intelligence can overcome those limits, so effectively there are no limits to growth of population or the economy. The debate between Malthusians and Cornucopians is embodied in the **Simon-Ehrlich wager** that scientist Paul Ehrlich made with economist Julian Simon in 1980. Ehrlich posited that population growth would increase demand on a limited supply of metals, causing the price of those metals to increase in one decade. Simon won the bet because the price of all five metals decreased due to market competition.*

Malthus predicted future food shortages because population grows faster than food supply. Population grows exponentially, while food supply grows arithmetically (linearly). Why the difference? A quantity grows exponentially when its increase is proportional to what is already there. In

* Simon lost a less-known wager that he made with David South of the Auburn School of Forestry in 1995. Simon wrongly bet that timber prices would decrease in five years.

exponential growth, the amount of increase rises from one period to the next, while in linear growth the amount of increase per unit time remains the same. Absent limits, everything that reproduces grows in number at an exponential rate because individuals added to the population also reproduce. Compound interest works the same way: interest added to the principal each year also draws interest in future years.

The rate of population growth is usually expressed as an annual percentage of the population. Currently global population growth is ~1.1% per year. The **doubling time** is the amount of time it takes to double the size of the population. It can be easily estimated by dividing 70 by the annual percentage growth rate. Thus, the current population doubling time is 70/1.1 = 64 years.

3.4.2 Trends in Human Population

For most of history, human population was small and grew slowly. **Leibig's Law of the Minimum** states that the resource in shortest supply will limit growth in a species' population, and for most of human history the limiting resource was food. In the nineteenth and twentieth centuries, however, the **Industrial Revolution** and the **Green Revolution** allowed for the production of ever-increasing amounts of food. This eliminated the main constraint on human population growth, so the population subsequently grew rapidly. Increased longevity due to improvements in medicine and hygiene has further quickened the pace of population growth. The last 100 years alone has seen a spectacular increase in global population from roughly one to seven billion people, with the population continuing to rise because births exceed deaths each year (Figure 3.7). To be sustainable humanity must achieve a steady state where the birth rate equals the death rate. The latest projections show that while the gap between births and deaths will likely decrease over the twenty-first century, it will not be closed until after the twenty-first century (Figure 3.7).

Because the total environmental impact of humans is proportional to human population (Equation 3.2), an increase of six billion people over the last 100 years has had an enormous effect on the environment. As of 2014, the human population is 7.2 billion, and there is an 80% probability that by the year 2100 population will increase to between 9.6 billion and 12.3 billion (Gerland et al. 2014). Let's split the difference and assume that population will increase to 11 billion, a 53% increase over today's population. Given that 7.2 billion people now have a global average EF of 1.4 Earths, then increasing the population to 11 billion without changing per capita EF would increase the total EF

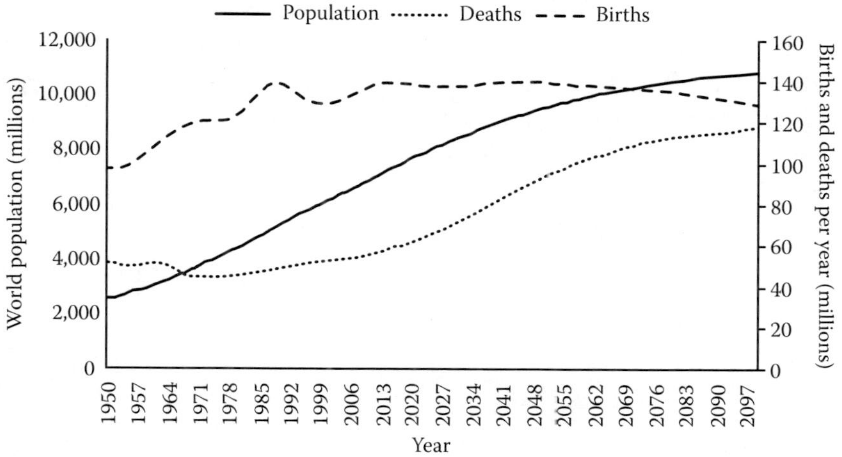

FIGURE 3.7 Historical trends in births, deaths, and population. Numbers from 2011 to 2100 are projections. (Data from UN World Population Prospects: The 2012 Revision. Retrieved March 28, 2015 from http://esa.un.org/wpp/Excel-Data/Interpolated.htm.)

to over 2 Earths,* an overshoot of more than 100%. Experts expect much of that growth to occur in developing countries that are least able to sustain it (e.g., India and many countries in Africa) (Friedman 2008).

Ecological studies of animal populations and computer simulations using system dynamics modeling lead to the generalization that systems that have physical limits (fixed amounts of available resources) and that experience exponential growth eventually become unstable and collapse (Meadows, Randers, and Meadows 2004). Stable systems are characterized by little change; conversely, systems undergoing rapid change are unstable, and large, rapid changes in one part of the system (e.g., human population) will cause large, rapid, and sometimes unpredictable changes in other parts of the system.

To reduce the risk of a Malthusian catastrophe, we need to reduce the human global environmental impact $I = P \times A \times C \times T$ by reducing population P, reducing consumption $A \times C$, or reducing inefficiency T. In the 1970s environmentalists like Paul Ehrlich advocated for decreasing the population, but this approach has fallen out of favor for several reasons. First, many people view childbearing as a fundamental human right, and coercive measures to limit **fertility** (number of children born per woman) in countries like China has led to widespread condemnation.† Second, population growth seems to slow naturally as countries develop. When the standard of living and life expectancy increase, average family size and birth and death rates decline, a phenomenon called the **demographic transition**. This transition results in a general trend of decreasing population growth rate with increasing per capita income (see the chart at www.bit.ly/1ybHrNK). Countries in Africa and in failed states like Haiti generally show the highest birth rates and population growth rates, resulting in a high fraction of young people (Figure 3.8).

Currently the world's developed countries have very low birth rates, sometimes below the **replacement level** of roughly of 2.1 children per woman, causing population to be steady or to even decline. However, the population is still growing at an exponential rate in developing countries. The population of these countries will not stabilize for a long time because of **population momentum**: a high proportion of the population in countries like China and Haiti is of child-bearing age (Figure 3.8), so the population will increase even if fertility drops (Diamond 2005). Because the amount of nonrenewable resources is fixed, and the renewal rate of renewable resources is relatively constant, exponential growth in population results in an exponential decline in the average amount of resources available to each person.

For systems as complex as the Earth and global society, it's impossible to predict if and when a global collapse will occur. Local collapses resulting from population pressures have already occurred in countries like Rwanda (Diamond 2005; Brown 2009). All we know is that the longer we stay in a state of ecological overshoot and the greater the degree of overshoot, the more likely a global collapse becomes.

History has shown that when agricultural output increased, population followed suit (Heinberg 2010). This occurred in ancient China and other civilizations, and more recently it happened globally as a result of the Green Revolution (de Vries 2013). Therefore, our goal should not be to figure out how to feed more people, but to adopt policies that decrease the population growth rate to zero.

3.4.3 Solution: Stabilize Population

> We have been god-like in the planned breeding of our domesticated plants, but rabbit-like in the unplanned breeding of ourselves.
>
> **Arnold Toynbee**
> *English Economic Historian*

Because we are already in ecological overshoot at 1.6 Earths, to slow the growth of human ImPACT we must decrease population growth to zero as quickly as possible. It is particularly important

* $11/7.2 = x/1.4$, $x = 2.1$ Earths.

† China rescinded this policy in 2015, see the case study in Section 16.2.

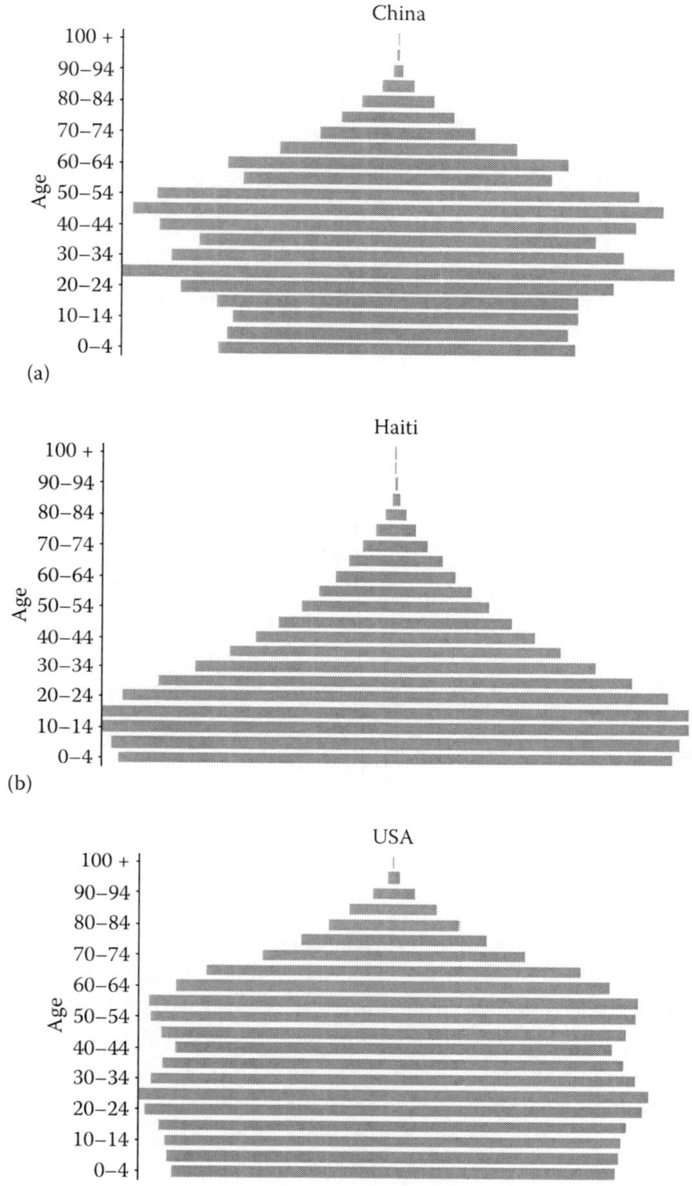

FIGURE 3.8 Population pyramids for selected countries: (a) China, (b) Haiti, and (c) USA. (2016 data from U.S. Census Bureau International Data Base, http://www.census.gov/population/international/data/idb/informationGateway.php, downloaded April 11, 2016.)

to limit growth in developed countries that have high per capita EF (Goodland and Daly 1996). However, population growth rates are highest in developing countries that have not yet achieved the demographic transition. Evidence shows that the demographic transition can be accelerated through family planning, education of women of child bearing age, economic growth, and urbanization.

Population growth ceases when the average fertility rate drops to the replacement level of 2.1 children per woman (the number is higher than 2 because of premature mortality). The fertility rate has already decreased to below 2.1 in developed countries that have passed through the demographic transition (Meadows, Randers, and Meadows 2004). To foster the success of population growth policies in developing countries, planners must use a comprehensive approach that recognizes the rights and needs of women and the poor while addressing the root causes of overpopulation. They must address family planning and reproductive health, mother and child health, women's rights, and women's education. For example, low birth rates correlate with high levels of literacy, school enrollment, and graduation (see the chart at www.bit.ly/1al8hYA). Not only do educated women have fewer children on average, but they also delay childbirth until later in life, which further contributes to a decline in population growth rates.

To stabilize the population, women need access to family planning services. The country of Iran began a remarkably effective family planning program that provides all contraceptives free of charge and requires couples to take a class on contraception before receiving a marriage license (Brown 2009). Between 1987 and 1994, Iran's population growth rate decreased 50%. An added benefit of providing contraceptives is that their use greatly reduces the number of abortions. Providing contraception to the 200 million women who do not have access could prevent 52 million unwanted pregnancies, 22 million induced abortions, and 1.4 million infant deaths (Brown 2009).

Birth rates in the developing world are high partly because having more children helps guarantee the care of aging parents and the continuation of the family lineage. Producing large numbers of offspring is a form of insurance. However, for parents with limited resources, a more sustainable, effective, and humane approach is to have fewer children to invest their resources in. Higher per-child investments mean more and better quality food and better education and healthcare, which increase health, longevity, and productivity.

Urbanization helps speed the demographic transition by providing more education and economic opportunities, particularly to women, resulting in a decrease in fertility rate (see the chart at www.bit.ly/1aKaptz). This is because women who move to cities gain access to birth control and education and acquire new economic opportunities. In villages children are an asset, but in cities they become a liability, and so women begin to invest more in fewer children. Even in squatter cities the fertility rate is markedly lower than in villages.

In developing countries urbanization is occurring most rapidly, and the fertility rate is falling fastest. For example, in Mexico the birthrate was 6.5 in the 1970s, but only ~2 in 2008, and still falling. However, high fertility rates are expected to persist in Africa, where the population is expected to increase from ~1 billion in 2014 to between 3.1 billion and 5.7 billion by 2100, partly because the supply of contraceptives has never kept pace with demand (Gerland et al. 2014).

The global population will continue to rise through the twenty-first century, but if urbanization continues 80% of those people will live in cities that occupy only 3% of land area. In the meantime, because population growth leads to economic growth, the global south will contain new cities with young people and high economic growth rates, while the global north will contain old cities with old people and low economic growth rates (Brand 2009). The best measure of the aging of a population and resulting average productivity decline is the **potential support ratio**, equal to the number of working age people ages 20 to 64 divided by the number of retirement age people older than 65 (Figure 3.8). In the United States, this rough measure of the number of workers per retiree is projected to decline from 4.6 in 2014 to 1.9 by 2100, which will likely lead to stagnation of the economy (Gerland et al. 2014).

Declining population growth rates likely will slow economic growth, which could leave less money for environmental protection, which would cause greater environmental impact (Kuznet's curve in reverse). However, because urbanization accelerates innovation, even cities with declining population might be able to maintain positive economic growth. If this happens, scientists and economists must learn how to decouple economic growth from environmental degradation. We must figure out how to make cities green while protecting newly emptied countryside. Adoption of the principles of **urban ecology**, **urban agriculture**, and **New Urbanism** may ultimately lead to the development of sustainable cities (Section 5.4.2 on urbanization).

There is debate in the literature about how much future environmental impacts can be reduced by focusing on slowing population growth. Some argue that demographic momentum is still so large that even rapid adoption of radical population control measures such as "one child" policies would not decrease global population by 2100, while others argue that such demographic scenarios are unrealistic, and furthermore that any decrease in population growth yields dividends in the form of reduced environmental impacts such as greenhouse gas emissions (see Bradshaw and Brook 2014 and rebuttals). Some argue for reducing the future environmental ImPACT by adopting policies that limit growth in *P*, while others maintain we should focus on reducing resource consumption $A \times C$. From a sustainability perspective, it seems prudent to take measures to reduce all of the factors that make human society unsustainable. In later chapters we will examine how individual actions and social policies can effectively accomplish this, but for now suffice it to say that to be sustainable, you should not have more than two children. (See World Population Clock: http://www.census.gov/ipc/www/popclockworld.html.)

3.5 CONSUMPTION

In developed countries, the demographic transition has greatly slowed the rate of increase of the human population. As a result, consumption growth has replaced population growth as the primary cause of environmental degradation and increasing ecological overshoot (Pearce 2009). Rapid increases in the per capita consumption rate in both the developed and developing world have magnified the environmental impact of population growth.

Consumption drives production, which in turn relies on natural resources. We use products and then return them to the environment as waste. Production and disposal, and sometimes product use, increase environmental stress. As stated previously, the ImPACT identity $I = \text{EF} = P \times A \times C \times T$ expresses the total environmental impact *I* of the human population as the ecological footprint. The effect of consumption is given by $A \times C$, the affluence or income *A* times the resource use intensity *C* (Waggoner and Ausubel 2002). The ImPACT identity tells us that a person's impact on the environment increases as his or her consumption rate increases.

Globally, per capita consumption expenditures almost tripled between 1960 and 2006. Prosperous nations in the developed world that have high consumption and low population like the United States have a greater negative impact on the environment than more populous countries in the developing world with low consumption and high population like India. The average American consumes 88 kg (194 lb) of resources daily (Assadourian 2010). Much of these resources now come from overseas, giving Americans the impression that we have limitless resources, when in fact we have already exceeded the capacity of our own country to provide much of the resources we consume (Adams 2006).

Some products that we consume cause more environmental damage than others. These bad actors can be remembered as the "ABC's of unsustainability": the use of autos and airplanes (A), beef (B), and coal (C). Use of these products is unsustainable because the environmental impacts are very high (sustainable transportation is discussed in Section 11.2.1.1, beef in Chapter 13, and coal in Section 9.3). Full-cost accounting using **Life Cycle Assessment** (see Section 3.6.2) shows that the true costs of these products are much greater than their benefits. The inclusion of autos and airplanes, beef, and coal as the ABCs of unsustainability is well supported. Many studies demonstrate that these products have the greatest environmental impact over their lifecycles (Tukker and Jansen 2006). Transportation, food, and housing together make up 70% of environmental impacts across all categories of consumption, and the specific products that cause the most environmental damage in those categories are autos and airplanes, meat (especially beef) and dairy, and electricity produced using coal. To become sustainable our society must decrease overall consumption, focusing on eliminating the most harmful forms of consumption such as the ABCs of unsustainability.

3.5.1 Problem: Excessive Consumption and Materialism

> Overconsumption is not a national religion or patriotic duty, but a highly destructive and often morally questionable pursuit.
>
> **David Wann**
> *2007, p. 26*

Consumerism has become a dominant theme in western culture. Consumers find "meaning, contentment, and acceptance through what they consume" (Assadourian 2010). Our consumer culture began to flourish when World War II ended. The enormous industrial capacity developed for defense during the war had to be converted to peacetime use. In 1955, Victor Lebow wrote in the *Journal of Retailing*, "Our enormously productive economy demands that we make consumption a way of life, that we convert the buying and use of goods into rituals, that we seek our spiritual satisfaction, our ego satisfaction, in consumption…We need things consumed, burned up, worn out, replaced, and discarded at an ever-increasing rate." Thus, corporations sought new markets for their products as their production capacity expanded. Consumer demand rose with increasing population and drove economic growth. Further increases in consumption required escalation in the per capita consumption rate. Corporations achieved this by expanding **advertising** to convince people to buy products they did not need, and by using **planned obsolescence** to increase the frequency of product purchases. After decades of being bombarded by advertisements, most Americans unquestioningly accept materialism as a way of life. The upsurge in per capita consumption rates has led to an increasingly unsustainable American lifestyle. Materialism is rampant in our culture. It is perhaps the most important social force in our society. It drives our economy, fuels our desires, and preoccupies our minds (as a bumper sticker reads, "Born to shop, forced to work").

Many studies have shown that, once an income level is achieved that satisfies basic human needs, further increases in income do not increase happiness or life satisfaction (Speth 2008). Income and environmental ImPACT measured by the EF are correlated (Figure 3.5); both can be used as measures of consumption. Using these measures, studies of developed countries show that people who consume little are just as satisfied as those who consume a lot (Simms 2008). Many countries in Europe such as Spain and Norway with relatively low per capita consumption rates (measured by EF) have high levels of well-being (measured by HDI) comparable to those in countries such as the

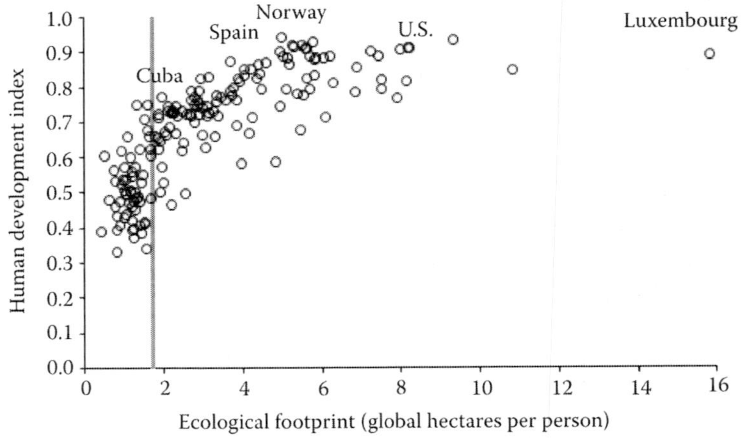

FIGURE 3.9 Ecological footprint (EF in hectares) versus the Human Development Index (HDI). The vertical line indicates global per capita biocapacity. (HDI data are 2012 values from UNDP Human Development Reports, 2015. EF Data are 2012 values ©Global Footprint Network 2016. National Footprint Accounts, 2016 Edition. Licensed and provided solely for noncommercial informational purposes. Contact Global Footprint Network at www.footprintnetwork.org to obtain more information or obtain rights to use this and/or other data.)

United States with higher per capita consumption rates (Figure 3.9). One might ask how people in EU countries live more sustainably by consuming half as much as Americans without sacrificing well-being. Rosenthal (2009) argues that it's not because Europeans are morally superior or care more about the environment. It's because the designs of their cities encourage sustainability. Most homes are small, which reduces energy required for heating and cooling. Many residents have no room for washers and dryers, so they line-dry their clothes. Most houses do not have air conditioning, but are designed for efficient passive heating and cooling. The cities were established hundreds of years before cars were invented, so parking is limited. As a result, most European city dwellers rely on mass transportation. The population density is so high that cities have little space to sacrifice for waste disposal; thus, Europeans are forced to reduce their waste production greatly.

Clearly, above a certain point, a higher income and more material possessions do not guarantee a higher level of well-being or greater happiness. Yet, despite this, our society continues to live in a materialistic manner. According to Bill McKibben (2007), "our single-minded focus on increasing wealth has succeeded in driving the planet's ecological systems to the brink of failure, even as it's failed to make us happier…we kept doing something past the point that it worked. Since happiness has increased with income in the past, we assumed it would inevitably do so in the future." Remember that the goal of sustainability is to increase well-being, and in the developed countries this can be accomplished without further increases in affluence.

3.5.2 Solution: Reduce Consumption and Waste

> Too many people spend money they haven't earned to buy things they don't want, to impress people they don't like.
>
> **Will Rogers**
> *American cowboy and humorist*

> You can't have everything. Where would you put it?
>
> **Steven Wright**
> *American comedian*

Exponential growth of both population and consumption has exceeded the capacity of the Earth to provide needed resources for the human population. Each individual consumes resources and produces waste; the more people there are, and the more each person consumes, the faster resources become depleted and waste builds up. Here we focus on dematerialization strategies that promote conservation of resources. The positive impacts of dematerialization are manifested at both the individual and national level. They include increased economic security, decreased environmental degradation, and in many cases increased well-being. Dematerialization success can be measured by decreases in the ecological footprint resulting from decreases in A, the amount of money spent on a material and C, the material intensity.

Dematerialization is an important goal for achieving sustainability. It is a general term that applies to all types of materials. Specific examples include decarbonizing by reducing carbon intensity; reducing energy use by reducing energy intensity; and reducing water use by reducing water intensity. All are special cases or subsets of material intensity. The goal must be to reduce the material intensities for all products, which is primarily achieved through conservation measures (Table 3.1).

Dematerialization strategies include buying less, avoiding purchases of disposable and shoddily manufactured products, repairing rather than replacing, reusing (including repurposing or **upcycling**) and recycling, buying small rather than large, and saving rather than spending money (see Chapter 14). Dematerialization is like dieting. Dieters lose weight by reducing their calorie consumption. To do this successfully they buy less food; avoid buying unhealthy foods that might tempt them; spend a little extra to buy healthy food rather than cheap junk food; eat slowly and savor the food they have rather than thinking about the food they want; and eat small amounts by eating only until they feel full. Likewise, to reduce material consumption consumers would buy less material by buying only what they need; avoid buying disposable products and products with short lifetimes; spend a little extra on high quality products with long lifetimes; and spend time enjoying the products they have rather than in constant pursuit of new products. In both cases, it helps to start with the "low-hanging fruit," the easy changes that have a big impact as measured by ecological footprint reduction.

One of the easiest dematerialization strategies to adopt is to avoid disposable products. For example, instead of buying disposable water bottles, buy a BPA-free water bottle, carry it with you wherever you go, and refill it for free. Keep a reusable coffee mug in your car to avoid using disposable cups and lids, and keep reusable grocery bags in the car trunk for grocery shopping. The average American produces about 4.5 lb of garbage every day (Chapter 14); avoiding disposable products can greatly reduce this figure.

Another very effective dematerialization strategy is to downsize, that is, to buy small rather than large. For example, the curb weight of a Hummer H1 is 3,559 kg, while a Toyota Prius is 1,310 kg, so the manufacture of a Hummer H1 requires more than three times the mass of resources as a Prius. Furthermore, over its lifetime the Hummer will consume far more than three times the amount of gasoline as a Prius. Smaller houses require fewer resources and consume less energy. Recently, the **small house movement** has gained traction in the United States, with a popular TV show called *Tiny House Nation* and a 2013 documentary called *Tiny: A Story About Living Small*. Adopters find that the "tiny" lifestyle frees them from financial worries (a tiny mortgage and tiny energy and maintenance bills) and consumes much less of their time, freeing them to do the things they truly enjoy.

Some environmentalists believe they can live sustainably without lifestyle changes simply by buying products labeled as "green." By now most people are aware of the practice of **greenwashing** in which products that aren't truly green are labeled as such. Apps such as **Goodguide** can help consumers find products that are more sustainable than others. But even if the average American consumer always chooses the most sustainable purchasing options, if they do not practice the dematerialization strategies listed above they will have an ecological footprint much greater than one.

3.6 TECHNOLOGY

The pace of technological innovation is accelerating for many reasons. First, technological advances breed more advances in a positive feedback loop. Second, synergies between technologies enable new, often unanticipated applications. For example, multiple new technologies are leveraged together in a smart phone, a tool that is greater than the sum of its parts. The economic engine of capitalism is also driving technological change: venture capitalists fund start-up companies in Silicon Valley that develop and apply new technologies. Finally, Silicon Valley and many other technology hubs would not exist without the knowledge production centers of higher academics. The transistor chip and many other breakthrough technologies were discovered at Stanford University, near Silicon Valley. And many other universities have developed new technologies (medical, electronic, mechanical, etc.) and new methods (e.g., natural resource exploration, mining techniques such as fracking, and resource processing) that have led to technology spin-offs, business incubation centers, and creation of new corporations. For example, Google was the outgrowth of an NSF-funded project at Stanford University. Some technological advances have solved one problem but created others, while many have done more good than harm. The impact of technology on the environment is embodied in the term T of the ImPACT identity, and it can be reduced by increasing the efficiency of resource use (Table 3.1). On balance, technology has great potential to increase global sustainability, but below we will see that its impact is smaller than changes in population and consumption.

3.6.1 Problems

Technology has always been a double-edged sword for humanity, both a blessing and a curse. For all of the amazing technological advances that have improved human welfare, there are almost as many that have caused problems that usually were not anticipated. For example, agricultural pesticides originally promised to eradicate crop-damaging insect species, but were soon found to cause collateral damage to humans and ecosystems, and have become less effective over time as insects have developed resistance. The pesticide DDT was found to inhibit the reproduction of eagles, and chlorofluorocarbons used as refrigerants were found to cause damage to the Earth's ozone layer. Yet technological innovation, aided by basic scientific research, has been the primary driver of economic growth, especially in the United States. Science and technology (the application of basic science) do not always solve our problems, and sometimes create new problems, but on balance they have greatly increased the material welfare of humanity.

While technology has increased societies' economic capital, it has caused a net decrease in environmental capital because it has magnified the ability of humans to degrade the environment. For example, coal use in Great Britain during the Industrial Revolution enabled the new technologies of the steam engine, but led to intense air pollution that caused many deaths (Gieré and Querol 2010). Yet we know from the ImPACT identity that technology can reduce environmental impacts by reducing resource use inefficiency (making the T term in Equation 3.2 smaller). In fact, we already have the technological knowledge needed for society to live sustainably, but we have to convince decision makers that these technologies are worth investing in, and the public to adopt them. Also, we must remember that while technology may sometimes provide an easy fix to environmental problems, it has limitations. New technologies often require decades to fully implement because they require new infrastructure construction. For example, plans for construction of wind and solar energy installations have been put on hold in many places in the United States because high voltage transmission lines must first be put in place. Adoption of electric cars is slow in the United States because of a shortage of charging stations, limiting the distances that people can travel.

One problem often encountered during attempts to use technology to reduce environmental impacts is that the impacts are often simply moved to another process or to a different geographic location. To solve pollution coming from car tailpipes, engineers design cars fueled by hydrogen or electricity, but both increase the demand for electricity that is usually produced by polluting coal-fired power plants.* There is also the issue of cost. We know how to produce fresh water through desalination, but the process is very energy-intensive, and that makes desalination expensive.

Because energy use is the greatest component of societies' environmental impact, much attention is focused on developing technologies to increase energy efficiency. The public prefers to rely on energy efficiency technology to reduce energy consumption because, unlike energy conservation, it does not require lifestyle changes. However, advances in technology that improve efficiency usually don't reduce energy consumption as much as expected. Often consumers spend the energy savings by increasing use, an effect known as **Jevons paradox** or the rebound effect (Herring 2008). For example, fuel-efficient vehicles make people willing to drive farther. LED TVs are much more energy efficient than older designs like plasma and LCD, but this has only caused people to buy larger TVs that consume just as much power. The electrical efficiency of personal computers doubled every 1.5 years from 1975 to 2009, but computing speeds demanded by consumers also doubled every 1.5 years, causing the two factors to cancel out and make the energy consumption of desktop computers nearly constant (Koomey et al. 2011). Increases in computer monitor size and hard drive capacity have also offset efficiency gains.

The rebound effect tells us that if producers use technology to decrease resource inefficiency T, consumers then feel free to increase their consumption $A \times C$ which has units of mass of resource/

* However, note that even when taking this into account using life cycle assessment, electric cars still have lower environmental impacts.

person. If economies of scale and technology developments lead to lower cost per mass of resource, consumers then increase their resource consumption to the level they can afford. For example, advances in lighting technology have greatly decreased the energy inefficiency of lights in units of Joules/lumen, where Joule is the SI unit for energy and lumen is the SI measure of the amount of visible light produced. To calculate the energy footprint of lighting in Joules:

$$I_{\text{lighting}}(\text{Joules}) = P \times A \times C \times T = \text{persons} \times \text{dollars/person} \times \text{lumens/dollar} \times \text{Joules/lumen} \quad (3.7)$$

Because consumers can get more lumens/dollar, they increase $A \times C$, the amount of lumens they consume (lumens/person). Thus, substitution of LED lights for incandescent lights simply leads consumers to use more lighting (Tsao et al. 2010). Fortunately, the rebound effect usually does not negate all efficiency gains. Estimates of the rebound effect range from 5% to 40%, meaning 60% to 95% of efficiency gains translate into reduced resource consumption (Greening, Greene, and Difiglio 2000).*

More important than the rebound effect is incomplete adoption of sustainable technologies and lifestyles, even when it is in a person's best interests. Homeowners don't implement energy efficiency measures in their homes even when upfront costs are low and potential financial savings are high. People eat unhealthy foods and avoid exercise even when those choices make them unhealthy and decrease their longevity. Individual barriers to adoption of sustainable behaviors include ignorance and skepticism, which can be overcome if consumers are effectively educated so that they make good decisions. Barriers to adoption of sustainable technologies include high up-front costs, bureaucratic obstacles, and lack of incentives. These problems and potential solutions are discussed for energy in Chapter 11.

3.6.2 Solutions

Technological advances are needed to reduce humanity's environmental ImPACT. For example, fossil fuels should be replaced by renewable energy sources that have lower environmental impacts per unit of energy (Chapter 11). More public funds should be invested in renewable energy research, and these technologies should be provided to developing countries at low cost. In many cases, the necessary technologies are already developed and simply need to be implemented on a larger scale. Technology transfer from the developed to developing countries is essential to achieving a global sustainability transition (Goodland and Daly 1996).

The goal of **green technology** is to minimize environmental ImPACT, primarily through reduction in resource use inefficiency *T*. Green technologies require less raw materials and energy and produce less waste. The new green economy offers many commercial opportunities and is growing rapidly. It is being driven by green entrepreneurs and by engineers who are hired to solve problems and design green products (Brand 2009).

Examples of new technologies that promote sustainability include the development of philanthropic organizations that use new online payment technologies and inexpensive cell phones to give money directly to the poorest of the poor (e.g., https://givedirectly.org/); the recent discovery of high concentrations of precious metals in sewage and new technologies for recovering them (Westerhoff et al. 2015); and the use of medical devices and apps for tracking health, which increase health and decrease health expenditures, resulting in increased social and economic capital.

To assess whether a product is green people often look only at the operating costs. For example, an energy-star rated refrigerator is considered green because it uses less electricity and therefore will cost less to operate. That doesn't necessarily mean it is the greenest choice. What if extraction

---

* As an example, let's say a new, more energy efficient lightbulb uses only 50% of the energy per lumen as a traditional bulb, so the maximum energy savings would be 50%. The rebound effect would reduce the energy savings; for a moderate rebound effect of 25%, the energy savings would be 0.5(1 – 0.25) = 0.375 or 37.5%.

of raw materials for that refrigerator damaged an ecosystem, resulting in loss of environmental capital? What if children manufactured it in a sweatshop that was an unsafe and unhealthy working environment, causing loss of social capital? What if it leaked poisons after disposal, causing further loss of environmental capital? To estimate the true cost of any product we have to look at its effect on economic, social, and environmental capital at every stage of its life. The **life cycle assessment** adds up the capital costs or environmental impacts incurred at each stage of a products' life, including raw materials, production, use, and disposal (the acronym RPUD is useful for remembering these four phases of the life cycle).

Life cycle assessment considers process and product design in the management of materials (Manahan 2006). It gives a complete picture of a product's environmental impacts. It shows which parts of the life cycle most negatively affect the environment and should be targeted for reduction. For the consumer it's useful for comparing the impacts of two competing products. Sometimes the result of a life cycle assessment is a resource footprint or emission footprint. For example, the carbon footprint is a measure of the mass of CO_2 emissions accumulated through the life cycle of a product (Hertwich and Peters 2009). The ImPACT identity can be used to calculate the resource footprint at each stage of the life cycle, and then the footprints are summed to obtain the total life cycle resource footprint.

Besides ceasing unsustainable practices that are wasteful, inefficient, or harmful to society, the economy, or the environment, we must create new, sustainable practices. McDonough and Braungart (2002) explain this concept in their influential book *Cradle to Cradle*. They argue that we can do more than just reduce pollution and waste. We must also create positive effects, and move from being "eco-efficient" to "eco-effective." McDonough and Braungart encourage businesses to use **sustainable design** to make products that will enhance economic, environmental, and social health. Products should be biodegradable, so that at the end of their life cycles they will become food instead of waste. Microorganisms can convert waste into the valuable resource soil and save landfill space. This is true recycling, not simply **downcycling** where we recycle the product into a degraded form.

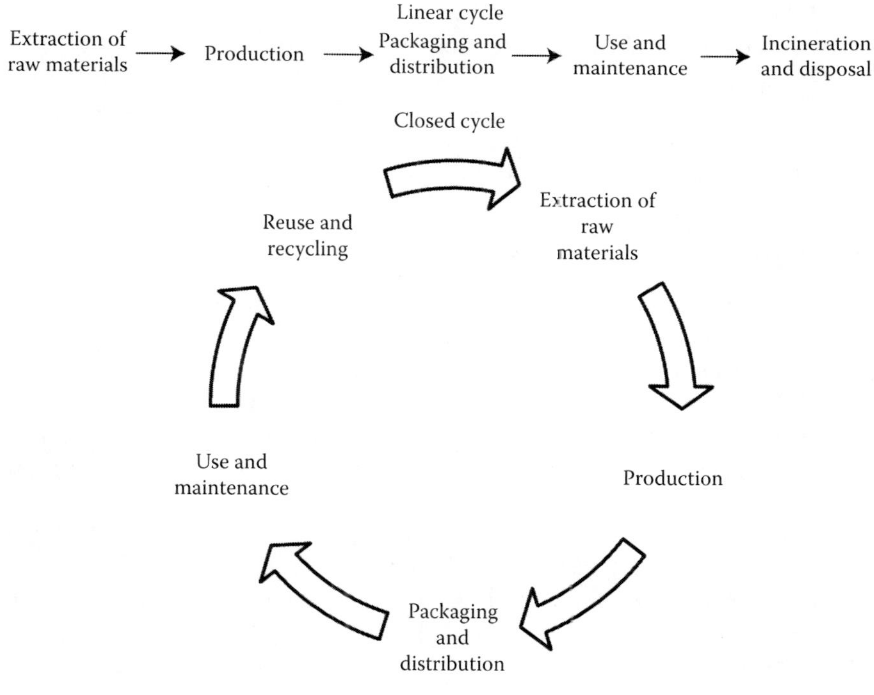

FIGURE 3.10 Comparison of linear or open material cycles and more efficient closed cycles.

The cradle to cradle approach also advocates that we design human industrial systems to mimic natural cycles in which nutrients (resources or materials) safely circulate in closed loops. Consider the life cycle of a resource (Figure 3.10). A linear cycle, or open loop, produces waste and ends at "incineration and disposal." Thus, rapid consumption of nonrenewable resources causes rapid depletion of the resource stock. Replacing "incineration and disposal" with "reuse and recycling" closes the resource loop and makes it more efficient. A completely closed loop with no leakage as waste is 100% efficient.

We can extend the lifetime of a resource by reducing the amount we consume and by closing the resource loop through reuse or recycling. If the resource loop for a nonrenewable resource is open, over time the production rate increases, peaks, and then begins to decline when supply cannot keep up with demand (more on this in Chapter 6). If we slow the rate of consumption by reducing resource intensity C through conservation, we delay the peak, and the decline is more gradual. The most sustainable approach is to both conserve and also reduce resource use inefficiency T through recycling, which closes the resource loop and further extends the usable lifetime of the resource.

The technological development of computers and the Internet in the late twentieth century has opened new opportunities for dematerialization. They may become the lifejacket that allows us to maintain our quality of life but with a greatly reduced ecological footprint. When people work online, socialize online, and obtain information and reading materials online, they greatly reduce their ecological footprints. Networking and the Internet have made communication and information transfer faster and easier and enabled new lifestyle choices such as **telecommuting** and the **sharing economy**. This has led to economic growth and reduced environmental impacts (Friedman 2006).

Miniaturization and other technological advances allow manufacturers to make more powerful devices using fewer resources. Furthermore, economic growth is increasingly driven by the information flow powered by the Internet rather than by the more environmentally and economically costlier physical flow of materials. This can lead to the desired decoupling between affluence and environmental impact.

Despite all of these improvements in energy efficiency over recent decades, the average energy consumption of Americans has remained relatively constant (see Chapter 11). Globally, increases in P and A have greatly outweighed reductions in C and T (Wilson 1998; Jackson 2008). Thus, the environmental impact of humanity as measured by the EF continues to grow despite improvements in energy conservation and efficiency. This suggests that technological advances alone are not effective at reducing environmental impacts, and that we are more likely to achieve sustainability by decreasing population P and consumption $A \times C$.

3.7 CONCLUSIONS

The global environmental impact is increasing rapidly due to increasing population and rising living standards in developing countries. Environmental impact is also increasing because people are migrating from developing to developed countries and adopting high consumption lifestyles (Diamond 2005). If humanity is to live sustainably, global population and consumption must stop increasing. Since we are already in overshoot mode, at least one must decrease (Engelman 2009). Failure to decrease the global environmental impact will eventually cause a decline in global human well-being.

Currently population is stable in the developed countries but rising rapidly in developing countries. Consumption rates are sustainable in most developing countries but not in the developed countries. Therefore, developing countries must focus on slowing population growth, and developed countries must focus on decreasing consumption. Using technology to increase efficiency is potentially an effective approach to reducing resource use and associated environmental impacts, but the benefits are often reduced by the rebound effect.

In this chapter we learned how to measure and reduce the environmental ImPACT. In the next chapter we will learn about the risks of not reducing it and how to reduce those risks.

WEB RESOURCES

- Consumption: The Story of Stuff with Annie Leonard: http://www.storyofstuff.com/
- Ecological Footprint Quiz by Redefining Progress: http://myfootprint.org/
- GoodGuide Ratings of Natural, Green and Healthy Products: http://www.goodguide.com/
- TED talk by Hans Rosling on population http://www.ted.com/talks/hans_rosling_religions_and_babies.html
- Tool for charting data related to sustainability: http://www.gapminder.org/

HOMEWORK PROBLEMS

1. Use www.gapminder.org to plot a trace of the population of your country over time to see if there is a correlation between:
   a. HDI and education
   b. Fertility and population density
   c. Longevity and income
2. Make a list of five examples of sustainable behavior/development and five examples of unsustainable behavior/development.
3. Propose a new sustainability index. How would you calculate it? Defend your choice of parameters and weightings. What are the advantages and disadvantages of your index?
4. If you were a scientist visiting a poor community in the developing world, what would you teach them that would significantly improve their lives and make them more sustainable?
5. Calculate your ecological footprint at www.myfootprint.org.
6. In 2016, the global biocapacity was 1.2×10^{10} hectares and human population was 7.4×10^{9}. What was the global per capita biocapacity in gha and in acres? How does this value compare to the average American ecological footprint? Your personal EF?
7. Calculate the doubling time if the annual growth rate is 3%.
8. If population doubled in 25 years, what is the annual growth rate in percent?
9. Use the calculator on the Sustainable World Initiative website (http://swinitiative.com/calculator/) to find a combination of the four factors of population, lifestyles, energy mix, and agricultural productivity that result in a sustainable world. Do so by adjusting the sliders until the number of Earths required to support the human population in 2050 decreases to 1.0. Record the values of the four factors in the following table:

Factor	Value
Projected population growth rate	
Projected change in average living standards (per capita ecological footprint) (%)	
Clean energy implementation (%)	
Agricultural yield changes (%)	

4 Risk, Resilience, and System Dynamics

4.1 INTRODUCTION

To this point, we have focused on sustainability from a resource management perspective, where resources are broadly interpreted as environmental, economic, and social capital. We discussed how in some cases human demands now exceed the ability of the environment to supply those demands. The risk is that environmental degradation can reduce the ability of the environment to supply ecosystem services while demand for those services continues to rise due to increasing population and consumption. Specifically, on a global scale, and in some places on a local scale, biocapacity is decreasing while the ecological footprint is increasing. In this chapter, we will look at sustainability from a different perspective: preservation of the things we value as individuals and as a society, which will lead us along a preferred future path. We will examine **risk** as the potential to cause deviation from that preferred path, and **security** as a form of insurance to prevent that deviation (terms defined in Table 4.1). We then look at **resilience** as a system property (where a system can be a forest, lake, a society, or the Earth) that describes how difficult it is to change a system state (terms defined in Table 4.2). Resilience is one component of **vulnerability**, which is a measure of susceptibility to harm or potential for change of a system when affected by a perturbation. Cyclical changes in systems are explored to gain a better understanding of long-term changes in dynamic systems and how "history repeats itself." We then examine system dynamics as a tool for understanding and modeling dynamic systems that can change between desirable and undesirable states. Finally, we will look at historical and modern examples of environmental and social collapse and briefly address strategies for avoiding collapse.

4.2 SECURITY AND RISK

All of us have experienced feelings of anxiety or dread when we feel vulnerable, or are confronted by great risk. Every day we are exposed to risks to our health, finances, and emotional security. Following Becker (2014), risks can be classified using the same categories we used to classify capital: environmental, economic, and social. There are three main types of environmental risks. *Hydrometeorological events* include floods, droughts, mudflows and avalanches, storms, heat waves, cold spells, and wildfires. *Geological events* include earthquakes and tsunamis, landslides, volcanic eruptions, and sinkhole formation. *Biological events* include disease, epidemics, and pests and invasive species. Economic risks to the individual include loss of uninsured property, job loss, and stock market losses, while on the societal scale financial crises can affect large numbers of people. Social risks include domestic accidents affecting individuals (e.g., falling down the stairs); industrial, infrastructure, and transport accidents; and environmental impact and pollution. Social risks from antagonistic events include conflict, terrorism, and crime. Many events involve multiple types of risk: the environmental hazard flooding can cause economic losses and the spread of water-borne diseases. All of these event types can reduce individual and collective human well-being.

The risk of a hazard is proportional to the magnitude of the consequence (severity) and to the probability that the event will occur (frequency). A plane crash has high severity because it causes many deaths and large economic losses, but the frequency of plane crashes is low. A car crash causes fewer deaths than a plane crash, but the frequency of car crashes is much greater. Which

TABLE 4.1
Terms Related to Risk

Term	Meaning
Capacity of response (adaptive capacity)	A system's ability to adjust to a disturbance, moderate potential damage, take advantage of opportunities, and cope with the consequences of a transformation that occurs.
Exposure	The degree, duration, and/or extent in which the system is in contact with, or subject to, the perturbation or risk agent. The presence of people, livelihoods, species or ecosystems, environmental functions, services, and resources, infrastructure, or economic, social, or cultural assets in places and settings that could be adversely affected.
Hazard	Threat to a system, comprised of perturbations and stress.
Perturbation	Major spikes in pressure outside the normal range in which the system operates, commonly originate beyond the system (exogenous).
Risk	Potential to cause deviation from a preferred path, or losing something of value. Depends on both the probability (frequency) and magnitude (severity of consequences) of an event.
Security	Degree of protection from harm.
Stress	Continuous or slowly increasing pressure on a system, within the range of normal variability and usually originating within the system (endogenous).
Vulnerability	Potential for change of a system when affected by a perturbation. Encompasses a variety of concepts including sensitivity or susceptibility to harm and lack of capacity to cope and adapt. Components include exposure and capacity of response, which includes resilience.

Sources: Gallopín, Gilberto C. 2006. "Linkages between Vulnerability, Resilience, and Adaptive Capacity." *Global Environmental Change* 16 (3): 293–303. doi:10.1016/j.gloenvcha.2006.02.004. IPCC 2014. "Summary for Policymakers." In *Climate Change 2014: Impacts, Adaptation, and Vulnerability. Part A: Global and Sectoral Aspects. Contribution of Working Group II to the Fifth Assessment Report of the Intergovernmental Panel on Climate Change*, edited by CB Field, VR Barros, DJ Dokken, KJ Mach, MD Mastrandrea, TE Bilir, M Chatterjee et al., pp. 1–32. Cambridge, United Kingdom: Cambridge University Press.

presents a greater risk? To make a fair comparison we must measure the risk of each event using the same scale. In this example, the comparison can be in terms of fatalities per person mile traveled:

$$\begin{aligned}\text{Risk} &= \text{severity} \times \text{frequency} \\ &= \text{average \# people killed/event} \times \text{\# events/person mile traveled}\end{aligned} \tag{4.1}$$

In the United States, between 2000 and 2009, risk in terms of passenger fatalities per billion passenger miles was 213 for motorcycles, 7.3 for cars or light trucks, 0.11 for buses, and 0.07 for commercial aviation (Savage 2013). Clearly, a person's choice of transportation has a large influence on the risk to which he or she is exposed. Other risk factors include whether the person is wearing a seatbelt or helmet and whether the driver's ability to drive is impaired.

Unfortunately, humans are not good at evaluating risk (more on this in Chapter 8). We tend to overemphasize the severity part of the equation, probably because we are not good at estimating probabilities (Kahneman 2011). Thus, many people are more afraid of plane travel than car travel, even though plane travel is safer because fewer people are killed per person mile traveled. The occasional plane crash that kills many people sticks in our minds more than the everyday car crashes.

An important component of human well-being is longevity. We all hope to live long, fulfilling lives. As individuals, we do our best to avoid unnecessary risks that may lead to premature death. We use many strategies to reduce our risk by decreasing vulnerability, our susceptibility to harm.

TABLE 4.2
Terms Related to Resilience

Term	Meaning
Actors or agents	The people who play a role in or have some influence on an SES.
Adaptability	The capacity of actors in a system (people) to manage resilience. This might be to avoid crossing into an undesirable system regime, or to succeed in crossing into a desirable one.
Adaptive capacity	Ability to adjust to change or damage, to take advantage of opportunities, or to cope with the consequences.
Adaptive cycles	The progression of SESs through various phases of organization and function. Four phases are identified: rapid growth, conservation, release, and reorganization. The manner in which the system behaves is different from one phase to the next with changes in the strength of the system's internal connections, its flexibility, and its resilience.
Diversity	The different kinds of components that make up a system. *Functional diversity* refers to the range of functional groups that on which a system depends. For an ecological system, this might include groups of different kinds of species like trees, grasses, deer, wolves, and soil. Functional diversity underpins the performance of a system. *Response diversity* is the range of different response types existing within a functional group. Resilience is enhanced by increased response diversity within a functional group.
Domain or basin of attraction	An attractor is a stable state of the system, an equilibrium state that does not change unless it is disturbed. The domain or basin comprises all of the system states that tend to change toward the attractor (think of a topographic basin).
Equilibrium	A steady state condition of a dynamic system where the interactions among all the variables or actors (e.g., species) are such that all the forces are in balance, and no variables are changing.
IPCC	The capacity of an SES to cope with a hazardous event or trend or disturbance and maintain the capacity for adaptation, learning, and transformation.
Resilience	The magnitude of a disturbance that can be tolerated before an SES moves to a different region of state space controlled by a different set of processes. Can be desirable or undesirable. Systems that are highly resilient are persistent.
Socio-ecological system (SES)	A system that includes societal (human) and ecological (biophysical) subsystems in mutual interaction. The SES can be specified for any scale from the local community and its surrounding environment to the global system constituted by the whole of humankind (the "anthroposphere") and the ecosphere.
State of a system	Defined by the values of the state variables that constitute a system. For example, if a rangeland system is defined by the amounts of grass, shrubs, and livestock, then the state space is the three-dimensional space of all possible combinations of the amounts of these three variables. The dynamics of the system are reflected as its movement through this space.
Sustainability	The likelihood an existing system of resource use will persist indefinitely without a decline in the resource base or in the social welfare it delivers.

Sources: Gallopín, Gilberto C. 2006. "Linkages between Vulnerability, Resilience, and Adaptive Capacity." *Global Environmental Change* 16 (3): 293–303. doi:10.1016/j.gloenvcha.2006.02.004. Walker, B, and D Salt. 2006. *Resilience Thinking: Sustaining Ecosystems and People in a Changing World*. Island Press. IPCC 2014. "Summary for Policymakers." In *Climate Change 2014: Impacts, Adaptation, and Vulnerability. Part A: Global and Sectoral Aspects. Contribution of Working Group II to the Fifth Assessment Report of the Intergovernmental Panel on Climate Change*, edited by CB Field, VR Barros, DJ Dokken, KJ Mach, MD Mastrandrea, TE Bilir, M Chatterjee et al., pp. 1–32. Cambridge, United Kingdom: Cambridge University Press.

Economic risk can be reduced by saving money in a secure financial institution and by buying various types of insurance. Health risk can be reduced by eating healthy, exercising, reducing stress (which can be accomplished by reducing risk), getting adequate sleep, avoiding toxins (including alcohol and cigarettes), and through regular health checkups. When our health is at risk, we need health insurance and an affordable and effective healthcare system to take care of us. We can reduce the social risk of domestic accidents by wearing helmets when riding cycles, wearing seat belts in cars, using crosswalks, and choosing safe methods of transportation. Environmental risks can be reduced by choosing a relatively safe place to live (outside a floodplain, and far from earthquake faults, volcanoes, and pollution sources).

One of the primary aims of sustainable development is to reduce risk, thereby increasing longevity and improving human well-being. Communities try to reduce public risk by adopting safety regulations (including environmental regulations), planning for disasters, and funding public health infrastructure (sanitation and hospitals). To achieve these goals, public policy changes should focus on minimizing years of life lost (YLL) at minimal economic cost by identifying the primary causes of YLL (Figure 4.1). Interventions should target primary causes of death that are easily preventable, and that disproportionately affect children, as childhood deaths result in greater person-year losses.

Of course, the risks we need to focus on reducing are the ones that kill or harm the most people. Figure 4.1 shows that most of the greatest risks are health-related. So why does this book emphasize environmental risks? Because health risks are often caused by or exacerbated by environmental factors. For example, respiratory infections often result from air pollution; diarrheal diseases are caused by improper waste management and resulting pollution; heart disease is caused in part by inadequate nutrition, which often results from environmental problems such as climate change; and many premature deaths result from environmental hazards including hydrometeorological, geological, and biological events (Becker 2014).

Three primary causes of YLL emerge as candidates for targeted public health programs. *Diarrheal diseases* can be eliminated at relatively low cost by providing water treatment facilities and sanitation. Improvements in these areas have been substantial: between 1990 and 2012, the percentage of the population using improved drinking water sources increased from 76% to 89%, and improved sanitation 47% to 64% (World Health Organization 2015). *HIV/AIDS* and *malaria* can be reduced through public health announcements that provide the information needed for people to reduce their risks and by providing inexpensive tools for combating the diseases (condoms for HIV/AIDS, mosquito nets for malaria). For these reasons, philanthropic organizations such as the Gates Foundation that use science to improve human well-being in developing countries focus on reducing the prevalence of these diseases. Successful implementation of preventive health measures is responsible for most of the increase in global average longevity from 64 years in 1990 to 71 years in 2013 (World Health Organization 2015).

It is also important to focus public health measures on contributing factors to premature death. Poverty may be the single largest contributing risk factor, as it contributes to many other risk factors. Poverty is linked to malnutrition in developing countries, obesity in developed countries, inadequate healthcare and education (especially health education), and generally unsafe living environments (high crime, unsafe homes, and transportation). Without money to buy health insurance, a safe car, or healthy food, poor people tend to have shorter lives. On the national scale, average longevity increases dramatically with increasing income until it plateaus at ~$30k (see the chart at www.bit.ly/1Jrlu2n), presumably because all important needs are met by this level of income. Sustainable development must ensure that minimum income approaches this level to reduce risks of premature death and lost productivity and social capital.

4.3 ENVIRONMENTAL RISKS

Here we will briefly discuss the environmental risks that are geological and hydrometeorological, which we collectively refer to as natural hazards (biological risks will be discussed in Chapter 15). Natural hazards have always existed, but increasing population and development have magnified their risks. Although not one of the leading causes of YLL, natural disasters frequently cause large economic losses, and sometimes large loss of life. As the global population and the value of infrastructure increase, and climate change increases the intensity of many hydrometeorological events (Chapter 7), the economic and social costs of all natural hazards will further increase.

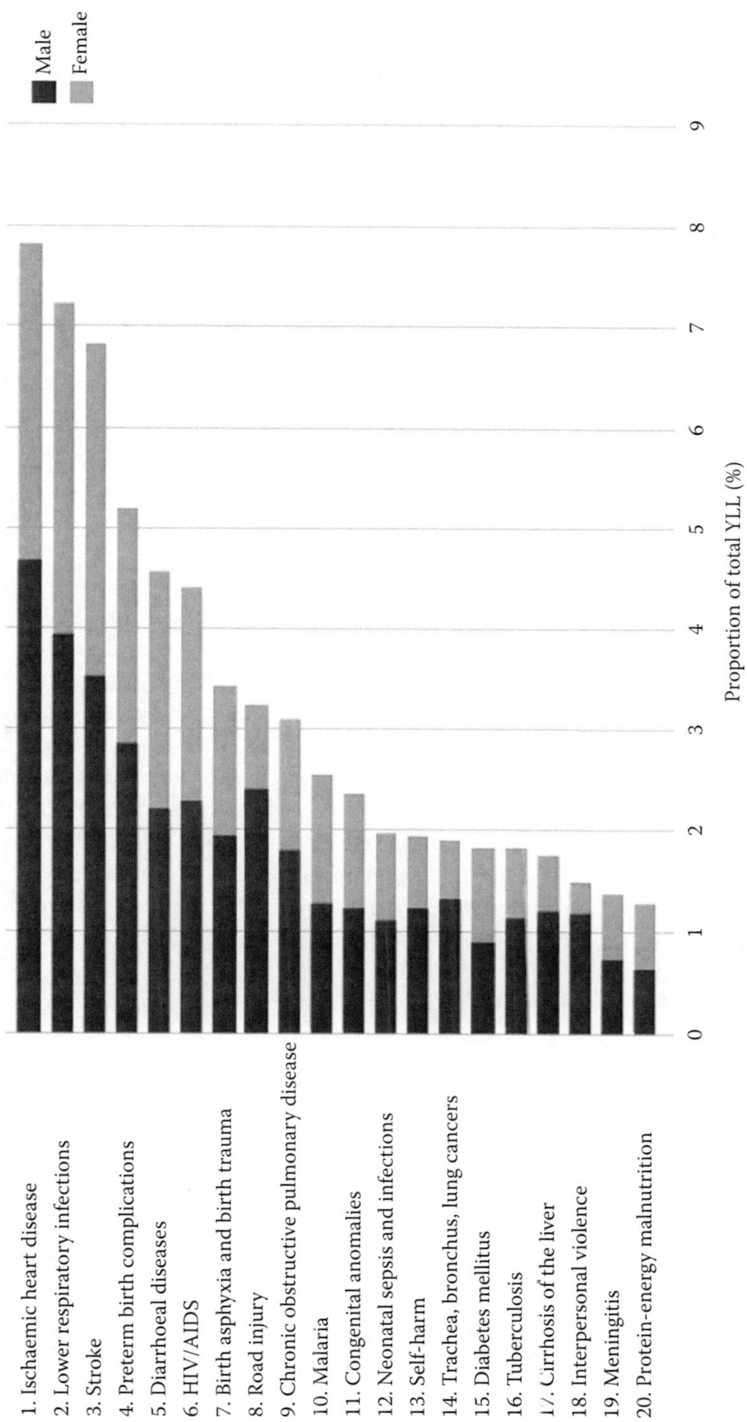

FIGURE 4.1 The 20 leading causes of years of life lost (YLL) due to premature mortality. (From World Health Organization, 2014. "World Health Statistics 2014." apps.who.int/iris/bitstream/10665/112738/1/9789240692671_eng.pdf.)

Dangerous natural processes include earthquakes, floods, volcanic activities, landslides, and storms. Many types of natural hazards tend to occur frequently in specific types of geographic settings. For example, earthquakes and volcanoes occur most frequently near active margins of tectonic plates such as the western coast of the United States. Landslides are also more likely to occur in these areas because they have high topographic relief and therefore steep slopes, and because earthquakes can trigger landslides.

Between 1900 and 2015, almost 14,000 natural disasters killed 33 million people (Figure 4.2). In developed countries, natural hazards cause fewer casualties but higher economic losses than in developing countries. For example, in southern California strict building codes make buildings safer but more expensive to build and repair. As a result, a magnitude seven earthquake in southern California would kill a much lower percentage of the population than it would in a poor, mostly undeveloped country like Haiti. Over time, smart development has reduced annual deaths from natural disasters, but economic costs have increased (Figure 4.2). A priority of sustainable development is to build infrastructure that is not vulnerable to natural hazards (Becker 2014).

We mentioned previously that risk is proportional to severity and frequency. These two quantities are typically inversely related. So, for example, low severity magnitude five earthquakes occur frequently, but high severity magnitude eight earthquakes occur infrequently. Low severity, high frequency hazards are not always destructive but common, while high severity, low frequency hazards are usually catastrophic but infrequent. Natural factors largely control frequency, whereas both natural and human factors control severity.

Most natural hazard risk arises from processes of moderate severity and frequency. Sustainable development in the form of proper land-use planning can reduce risks associated with these processes. Responsible developers should not build housing developments, hospitals, or schools on active faults, in floodplains, or on steep cliffs prone to erosion. These are also situations to avoid when choosing a place to live. Choose a location with solid bedrock, fertile soils, nearby bodies of fresh water, on high ground but without steep slopes. Following these guidelines will reduce your vulnerability to natural hazards.

The four steps in the disaster cycle are prevention (mitigation), preparedness, response, and recovery. Prevention involves making sure that structures such as bridges, river levees and dams, and buildings have sufficient structural integrity to withstand recurrent natural disasters. Preparedness involves identifying and assessing risks and then implementing plans to reduce those risks. Risk assessment identifies system components that are necessary to the response and recovery steps but are vulnerable to disruption, such as communication networks and provision of food, water,

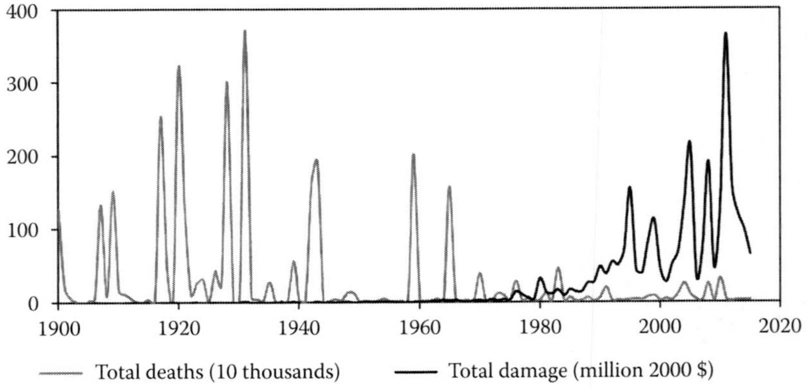

FIGURE 4.2 Global deaths and economic losses from natural disasters from 1900 to 2015. (Data from D. Guha-Sapir, R. Below, Ph. Hoyois—EM-DAT: The CRED/OFDA International Disaster Database, www.emdat.be, Université Catholique de Louvain, Brussels, Belgium.)

and medicine, and implements plans. Preparedness plans also provide for early warning networks and evacuation routes, which can save many lives when hurricanes or tsunamis strike. Response involves disaster relief efforts including rescue operations, provision of medical care and essential resources, and assistance with relocation. Recovery includes provision of food and shelter for disaster victims, as well as damage assessment and repair. This step can take many years even in developed countries, as shown by Hurricane Katrina.*

4.4 RESILIENCE AND SOCIO-ECOLOGICAL SYSTEMS

> What is perhaps most intriguing in the evolution of human societies is the regularity with which the pattern of increasing complexity is interrupted by collapse.
>
> **Joseph Tainter**
> *Sustainability of complex societies, Futures, 1995*

> Unlike sustainability, resilience can be desirable or undesirable…sustainability is an overarching goal that includes assumptions or preferences about which system states are desirable...confusion can be avoided by answering the question, "Resilience of what to what?"—that is, over what time period and at what scale.
>
> **S. Carpenter et al.**
> *From metaphor to measurement: Resilience of what to what? Ecosystems, 2001*

When a disaster strikes, some communities recover quickly, some slowly, and some not at all. **Resilience** is the capacity of a system to survive disturbances and still retain its basic structure, function, and internal feedbacks, that is, its identity. It is related to the capacity of response of a system, and therefore is one component of vulnerability (Gallopín 2006). As an example, before Hurricane Katrina struck in 2005, the city of New Orleans was vulnerable to hurricane risks because many areas were at elevations below sea level, protected by crumbling levees. Steps taken to reduce risk from frequent events, such as building levees to protect against flooding from Category 2 hurricanes, made New Orleans more vulnerable to large, rare events such as Category 5 hurricanes (Kates et al. 2006). Furthermore, by reducing flooding frequency, the levee system caused the sinking delta to lose the sediment source that sustained it.† Thus, many coastal wetlands that previously protected the coastline from storm surges either sank below sea level or were drained and developed. Poor planning therefore magnified the negative impacts of Katrina. As a result, it has taken more than a decade for the city to recover from the damage, and some people remain unable to repair their homes.‡

Resilience science accepts that change is inevitable and that attempts to resist change or control it are doomed to failure. System change can be a continuous or episodic process. Large-scale disasters cause episodic changes. Variable rates of change lead to uncertainty in how to manage systems that provide resources such as food and water. As we'll see, global climate change is affecting many of these systems in ways that are hard to anticipate (Chapter 7), making it more difficult to manage provisioning systems, leading to reduced water security (Section 12.3) and food security (Section 13.2.2).

A **socio-ecological system (SES)** is a complex adaptive system that has multiple possible regimes or stable states (Table 4.2). These regimes are separated by thresholds defined by changes in feedbacks. Over-stressed SESs gradually lose their resilience, which can cause an SES to cross a threshold into a new, sometimes undesirable regime. For example, the Florida Everglades are an

---

* http://www.directionsmag.com/articles/gis-helps-support-world-food-programs-food-security-program/355268.

† Restoring the River, Megan Sever, *Geotimes*, August 2007, http://www.geotimes.org/aug07/article.html?id=feature_river.html, retrieved June 13, 2016.

‡ New Orleans' Recovery From Hurricane Katrina Leaves Some Behind, Daniel A. Medina and George Itzhak, August 29, 2015, http://www.nbcnews.com/storyline/hurricane-katrina-anniversary/new-orleans-recovery-hurricane-katrina-leaves-some-behind-n417751, retrieved June 13, 2016.

SES that is shifting from a sawgrass-dominated regime to a cattail-dominated regime as a result of anthropogenic phosphorous inputs (Walker and Salt 2006).

The concept of alternate regimes separated by thresholds is usually depicted using a ball-in-basin metaphor (Figure 4.3). The ball represents the current state of the SES, and each basin is a possible system state. The instability or energy content of the system is plotted on the *y*-axis. In our Everglades example, one basin is the sawgrass regime and the other the cattail regime. The resilience of the original sawgrass regime is represented by the distance of the ball from the threshold, and by the depth of the basin of attraction (Figure 4.3a). Changes in inputs or the environment will make some regimes less stable, which is represented by the regime basin becoming smaller or less deep (Figure 4.3b). Usually SESs tend to move toward the bottom of the basin, which represents the most stable or equilibrium state. However, the bottom and the walls of the basin keep moving, and processes that undermine resilience may move the SES represented by the ball up the wall of a basin. Eventually, the SES may cross a threshold or **tipping point** and slide down into a different basin that represents a new regime, in our example one dominated by cattails. Once a system crosses a threshold into a new regime it becomes difficult or impossible for it to return to the original regime. For example, if the SES in Figure 4.3 transitions to undesirable regime 1, it would take a great deal of energy to push the system back up and over the threshold to return it to desirable regime 2.

The constancy of change is evident in the **adaptive cycles** model of resilience science (Figure 4.4). Ecologists recognized that ecological systems often pass through repeated cycles of rapid growth, conservation, release, and reorganization. This cycle has been observed in many different types of systems, including economic and social systems, and we will see many examples throughout this book. Forests grow, are destroyed by fire or disease, and then regrow. Human societies flower, grow in complexity, collapse, and then sometimes are replaced by new societies. The theory of adaptive cycles forms the basis for **adaptive management** of systems that aims to reduce risk over time. Adaptive management aims to keep desirable SESs away from thresholds by building system resilience.

Complex adaptive systems usually spend most of their time in the growth and conservation phases of the adaptive cycle (Figure 4.4). The release and reorganization phases are generally short-lived. Accumulating economic, environmental, and social stresses may move an SES toward a threshold or tipping point beyond which change becomes irreversible. Resilient communities can experience high stress levels without experiencing irreversible change; they retain their structure and internal feedbacks.

Ecological succession in forests provides a good example of adaptive cycles (Figure 4.5). After a forest fire (collapse Ω) opportunistic grasses quickly spread and grow to form a meadow

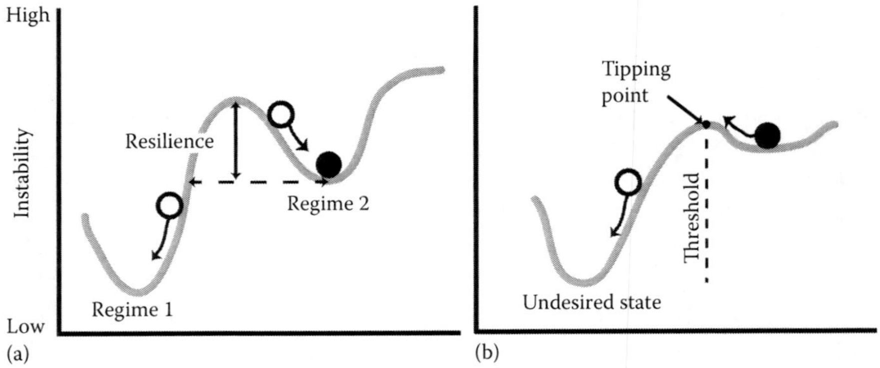

FIGURE 4.3 Ball-in-basin model of ecological resilience and tipping points. The ball represents a socio-ecological system (SES). Resilience corresponds to the depth of the basin. (a) High resilience keeps the system in regime 2. (b) Reduced resilience allows a small perturbation to push the system past a tipping point, causing the SES to shift from regime 2 to regime 1.

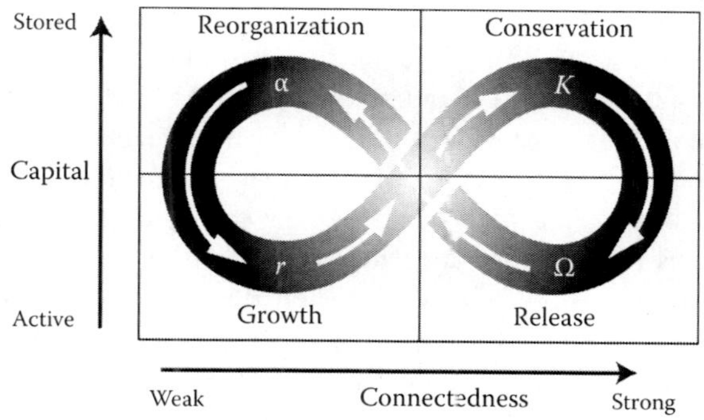

FIGURE 4.4 Adaptive cycles of ecological systems. The phases of the cycle are renewal/reorganization α, rapid growth *r*, conservation *K*, and collapse or release Ω. (After *Panarchy* by Lance Gunderson and C.S. Holling. Copyright ©2002 Island Press. Reproduced by permission of Island Press, Washington, DC.)

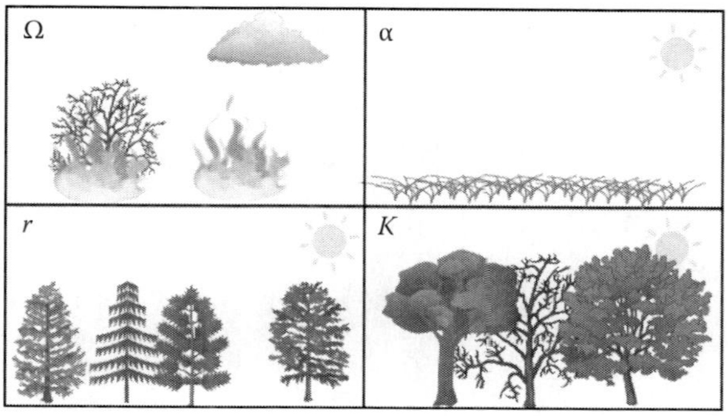

FIGURE 4.5 Ecological succession in a deciduous forest in a temperate climate. (After https://commons.wikimedia.org/wiki/File%3ASecondary_Succession.png by Katelyn Murphy (own work) [CC BY-SA 3.0 (http://creativecommons.org/licenses/by-sa/3.0)], via Wikimedia Commons. Symbols courtesy of the Integration and Application Network, University of Maryland Center for Environmental Science [ian.umces.edu/symbols/].)

(reorganization α). During the growth phase *r*, trees slowly grow to replace grasses and form a forest with less diversity, more specialization, and less efficiency. In the conservation phase *K*, biomass reaches a maximum, and deadwood accumulates until a threshold is reached and collapse occurs (release phase): another forest fire decimates the forest, but it sows the seeds for rebirth (the reorganization phase). Trying to maintain a forest in the conservation phase by preventing wildfires only makes the system less resilient over time; greater amounts of deadwood accumulate, and as a result, the eventual wildfire is larger, more intense, and harder to control. Thus, collapse is a necessary part of the adaptive cycle of complex systems, and it provides new opportunities. The positive message is that while collapse can be painful (e.g., the economic collapse that caused the U.S. Great Depression), it is always followed by a stage of renewal (Homer-Dixon 2006). Furthermore, humans can change their behavior much more rapidly than evolution can cause species adaptation; our intelligence gives us the ability to adapt quickly to changing circumstances, which should minimize damage during the collapse stage (Greer 2009).

Resilient systems can sustain their structure and function over long periods. Thus, sustainability requires resilience. Previously we discussed how natural resources could be sustainably managed if the harvesting rate is kept below the regeneration rate. Resilience science tells us that this approach may not always work because complex adaptive systems change over time, so that the regeneration rate is a moving and usually unknown target. Water inflow to a lake or groundwater aquifer may be high one year and low the next. This introduces uncertainty into the management of the shared resource. Furthermore, managing a resource system by optimizing on only one variable generally leads to decreased system resilience (Walker and Salt 2006). For example, river water may be diverted from groundwater recharge to maintain the level in a surface water reservoir, but this reduces the ability of the aquifer to provide water during subsequent droughts. This illustrates that increasing resilience in one part of the system may decrease resilience in other parts.

As another example, consider marine fisheries. In the early stages of resource use, the harvesting rate was lower than the regeneration rate. Most countries did not limit the amount of fish that could be harvested, so over time more boats were built to harvest the fish and total yields increased. Eventually the global harvest plateaued as the harvesting rate reached the regeneration rate, and the global fishery reached its maximum sustainable yield (Figure 4.6). Some fishery managers recognized this limit and instituted quotas to ensure the maximum sustainable yield was not exceeded, which would have caused their fisheries to collapse. Yet in some cases, even when a strictly enforced quota system prevented fish overharvesting, fisheries still collapsed. Why? For a variety of reasons. First, the maximum sustainable yield can vary from year to year. Sometimes the changes can be incremental and linear, and be recognized as consequences of cause and effect (e.g., an unusually cold breeding season decreases the fish regeneration rate). However, some changes are catastrophic, meaning they are large and occur suddenly. These nonlinear changes in a system, such as a storm destroying a primary breeding ground for a fishery, can push a system past its tipping point, leading to collapse. Local collapse of multiple marine fisheries has caused the global annual fish catch to level off despite intensification of fishing practices (Figure 4.6).

Walker and Salt (2006) argue that there is no sustainable "optimal" state of an ecosystem or natural resource supply system. The more we try to "control" a system to produce an optimal yield, the more undesirable outcomes are produced. The more order we try to impose, the more energy is required to maintain that order and prevent the tendency of systems to move toward increased entropy. The sustainability approach to managing systems is to enhance resilience rather than to optimize isolated parts.

Industrial agriculture is a good example of unsustainable management practices because it focuses solely on maximizing crop yields by adjusting various inputs of energy, fertilizers, pesticides, and engineered seeds, without considering how those inputs will affect the ecological system, including soil and water. This simple optimization approach is doomed to failure in the long term. A holistic approach is needed that balances crop yields against environmental degradation and that maintains or builds the resilience of the agricultural system, including its surrounding environment. Sustainable management of complex systems requires optimizing many variables, not just one.

So what are the necessary components of resilient systems? A resilient world or SES would have all of the following properties (Walker and Salt 2006; Mazur 2013):

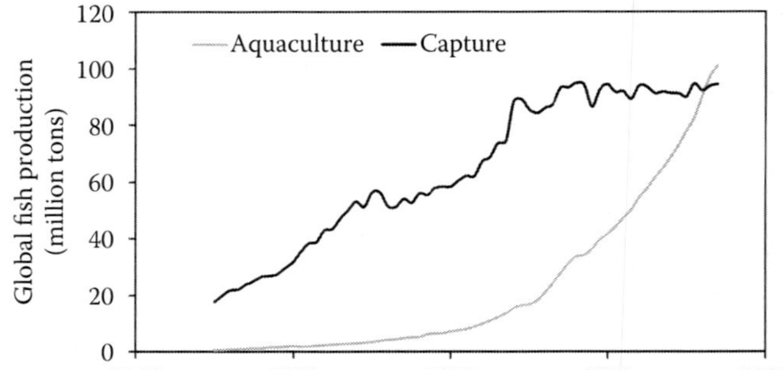

FIGURE 4.6 World wild fish catch and farmed fish production per person, 1950–2010. (Data from Brown, Lester R. 2012. *Full Planet, Empty Plates: The New Geopolitics of Food Scarcity*. WW Norton.)

Diversity: Diversity refers not only to the number of elements in a system (number of different species in an ecosystem, number of different crops in a farm field), but also to the diversity of functions in our communities (e.g., number of different types of businesses), diversity of land uses (vegetable gardens, fruit orchards, aquaculture, nut trees, rainwater harvesting and storage), and diversity of potential responses to challenges. Having many different components allows for many different responses to a disturbance, so we should promote and sustain diversity of all forms (biological, landscape, social, and economic). Increasing all of these types of diversity can lead to a more resilient and sustainable community, although high diversity alone does not make an SES sustainable (McCann 2000).

Currently biodiversity is declining (Chapter 15): one-eighth of birds, one-quarter of mammals and conifers, one-third of amphibians, and one-half of turtles and tortoises are at risk of becoming extinct (Walker and Salt 2006). Diversity in crops and livestock is also decreasing (Chapter 13): our food security completely depends on a very narrow genetic base that has been selected solely for maximizing production in a limited set of conditions (conservation phase). Environmental change, including climate change, may change the equilibrium state, causing decreasing yields.

Redundancy: A resilient SES has many different ways to perform basic functions; failure of one component does not cause collapse. For example, the Internet has many network nodes (routers); failure of one simply redirects traffic to other routers. In a resilient network, each node has many connections; in efficient networks most nodes have only one connection to a central node, making them less resilient (Figure 4.7).

Modularity: Individual components still function if disconnected from larger networks. For example, electrical "microgrids" are designed to be diverse and modular; they have a variety of electricity sources that can attach to the national grid or operate independently, ensuring that they provide electricity even during regional brownouts. Subecosystems can survive geographic isolation resulting from mountain formation or submergence of a land bridge if they are sufficiently modular and have the other requirements for resilience.

Reserves: A resilient SES has large reserves of environmental, economic, and social capital to draw on during emergencies when resources are scarce.

Agency and inclusiveness: Individuals in a resilient community are empowered to make choices and take action to protect their own well-being and their community. They are involved in community decision making and, importantly, have a sense of control over their own destiny. In resilient communities, power is distributed rather than concentrated.

Tight feedbacks: How quickly and strongly are changes felt in other parts of a system? Feedbacks are responses to perturbations, information about which is usually transmitted over some type of network. For example, information about an interest rate cut by the Federal Reserve is transmitted by various communication networks to traders. The initial perturbation of the financial system provokes a response: stock prices rise in anticipation of resulting economic growth.

Individuals in a resilient system quickly see the consequences of their actions, which allows them to conceptually link cause and effect and to modify their behavior to achieve desired outcomes. Through most of human history, people saw that when they overfished or overgrazed they undercut

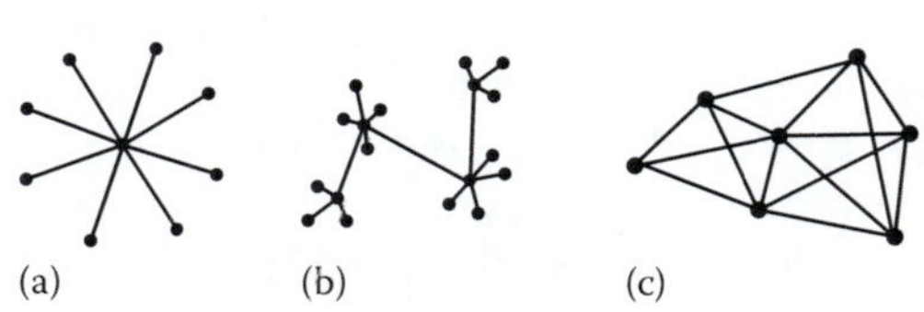

FIGURE 4.7 (a) Centralized networks have few connections per node and one central node. They are efficient but have low resilience. (b) Decentralized networks also have few connections per node, but are more resilient because there are multiple central nodes. (c) Distributed networks have more connections per node. The high redundancy makes them less efficient but gives them the greatest resilience.

their ability to obtain food from their local environment, so they reduced their harvests to sustainable levels. Today people in developed countries often don't know where their food comes from, and don't see the consequences of the unsustainable practices used in developing countries to provide that food. In a resilient system, information about destructive practices would become rapidly available to the people driving those changes, which may provoke a negative feedback that reduces or eliminates those practices.

Innovation: Complex systems always change. The biosphere changes slowly by evolution; human societies evolve comparatively rapidly through innovation. Innovation can provide solutions to difficult societal problems and in the process make societies more resilient. For example, the Green Revolution increased food security throughout the world (Chapter 13).

Today's global society seems to lack resilience because it does not possess all of these properties. We rely on complex technological systems to provide critical resources "just in time." These globalized systems are highly efficient and therefore profitable, but they generally lack diversity, redundancy, modularity, and tight feedbacks. Recent examples of socio-ecological systems that lacked some of these components of resilience, so that they had trouble returning to normal after a disaster, include Haiti following the 2010 earthquake in Port-au-Prince (Section 4.8); New Orleans following Hurricane Katrina in 2005; Fukushima, Japan following its nuclear disaster triggered by an earthquake and tsunami in 2011 (Section 9.5.3); and the **Great Recession** that was triggered by a housing market crisis in the United States in 2007–2008. On a global scale, combine more people and fewer resources with a homogenizing globe that is rapidly losing biodiversity and you end up with shrinking options. There are fewer places to go and less diversity with which to respond to new challenges and unforeseen crises. We need to make our SESs resilient so that they adapt as the world changes while still maintaining their structure and functionality. Resilient systems are forgiving of management mistakes and miscalculations.

Maintaining resilience has **opportunity costs**. Farmers may increase the resilience of their agricultural systems by shifting from a monoculture of highly productive corn to heirloom varieties. A disease that strikes one variety may leave the others untouched, protecting the farmers' economic return on investment, but efficiency and yields will be lower. Viewed from this perspective, the different components of resilience are different forms of insurance; investing a share of profits in insurance may decrease overall profits but it increases economic security. Managing resilience involves balancing short-term profit losses from maintaining or enhancing resilience against long-term benefits of avoiding collapse. Just as it is wise to increase economic security by insuring financial capital, it is also wise to increase other forms of security (food, water, energy) by adopting appropriate forms of insurance that increase the resilience of resource provisioning systems.

The Great Lakes district of North America provides a good case study of resilience (Carpenter et al. 2001). These glacial lakes provide a range of ecosystem services that fall into two categories: agricultural production and aquatic ecosystem services, which include pollution dilution, municipal water supplies, and recreation. Over time, intensification of agricultural production has come at the expense of aquatic ecosystem services. Much of the problem can be traced to the geochemical behavior of phosphorous, which is the limiting nutrient for plant growth in the region. Initially, a clear-water oxygen-rich lake, agricultural perturbations decreased the resilience of the lake and eventually pushed it into a new state of a turbid, oxygen-poor lake that could not support fish. Unfortunately, this new undesirable state is resilient, so intensive management is required to return the lake to the desired clear-water state. Because it is the critical control on the lake state, phosphorous concentration can be used to measure the resilience of both lake states (clear and turbid).

The geochemical controls on phosphorous behavior include the influx of phosphorous to the lake, which depends on erosion rates and phosphorous concentrations in eroded soil and sediment, all of which are inversely related to the resilience of the clear-water state. In the initial clear-water state when phosphorous entered the lake, much of it was removed through adsorption onto the surfaces of iron oxide minerals in the lake sediment, a negative feedback. However, agricultural activities such as fertilizer application, tilling, and livestock production led to rising inflows of phosphorous to the lake, leading to rising phosphorous levels in lake water. Rising nutrient levels led to algae blooms, which led to accumulation of dead plant matter on the lake floor, which consumed oxygen during decomposition, which led to the iron oxide minerals becoming unstable, which led to release of more phosphorous to the water. This positive feedback either eliminated the clear water basin of attraction or pushed the lake system past a tipping point so that it entered a new basin of attraction, the turbid lake system. Most lakes have transitioned from the *r* phase (rapid development and rise in *P*) to the *K* phase, followed by rapid post-WWII development that pushed the system past the tipping point causing collapse Ω and a permanent transition to a degraded turbid state with low utility (Figure 4.4). Multiple adaptive management cycles have resulted in rises and falls of the resilience of the clear-water state, measured as the inverse of *P* concentration. Most of these attempts at managing phosphorous levels in the Great Lakes district have failed (Carpenter et al. 2001).

A sustainable society must preserve or increase economic, social, and environmental capital and *resilience*. For example, a community can increase its resilience by "relocalizing," that is, reducing its dependence on imported resources, which makes it less vulnerable to transportation disruptions. Relocalization also brings ecological advantages. For example, depositing urban organic compost on nearby farms and forestland would close the nutrient cycles broken by the current spatial separation of rural ecosystems and urban populations. Resilience thinking provides a foundation for achieving sustainable patterns of resource use.

4.5 SYSTEMS THEORY

To understand the risks of exponential growth, we will use the concepts of **system dynamics**. System dynamics is a powerful approach to modeling complex systems that exhibit nonlinear behavior using stocks and flows. It can account for **feedback loops** and time delays, which we will discuss in this chapter. System dynamics is also a useful way to think about, visualize, and understand complex systems, and even to make accurate predictions about their behavior over time. I strongly recommend the book, *Thinking in Systems: A Primer* by Donella Meadows, who was one of the leading practitioners of systems modeling and was the lead author on the *Limits to Growth* books.

According to Meadows (2008), a system is "a set of elements or parts that is coherently organized and interconnected in a pattern or structure that produces a characteristic set of behaviors, often classified as its 'function' or 'purpose'. Examples include the human body, ecosystems, the atmosphere, a drainage basin, and a nation. Systems can change, adapt, respond to events, seek goals, mend injuries, and attend to their own survival in lifelike ways, although they may contain or consist of nonliving things. Systems can be self-organizing, and often are self-repairing over at least some range of disruptions. They are resilient, and many of them are evolutionary." The Earth is a large, complex system that consists of many smaller, simpler systems including ecosystems, economic systems, and social systems.

The *elements* of a system are easy to identify, for example, a forest ecosystem has trees, soil, and animals. The *connections* or relationships among the elements are harder to see. Connections consist of flows of capital (resources, information) and the rules and laws that govern those flows (de Vries 2013). Examples include the uptake of nutrients from soil by trees, the harvesting of fruit from trees by animals, and so on. The *purpose* of human systems is usually easy to discern, for example, the purpose of a school system is education. For animals, plants, and ecosystems we can say that the purpose is to survive or sustain.

An environmental system is a part of the Earth we wish to study, and can range in scale from microscopic to the entire planet. It contains component parts that interact with each other to cause change. Ecosystems contain local groups of living creatures and the environment they inhabit, which are all interdependent. Boundaries between geographic or climatic provinces usually set the spatial limits of ecosystems. Ecosystems are dynamic because material and energy are constantly moving within them. If we can quantify the rates at which renewable resources move within systems, we can estimate sustainable rates of extraction of resources such as water to ensure we don't exceed those rates. System dynamics helps us do this and to predict the effects of anthropogenic changes to natural systems.

In an **open system**, the amounts of matter and energy can vary through exchange with the universe outside the system. A **closed system** allows for the exchange of energy but not matter. While a lake acts as an open system, Earth behaves as a closed system because the total amount of matter is roughly constant; although small flows of hydrogen escape from the Earth's atmosphere into space, and although meteorites and comets entering the Earth's atmosphere can add matter, these amounts are minuscule compared with the total mass of the Earth. The Earth is not an isolated system because it receives electromagnetic energy from the sun and loses heat energy to space. We can also treat small portions of the earth such as ecosystems as closed systems, especially over short timescales. Thus, for our purposes, we can safely assume that the mass of any element in the systems we will consider is constant.

To illustrate the workings of a complex, dynamic Earth system, we will examine a system map of the **water cycle** (Figure 4.8). Many natural materials cycle repeatedly through closed loops. As discussed in Section 2.2, we use two types of variables to characterize these cycles. *Stocks* are accumulations of capital characterized by how much capital they hold at a given time. In the water cycle, water is a form of environmental capital, and we can define our system by specifying the present amount (mass) of water in each stock, for example, the oceans, icecaps, groundwater, and so on. In a dynamic system, *flows* transfer capital between stocks. Flows are rates measured as amount per unit of time; the higher the flow rate, the faster water moves from one stock to another, for example, from a river to the ocean.* In closed systems like Earth's water cycle, mass is conserved, so the total amount of water is constant and flow distributes the water between the various stocks. Energy from the sun drives evaporation, which transfers water from surface bodies of water to the atmosphere. Condensation and precipitation then move water back to surface water bodies. Thus, water constantly cycles between stocks. Our illustration of the water cycle is an example of a **stock and flow** diagram, and it specifies the size of each stock (in Exagrams [Eg] [= 10^{18} grams] of H_2O) and the flow rate associated with each process (in Eg H_2O per year).

The size of the stock will change when inflows do not equal outflows (Figure 2.2). For our lake example, the time rate of change of stock size can be calculated from an equation of the form:

$$\Delta M_{\text{water}} / \Delta t = \text{Sum Inflows} - \text{Sum Outflows} \tag{4.2}$$

where ΔM_{water} is the change in mass of the water. This is called a **difference equation**, and each term has units of kg/year.† When the difference between water inflow and outflow is zero, then the mass of water in the lake does not change. A dynamic equilibrium called a **steady state** exists because the inflows exactly balance the outflows. A steady state is desirable because it is sustainable.

Stocks can only be changed by changing their inflows and outflows, and cannot change faster than the difference between inflow and outflow. This gives the system inertia, that is, stocks are a source of time delays in system responses to change. Many people have trouble understanding this important system property, and this has led to poor public policies. For example, air pollutants may persist long after we stop adding them to the atmosphere if the difference between inflows and outflows is low.

In a system at steady state, the **residence time** represents the average amount of time an atom or molecule stays in the stock and is equal to the stock mass divided by the total inflow or outflow:

$$t_{\text{residence}}(y) = \text{Mass }(g)/\text{Flow }(g/y) \tag{4.3}$$

---

* Note that stocks are equivalent to integrals and flows are time derivatives. See de Vries (2013) for an in-depth discussion.

† When $\Delta t \rightarrow 0$, $\Delta M_{\text{water}}/\Delta t$ becomes the **derivative** $\Delta M_{\text{water}}/\Delta t$ and Equation 4.2 becomes a **differential equation**.

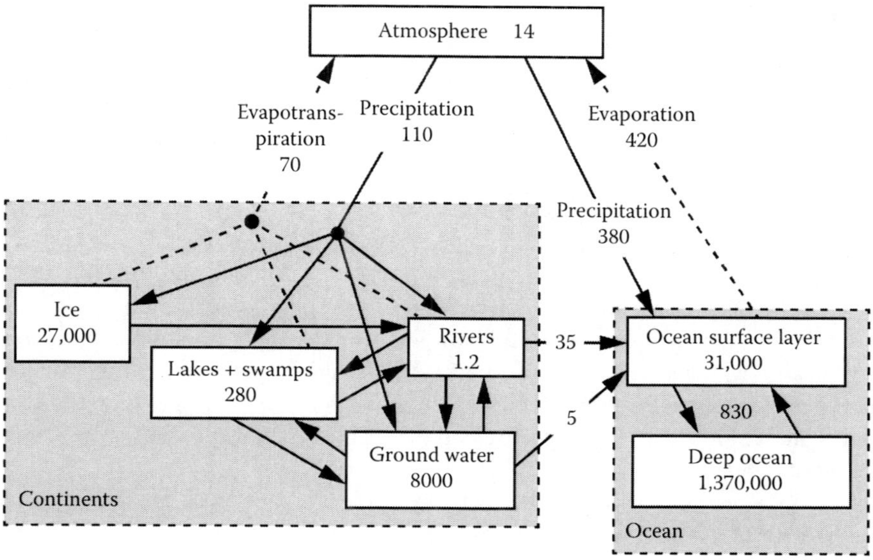

FIGURE 4.8 The water cycle. Water reservoirs are displayed as boxes with values given in units of Exagrams. Arrows represent annual fluxes with values of Eg/year. (Modified after Walther, 2009, *Essentials of Geochemistry*, Jones & Bartlett Learning, Burlington, MA, www.jblearning.com. Reprinted with permission.)

To understand this, imagine a small pond (low mass) with an input stream and an output stream having equal and high flow rates. The residence time of water in the pond is small because the mass of water is small, and rapid flow quickly replaces the water.

Because the water cycle is approximately in a steady state, we can calculate the residence time of water in each stock. For example, the total inflow of water to the atmosphere equals the sum of the inflows (numbers from Figure 4.8):

$$\begin{aligned}\text{Atmosphere inflows (Eg/year)} &= \text{evaporation from ocean}\\ &+ \text{evaporation from lakes and rivers} = 420 + 70 = 490\end{aligned} \tag{4.4}$$

The residence time for water in the atmosphere is then equal to the stock (mass of H_2O in the atmosphere) divided by the total inflow:

$$\begin{aligned}t_{\text{residence}} &= \text{mass } H_2O/\text{inflow}\\ &= 14\ (\text{Eg})/490\ \text{Eg/year} = 0.03\ \text{years} = 10\ \text{days}\end{aligned} \tag{4.5}$$

Now consider a much larger body of water: the ocean. The residence time of water in the ocean is 3300 years, which means that when we add water-soluble pollutants to the ocean, it stays polluted for a long time.

For a system that is not in a steady state, the size of a stock changes over time because inflows do not equal outflows (Figure 2.2). Here we call the residence time the **response time** because it no longer represents the average time a molecule spends in the stock. The response time is a measure of how quickly the mass of a substance in a stock changes. Rearranging Equation 4.2:

$$t_{\text{response}} = \text{Mass (g)}/(\text{inflows} - \text{outflows})\ (\text{g/y}) \tag{4.6}$$

The response time is large if the mass of the substance in the stock is large and if the difference between inflows and outflows is small. Most of the time the water cycle is in a steady state, but climate change and human activity have perturbed the system and thrown it out of equilibrium. Currently inflows to the ocean exceed outflows, not only because the rate of glacier melting has increased in response to global warming, but also because humans are pumping water out of ancient aquifers and dumping it into the ocean (Sahagian, Schwartz, and Jacobs 1994).

A key concept in system dynamics is **feedback**, which is a system response that can amplify or diminish the effects of system perturbations. For example, audio feedback occurs when a microphone is placed in front of a speaker. The microphone detects a sound and sends it to the amplifier and speaker. The speaker plays an amplified version of the sound, which the microphone detects. The sound keeps cycling through the system, growing increasingly loud. This is an example of positive feedback, in which a feedback loop amplifies the original signal or perturbation. Feedbacks occur in all types of systems, including economic systems. For example, a growing economy causes an increased demand for energy, which causes oil prices to rise, which causes a negative feedback that slows or stops economic growth.

Feedback loops are essential components of dynamic systems. In a positive feedback loop, exponential growth of a stock occurs at a rate that is proportional to the size of that stock. This leads to exponential growth that is unsustainable. Positive feedback loops can destabilize systems (de Vries 2013). In a balancing or negative feedback loop, a stock shrinks at a rate that is negatively proportional to the size of the stock. In dynamic systems, the size of a stock will tend to decline to a constant K. If the human population becomes too large, then the death rate will exceed the birth rate, and population will decline until it reaches the carrying capacity K. Negative feedback loops tend to stabilize systems.

Most systems have both positive and negative feedback loops. If positive feedback loops dominate, then growth occurs, and if negative feedback loops dominate, then a stock shrinks. Because feedbacks can lead to exponential growth and decline, a small change in one system variable can cause big changes in the system over time. This nonlinear behavior of feedback loops can lead to **tipping points** where the structure of a system changes. These features of dynamic systems (feedback loops and tipping points) make it very difficult to make predictions about their future behavior.

The systems diagram for population is key to understanding what we need to do to transition to a sustainable society. Exponential growth in population results from a positive feedback loop (Figure 4.9): as population increases, the number of people who can give birth increases, and for a given fertility, the number of births increases, thereby further increasing the population. This positive feedback loop

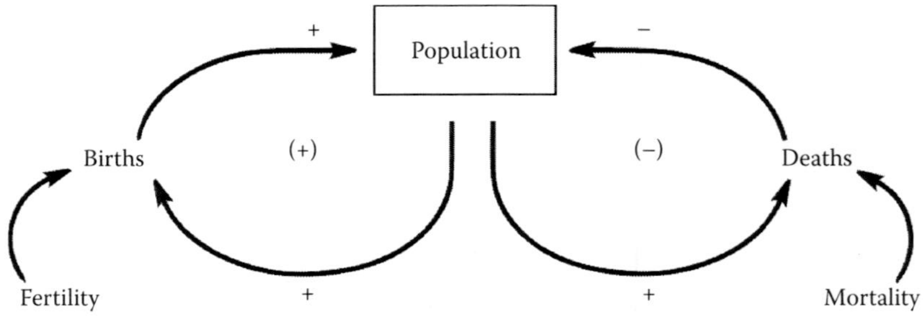

FIGURE 4.9 Causal loop diagram for population. Arrows originate at causes and end at effects. The sign above the arrow describes whether an increase in the causal factor causes an increase (arrow labeled with a "+") or decrease (arrow labeled with "–") in the effect. If the number of negative signs in a loop is odd, the cycle leads to negative (stabilizing) feedback denoted by "(–)" in the middle of the loop, so that the initial perturbation is damped; if the number of negative signs is zero or even, it is a positive (reinforcing) feedback loop denoted by "(+)," and the initial change in the causal factor gets amplified. (Reprinted from *Limits to Growth*, copyright 2004 by Dennis Meadows, used with permission from Chelsea Green Publishing, www.chelseagreen.com; de Vries, B J M. 2013. *Sustainability Science*. Cambridge University Press.)

is counteracted by a negative feedback loop (Meadows, Randers, and Meadows 2004): deaths decrease the population, which therefore decreases the birth rate. Currently the effect of the positive feedback loop outweighs that of the negative feedback loop, so that global population is increasing at an exponential rate. The preferred approach to stabilizing population is to decrease the positive feedback loop by decreasing fertility rather than increasing the negative feedback loop by increasing mortality (decreasing longevity). Once these two feedback loops are in balance, population will be in a sustainable steady state.

System dynamics and stock and flow diagrams can help us understand complex systems. Sometimes we can use them to predict the consequences of human perturbations to natural systems. However, some natural dynamic systems have many feedback loops, some of which are poorly understood, which makes it difficult to predict the effects of human perturbations (e.g., the global climate system, see Chapter 7). In these cases, we should follow the **precautionary principle**, which states that an action should not be taken, or policy enacted, unless scientific consensus or burden of proof establish that it will not cause harm to people or the environment. This means that to minimize risk, we should change the natural system as little as possible, and allow natural cycles to remain close to their steady state.

4.6 ECOLOGICAL OVERSHOOT AND COLLAPSE

Many systems have similar underlying structures that make them function in a similar way even though there are no obvious similarities, for example, the elements of the systems may be very different. There are several structures called "archetypes" that are important because they appear in many types of systems. One archetype is the physical and social "limits to growth" that ends exponential growth (de Vries 2013).

As mentioned previously, exponential growth results from one or more positive feedback loops. Eventually exponential growth makes a stock so large that new negative feedback loops kick in and slow the rate of growth until it reaches zero. This results in a special archetype called "logistic growth" in which a stock initially grows exponentially so that the absolute growth rate continues to increase until a negative feedback loop kicks in. Then the absolute growth rate increases more slowly until it reaches a maximum before it decreases to zero. This results in a smooth transition to a sustainable state of zero growth. Human population is usually modeled by a logistic growth equation, which assumes that humanity will transition smoothly to an optimal population, Earth's carrying capacity K. Figure 3.7 shows how the stock (population) size increases over time such that it forms a sigmoid or "S" shape. The annual population relative growth rate (note that relative rates are expressed as percentages, while absolute rates are in number per year) peaked near the mid-1960s at 2.2% and has shown a general decline ever since to 1.1% in 2013.*

In a more pessimistic scenario, the global population overshoots the carrying capacity K during the conservation phase of an adaptive cycle. The further the population grows beyond the dynamic equilibrium point (the attractor K), the stronger the negative feedback loops become. This can result in an unstable system that collapses when the death rate skyrockets. In order to avoid this catastrophic scenario, which would entail much human suffering, we should estimate the Earth's carrying capacity and then take steps to slow population growth gradually until it reaches zero when K is reached. Unfortunately, the best estimates of Earth's carrying capacity based on the ecological footprint suggest that we are already in overshoot, with K being near 4 billion while the current population is over 7 billion (Section 3.3). The UN projects that population will not plateau until after the year 2100, when it will likely exceed 10 billion (Gerland et al. 2014), assuming overshoot is not great enough to trigger a population collapse.

A simple model that illustrates the system dynamics view of exponential growth and resulting overshoot and collapse should suffice (Meadows et al. 2004). When we make alcoholic beverages, we make use of yeast. The yeast feeds on sugars and converts them to alcohol, a process called fermentation. As long as abundant food and no toxic waste are present, the yeast will multiply at an exponential rate. This is a characteristic of all living organisms that self-replicate. Let's assume we start with one yeast cell

* CIA World Factbook, https://www.cia.gov/library/publications/the-world-factbook/geos/xx.html.

that can reproduce every 10 minutes. At the end of each 10-minute interval over a period of two hours we will have 2, 4, 8, 16, 32, 64, 128, 256, 512, 1024, 2028, and 4096 yeast cells! The rates at which the yeast consumes sugar and produces waste alcohol increase in parallel with the growth in yeast population. Exponential growth occurs when the increase is proportional to what is already there; thus, there is a positive feedback similar to the positive feedback loop in Figure 4.9. Exponential growth is much faster than linear growth, where the increase is constant over a given period. As the yeast multiplies at an exponential rate, the amount of alcohol it produces increases at an exponential rate. The waste alcohol is toxic to yeast; the yeast cannot survive above 10% to 14% alcohol, which is the upper limit of alcohol content of nondistilled alcoholic beverages. Thus, the rising levels of waste alcohol result in a negative feedback loop (Figure 4.9) that causes the yeast to die more rapidly. Eventually the alcohol concentration becomes high enough to cause a system collapse in which all of the yeast dies.

The logistic archetype can also explain exponential decline. One example is a nonrenewable resource. Figure 4.10 shows a hypothetical case of depletion of a nonrenewable resource. The resource consumption rate increases from year 1900 to 2000, when it reaches a maximum. During that time, the amount of the resource in the ground decreases exponentially, but in the year 2000, there is an inflection in the curve. After 2000, the production rate decreases and the rate of depletion slows. Estimating the timing of the peak in the production rate is important because after that time the production rate decreases and the resource becomes "scarce" as the supply lags behind demand, causing prices to increase (more on this in Chapter 6).

The increasing demand on resources has led to a shortage in supply of many nonrenewable and renewable resources. For example, the harvest of the renewable resource marine fish peaked in the mid-1980s (Figure 4.6). Recent declines in the wild catch suggest that some fisheries have been overharvested so that the stock of fish has declined, even to the point of complete collapse where the stock plummets to zero. Such trends of decreasing environmental and economic capital are unsustainable.

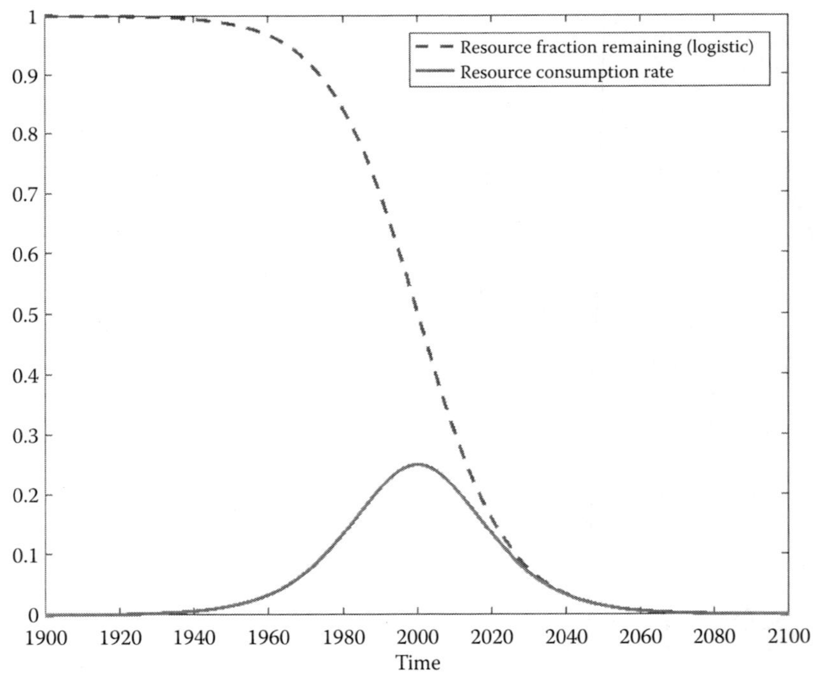

FIGURE 4.10 Logistic decline curve for a hypothetical nonrenewable resource (dashed curve). The solid curve is the resource consumption rate, which equals the absolute value of the first derivative of the resource stock size.

The most famous system dynamics model is World3, which formed the basis of the Club of Rome book *Limits to Growth* (Meadows et al. 1972) and subsequent updates (Meadows, Randers, and Meadows 2004). World3 has five variables that increase exponentially over time: world population, industrialization, pollution, food production, and resource depletion. In the *Limits to Growth* studies, scenarios were created making various assumptions about future changes in population, pollution, abundance of nonrenewable resources, agricultural yields, and so on. The original study found that exponential growth of human population and the economy were stressing the environment, and that humanity was approaching the limit of Earth's capacity to provide needed resources such as food.

The 30-year update of that book concluded that we were reaching those physical limits (Meadows et al. 2004). This is reflected in the fact that the global ecological footprint now exceeds the Earth's global biocapacity by ~40%, and that humans are changing the composition of Earth's atmosphere. A study of nine planetary systems (Rockstrom et al. 2009) found that humanity has already exceeded sustainable limits in three categories: rate of biodiversity loss, climate change, and human interference with the nitrogen cycle. The rapid rise in the global number of failed states, the ozone hole, global warming, rising species extinction rates, the recent global recession, increasing national debts, increasing conflicts over resources, the growing gap between rich and poor, and "natural" disasters becoming more frequent and more severe because of less resilience in socioecological systems are all consistent with the conclusion of the *Limits to Growth* study that we are currently in ecological overshoot (Meadows et al. 2004). However, an updated version of the model suggests that collapse is not imminent or likely, and the dates it forecasts as the turning points, when resource availability and quality of life begin to decline, have been pushed farther into the future; for example, food availability is forecast to begin declining around the year 2040 (Randers 2012).

The physical limits to growth include resources, environment, space, and food, while the traditional approaches to solving growth-related stress include migration, expansion, economic growth, and technology (Myrtveit 2005). However, these traditional approaches are no longer effective at relieving growth-related stress. The associated social problems seem intractable; no political approaches seem to solve them. If the world system stays in ecological overshoot too long and the magnitude of the overshoot becomes too great, it will eventually collapse. An alternative, new approach is needed to return the world system to sustainability; we must apply self-restraining policies that switch the development from growth to equilibrium, that is, we must change to a steady-state economy. In a steady state economy, the physical throughput (mass of resources consumed and mass of waste produced) is constant, but wealth measured as total and per capita GDP are allowed to increase.

Can a sustainable society still grow in population? A sustainable global society must operate below the carrying capacity of the Earth K where

$$K = \mathrm{EF}_{\max} = B = P \times A \times C \times T \tag{4.7}$$

where $\mathrm{EF}_{\max}$ is the maximum sustainable ecological footprint of humanity, B is global biocapacity, P is population, A is affluence, C is resource use intensity, and T is resource use inefficiency. For example, humanity may choose to operate at a level not greater than 90% of global biocapacity B. This would leave a safety margin that can compensate for uncertainty in the size of global biocapacity and for fluctuations in biocapacity over time. Once humanity sets that limit, it will operate at and not below that limit because that is human nature. When humanity operates at its sustainable limit, it cannot grow in one area without shrinking in another. If population P grows, then humanity would need to decrease the per capita environmental impact represented by $A \times C \times T$. Thus, sustainability requires that these components of EF be balanced so that there is no net growth in EF.

Due to current exponential growth in population and material wealth, future societies will be operating near their sustainable limits. To have global sustainability, ecological overshoot in one society would have to be balanced by undershoot in another. Since it is unlikely that any society would voluntarily agree to reduce its EF through reductions in affluence or resource use intensity to compensate for the overshoot of other societies, each society would need to stay at or below its sustainable limit to avoid local overshoot, and every society must be sustainable to avoid global collapse. In addition, intra-generational or distributional equity demands that each society must operate at a sustainable level without placing demands on other societies. Thus, global sustainability will require local sustainability.

As an example of local sustainability, imagine a sustainable village in the future that wants to build a new, better neighborhood. The village population is stable, so it has the freedom to increase A or C or T, if that increase is compensated for by decreasing the product of the other two variables. The village has maintained the area of surrounding forests at its lowest sustainable level to maximize the space available for use in the village (again, human nature). If it cuts down trees in one area for the new neighborhood, it must restore forest in another area of equal size by replanting. The new neighborhood would place demands on renewable resources such as water; those demands must be met by reducing demand in other areas. The situation might be a simple one where an old neighborhood is torn down and replaced by a higher technology, more efficient neighborhood. Village leaders would need to devise plans that would not increase $A \times C \times T$. Since the new neighborhood would have a lower T, it could have a higher $A \times C$. This means it could raise its consumption level by, for example, building larger homes, but only to the level defined by the current sustainable value of $A \times C \times T$, that is, without net growth in EF.

Because future societies will likely operate near their physical limits, that is, near the maximum sustainable EF equal to the biocapacity, their choices for development will be constrained. However, they will always have the choice of increasing quality of life. Future sustainable societies could develop, for example, by increasing their social capital without increasing their EF. However, unlike societies of the past, they will not have the freedom to choose unrestrained population growth and economic growth. Growth will forever be constrained. Thus, when future planners in sustainable societies consider growth in one area, they should plan shrinkage in another, and they should ask questions such as (Meadows et al. 2004):

- What is the growth for?
- Who would benefit?
- What would it cost?
- How long would it last?
- Would it serve important social goals while enhancing sustainability?
- Could the Earth's sources and sinks accommodate the growth?

These are questions that today's planners usually do not ask, but they must start asking if we hope to achieve sustainability.

It is also important to remember that the EF is an aggregate sustainability indicator comprising many components such as consumption of energy and renewable resources and production of waste. Thus, it will not be enough for sustainable societies to maintain their ecological footprints below their biocapacities; they must also limit growth in demand on every resource and growth in emissions of every pollutant. Every type of footprint, including the carbon, water, and agricultural footprints, will have a limit that a sustainable society cannot exceed. This means that a sustainable society must observe the Laws of Sustainability strictly and adhere to strong sustainability and the precautionary principle (Section 2.2).

Currently our society is driving up a mountain of risk with a cliff at the top representing a tipping point. The higher we climb, the greater the degree of ecological overshoot, and the higher the risk associated with falling off the cliff. However, because there is uncertainty about the location of the cliff, we don't know how close we are to it; it's like driving in a fog. A prudent driver would slow down or stop, and collect more information by consulting maps or a GPS to determine the location of the cliff, and then take steps to avoid it. However, society is not waiting for scientists to collect enough information to determine where our cliff is, and it is not slowing down. Instead, society is speeding up, increasing our ecological overshoot at an ever-quickening pace, driving blind toward a cliff but unable to take its foot off the gas pedal.

4.7 EXAMPLES OF COLLAPSE: UNSUSTAINABLE SOCIETIES PAST AND PRESENT

> Throughout history, we've lost an Easter Island here, a Roman Empire there, but now we face major ecological and economic disruptions at the planetary scale—the whole ball of wax, so to speak.
>
> **David Wann**
> *2007*

Many societies today and in the past have partially or completely collapsed because they could not provide the basic human needs of shelter, water, and food. Modern societies use energy to help provide these resources. We use energy to construct homes and to heat and cool them; to pump water; and to produce and transport food. When any of these resources become unavailable in a given location, we must either use energy to transport them to us or move to a new location where they are readily available. For most of human history, we operated as hunter-gatherers, staying in one area until we depleted the food and then moving on. However, in the modern world it is not possible to move an entire civilization or city when resource shortages arise.

Shortages in essential resources and subsequent societal collapse have many potential causes. In *Collapse*, Jared Diamond gives five reasons for the collapse of ancient societies: climate change, hostile neighbors, changes in friendly trading partners, environmental damage, and society's responses to environmental problems. As an example of environmental damage causing collapse, natives of Easter Island chopped down all of the trees on the island for fuel and other uses. Forest removal reached a peak in the 1400s, and was complete by the 1600s (Diamond 2005). Deforestation led to increased soil erosion, widespread starvation, and a population crash. In essence, societies like Easter Island's committed ecological suicide or **ecocide**.*

The specific types of environmental damage that led to the downfall of ancient societies included deforestation and habitat destruction, soil problems (erosion, salinization, and soil fertility losses), water management problems, overhunting, overfishing, the introduction of non-native species, human population growth, and increased human environmental impact (Diamond 2005). Diamond observes that it has always been difficult to manage environmental resources sustainably. People overexploit environmental resources because they initially seem inexhaustibly abundant. They don't recognize the signs of incipient depletion because they are masked by fluctuations in resource levels over years and decades (e.g., rainfall). People find it impossible to predict the consequences of their impacts on the environment because ecosystems are so complex. Even when depletion becomes apparent, people often can't agree to exercise restraint in harvesting a shared resource (see Section 5.4.3).

Past societies that collapsed often did so swiftly after attaining peak population numbers and power. As noted by Diamond (2005) "... one of the main lessons to be learned from the collapses of the Maya, Anasazi, Easter Islanders, and those other past societies (as well as from the recent collapse of the Soviet Union) is that a society's steep decline may begin only a decade or two after the society reaches its peak numbers, wealth, and power...The reason is simple: maximum population, wealth, resource consumption, and waste production mean maximum environmental impact, approaching the limit where impact outstrips resources." In the good years, adequate food and water supplies often caused the population to grow, which led to deforestation and the geographic spread of people to marginal lands. Societies became increasingly complex, interdependent, and no longer locally self-sufficient, all of which decreased their resilience. Collapse occurred in the bad years

* Recent research suggests that rats introduced by island settlers may have decreased tree populations; see "What Happened On Easter Island—A New (Even Scarier) Scenario," http://www.npr.org/sections/krulwich/2013/12/09/249728994/what-happened-on-easter-island-a-new-even-scarier-scenario, retrieved October 3, 2016.

such as during a drought when the environment became incapable of supporting large populations in the long term.

This chain of events is well illustrated by the Anasazi, who arrived in the U.S. southwest by 11,000 B.C. (Diamond 2005). They existed as hunter-gatherers until import of domesticable plant and animal species from Mexico at ~2000 B.C. started agriculture. The fundamental problem faced by the Anasazi is that the U.S. southwest is a fragile and marginal environment for agriculture: it has low and unpredictable rainfall, quickly exhausted soils, and low rates of forest regrowth. Rises and falls of Anasazi population closely mirrored rises and falls of calculated annual corn harvests. In one well-studied Anasazi settlement in Chaco Canyon in modern-day New Mexico, water management and deforestation became problems for the growing population by ~1000 A.D. The rate of tree regrowth was too slow to keep up with the rate of logging. Although the Anasazi of Chaco Canyon had survived earlier droughts, they could not survive the drought in ~A.D. 1130 because too many people depended on outlying settlements for food and there was no unoccupied land. Civil war and cannibalism followed, and the Anasazi abandoned Chaco Canyon between 1150 and 1200 A.D.

Not just ancient societies but also modern societies can collapse, leading to the formation of **failed states**. Global collapse has not occurred, and may never occur, but many countries have already collapsed and are now failed states. The failed state most familiar to Americans is Haiti (see case study below).

Many parts of the modern developed world are unsustainable. For example, Diamond (2005) argues that the human population in Montana is unsustainable. Its three main industries have been mining, timber, and agriculture. However, mining was unsustainable because it caused massive environmental damage and because Montana could not compete with mining in the developing world. Likewise, the state cannot be economically competitive in timber or food production because of its cold climate and poor soil. The one growing industry is real estate development fueled by wealthy outsiders buying vacation homes (Diamond 2005), a practice that is unsustainable because it involves excessive consumption and is incompatible with the three main industries: none of the wealthy outsiders wants a mine, clear-cut forest, or a foul-smelling farm in their backyard. As a result, Montana relies on receiving $1.50 from the federal government for every dollar they contribute in taxes. Without federal subsidies, most Montanans would be forced to leave Montana.

Commonly the root cause of societal collapse boils down to resource shortages. Some cases labeled "environmental damage" involved the loss of basic resources, such as trees on Easter Island. For example, many western U.S. ghost towns became so when the gold ran out. Although gold itself is not an essential resource, it allowed for the purchase of essential resources. Strictly speaking, people did not leave town because the gold ran out; they left because they lost their ability to purchase what they needed for survival. Such "boomtowns" lacked resilience.

4.8 CASE STUDY: HAITI

4.8.1 Local Problems

Haiti is considered the "basket case" of the western hemisphere, being the poorest country and among the most overpopulated countries in the New World (Table 4.3). The enormous problems Haitians face have persisted despite decades of international aid (Diamond 2005). Haiti has the lowest HDI in the world outside Africa. Environmental problems include deforestation, soil erosion and river siltation, lack of uncontaminated drinking water, frequent large earthquakes, and lower agricultural productivity than in the temperate zones.

Yet Haiti's problems are not simply a result of its geographic setting. Despite being on the same island of Hispaniola, citizens of the Dominican Republic (DR) fare much better than do Haitians. Both sides of Hispaniola were originally forested; now Haiti is only 1% forested, while 28% of the DR remains forested. As shown in Table 4.3, the DR has a much higher literacy rate, life expectancy,

TABLE 4.3
Comparison of Sustainability Indicators for Haiti, the Dominican Republic, and the United States

Country	Literacy Rate (%)	Fertility Rate (# per woman)	Fragile States Index	Life Expectancy at Birth (y)	Gini Index	GDP at PPP per Capita ($)
Haiti	49	3.2	104.5 (High alert)	63	60.8	749
DR	90	2.5	71.2 (Warning)	74	45.7	6086
United States		2.1	35.3 (Highly stable)	79	41.1	49,725

Source: GDP data are 2012 values from the International Monetary Fund World Economic Outlook Database, published 2015. GINI index (World Bank estimate) values for 2012–2013 from http://data.worldbank.org/indicator/SI.POV.GINI/, retrieved June 9, 2016. Fragile States Index from Messner, J.J., Nate Haken, P. Taft, H. Blyth, K. Lawrence, S.P. Graham, and F. Umana. 2015. "Fragile States Index 2015." *The Fund for Peace*. http://library.fundforpeace.org/library/fragilestatesindex-2015.pdf. Literacy rate, fertility rate, and life expectancy (average of men and women) are 2012 values from the United Nations Statistics Division, retrieved from http://unstats.un.org/unsd/demographic/products/socind/ on June 14, 2016.

and per capita GDP, and much lower fertility rate and fragile states index than Haiti. The population pyramid for Haiti (Figure 3.8b) shows that few Haitians reach old age. Every sustainability indicator shows that the Dominican Republic is in better shape than Haiti.

In the nineteenth century, Haiti was richer and more powerful than the DR. So why is the DR better off today than Haiti? The different outcomes are not due to different environments alone, referred to as **environmental determinism**. Rather, the DR responded better than Haiti to environmental problems because it had more social capital. Jared Diamond (2005) argued that social and political differences gave the DR a larger advantage over Haiti than environmental differences. Higher population density led to more rapid deforestation in Haiti. The DR spared wood by importing fossil fuels for cooking, while Haiti still relies on wood to produce charcoal for cooking. Haiti forbids foreigners to own land, which has discouraged external investment. The DR used a mix of bottom-up and top-down approaches to solve their environmental problems, but Haiti's response has been hampered by an extremely unequal distribution of wealth and a perennially corrupt government that offers minimal public services. All of these problems have led to reduced environmental, economic, and social resilience in Haiti, making it vulnerable to natural disasters.

Haiti is an extreme example of the negative effects of inequitable wealth distribution. Haiti has one of the highest levels of income inequality in the world, as measured by the **Gini coefficient** (Table 4.3). It is the wealthy, not the government, that control nearly everything that happens in Haiti. Any action taken by the government or by foreign aid agencies must first be approved by the wealthy, as they control the businesses that accept foreign aid to rebuild Haiti. They are thus the primary beneficiaries of foreign aid. The situation in Haiti won't improve until the rest of Haiti can benefit from foreign aid and have enough money to meet their basic needs. Haiti cannot become sustainable until the people have enough power, exercised by a strong democratic government, to improve their situation and shape their future. This can only happen if Haiti's wealth is distributed more equitably, and more money is invested in effective strategies to build environmental and social capital.

The catastrophic magnitude 7.0 earthquake that struck 25 km (16 miles) southwest of the capital of Port-au-Prince, Haiti, on January 12, 2010 caused 230,000 deaths and left 1,000,000 homeless. Already a failed state with no resilience, Haiti was unable to help itself in the aftermath. Its weak and ineffective government was further hobbled by damage to administrative buildings including the Presidential Palace and the National Assembly building. Even the headquarters of the United Nations Stabilization Mission in Haiti collapsed, killing many people. At least 52 aftershocks measuring 4.5 or greater occurred by January 24, making rescue and humanitarian efforts more difficult.

The city of Port-au-Prince had been completely destroyed by earthquakes at least twice previously, in 1751 and 1770. Earthquakes occur frequently in Hispaniola because it sits at the boundary between the Caribbean and North American tectonic plates. Haiti is susceptible to earthquake damage because its buildings are not earthquake-resistant, and its emergency response infrastructure is nonexistent. In a four- to five-year period preceding the earthquake, several statements and studies by geologists concluded that the Port-au-Prince region was at high risk of a large earthquake because the geologic fault near the city had been locked and accumulating stress for at least 40 years. It was located in a **seismic gap**, meaning that earthquakes had occurred recently on other segments of the fault, and that the gap was overdue for an earthquake. It would be wise to move the capital to an area with lower seismic risk so that future earthquakes would not paralyze the government as it was in 2010.

4.8.2 Local Solutions

Haiti collapsed because it was in severe ecological overshoot, that is, it plots in the "unsustainable" region in Figure 2.3 because its ecological footprint is greater than its biocapacity. Other countries like Kuwait and the UAE have higher *EF/B*, but they have not collapsed yet because they have sufficient money to compensate through trade. This is a form of substitution; money combined with trade is substituting for natural resources. Thus, weak sustainability works when there are external sources of needed capital that can be traded for.

Longevity is a good indicator of societal collapse. The Gapminder chart at www.bit.ly/1O0CbSr shows examples of collapse causing stagnation or dramatic drops in longevity for Afghanistan, Haiti, and Lesotho between 1900 and 2013. All three countries show drastic drops in longevity in 1918 due to the global flu pandemic that killed 3% to 5% of the global population. Longevity in Lesotho rose from ~33 years in 1945 to 61 in 1991 before dropping dramatically to 44 in 2006 due to political instability and the spread of HIV/AIDS. Also shown is the drastic decrease in Haiti in 2010 caused by the Port-au-Prince earthquake.

Collapse as modeled using system dynamics, observed in modern states such as Haiti, or inferred for ancient civilizations (Diamond 2005) is by definition a sudden event that can be thought of as a societal tipping point. Tipping points occur in complex systems that are nonlinear due to feedbacks. If a tipping point reinforces the change that caused it, it may push the system to another tipping point, and so on. Societal collapse may result from human systems passing multiple tipping points due to positive feedbacks. In Haiti, the removal of too many trees led to a tipping point where the irreversible loss of soil commenced. Now in many places there is not enough soil to grow trees. The loss of soil and vegetation led to another tipping point, the irreversible loss of stored fresh water. All of these reinforcing changes led to an irreversible loss of biocapacity that makes it impossible for Haitians to grow enough food, causing a collapse of Haitian society, which cannot be maintained without foreign aid and is therefore unsustainable.

Why is it essential to provide international aid to prevent countries from becoming failed states? Because societal collapse involves a successful state passing a tipping point and becoming a failed state, and it is very difficult to reverse the process because the structure of the system has been destroyed, and it takes decades to restore. For example, Haiti has been a failed state for decades, and every foreign aid worker in Haiti uses the word "hopeless" to describe Haiti (Diamond 2005). The SES of Haiti lacks many of the components of a resilient system: diversity, redundancy, reserves, agency and inclusiveness, tight feedbacks, and innovation. Haiti is so poor, and so deficient in natural resources and in trained or educated human resources, that it is difficult to see what might bring about improvement. It even lacks the capacity to utilize outside assistance effectively. Wealthy countries should provide humanitarian aid to prevent countries from collapsing to the point where investments become ineffective, and then wean the countries off the aid so that they can become self-sustaining.

The two biggest problems in Haiti are deforestation and lack of clean water. Deforestation primarily results from poor people chopping down trees to make charcoal to fuel their stoves. Women often spend many hours every day collecting wood to make charcoal. A simple solution is to provide solar cookers with instructions to the women in each household. Haiti has abundant sunshine, and to become sustainable the Haitian people need to make use of this valuable, free resource. Solar cookers eliminate the need to cut down trees for charcoal. The time saved could be used by women and girls to improve their situation, perhaps through education. An additional benefit is that solar cookers can be used to pasteurize water, thereby preventing water-borne diseases, which are a serious problem in Haiti. Solar cookers are an extremely cost-effective solution to the problems of deforestation and water contamination. Solar Cookers International* has an aid program to distribute solar CooKits, pots and water pasteurization indicators in Haiti. This is an example of high-impact philanthropy, where charities maximize benefits by leveraging existing resources such as abundant sunshine.

4.9 A PRELIMINARY LOOK AT GLOBAL SOLUTIONS

Lester Brown's Plan B to prevent the global collapse of society involves four parts: stabilizing population, eradicating poverty, restoring the Earth's natural support systems, and stabilizing climate (Brown 2009). The basic social goals of stabilizing population and eradicating poverty could be accomplished by instituting universal primary education and school lunch programs, eradicating adult illiteracy, and providing universal basic health care, reproductive health care and family planning services, and aid to women, infants, and preschool children, all at an annual cost of \$75 billion. Restoring the Earth would involve planting trees, protecting topsoil on cropland, restoring rangelands, restoring fisheries, stabilizing water tables, and protecting biological diversity at an annual cost of \$110 billion. Stabilizing climate would require cutting global net CO_2 emissions 80% by 2020 by raising energy efficiency and restructuring transportation, replacing fossil fuels with renewables, and ending net deforestation and planting trees to sequester carbon, all to prevent global atmospheric CO_2 concentrations from exceeding 400 parts per million so as to minimize future temperature rise. In this book, we will look closely at these and other potential solutions that can make society more sustainable. It is important to keep in mind that human welfare can increase without limits, but only as long as the carrying capacity is not exceeded. In the next chapter, we will explore sustainable development as an effective approach to preventing local and global collapse.

HOMEWORK PROBLEMS

1. Update the charts and tables in this chapter with more recent data. Do the new data lead to conclusions different from those in the text?
2. Use data from the Internet (provide sources) to estimate the average global risk associated with shortages of food, safe drinking water, infectious diseases, air pollution, and (if possible) global climate change.
3. Identify one society not discussed in the text that collapsed in the past. What do published studies have to say about the cause of the collapse?
4. Describe an example of an ecosystem or past society that went through an adaptive cycle. Describe each of the four phases.
5. Draw a causal loop diagram illustrating each of the following dynamic processes: forest growth, fish population in a fishery, or fossil fuel depletion. Why is the diagram different for fossil fuel depletion?
6. Define positive feedback loop and negative feedback loop in your own words, and give an original example of each.
7. Calculate the residence time of water in the ocean using data from Figure 4.8.

* http://www.solarcookers.org/.

5 Sustainable Development: How to Avoid Collapse and Build a Better Society

We could have saved the earth, but we were too damn cheap.

Kurt Vonnegut, Jr.
American author

In the previous chapter, we looked at factors that affect the sustainability of societies, including vulnerability and resilience. We learned that system dynamics models can be used to explore factors that cause a socio-ecological system (SES) to go into ecological overshoot and then collapse, and we looked at examples of collapse of ancient and modern societies. This chapter will be a bit more uplifting, as we will examine approaches to making an SES sustainable. The focus will primarily be on developing countries, what influences their development, and how they can develop sustainably.

5.1 GROWTH VERSUS DEVELOPMENT

For our purposes, the size of the economy is defined as the physical throughput, that is, the rate at which raw materials are harvested, produced, used, and disposed of (the four stages of the life cycle—RPUD). The human population and the global economy are increasing in size at exponential rates due to positive feedback loops (see the chart at http://www.bit.ly/1hGm5Bc). Exponential rates of change in a system are unsustainable and can lead to instability and collapse. To achieve sustainability we must slow the rate of growth of the population and the economy and promote sustainable development, which will allow us to leave future generations the capacity to live as well as we do today (WCED 1987). True sustainability requires a steady state population and economy. Organizations such as The Center for the Advancement of the Steady State Economy* advocate the slowing of economic growth because it can lead to depletion of natural resources and environmental degradation.

To grow means to increase in size while to develop means to expand, bring out capabilities, and to advance from a lower to a higher state. Development is sustainable but continuous economic growth is not because it is limited by the physical environment. The Brundtland commission used "development" in this sense, since they discussed sustainable development in terms of meeting human needs. In contrast, economists tend to focus on material success rather than human development, on GDP rather than human well-being.

From an environmental science perspective, sustainable development is defined as "development without growth in throughput of matter and energy beyond regenerative and adsorptive capacities" (Goodland and Daly 1996). Sustainable development improves the human condition by meeting human needs in both the short and long term, but a country is truly sustainable only if its environmental impact, best measured by the ecological footprint, is less than its biocapacity.

Examples of unsustainable practices that our society should abolish in order to become sustainable include use of disposable or inefficient products; deficit spending; reliance on nonrenewable

* http://steadystate.org/.

resources such as fossil fuels; overuse of renewable resources such as water; pollution; planning development without considering environmental impacts or environmental change; focusing only on growth of economic capital, while ignoring social and environmental capital; and wasting critical resources including food, water, or energy. In contrast, a sustainable society maximizes and maintains social, economic, and natural capital. It increases security and survival rates—and therefore longevities—by securing multiple reliable, sustainable sources of water, food, and energy. It conserves these resources and uses them efficiently. It strives to make each community autonomous and self-reliant and therefore resilient. We will examine strategies for societies to effectively use, protect, and diversify social, economic, and environmental capital to increase security.

The goal of society is to improve human well-being, and in developing countries increasing GDP and material consumption is a means to that end. However, rising income shows diminishing returns on well-being and happiness, with the latter actually showing a stronger dependence on health and marital status (Easterlin 2003). Related to this, as a measure only of economic capital, the GDP does not account for the many positive contributors to well-being such as environmental and social capital. Better measures of the contributions of economic activity to promoting human well-being, ones that are consistent with strong sustainability, are necessary. One proposed measure is the Genuine Progress Indicator (GPI), which starts with GDP, deducts the environmental and social costs incurred by economic activity, and adds the benefits of nonmarket work such as housework and volunteer work. Costanza et al. (2013) distinguish between three current economic models and their objectives:

- **Current "business as usual" model**: More economic growth as measured by GDP. Markets can overcome resource limits through substitution and new technologies.
- **Green economy model**: More economic growth but with lower environmental impact through decoupling.
- **Ecological economics model**: More human development through growth in sustainable human well-being, as measured by GPI.

In terms of constant dollar per person in the United States, from 1975–2005 GDP has increased steadily but GPI has remained flat, suggesting that the business as usual approach to economic growth is unsustainable (Costanza et al. 2013).

5.2 PROPERTIES OF SUSTAINABLE SOCIETIES

We start by examining two societies that meet many of the objectives of sustainability. In *Collapse* (2005), Jared Diamond points to Japan as a successful model of the move toward sustainability. Japan has experienced resource limitations for centuries because of its high population density, small land area, and island geography. Japan's resource limitations forced its citizens to live more sustainably and its social organizations (governments and corporations) to operate more sustainably. Its average per capita ecological footprint is ~4 global hectares (gha), less than the U.S. value of ~7 gha though higher than the average global biocapacity of ~1.8 gha, a truly sustainable level (McLellan et al. 2014). Its homogeneous culture and strong leadership was well-prepared to adapt to resource limits. For example, Japanese leaders used a top-down approach to develop sustainable forestry management in response to wood shortages that started in the mid-seventeenth century (Diamond 2005). More recently, Japanese companies dealt with resource shortages by emphasizing quality over quantity. Its early lead in adapting to resource shortages gave Japan an economic edge, and may explain why their high-quality, efficient automobiles have outsold American cars for the last few decades. Their sustainable approach also explains the resilience of Japan in the aftermath of the Fukushima nuclear accident in 2011, and the fact that Japanese citizens have the highest average longevity in the world.

As another example of sustainable communities, Amish communities in the United States have achieved autonomy and nearly complete sustainability. In traditional Older Order communities in

Pennsylvania, Ohio, and Indiana, the Amish live simple lives and reject the use of many modern technologies such as electricity. Education is limited to the eighth grade, and they spend most of their lives in manual labor. They use horses and buggies for transportation and wear spare clothing that they make themselves. The Amish do not rely on centralized sources of energy, water, or food, and they are pacifists. They use traditional organic farming methods and obtain crop yields that are 50% to 75% of those obtained by industrial agriculture (Heinberg 2004). The Amish refuse to buy insurance, instead relying on neighbors and their church for support. If a storm destroys a barn, or if a new family needs a barn, the community holds a barn raising and builds a new barn in a single day. Unlike the rest of the United States, the Amish have little to fear from future resource shortages or economic crises: they have much smaller ecological footprints than the average American, and their communities are sustainable and resilient.

We don't have to live like the Amish to be sustainable. Perhaps the best way to think about the Amish lifestyle is as a starting point: how can we incorporate some of their sustainable practices into our own lives, while retaining elements of our culture such as higher education and wealth that promote human development?

5.3 FACTORS THAT INFLUENCE THE SUSTAINABILITY OF DEVELOPING COUNTRIES

5.3.1 Poverty

Perhaps the largest obstacle to sustainable development is poverty. Poverty is a lack of economic capital, and it is largely inherited. In general, people with low incomes are less productive because they are poorly educated, have poor health, and live shorter lives (www.bit.ly/1Jrlu2n). According to the Millennium Ecosystem Assessment (2005):

- 1.1 billion people survive on less than $1 per day.
- Over 850 million people were undernourished in 2000–2002, up 37 million from the period of 1997–1999.
- Some 1.1 billion people still lack access to improved water supply, and more than 2.6 billion lack access to improved sanitation.
- Global improvements in levels of poverty are skewed by rapid economic growth in India and China; poverty elsewhere (especially in sub-Saharan Africa) is profound and persistent.

Poverty is a problem even in wealthy countries such as the United States. In 2010, over 14% of Americans lived below the annual income poverty line of $10,500 per person or $21,500 for a family of four. The most common cause of bankruptcy in the United States is poor health. Illness depletes financial resources and often deprives people of the ability to earn a paycheck, which is why universal healthcare has been intensely debated in recent years.

The "war on poverty" is a primary goal of economic development policies of the UN, the World Bank, and other organizations because collective social well-being improves as the proportion of the population living in poverty decreases. The objective is to narrow the equity gap between rich and poor, not by lowering the top but by raising the bottom, that is, enriching the poor (Goodland and Daly 1996). The good news is that poverty rates are declining in most developing countries. The poverty rate is decreasing for the populous BRIC countries (Brazil, Russia, India, and China) other than Russia, where the rate is stable (see the chart at www.bit.ly/1Kg4LxQ). The global poverty rate is also declining due to globalization and associated economic development (see below).

We cannot achieve global sustainability without eliminating poverty, which is why "eradicate extreme poverty and hunger" was the first of the eight Millennium Development Goals (UN 2000). People enduring poverty and hunger must often take desperate measures to survive, which in turn can cause environmental degradation. Starving people cannot afford to think about the future: they eat chickens that could lay eggs and cows that could produce milk in the future. They might cut down fruit and nut trees for cooking and heating, eliminating a future food source.

People living in poverty often degrade the environment because they have no choice. In countries like Haiti, the poor deforest surrounding areas for wood for cooking, causing soil erosion and desertification. In many areas of the world such as India, poor people do not have proper sanitation, so they pollute surface water bodies with their waste. When they need water, they pump aquifers until they are dry, or they use water from unclean sources and contract diseases, placing an economic burden on public health facilities.

The loss of environmental capital makes it harder to build economic and social capital: how can your construction business make money building homes if people have used the supply of your main building material—wood—to fuel their stoves? How can you grow a prosperous city if there is no fresh water left? Poverty can also cause further loss of economic capital because lost environmental support services must be replaced using economic capital. For example, when groundwater over-pumping or surface water pollution causes the loss of fresh water, people are forced to buy fresh water that they previously obtained for free, or they spend more money on pumps and fuel to pump groundwater from greater depths. The initial lack of economic capital causes loss of environmental capital, which causes further loss of economic capital. This positive feedback that results from competition for scarce resources makes the poor poorer and also tends to make the rich richer. Using the language of systems theory (Meadows 2008), "According to the competitive exclusion principle, if a reinforcing feedback loop rewards the winner of a competition with the means to win further competitions, the result will be the elimination of all but a few competitors, i.e., *the rich get richer and the poor get poorer*."

How do we break this vicious cycle? One way is to adopt policies that fuel economic growth in the short term, which increases economic capital and raises people out of poverty if the distribution of new wealth is equitable. In developing countries that are not already in ecological overshoot, equitable economic growth can also increase social capital and preserve environmental capital (for the reasons outlined above), thereby making society more sustainable.

So what policies and approaches are most effective at eliminating poverty and improving human well-being? The UN's Human Development Index HDI is one of the best measures of social capital and quality of life (UNDP 2015). HDI increases strongly with increasing income (Figure 5.1) until it levels off at an annual income of ~$30,000. Thus, the most effective approach to improving quality of life in developing countries in the short term is to promote economic growth until per capita GDP reaches an annual level of ~$30,000.

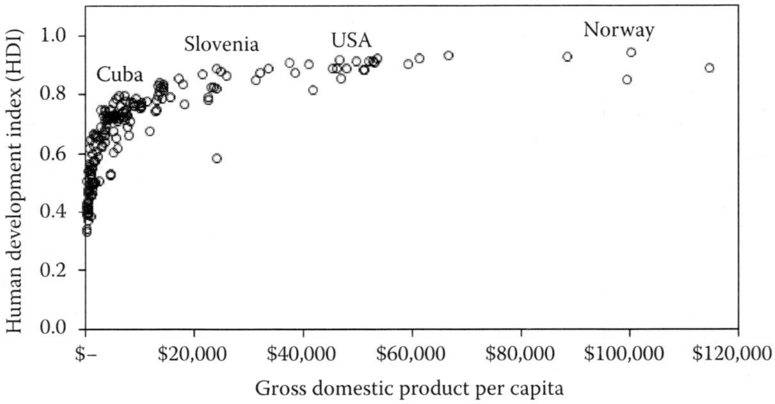

FIGURE 5.1 Gross domestic product per capita in 2012 versus the UN Human Development Index (HDI) for most countries. (GDP data are 2012 values from the International Monetary Fund World Economic Outlook Database, Published 2015. HDI data from http://hdr.undp.org/en/data, retrieved April 12, 2016.)

Most economists and the World Bank advocate economic growth as the primary tool for ending poverty. Some, like Libertarian John Stossell (2004), argue that we should eliminate all barriers to wealth creation, including environmental regulations. However, that would create economic capital at the expense of environmental and social capital. Governments should institute policies that promote growth in economic capital without reducing environmental or social capital, consistent with strong sustainability and either the ecological economics or green economy model.

Can the objective of the green economy model of decoupling income from environmental ImPACT be achieved? Some evidence suggests that it can. Figure 3.5 shows that some countries have relatively low EF and high GDP. Compared to the United States, Brunei has a much higher GDP but lower EF (Figure 3.5). Brunei has done a better job than the United States of decoupling income growth and environmental impacts. However, no country with an income higher than $12,881 has an EF less than Earth's biocapacity. This suggests it is not currently possible for all countries to achieve a maximum HDI by increasing per capita annual GDP much higher than ~$30,000 (the goal defined by Figure 5.1). A goal that is more realistic, because it is potentially sustainable, is to lift the income of the poorest to the level of a country like Tonga, which in 2012 had an average income of $4,331 and HDI of 0.72, or Cuba with an average income of $5,176 and HDI of 0.77 (Goodland and Daly 1996). Norway has the highest HDI of 0.94 but an average income of $100,172 and EF of 5.0 gha, which is not sustainable.

Supplementary approaches to reducing poverty include slowing population growth, improving education and health, and rescuing failing states. No developing country has successfully modernized without slowing population growth through the demographic transition (Brown 2009). For example, part of the reason for China's rapid economic growth is its efforts to slow population growth, which has paid a **demographic dividend**, defined by the United Nations Population Fund as "the economic growth potential that can result from shifts in a population's age structure, mainly when the share of the working-age population (15 to 64) is larger than the non-working-age share of the population (14 and younger, and 65 and older)." The population pyramid for China in Figure 3.8a shows that a very high proportion of China's population is of working age. This is a result of China's one-child policy instituted in 1979. The demographic dividend leads to an increase in average per capita productivity, savings, and investment, and thus fosters economic growth in the short term. The demographic dividend lasts for a few decades and gives countries a chance to modernize. The downside is that China's population is aging, and in the future, the percentage of Chinese who are of working age will begin to decline, and economic growth will slow. Furthermore, there will be fewer people to care for Chinese senior citizens. These demographic changes will strain the Chinese economic and social systems.

We saw in Section 3.4.3 that education of girls decreases the fertility rate, and with fewer children, it is easier for women to lift themselves and their families out of poverty. Another way to reduce poverty is by improving health. In developing countries, infectious diseases (diarrhea, respiratory illnesses, tuberculosis, malaria, measles, and AIDS) are the overriding health threats (Figure 4.1). Childhood immunization is a highly cost-effective approach to reducing the prevalence of infectious diseases. It has eradicated smallpox and nearly eradicated polio. Programs that discourage use of harmful products like tobacco can also greatly improve public health. In 2005, 5.4 million people died of tobacco-related illnesses, more people than all infectious diseases and more than from all other air pollutants combined (3 million) (Brown 2009).

A final approach to decrease poverty is to rescue failed states such as Haiti. In failed states, government is weak or nonexistent and therefore unable to provide basic public services such as education and police protection. The number of failed states is increasing over time (Brown 2011), which is leading to civil wars, uncontrolled immigration, and rising terrorism, and if this trend continues unabated, global civilization may be at risk of failing. Systemic failure requires a systemic response, but the United States does not have a coherent process for aiding failing states. Brown (2009) advocates for the formation of a Department of Global Security to address these problems. He argues that threats to national security are becoming less from military power and more from trends that undermine states (rapid population growth, poverty, deteriorating environmental support systems, and spreading water shortages). These development trends fuel discontent, resulting in growth of terrorism aimed at wealthy countries. Wealthy countries like the United States serve their own interests by adopting policies that aid citizens of developing countries.

5.3.2 Globalization

> Outsourcing and globalization of manufacturing allows companies to reduce costs, benefits consumers with lower cost goods and services, causes economic expansion that reduces unemployment, and increases productivity and job creation.
>
> **Larry Elder**
> *American Libertarian radio and television personality*

> The negative side to globalization is that it wipes out entire economic systems and in doing so wipes out the accompanying culture.
>
> **Peter L. Berger**
> *American sociologist and Lutheran theologian*

Another force that profoundly influences sustainable development is **globalization**, which is an ongoing process by which regional economies, societies, and cultures have become integrated through a globe-spanning network of communication and trade (Friedman 2006). Growth in human population and consumption has fueled global economic growth, while advances in transportation and communication have accelerated the transfer of physical goods and information, transforming the world in the last 200 years.*

Globalization is partly a result of increasing international trade made possible by cheap transportation fuels. Free trade agreements removed trade barriers, allowing goods to flow freely from producer to consumer nations. This resulted in greater efficiency and lower prices because each country can specialize in producing only products or services for which they have a comparative advantage over other countries. Countries with specialized production then obtain all other goods and services through trade (Rees 2010). For example, having the raw materials for a product that other countries do not have gives a country a competitive advantage; a country with large iron deposits and cheap labor is likely to produce and export steel. However, specialization makes nations less resilient, as they must rely on other nations for critical resources. In *Collapse*, Diamond (2005) notes that the loss of a trading partner was one of the influential factors contributing to the collapse of some ancient societies, particularly those in the very resource-limited Pacific Islands. For example, if country B relies on country A for food, but food production in country A declines because of climate change, then country A may stop exporting to country B so that it can feed its own people, and people in country B will starve. Thus, increasing specialization leads to decreasing self-reliance and increasing vulnerability. Many countries may depend on a single supplier of a product, so if that country stops exporting that product many countries are adversely affected and the economic impact is magnified. In a globalized world, countries are increasingly interdependent, so that trade disruptions can become global economic crises.

Globalization corresponds to the conservation phase of the adaptive cycle (Walker and Salt 2006). Rapid economic growth and diversification in the nineteenth and early twentieth centuries transitioned to increased industrial specialization and efficiency. Increasing competition caused many small, local manufacturers to be replaced by fewer, larger global manufacturers. Economies of scale led to lower prices, and specialization led to increasing international trade. However, increasing efficiency leads to decreasing flexibility, and increasing interconnectedness makes the system become more rigid, causing resilience to decline. This means the economic system is more stable, but over

* See Hans Rosling's video, "200 Years that Changed the World": http://www.gapminder.org/videos/200-years-that-changed-the-world/.

a narrower range of conditions; it is increasingly vulnerable to disturbances. The longer a system stays in the conservation phase, the smaller the disturbance needed to end it through collapse. For example, the **Great Recession** started with the housing crisis in a single country, the United States.*

The conservation phase of the adaptive cycle is called the "K" phase because "K" is the symbol for carrying capacity (Figure 4.4). It is the global carrying capacity, the maximum human population that the global economic system can support, that is the ultimate limiting factor for the conservation phase. Sometime after the carrying capacity is exceeded, the economic system will likely collapse and enter into the release or "omega" phase of the adaptive cycle. Economic chaos will cause many large corporations to go bankrupt and lead to a global recession or depression. The good news is that this process will release economic capital, creating new opportunities for creative inventors and entrepreneurs in the reorganization or "Alpha" phase (Walker and Salt 2006). Thus, the death symbolized by omega is followed by rebirth and a new beginning in the alpha phase. For example, the United States emerged from the Great Depression of the 1930s to become in the late twentieth century the world's most powerful country with the largest economy. Once we recognize that, in complex adaptive systems, collapse is always followed by phases of new opportunities and growth, it becomes easier to accept that collapse may occur (Homer-Dixon 2006). The process of collapse is least painful to those who anticipate it and prepare for it.

Globalization is raising people out of poverty faster than at any time in human history. However, by raising living standards, it has rapidly increased consumption rates, leading to more rapid environmental degradation, decreasing resilience, and increasing vulnerability (Friedman 2008). Governments around the world have adopted policies that promote economic growth at all costs. This scorched earth approach has made the global socio-ecological system vulnerable to collapse (Rees 2010). Sustainable development aims to lift people out of poverty without causing environmental degradation. The effects of globalization on sustainability have to be considered when formulating policies for sustainable development.

5.3.2.1 Effects on the Economy

By removing barriers so that goods, services, people, and ideas can freely move from place to place, globalization has leveled the global economic playing field (Friedman 2008). Removing trade barriers reduces costs and increases efficiency. Thus, globalization spurs economic growth, as found in a study that compared economies of countries (Kwong 2005). It also narrows the economic gap between poor and rich countries; the world is becoming "flat," meaning that per capita income in developing countries like China and India is approaching that in developed countries (Friedman 2006). Until 2015, globalization allowed China's economy to grow at an astounding annual rate of 10% over a period of 30 years, making it the world's second largest economy. Eventually, the United States may not always be able to outbid other countries for resources that are scarce because poor countries are becoming richer and more economically competitive, as has happened to China. Another long-term implication is that outsourcing may become more expensive as labor costs in developing countries increase, making it a less attractive option.

Globalization works only in the presence of economic freedom. Economic freedom promotes long-term economic growth *and leads to more efficient resource use*. The three key ingredients of economic freedom are (1) free and open markets, (2) clearly established property rights, and (3) enforced laws (Kwong 2005). These ingredients hold people accountable for their actions and reward them for positive behavior. For example, they discourage theft of land and resources by defining what constitutes theft (property rights) and by punishing thieves (enforced laws, meaning no corruption). Combining economic freedom with free trade agreements that include environmental protections and compensation to locals can lead to economic growth that does not reduce social or environmental capital. Developing countries must carefully design and implement economic development policies in ways that do not enhance social and economic inequality.

---

* https://www.britannica.com/topic/Financial-Crisis-of-2008-The-1484264, retrieved August 15, 2016.

The study of societal collapses of the past illustrates the economic risks associated with modern globalization. For example, Polynesians on Pitcairn and Henderson Islands were completely dependent on trade with the island of Mangareva between 1000 and 1450 A.D. (Diamond 2005). The islands of Pitcairn and Henderson are very isolated and ecologically fragile, and lack many critical resources including most Polynesian foodstuffs. The larger western island of Mangareva was also ecologically fragile, and overpopulation and deforestation led to soil erosion and resulting declines in agricultural and fish yields. Mangareva society slid into civil war, chronic hunger, and cannibalism as recalled by modern islanders, and all exports to Pitcairn and Henderson ceased by 1500 A.D. Residents of Pitcairn and Henderson were trapped on their small islands because they had no trees for making canoes. Henderson's population ceased to exist by 1606 A.D., and Pitcairn's by 1790 A.D. Many modern countries are completely dependent on international trade to support their populations, and the loss of trading partners could be devastating, as it was for the residents of Pitcairn and Henderson islands.

In the United States, globalization in general and increased imports resulting from international free trade agreements such as **NAFTA** and entry of China into the **World Trade Organization** have resulted in manufacturing job losses in the United States, which in part has led to rising social and income inequality and political radicalization and polarization.* However, most economists believe the benefits of free trade greatly outweigh the costs. In general, free trade benefits the poor the most because it lowers the cost of consumer goods. However, without effective policies to aid workers who lose their jobs due to decreased trade barriers, the social changes induced by globalization can lead to social and political instability.

The interconnectedness of countries resulting from globalization means that problems that develop overseas automatically become problems for the United States. In the past, economic problems in the United States or Europe have caused global economic crises, and in 2015–2016 it became apparent that China's economy has global influence, as U.S. stock market indices declined as a result of slowing economic growth in China.† Even the debt crisis in Greece affected the global stock market in 2014–2015. The economies of countries are highly interconnected, and the collapse of economies and societies in foreign countries adversely affects U.S. interests and decreases our national security, so it is in the best interests of the United States to provide foreign aid to keep foreign economies from failing.

5.3.2.2 Effects on Society

> Globalization, as defined by rich people like us, is a very nice thing...You are talking about the Internet, you are talking about cell phones, you are talking about computers. This doesn't affect two-thirds of the people of the world...If you're totally illiterate and living on one dollar a day, the benefits of globalization never come to you.
>
> **Jimmy Carter**
> *39th U.S. President, 2002 Nobel Peace Prize recipient*

Having established that globalization increases economic capital and improves living standards for many of the poor in developing countries, we want to know whether globalization increases or decreases social capital. Much of the backlash against globalization stems from a fear that it will lead to a homogenization of culture. This process has operated throughout history, but electronic media and global transportation have accelerated the process because they have removed barriers to the exchange of information. Without barriers, random processes cause the entropy of the global social system to increase, eventually leading to homogenization. It's like the classic experiment that explains entropy and diffusion. Divide a box into two chambers and fill each with a different gas.

---

* "Where jobs are squeezed by Chinese trade, voters seek extremes," Nelson D. Schwartz and Quoctrung Bui, April 25, 2016, *The New York Times*, http://mobile.nytimes.com/2016/04/26/business/economy/where-jobs-are-squeezed-by-chinese-trade-voters-seek-extremes.html?.

† "Why China is still the biggest threat to U.S. stocks," Matt Egan, CNN, http://money.cnn.com/2016/01/04/investing/china-biggest-threat-to-us-stocks/, retrieved June 21, 2016.

When you remove the divider, gas A molecules begin to diffuse into the gas B chamber, and vice versa. The entropy, or disorder, of the system increases as the two chambers change composition from pure gas to increasingly similar mixtures of A and B molecules. When the process is complete, the two chambers have the same compositions. Removing the barrier resulted in an increase in entropy and stability.*

Likewise, geographic and communication barriers have historically divided world cultures. A diverse array of cultures developed in isolation, which led to decreased stability and increased conflicts. The modern removal of communication barriers inevitably reversed the process of cultural divergence by increasing the efficiency of information exchange and removing cultural obstructions. Diffusion of ideas, cultural objects, cuisines, music, and movies has had large impacts on the food and entertainment industries.

Just as a homogeneous mixture of two gases is more stable than the segregated pure gases, cultural homogenization should promote stability. Removal of cultural differences and barriers increases understanding, which decreases fear and hatred, which increases stability.† Theoretically, cultural diffusion and the resulting cultural "blending" will ultimately (over long periods of time) lead to cultural homogenization and societal stability, but in the short term the process can be disruptive and painful (e.g., the many countries that have had a recent **ethnic conflict**: Rwanda, Chechnya, Yugoslavia, etc.). However, the forces driving cultural homogenization are relentless. As long as humanity has affordable global travel and digital communication, the only way to slow or prevent global cultural homogenization is to slow or stop the exchange of information, which is neither desirable nor acceptable.

Instability in foreign countries can also lead to mass migration. Immigration from Latin American countries, particularly Mexico, has led to much hand-wringing in the United States. Since 2015, mass migration from countries destabilized by ISIS (Syria, Iraq, and Afghanistan) to Europe has been causing social upheaval and the deaths of many migrants. Such migration events often cause the rise of right-wing political groups opposed to immigration in destination countries. Many of these social changes are at least partly caused by globalization.

5.3.2.3 Effects on the Environment

If globalization spurs economic growth, what effect does it have on the environment? We have argued that consumption increases with income, and that environmental degradation increases with consumption, so logically, environmental degradation must increase with income. This is in fact what we observe in developed countries: per capita and total EF increases with per capita GDP (Figure 3.5). We can expect that as developing countries become developed countries, globalization will lead to further increases in average income and further environmental degradation in those countries, too.

In contrast, most economists argue that globalization is *good* for the environment because it encourages economic growth, which, for pollutants that obey the environmental Kuznets curve (Figure 3.2), eventually leads to decreasing pollution. In developed countries that have passed the

* In chemical systems, stability is enhanced when entropy increases, which decreases the Gibbs free energy. A system is most stable, that is, is at chemical equilibrium, when it has the lowest possible Gibbs free energy.

† However, we also previously argued that decreased diversity leads to decreased resilience. A system is most resilient when diversity is at a maximum. For example, ecosystems with high biodiversity are more resilient than those with low biodiversity. In a farm or garden, a polyculture is more resilient than a monoculture. Reasoning by analogy, high cultural diversity corresponds to greater resilience. Cultural diversity makes it more likely that society will find solutions in the face of global threats such as global warming. In the past, some cultures were better prepared to deal with adversity, while other less adept civilizations collapsed. For example, in contrast to the Easter Islanders who practiced unsustainable logging practices until no trees remained, Japanese leaders successfully dealt with timber shortages in the mid-seventeenth century. They invoked Confucian principles of limiting consumption and accumulating reserve supplies to develop sustainable forest management (Diamond 2005). In our global society, one culture may provide the seed of knowledge or understanding that will lead to the preservation of global civilization. What if that culture was wiped out during cultural homogenization? It seems that global homogenization of culture may decrease the resilience of humanity.

turning point, further economic growth results in reduced environmental degradation, suggesting that countries can grow out of pollution problems by increasing wealth (Kwong 2005). For example, since the 1970s wealth in the United States increased while the extent of many forms of environmental degradation decreased, primarily due to adoption of the **Clean Air Act** and **Clean Water Act** and subsequent additional environmental laws and regulations such as the **Resource Conservation and Recovery Act**. The explanation is that wealthy countries like the United States can afford to invest in restoring the environment. Countries currently below that threshold find that environmental degradation is increasing with economic growth, which is why some of them (such as Afghanistan, India, and China) are so polluted.

However, the environmental implications of the Kuznets curve have been overgeneralized. First, the curve seems to apply only to capitalist economies, which "use fewer resources to produce the equivalent level of output and hence do less damage to the environment" (Kwong 2005), and does not apply to socialist or communist countries. Second, data suggest that the Kuznets curve does not apply to many pollutants such as carbon dioxide, and has not been shown to be applicable to other forms of environmental degradation such as deforestation, erosion, biodiversity loss, and so on. (Dinda 2004). Third, the decline in pollution in developed countries may be only a local and not a global decline, as polluting industries are exported from developed to developing countries. Finally, the income levels required to reverse negative environmental trends may be unsustainable (Figure 3.5).

Environmentalists argue that globalization and free trade provide a license to pollute and spur a "race to the bottom." The UN's Brundtland Commission argued in 1987 that economic development was unsustainable because the number of people living in poverty was increasing (but note that the percentage of people living in poverty has been decreasing). Also, in most places the environment was degrading (Friedman 2008), meaning that it would support fewer people in the future.

Figure 5.1 suggests that the most effective approach to improving human well-being is to increase income by growing the economy, as economists frequently argue. Furthermore, global measures of human well-being are increasing despite increases in environmental degradation and loss of ecosystem services (Millennium Ecosystem Assessment 2005b). The most plausible explanation is that there is a time lag between ecosystem service losses and resulting declines in human well-being. Such time lags are characteristic of complex systems with nonlinear feedbacks, and are predicted by system dynamics models. Other possible explanations are that technology has decoupled well-being from nature, and that well-being mostly depends on food services, which are increasing (Raudsepp-Hearne et al. 2010). Many questions about the relationship between human well-being and ecosystem services remain unanswered. Is the current economic growth-fueled improvement in human well-being a short-term benefit that comes at the expense of the environment? In the long term, will it result in loss of ecosystem services and a reversal of human development? Are globalization and associated economic growth unsustainable?

One development that could work against economic globalization is future increases in transportation costs caused by Peak Oil (Chapter 6) or new carbon taxes. Globalization requires global economic trade, which in turn requires long-distance transportation of goods. Cheap energy makes this possible. Will the trend of globalization stop and perhaps even reverse when energy for transportation becomes too expensive? Or will we transition to transportation based on renewable energy sources before this happens?

5.3.2.4 Problems and Solutions

> Instead of saying that globalization is a fact, that it's inevitable, we've also got to demonstrate that while the growing interdependence of the world economy is indeed a fact, it's not uncontrollable.
>
> **Peter Mandelson**
> *Labour Party politician in the UK, member of Parliament 1992–2004*

> We must create a kind of globalization that works for everyone...and not just for a few.
>
> **Nestor Kirchner**
> *Former president of Argentina*

From the perspective of sustainability, we want to know whether globalization can be good for both society and the environment and whether it can lead to more sustainable lifestyles. We also want to know whether globalization itself is sustainable.

In his book *Hot, Flat, and Crowded*, Friedman (2008) argues that the convergence of globalization, climate change, and population growth has brought us to the edge of a global crisis that marks the beginning of the Energy-Climate Era. His book addresses five key problems we face in this era:

- The growing demand for ever scarcer energy supplies and natural resources
- A massive transfer of wealth to oil-rich countries and their petrodictators
- Disruptive climate change
- Energy poverty, which is sharply dividing the world into electricity "haves" and "have-nots"
- Rapidly accelerating biodiversity loss, as plants and animals go extinct at record rates

However, there is room for hope. In *Global Sustainability: A Nobel Cause* (Schellnhuber et al. 2010), Ian McEwen describes positive aspects of globalization: "Globalization, while it has unified economies, increased production and raised carbon dioxide levels, has also created global networks of expert opinion and citizen's demands that are placing pressure on governments to take action." These positive aspects arise from the increased access to information provided by computers and the Internet, which in turn has increased transparency and the ability of the masses to organize and convey their opinions to their political representatives and other citizens.

The hope is that globalization will improve education and, as a result, decrease birth rates in the developing world. This, combined with education on sustainable practices, could stop and even reverse environmental degradation, which would lead to preservation and restoration of ecosystem services. If we adopt international trade agreements that promote sustainability so that economic capital is built without reducing environmental and social capital, then globalization could help raise the poorest out of poverty, thereby drastically increasing longevity and quality of life, without causing environmental degradation.

5.4 SUSTAINABLE DEVELOPMENT

The goal of sustainable development is that developing countries will manage their resources in a way that maximizes human well-being in both the short and long term (Vallero and Brasier 2008). The goal should be to preserve or grow environmental, economic, and social capital while promoting human development. Sustainable development thus involves the simultaneous pursuit of social equity, economic prosperity, and environmental quality, collectively referred to as the **triple bottom line**, in a way that moves a society toward a sustainable state (Ott 2003). Achieving sustainability requires sustainable development, but once we achieve it, we have to keep working to maintain it.

We can draw a stock and flow diagram for each renewable resource in an SES. For the SES to maintain sustainability (long-term stability), it must maintain a steady state for each resource stock in which the annual outflow (consumption) of each capital stock is equal to the inflow (renewal rate). For example, if our capital stock is a forest, the number of trees harvested per year must not exceed the number of trees planted per year. The water withdrawn from the lake or aquifer must not exceed the amount replaced by recharge. True, in years of drought the withdrawal of water from the aquifer may exceed the recharge by necessity, but when the drought ends the amount withdrawn must be less than the amount recharged until we pay the deficit.

Communities and countries can achieve long-term stability through sustainable development followed by sustainable maintenance. Use of nonrenewable energy resources such as oil can temporarily fuel economic growth, which in turn can finance sustainable development. However, consumption of nonrenewable resources such as fossil fuels must eventually dwindle to zero, usually through substitution of renewable resources such as solar or wind energy, so that the community or country can persist indefinitely in a steady state after the nonrenewable resources have run out.

Unfortunately, some countries, including the United States, are in ecological overshoot and are practicing deficit spending both in economic terms but also in terms of the environment. To achieve sustainability, developed countries like the United States must either curtail economic growth (but not human development), or find a way to decouple income growth and environmental degradation.

The United Nations has long been aware of the need for sustainable development. In 2000 the Millennium Declaration was adopted by 189 nations. It listed eight Millennium Development Goals for countries to achieve by 2015. Progress toward achieving these goals was measured using the Millennium Development Goals Index, which unsurprisingly shows a high positive correlation with per capita income. Progress on achieving many of the Millennium Development Goals was slow (Sachs and McArthur 2005), and many developing countries never achieved their goals. At the UN Sustainable Development Summit in 2015, the eight Millennium Development Goals were replaced by 17 Sustainable Development Goals, with the objectives being "to free the human race from the tyranny of poverty and want" and "to heal and secure our planet" by 2030.* While the Millennium Development Goals sought to cut poverty in half, the Sustainable Development Goals aim to eliminate poverty. A major problem with the Sustainable Development Goals is that no metrics have been provided to measure progress toward achieving many of the goals.

5.4.1 Case Study: Uganda

Dr. Moses Musaazi from Uganda is an excellent example of how people in the developing world can promote sustainable development.† He is an electrical engineer at Makerere University in Kampala, Uganda. Musaazi has designed and produced many products to improve the health and well-being of Ugandans.‡ For example, studies had shown that an obstacle to educating teenage girls in Uganda was that they could not attend school during their menstrual cycles, as they could not afford to buy sanitary napkins. To address this problem, and with the help of the Rockefeller Foundation, Musaazi developed the "Makapad," a sanitary napkin made primarily from native papyrus and paper waste. The pads cost 50% less than commercial pads and are manufactured by area workers. Nongovernmental organizations purchase the pads and distribute them to women in need, allowing them to attend school.

To promote sustainable development in Uganda further, Musaazi designed another sustainable product, curved, interlocking soil bricks. Traditionally, Ugandans constructed their homes using oven-fired bricks. To produce the bricks, locals would destroy a local swamp and convert it to a pond where they could collect clay. Then they would cut down local trees to fuel the oven that fired the bricks. After completing the homes, they left the land barren, and the soil turned to dust and blew away. Musaazi designed bricks that require no firing. To make a soil brick, the homebuilder simply needs to mix soil and inexpensive Portland cement together and then use a lever or hydraulic ram to squeeze them into a mold. The soil bricks interlock, producing sturdy structures that do not rely on mortar. As additional benefits, soil bricks reduce the time required to construct each house, and they don't damage the environment.

Musaazi also uses curved ISSBs to make cylindrical water tanks (cisterns) for storing harvested rainwater. Because human waste often contaminates groundwater in the area, locals must sometimes abandon the practice of digging wells or collecting water from springs. Instead, they can use gutters to collect clean rainwater from roofs and then drain it into cisterns equipped with valves for water dispensing.

Intelligently designed, sustainable products like these can greatly improve people's lives. In fact, many products like these are available in different parts of the world, but few people know they

---

* http://blogs.worldwatch.org/sustainable-development-goals-will-the-world-stick-to-its-new-years-resolutions/ and https://sustainabledevelopment.un.org/?menu=1300.

† I met Dr. Musaazi when he gave a talk at Vanderbilt University and corresponded with him thereafter.

‡ See http://www.t4tafrica.com/.

exist. One of the most effective ways to help people in the developing world is to identify well-educated, caring, and dedicated people like Dr. Musaazi and give them resources they can use to transfer knowledge to their communities.*

5.4.2 Urbanization

Urbanization, the movement of people from rural to urban settings, is the dominant economic event of the first half of the twenty-first century, and has dramatically affected developing countries. Urbanization accelerated dramatically in the twentieth century. In 1800, the global population was 3% urban, increasing to 14% in 1900 and 50% in 2007 (Brand 2009). Urbanization can increase economic, social, and environmental capital by reducing the costs and impacts of transportation and by improving opportunities in education, employment, and housing. Other positive impacts include the movement of rural dwellers who practiced environmentally destructive subsistence farming to cities, and increased education and decreased fertility of women. However, urbanization can also cause negative impacts such as increased social inequality, high crime, and concentration of pollutants. Sustainable development can reduce the negative environmental and social impacts of urbanization.

Urbanization has the potential to increase the human carrying capacity of the Earth by making life support systems more efficient. Compared to rural settings, cities benefit from **economies of scale** on many levels. For example, consider the per capita cost of supplying fresh water in a city with one million people using one large reservoir versus the same number of people in a rural area covering several different watersheds that would necessitate the use of 100 reservoirs, each serving 10,000 people. First, to build one large reservoir that could supply water to one million people would cost less than to build 100 reservoirs that each serve 10,000 people. Also, it would cost less to build the water delivery system in a city with high population density: the total number of connections and the total length of pipe would be less. The costs would be lower because less building material would be required, and on average the materials would be transported shorter distances to construction sites. Furthermore, fewer people would be needed to maintain the city water supply system.

Recent work has shown that population size has the greatest influence on the character and sustainability of cities (Bettencourt and West 2010). On average, when a city increases in population by 100%, it requires only an 85% increase in infrastructure (roads, water pipes, electrical cables, etc.), meaning it becomes 15% more efficient. Communities of other species such as ants and bees exhibit similar economies of scale. Urbanites also tend to have smaller homes and to rely more on public transport, which is more energy-efficient than transportation by personal automobile.

Cities also drive economic growth because they benefit from **economies of agglomeration**, that is, high population density accelerates economic activity by placing consumers closer to producers. According to urban theorist Richard Florida, the world's 40 largest megaregions are home to 18% of the world's population but account for 66% of global economic activity and 86% of patented innovations.† A UN report concluded, "Cities are so much more successful in promoting new forms of income generation, and it is so much cheaper to provide services in urban areas, that some experts are actually suggesting that the only realistic poverty reduction strategy is to get as many people as possible to move to the city" (UN-Habitat 2004). Urbanization increases human population density, which in the absence of a population increase allows more land to be returned to a natural state. Urbanization decreases fertility to below the replacement rate, which explains why some European

* For more information about Dr. Musaazi, see http://www.inc.com/magazine/201205/leigh-buchanan/uganda-moses-kizza-musaazi-never-stops-innovating.html.

† "Megaregions: The importance of place," *Harvard Business Review*, March 2008, https://hbr.org/2008/03/megaregions-the-importance-of-place#, retrieved June 21, 2016.

countries with high population density have decreasing populations. Other benefits of cities include lower transportation costs and greater opportunities for jobs, education, housing, and transportation.

Of course, urbanization has negative effects on society. While a population doubling increases efficiency by 15%, it also increases negative factors like social inequality, traffic congestion, crime, and certain diseases by 15% (Bettencourt and West 2010). The concentration of waste in cities can adversely affect health, as can increased stress levels resulting from the faster pace of city living, a decreased sense of security, and, paradoxically, social isolation. Finally, people living in cities generally have higher affluence and consumption levels, which in some cities has more than offset efficiency gains and led to a higher per capita environmental ImPACT than for people in rural areas (Kennedy, Cuddihy, and Engel-Yan 2007). The challenge then is to find ways to increase the sustainability of cities by reducing social inequality, crime, and waste production and increasing social capital for city dwellers.

A recent analysis of material and energy flows in 27 **megacities** (cities with populations exceeding ten million) revealed that they hold 6.7% of the world's population but account for 14.6% of GDP, 12.6% of waste disposal, 9.9% of gas use, and 9.3% of electricity use, but only 3% of water use (Kennedy et al. 2015). The higher per capita energy consumption and waste production result from higher per capita consumption levels associated with high income levels. So the good news is that cities produce wealth, and their environmental impacts are lower per unit of GDP than suburban or rural areas. However, the higher wealth they produce results in higher per capita consumption levels and higher per capita environmental impacts. The efficiency of cities is highly variable and depends on many factors including its urban form (population density, per capita building area), average annual temperature, transportation efficiency, and so on. The U.S. cities New York and Los Angeles are less efficient than most megacities. Still, citizens of Manhattan have the lowest ecological footprint in the United States (Brand 2009).

Well-planned cities have lower per capita ecological footprints than suburban and rural areas. For example, Germany, France, and most other European countries have lower average ecological footprints than the United States (Figure 3.6) primarily because they have higher population densities (Rosenthal 2009). We also saw in Chapter 3 that urbanization results in a decline in fertility and population growth rates (see the chart at www.bit.ly/1aKaptz) (Casterline 2010). Thus, urbanization can slow growth of the overall environmental ImPACT by decreasing population growth and the per capita EF. Together these reinforcing effects can greatly slow growth of the environmental impact of societies over time. And since more than half of the world's population now lives in cities, it makes sense to focus on making cities more sustainable.

Some cities in North America have been at the forefront of planning for sustainability. One of the greenest cities in the world is Vancouver, British Columbia. Most of Vancouver's residents live downtown in high-rises and compact communities (Steffen 2006). Planners designed the city for pedestrians and bicycles, and many residents have given up their cars. By avoiding urban sprawl, Vancouver has become one of the world's most livable cities, with low levels of traffic congestion and air pollution. San Jose also outperforms most cities in many metrics due to smart planning and development (Bettencourt and West 2010).

Planners have developed many living models in the last 100 years to improve on our current unsustainable models for urban and suburban living. **New Urbanism** tries to design cities to be more pedestrian-friendly and to foster community by incorporating shared central spaces like parks.* Sustainable cities can produce food through **urban gardening** techniques (Section 13.3.3) or by relying on farms just outside the city limits that deliver food to consumers at farmers' markets. **Ecovillages** take a step farther by considering not only human needs but also ecosystem needs. Ecovillages are designed to have minimal environmental impact. They are sustainable, intentional communities that are self-reliant, meaning that they contain all basic services (shops, schools, clinics, churches, etc.). Ecovillages aim to increase biodiversity by incorporating organic gardening

* See Congress for the New Urbanism: http://www.cnu.org/.

and green roofs into their design (Bates 2006). The Global Ecovillage Network* includes close to 20,000 villages worldwide.

A successful example of an ecovillage in the developing world is Gaviotas in the llanos (grasslands) of Colombia (Weisman 1999). Its founder, Paolo Lugari, had the foresight in 1971 to use a team of scientists and engineers to tackle the problems of sustainable living. This team came up with many novel solutions, including a special water pump that could extract groundwater from greater depths and with less effort than with traditional pumps. They connected the pump to a seesaw to put the energy of children's play to good use. They also developed solar water heaters, the sale of which became a major source of income. Finally, the planting of 1.5 million trees returned part of the llanos back to its preexisting state of tropical jungle by trapping the moisture in a microclimate. The villagers of Gaviotas tap the trees and sell the resin. The genius of the residents of Gaviotas enabled them to succeed in a harsh climate in a country that has had many social problems.

For those wishing to live sustainably, ecovillages seem to offer an effective solution. However, many attempts to start ecovillages fail because aspiring ecovillagers have a vision of what their life and ecovillage should be like, but don't realize that others who they attempt to recruit to their ecovillages have their own visions. Successful design and development of an ecovillage is a community project that requires compromise. To be sustainable (successful), the community must design, develop, and maintain its ecovillage holistically, paying attention to all three pillars of sustainability: environment, economy, and society. This is also true for growing cities. If done properly, urbanization can increase social, environmental, and economic capital, making cities more sustainable than rural settings.

5.4.3 The Commons Approach to Managing Shared Resources

Sustainable development requires effective management of resources, especially **commons**, which are shared natural resources such as bodies of water, fisheries, forests, and the atmosphere. For most of history, people and communities shared resources. For example, townspeople were free to chop down trees in a nearby forest, which was a commons. Because wood is a renewable resource, and there weren't many people compared to the number of trees, the harvest rate was always less than the rate of regrowth, and no rules were necessary to keep harvesting sustainable.

Once the population became large enough that its wood demand could exceed a sustainable harvesting rate, a community was faced with a decision on how to manage the shared resource. One issue plaguing the commons approach is the **free rider problem**, because freeloaders can remove resources from the commons without maintaining or replacing those resources. When businesses act as free riders they externalize their costs, meaning they profit from use of the resource while someone else bears the costs of maintaining the resource.

Until two centuries ago, the community management approach was to share the resource, make everyone responsible for maintaining the resource, and rely on informal sanctions to keep community members from taking more than their fair share. More recently, Garret Hardin's **Tragedy of the Commons**, which maintained that sharing of the commons inevitably leads to their degradation, was used to argue that property rights was the only efficient approach to managing the commons (Hardin 1968). Privatization of renewable resources like wood meant that they could no longer be taken from the forest for free; people had to pay for their use. This system is now used almost exclusively in the developed world, so much so that most people think that it is the only approach to managing the commons. In most cases, society now relies either on the "command and control" approach of centralized government, a market system where shared resources are appropriated by corporations and commodified (turned into commodities, or products for sale), or a combination of the two. This often leads to unsustainable resource use. For example, in 1988 a survey of African countries found that governments leased land to private loggers for fees that ranged from 1% to 25%

* www.gen.ecovillage.org.

of the cost of reforestation, amounting to a substantial erosion of the public commons (Goodland and Daly 1996).

Recently, social theoreticians, like Elinor Ostrom, have shown that the commons approach to managing shared resources can work effectively (Ostrom 1990). In some communities, people work together to manage common resources, such as meadows or lakes, because they have a shared sense of purpose and look out for one another. This leads to the development of social control mechanisms that inhibit the exploitation of the commons for individual gain. A community member who acts only in his own self-interests is likely to be the focus of intense social pressures that eventually will cause him to defer to the common good. Few can bear being ostracized by their community.

Ostrom showed through historical studies and case studies of modern communities that social contracts provide a very effective approach to managing the commons (Ostrom 1990). For example, farmers in a Swiss village have shared a meadow for centuries without problems of overgrazing, all because of a rule established in 1517 that no farmer could have more cows than they could care for over the winter. Hardin later wrote that he should have titled his essay "The Tragedy of the Unmanaged Commons" (Hardin 1998).

Today's socioeconomic paradigm is privatization and the free market. Part of the reason these ideas rose to prominence was the development of increasingly large and centralized sociopolitical and socioeconomic units. Gradually personal relationships became de-emphasized. People no longer knew everyone in their units, and with the loss of sense of community, they lost the motivation to act for the common good, and started to act only for their own good. This led to the development of the Tragedy of the Unmanaged Commons, where people often acted solely in their own interests without concern for the interests of their neighbors or of future generations. Commonly shared resources were appropriated and exploited by individuals for their own benefit, even when their actions damaged the well-being of their neighbors. The social contract gradually disintegrated. In this atmosphere, politicians argued that they could manage the commons for the common good only if the government privatized or regulated it. When government regulation became demonized in the United States and United Kingdom in the 1980s, it seemed that privatization was the only option for managing the commons. However, too often privatization led to the growth of individual wealth rather than the commonwealth. **Market failure**, which is the inefficient allocation of goods and services, often caused selling of the commons at prices well below its true value, as shown previously when African governments provided leases to timber forests at costs far below their value (Goodland and Daly 1996).

One of the root causes of our society's culture wars is the artificial dichotomy between market and state. Conservatives don't trust government and therefore favor deregulation and the operation of a free market. They generally accept government regulation only in cases of market failure. In contrast, liberals are suspicious of corporations and therefore favor government management. This dichotomy is based on the false assumption that there are only two types of goods—private and public. As shown in Table 5.1, goods actually fall into four categories, and the boundaries between them are gradational. Goods are classified using two characteristics. "Subtractability" means when you use a good, less remains for others (economists refer to these as "rivalrous" goods). "Excludability" means that you can exclude others from using a good unless they pay for it.

A free market is an effective mechanism for efficiently supplying and managing pure private goods because they are subtractable and excludable. Government regulation is usually necessary to protect public goods because they are not subtractable or excludable, meaning there is no way to make money from them. However, other types of resource management are required to manage toll (club) goods and **common-pool resources** (CPRs). Many environmental resources are CPRs, and so we focus on the management of CPRs.

Since CPRs and public goods are shared resources, they are both part of the commons. The defining feature of commons is therefore low excludability. An example from the United States is public roads, which are a public good. Public roads run between private properties, but anyone can use them, whether for driving a car, riding a bicycle, or walking. Occasionally wealthy enclaves

TABLE 5.1
Four Types of Goods

		Subtractability	
		Low (Nonrivalrous)	**High (Rivalrous)**
Excludability	Low	**Public goods**: peace and security of a community, national defense, knowledge, fire protection, public roads, weather forecasts, the atmosphere, etc.	**Common pool resources (CPRs)**: groundwater basins, lakes, irrigation systems, fisheries, forests, etc.
	High	**Toll or club goods**: theaters, private clubs, daycare centers	**Private goods**: food, clothing, automobiles, etc.

Source: Ostrom, E. 2010. "Beyond Markets and States: Polycentric Governance of Complex Economic Systems." *American Economic Review* 100 (June): 1–33.

aiming to decrease traffic or increase their privacy attempt to reduce or eliminate public access to roads in their neighborhoods. This is becoming increasingly common due to increasing traffic in neighborhoods caused by navigation apps such as Google Maps that route drivers through side streets to avoid traffic congestion. However, these neighborhoods do not own the roads that pass through them, and furthermore the public they attempt to exclude paid taxes for construction of those roads. As a result, these attempts to privatize public goods are usually thwarted in the courts.

The property of low excludability makes both CPRs and public goods susceptible to the free rider problem. However, effective management of CPRs is even more important than for public goods because their high subtractability can lead to resource depletion. Let's look at water as an example of a CPR. When water is abundant, it has low subtractability: Your use of water from a large lake does not affect the availability of water for others. Large water bodies also have low excludability and are therefore public goods, meaning everyone who needs water can obtain it. However, when the lake is small and shrinking and the number of users is large it becomes a CPR with high subtractability, and some type of management scheme is needed to ensure that everyone uses the lake water sustainably.

The commons provide many of our critical ecosystem services, especially provisioning services. Shared water resources and other CPRs are always vulnerable to the Tragedy of the Commons. Those who favor water privatization view water as a private good, while opponents of privatization view water as a public good. However, because water is a fugitive (moveable) resource, it is harder to privatize than land. The pressure to privatize water usually increases when water becomes scarce and therefore valuable. However, as demonstrated in case studies provided by Ostrom (1990) and elsewhere (Trawick 2001) for CPRs that are renewable and scarce and for situations in which users can harm one another but not harm nonusers (those outside the community), users can self-organize and benefit by treating the CPR as a managed or internally regulated commons. Self-organized, self-governed CPR institutions created to manage water resources include **acequias** in Spain and Spanish colonies (e.g., southwestern United States) and alpine irrigation in the arid mountains of Peru (Trawick 2001).

To manage resources successfully, it is critically important to choose the appropriate spatial scale of governance. Shared resources that cover small areas have small resource stocks and therefore tend to be subtractable and can be effectively managed by communities as CPRs. Larger resources such as inland seas or large forests have large resource stocks and lower subtractability, and so can be managed by government as public goods.

The greater the spatial scale of a shared resource, the larger the scale of government that is required to manage it effectively. Essentially the resource has to fit within the governed space. Resources that are contained within one state can be more effectively managed by the state than the federal government. However, intergovernmental treaties are required when shared resources such as rivers, lakes, and the atmosphere overlap or intersect with multiple administrative boundaries. The greater the number of affected administrative units, the more difficult it is to negotiate treaties and enforce regulations. Shared resources are most efficiently and effectively managed by one governing body that encompasses the resources. For this reason, the U.S. federal government, specifically the EPA, must protect resources that cross state borders. For international resources, there is no global government. The United Nations serves as a necessary forum for nations to negotiate treaties to protect global commons such as the oceans and atmosphere.

As an example of the relationship between the spatial scale of a problem and the required scale of governance, Figure 5.2 indicates the scale at which a specific atmospheric pollutant should be managed. Note that, when considering pollutants, it is not the entire atmosphere we are concerned with, but that portion of the atmosphere that is affected by the release of a pollutant at a specific location. Pollutants with long atmospheric residence times such as the greenhouse gases CO_2 and CH_4 affect larger areas. Once they are added to the atmosphere by a country, they will stay there for centuries and therefore be spread across the globe, and so must be managed at the global scale. In contrast to global scale pollutants, some components of acid rain such as SO_2 can be effectively managed at the local or regional scale. Note that the pollutant absorption capacity of the atmosphere is renewable, so each country must set a limit to the rate at which they add the pollutant to the atmosphere, so that the sum over all countries does not exceed the removal rate of the pollutant from the atmosphere.

Rule enforcement is most effective on the local scale when affected parties are involved in designing the rules and monitoring compliance. Compliance is more likely to remain high if sanctions are modest for first offenses and then increase in severity (Ostrom 1990). At larger spatial scales, regulators often call for the state "command and control" approach of government regulation or for market-based mechanisms. Government regulations are only effective if resources for monitoring and enforcement are sufficient and if fines are high enough to discourage rule-breaking. The market-based approach sets caps on withdrawals or emissions and issues Tradeable Environmental Allowances, which are harvesting or emissions permits that can be bought or sold to stay under the cap (Section 8.3.2).

It is especially difficult to manage resources that lie outside any national jurisdiction, such as marine fisheries and seafloor mineral ores (de Vries 2013). Both are subtractable and nonexcludable, meaning they are CPRs. For each of these resources there are limits to the amounts we can use sustainably. For nonrenewable resources, such as mineral ores on the seafloor, the global consumption

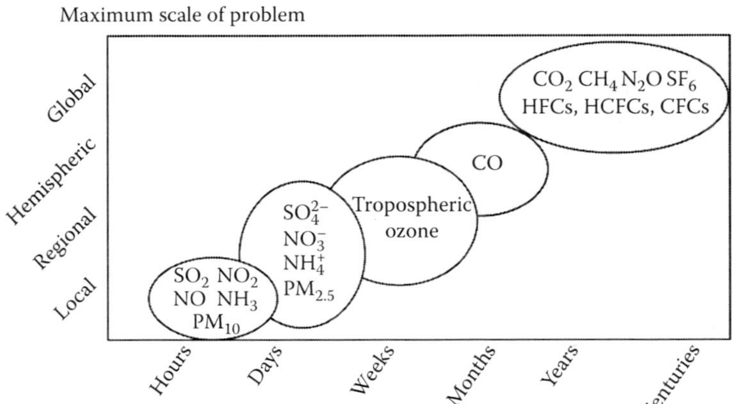

FIGURE 5.2 Selected pollutants, their average residence times in the atmosphere, and maximum extent of their impact. (After UNEP. 2007. Global Environment Outlook: Environment for Development (GEO-4). United Nations Environment Programme. http://www.unep.org/geo/geo4/report/GEO-4_Report_Full_en.pdf.)

rate cannot exceed the global rate at which substitutes are produced. For renewable resources such as marine fish, the global consumption rate cannot exceed the renewal rate.

Not only the spatial scale but also the rate of change must be considered when managing the impacts of environmental change on natural resource availability and provision of ecosystem services (Figure 5.3). For changes to the commons such as fishery depletion and sometimes deforestation, the rate of change is low enough that adaptive management may be used while the system is changing to reduce the magnitude and impacts of the change. When change is rapid, such as during flood events, most actions to reduce vulnerability must be taken before the event begins. Effective management of the commons can minimize the impacts of environmental changes, be they anthropogenic or natural.

New legal tools have been developed to protect the commons and preserve them for future generations (Walljasper 2010). **Trusts** can be used to hold and manage property for beneficiaries; trustees are legally obligated to protect the interest of beneficiaries (Barnes 2006). A model example is the National Trust, a charity in Britain with over 3 million members that owns over 600,000 acres of land and 200 historic buildings and gardens. In the United States there are more than 1,500 trusts that protect over 9 million acres. People can become trust members by donating money, which gives them access to the trust assets and allows them to vote for trustees. Another option is a **community land trust**, which can acquire land by gift or purchase and then lease it for purposes consistent with the trust's objectives, such as providing affordable housing. Leaseholders can own and sell buildings, but not the land, and capital improvements are limited to replacement costs. Over 200 community land trusts now exist in the United States. A **conservation easement** is a voluntary agreement between a landowner and a trust that protects the interests of the landowner in perpetuity. A trust acquires the easement and enforces its restrictions on current and future landowners. This legal arrangement seems the most appealing because it allows landowners to continue owning and using the land and to pass the land on to heirs, but it places permanent restrictions on land use. The owner and future owners give up certain rights, such as the right to clear-cut a forest or build additional buildings. However, the trust may compensate the current owner by purchasing the conservation easement. The Pacific Forest Trust protects ~50,000 acres of forest in the Pacific Northwest. Other types of trusts include watershed trusts, groundwater trusts, and airwaves trusts (Barnes 2006).

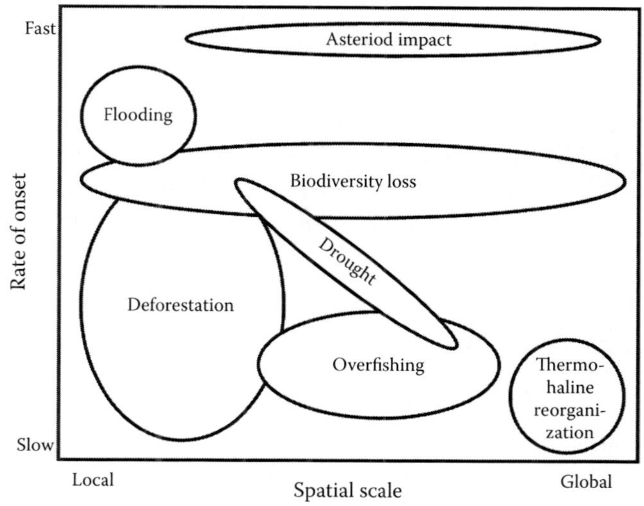

FIGURE 5.3 Environmental changes as a function of spatial scale and rate of onset. Thermohaline reorganization involves changes in oceanic circulation patterns (Chapter 7). (After National Academy of Sciences. 2012. *Ecosystem Services: Charting a Path to Sustainability.* The National Academies Press. http://www.nap.edu/openbook.php?record_id=13331.)

A shared resource that is becoming an increasingly important component of our lives and the economy is the Internet. Originally the Internet was a toll good because its subtractability was low and excludability was high, that is, you had to pay an Internet service provider for access. As its reach expanded and its availability became essential as a knowledge and information tool for all citizens, the government and some private corporations such as Google began to provide free Internet access to low-income homeowners and in public places such as community libraries.* In the ideal situation, this makes the Internet a public good, as its excludability changes from high to low, and its subtractability remains low.

E-commerce depends on the Internet to be accessible at all places at all times, which means that the network has to be extremely resilient and robust. Recent **cyber-attacks** by hackers have at times made the Internet a CPR by using **denial of service attacks** to increase network traffic to levels that overwhelmed network servers, effectively changing the subtractability of the Internet from low to high. Rapid rollout of **Internet Of Things** devices with inadequate security has allowed hackers to use these devices as a robot network or **Botnet** to launch coordinated attacks against network servers, which has compromised the resilience of the Internet.† In October 2016, a **denial of service attack** used Internet Of Things devices to shut down e-commerce at several large U.S. online retailers and banks using, causing large economic losses.‡

Like other public goods, the Internet must be protected by government regulation. The Internet Of Things has developed so rapidly that government regulation could not keep pace, so that current regulations are inadequate for ensuring Internet security. In 2016, the vulnerability of the Internet in the United States was exposed by Russian hackers who attempted to influence the outcome of the Presidential election by leaking information stolen from the Democratic Party and providing it to the website WikiLeaks.§

In summary, the commons approach is a "third way" to manage shared resources. It favors social cooperation over free market competition or centralized government regulation, and requires a shift in thinking "from the prevailing YO-YO ethic (You're on your own) to WITT (We're in this together)." The commons approach relies on "a shared understanding that some things belong to all of us and must be used in a sustainable and equitable way (Walljasper 2010). It is effective for managing CPRs, but government regulation is often necessary to protect public goods such as the atmosphere and the Internet.

5.4.4 Comparison of the Current Sustainability Status of Countries

Comparing the sustainability performance of countries helps us to identify areas where each country needs to improve, but also highlights which countries are performing well and can serve as models for other countries to emulate. Unfortunately, there is no single metric that summarizes sustainability performance, since sustainability has three components (economic, social, and environmental), each with a large number of subcomponents. Furthermore, many subcomponents have not been measured to date.

We previously used a global map to compare countries' environmental impacts, expressed as average per capita ecological footprint normalized to global per capita biocapacity (Figure 3.6). It is also instructive to compare average human well-being in countries using the HDI (Figure 5.4). We

* Google Fiber Is Coming To Public Housing For Free, http://www.informationweek.com/government/leadership/google-fiber-is-coming-to-public-housing-for-free/d/d-id/1321336, retrieved October 24, 2016.

† "Smart" or "connected" devices usually contain sensors that use wireless connections to upload data that can be used for remote operation or data collection. The proliferation of these devices is rapidly increasing the number of devices connected to the Internet and the vulnerability of that network.

‡ Hacked Home Devices Caused Massive Internet Outage, http://www.usatoday.com/story/tech/2016/10/21/cyber-attack-takes-down-east-coast-netflix-spotify-twitter/92507806/, retrieved October 24, 2016.

§ U.S. Says Russia Directed Hacks to Influence Elections, http://www.nytimes.com/2016/10/08/us/politics/us-formally-accuses-russia-of-stealing-dnc-emails.html, retrieved October 24, 2016.

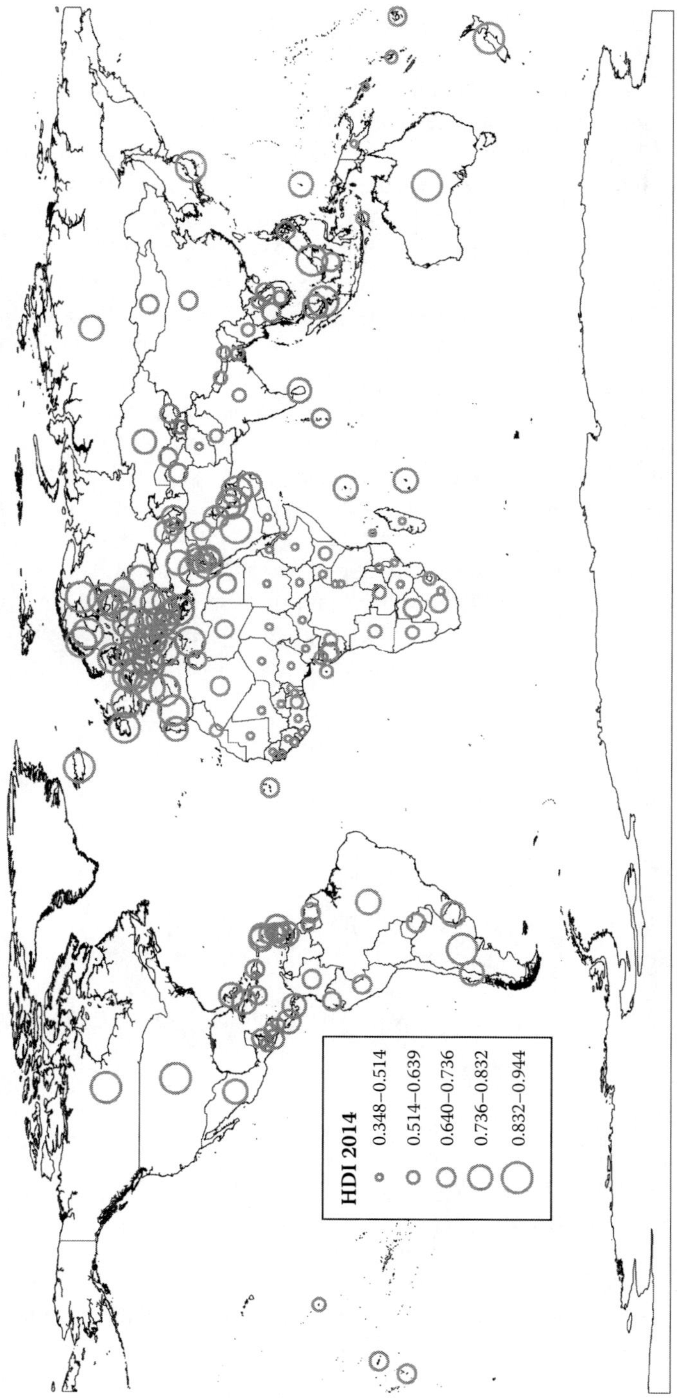

FIGURE 5.4 World map of Human Development Index (HDI) by country in 2014. (Data from http://hdr.undp.org/en/data, retrieved April 12, 2016.)

find that developed countries generally have high HDI and developing countries low HDI. This is because developing countries often have high birth rates, high poverty rates, low education levels, and low human longevity. Citizens of these countries must often take desperate measures to feed themselves, and in the process they often undercut the ability of their environment to meet their needs. Countries with low HDI scores are at risk of becoming or have already become failed states. Failed states remain so for decades or indefinitely. Developed countries like the United States also have problems. They are often straining their environment, as reflected by high EF/B (Figure 3.6). These countries have high consumption levels, and often demand more resources than their environments can provide. They compensate for this by using their wealth to purchase resources from less wealthy countries. Furthermore, high consumption levels in developed countries have many negative impacts, including global climate change resulting from greenhouse gas emissions.

The United States lags in some sustainability indicators (e.g., high EF/B) because it has already developed unsustainably. It's easier to start from scratch and develop sustainably than to fix all of the mistakes we've made. It will be very expensive to rebuild our infrastructure to be more sustainable. **Leapfrogging** allows developing countries to skip the implementation of old, inefficient technologies used in developed countries, thereby avoiding the environmentally harmful stages of economic development (e.g., the rising limb of the environmental Kuznets curve). For example, developing countries can build mobile phone networks that require much less raw materials and construction than land lines. They can deploy decentralized renewable energy systems such as PV panels, while in the United States we will be stuck for some time with old coal-fired power plants and an aging, inefficient electrical grid (Caldecott 2015). U.S. cities and suburbs were built when energy was cheap and there were no concerns about global warming, so we are forced to drive great distances to accomplish everyday tasks. Developing countries can concentrate development in cities where people can live more sustainably, and preserve green space around the cities. We can only hope that planners in developing countries learn from the mistakes of the developed countries.

WEB RESOURCES

- Consilience—Journal of Sustainable Development: http://www.consiliencejournal.org/blog/
- Foundation for Sustainable Development: http://www.fsdinternational.org/
- Sustainable Development—Earth Charter Initiative: http://earthcharter.org/
- Sustainable Development 101: http://youtu.be/Oa5dPsjrkik
- Sustainable development goals adopted by the UN in 2015: https://sustainabledevelopment.un.org/sdgs
- The Lazy Person's Guide to Saving the World: http://www.un.org/sustainabledevelopment/takeaction/

HOMEWORK PROBLEMS

1. Think of or search for one example of unsustainability in a developing country of your choice. What innovations are being/have been put into place to remedy this situation?
2. Update Figure 5.1. Does the update change any conclusions?
3. Choose a city that you think exemplifies unsustainability, one that is likely to become a ghost town or at least be economically depressed in the future and with a falling population. Give statistics to support your choice.
4. Group project: Devise a plan to build a sustainable city from scratch. Describe the guidelines you would use for choosing a suitable site, and the designs for your sustainable food, water, energy, and waste recovery/disposal systems. What criteria or metrics would you use for determining whether each system and the city as a whole are sustainable?

5. Many social experiments are being conducted in communities that aspire to be sustainable. Choose one of the following topics to research and give a presentation on (give references):
   a. Ecovillages
   b. Transition towns
   c. New Urbanism
6. You work for the Gates Foundation, and your budget for the year is $1 billion. Choose a country in the developing world to spend that $1 billion to promote sustainability, that is, improve the long-term well-being of its citizens. You should choose a country where that money will have the most "bang for the buck." Specify how the money would be spent: What would the objectives be and how would spending the money accomplish them? Defend your choice of country and spending decisions.

6 Nonrenewable Resources: Oil and Minerals

The intertwined problems of population growth, water shortages, and food shortages have been recognized since Malthus published his "Essay on the Principle of Population" in 1798. However, until very recently, much of the public was unaware of the related problems of **peak oil** and **global climate change**. In this chapter, we will examine the peak oil problem in detail, as it makes for a good case study of resource depletion, including uncertainties associated with the timing of peak resource production. This will also serve as an entry point for a discussion of global climate change, for which burning of fossil fuels like oil is the primary cause.

We will also discuss nonrenewable mineral resources and the problems posed by their finite supplies. There is concern that many nonrenewable resources may soon become scarce. First, demand is rising faster than ever. Second, prices are rising: Between 2002 and 2010, the average prices for 86 metals rose 6.4% annually, indicating that supply has not been able to keep pace with demand (Gardner 2013). Second, metal ore grades have been declining for over a century; this means higher environmental impacts and more energy required to mine and process the ore. Rising energy prices compound the cost problem. Maintaining adequate supplies of these resources is critically important to the economy. For example, 95% of the material flowing through the U.S. economy is nonrenewable (Gardner 2013). So it is crucial to find substitutes or use technologies, including recycling, to extend the lifetimes of valuable nonrenewable resources.

6.1 OIL

Oil has been the world's most important source of energy since the mid-1950s.* In the United States in 2014 it supplied almost 50% of the energy consumed, and 97% of the energy used in transportation, mostly in the form of gasoline produced by distilling oil (EIA 2016a). Oil is preferred for transportation because it is easily transported as a liquid and because it has a high energy density, meaning that a small volume or weight carries a large amount of energy. Vehicle owners prefer fuels with high energy density so that they can travel farther before they must refuel; electric batteries currently have lower energy densities than oil-derived gasoline, which results in smaller vehicle ranges for electric cars and lower consumer adoption rates. If oil becomes scarce, will we have an acceptable substitute for transportation?

To create sustainable resource use policies, we need to estimate how fast we are depleting resources like oil. We define a **resource** as a substance that can potentially be mined or harvested at a profit; the term refers to the total amount of the material in the Earth's crust, which for nonrenewable resources can only decrease (Meadows et al. 2004). In contrast, a **reserve** is an identified resource that can currently be legally extracted at a profit (Keller 2011). Reserves can increase with discoveries, price increases, and technology improvements.

Oil, and the gasoline it is used to produce, is effectively a nonrenewable resource because it forms much more slowly than we consume it. The consensus is that oil is the most limited of the important fossil fuels (Meadows et al. 2004). Because oil is a nonrenewable resource, global oil production will eventually reach a maximum, an event called **peak oil**, and then begin to decline. However, the demand for oil will continue to rise, and the growing gap between declining supply and growing demand could result in oil shortages and price increases (Figure 6.1). Thus, when considering future

* Note that we use the term "oil" synonymously with petroleum.

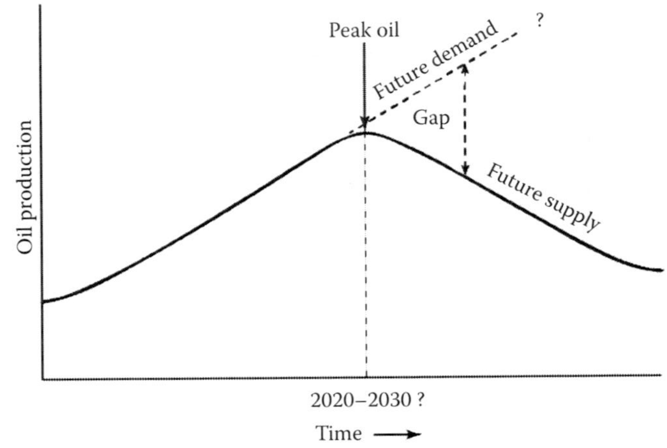

FIGURE 6.1 Peak oil and the supply–demand gap.

availability of nonrenewable resources such as oil, the timing of peak production is more important than the time of ultimate reserve depletion. In the long term, energy shortages caused by peak oil will only be avoided if, according to the second law of sustainability (Section 2.2), we transition to a renewable substitute faster than oil is depleted. Otherwise, we can expect transportation costs to rise substantially, which will force those who cannot afford the price increases to change their lifestyles. Finally, since we will have to leave much fossil fuel in the ground to avoid catastrophic climate change (Chapters 7 and 8), the amount that's left may not be the limiting factor on oil consumption; climate change mitigation regulations, not peak oil, will likely limit oil production in the future.

Peak oil is exactly the type of resource shortage predicted by the original Limits to Growth study (Meadows et al. 1972). That study enumerated physical limits in resource sources and sinks, corresponding to the beginning and ending of a linear material cycle. Peak oil is a source problem. Burning of oil and other fossil fuels has also caused a sink problem, namely the accumulation of the greenhouse gas carbon dioxide CO_2 in the atmosphere and resulting global climate change, a problem we will turn to in the next chapter.

6.1.1 Oil Supply and Resource Depletion Estimates

To understand why the amount of oil stored in the ground is finite, and the amount that we can retrieve is even smaller, we need to review how oil forms and how we recover it from the ground. Oil forms when plants use photosynthesis to store the sun's energy, die, are rapidly buried, and then transformed under heat and pressure into oil. This process can take millions of years, so the energy stored in oil molecules is ancient trapped sunlight. Oil can form from buried plants only under special conditions in the *oil window* at approximately 3 to 6 kilometers (2 to 4 miles) depth, and only when oxygen is not present to react with the carbon to form carbon dioxide (respiration). Oil is usually found only in **sedimentary rocks** that are less than 500 million years old because land plants did not exist before that time.

Because oil is "liquid gold," oil companies have used billions of dollars from oil profits to perfect oil exploration and recovery techniques. Over time, exploration shifted from the surface to the subsurface. Each drilled well provided information about the subsurface that could be used to locate more oil. To improve their oil-finding capabilities, oil companies developed new technologies that greatly increased the success rate of expensive drilling and allowed exploration geologists to find small patches of oil at great depth. These techniques greatly lowered the costs of exploration; they also greatly increased the amount of oil delivered to the market. Both factors helped to keep the price of oil low. These techniques were so effective that oil discoveries skyrocketed until 1965 but have fallen ever since, suggesting that most abundant oil supplies have already been found (Hall and Day 2009).

To understand better why we can expect to have future shortages of nonrenewable resources such as oil, we refer to Figure 6.2, which plots hypothetical production rates of renewable and nonrenewable resources as a function of time. For nonrenewable resources with open material cycles, that is, ones that are not or cannot be recycled, resource production curves are bell-shaped. These **Hubbert curves** are named after geophysicist M. King Hubbert, who developed the **Hubbert peak theory** in the 1970s.

Bell-shaped oil production curves have been observed for single wells, for oil fields, and for regions, so it seems likely that global oil production will show a similar behavior. To calculate a Hubbert peak curve, one only needs to know the ultimately recoverable resource amount (Cavallo 2004). When peak production is reached, exactly one-half of the ultimately recoverable resource has been used because the curve is symmetric about the point of maximum annual production. The total amount of a resource that is available is equal to the area under the curve, which can only be estimated. The timing of the peak depends primarily on the size of the resource and the production rate, which in turn depends on economic factors that are affected by politics and natural disasters. Uncertainties associated with these factors make it impossible to accurately predict exactly when oil production will peak (Gleick and Palaniappan 2010). It is worth noting that system dynamics models of nonrenewable resource consumption yield resource production curves similar to a Hubbert curve and to what has been observed repeatedly when natural resources are used unsustainably (Meadows 2008).

For every nonrenewable resource such as oil, production and consumption inevitably lead to reserve depletion. Nonrenewable resources that are not abundant and that we use rapidly run out quickly so that their resource production curves are very narrow. Those that we use slowly or that

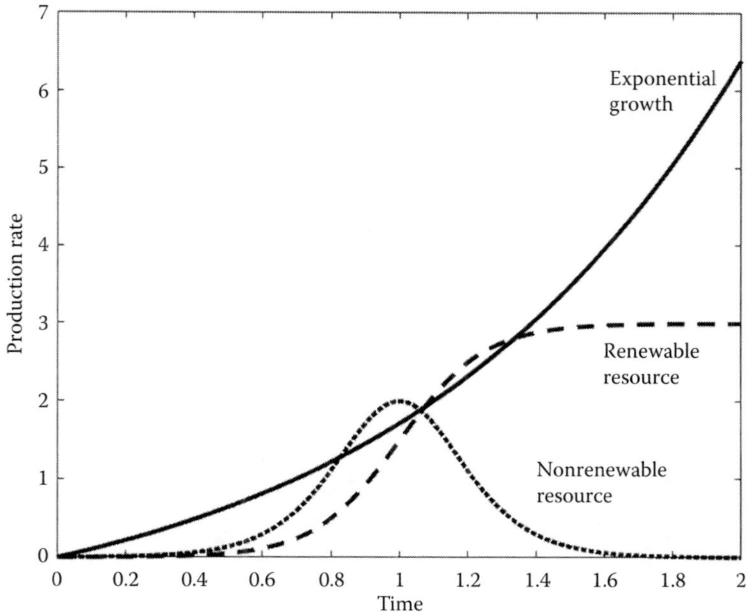

FIGURE 6.2 Hypothetical production rates as a function of time. Curves calculated in MATLAB 2015. (After Hubbert, M. King. 1974. "Statements on the Relations between Industrial Growth, the Monetary Interest Rate, and Price Inflation." *Hearing before the Sub-Committee on the Environment, Committee on Interior and Insular Affairs, House of Representatives, Serial*, no. 93–55.)

are abundant last much longer, so their curves are wide and do not peak until well into the future. Note that on the rising limb of the Hubbert curve demand drives supply: oil companies could simply increase production to meet rising demand. However, after passing the peak, supply cannot keep pace with demand (Figure 6.1), the resource becomes scarce, and prices thus rise steeply (Hopkins 2008). This rapid rise in cost acts as a negative feedback by reducing consumption (the poor are priced out).

In Figure 6.2 the "exponential growth" curve can represent human population and therefore resource demand. Initially the curves representing demand and supplies of both renewable and nonrenewable resources are close together, so supply can meet demand. But when human population approaches Earth's carrying capacity and nonrenewable resources reach peak production, the curves begin to diverge so that supply can no longer meet demand (Hubbert 1987).

The Hubbert peak approach is useful for estimating the time at which a nonrenewable resource will become scarce and therefore expensive due to physical depletion/scarcity. In developed countries with free markets, transparency, and lack of corruption, the physical amount available determines resource scarcity. However, in developing countries resource scarcity more often results from ineffective management and governance (Cordell and White 2014). We will discuss this issue as it pertains to food in Chapter 13.

Hubbert was mostly ignored when he predicted in 1956 that oil production in the United States would peak in the early 1970s. However, when production peaked in 1970 as he predicted (Figure 6.3), many scientists accepted his approach as legitimate. Most people forget that until the early 1970s the United States was, like Saudi Arabia of the 1980s and 1990s, the largest oil producer in the world. However, between 1972 and 2008 U.S. oil production decreased, and the United States became increasingly dependent on foreign countries like Saudi Arabia for oil. The Hubbert peak theory approach for predicting future oil production rates seemed reliable, as U.S. oil production seemed to follow a bell-shaped curve until 2008.

However, beginning in 2008 the United States reversed a long-term trend of declining oil production by developing new methods of **horizontal drilling**. These were combined with established technologies of **hydraulic fracturing** to make previously uneconomic shale oil deposits economically viable, that is, these methods of enhanced oil recovery converted oil resources into reserves. These techniques of **enhanced oil recovery** caused the reversal in U.S. oil production starting around 2008.

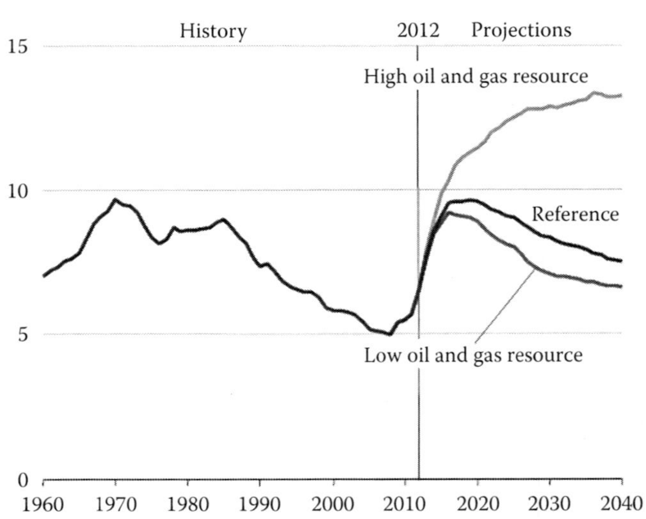

FIGURE 6.3 Historical and projected U.S. oil production in million barrels per day. (From U.S. EIA Annual Energy Outlook 2014. http://www.eia.gov/outlooks/aeo/0383(2014).pdf.)

Production of a natural resource will follow a Hubbert curve as long as the recovery technology and its efficiency remain unchanged. If resource recovery efficiency is increased, then the total recoverable amount of the resource represented by the area under the Hubbert curve increases. Depending on the magnitude of the efficiency increase and the implementation timing of the new technology, the resource recovery curve may display an additional production peak, or the single-peaked curve may simply extend farther into the future. The adoption of new oil and gas recovery technologies in the United States resulted in a second peak in their production curves (Figure 6.3). These new technologies have made it more difficult to predict future oil production rates.

Another approach to evaluating the scarcity of a nonrenewable resource is to calculate the reserve depletion time. In this approach, the size of a reserve ***R*** is used with the production rate ***P*** to estimate the amount of time before a reserve is depleted, assuming the production rate remains constant. For example, BP estimated that in 2014 proven world oil reserves totaled 1700 billion barrels. The production rate in 2014 was 32 billion barrels per year. If oil production remained constant at 30.5 billion barrels per year, the oil reserves would be depleted in ***R/P*** = 1700 bbl/32 bbl per year = 53 years (BP 2015).

In general, the *R/P* approach yields estimates of depletion dates that are later then Hubbert peak estimates because they assume that all of a reserve can and will be used, while the Hubbert peak estimates when half of the resource will be used. Both estimates are likely to prove inaccurate because they are static estimates that do not take into account changes in dynamic factors such as technological advances, substitution, and changes in cost and demand. However, they are still valuable as early warning indicators of physical resource depletion (Cordell and White 2014).

Reserve depletion times calculated using *R/P* can be inaccurate because both *R* and *P* can change over time. The size of a reserve *R* changes with technology and economics. New technologies may be developed that increase the reserves by allowing more oil to be extracted economically. Also, as the price of oil increases, more oil becomes economically mineable. The production rate *P* also changes over time, increasing until peak production is reached and then steadily declining. Thus, the *R/P* ratio often does not provide an accurate estimate of how long the supply of a nonrenewable resource will last (Bardi 2009). If production rate continues to increase at the same rate of 1.1% that it increased between 2000 and 2011, oil depletion would occur in 42 years.* The peak in oil production would occur much sooner than that.

Reserve estimates have been increasing steadily for decades due primarily to rising oil prices and development of new extraction technologies. Between 1980 and 2011, *R* increased 142%, while *P* increased only 33%. Because *R* has been rising faster than *P*, *R/P* has been steadily increasing (Figure 6.4), which seems inconsistent with the claim that peak oil is near at hand. Peak oil advocates argue that Middle Eastern countries inflate oil reserve estimates because their permitted production rates are tied to their reserve estimates. The higher Middle Eastern countries set their reserve estimates, the more they are allowed to produce, and the faster they can make money. For example, documents released by WikiLeaks in 2011 purportedly showed that Saudi Arabian reserves were overestimated by 40%.†

---

* A better estimate of the reserve depletion time is given by the following formula:

$$\text{time remaining } (y) = \frac{\log\left(1 - (1-g) \times \dfrac{R}{P}\right)}{\log(g)} - 1 \qquad (6.1)$$

where R = reserve, P = initial annual production, and g is the growth rate = $1 + f$, where f is the annual growth rate expressed as a fraction. For example, from 2000 to 2011, oil production increased an average of 1.1% annually. Assuming the consumption rate is equal to the production rate, $f = 0.011$, and $g = 1.011$. As calculated previously, in 2011 R/P = 54.2 years. Substituting these values into Equation 6.1 we obtain a revised oil depletion time estimate of 41.7 years.

† "WikiLeaks Cables: Saudi Arabia Cannot Pump Enough Oil to Keep a Lid on Prices," John Vidal, *The Guardian*, February 8, 2011, https://www.theguardian.com/business/2011/feb/08/saudi-oil-reserves-overstated-wikileaks, retrieved June 22, 2016.

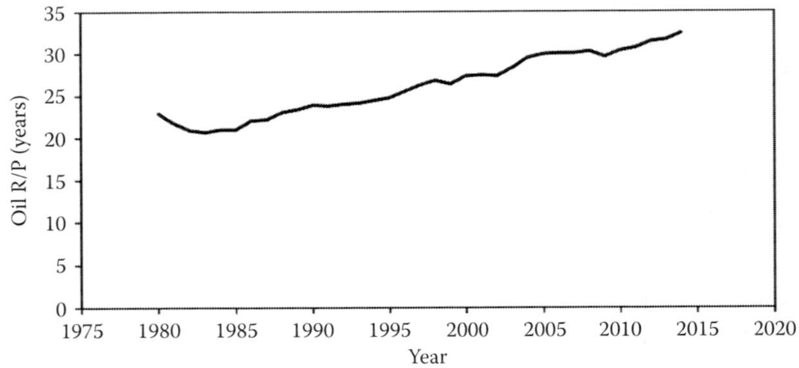

FIGURE 6.4 Global oil reserves *R* divided by world oil production rate *P* from 1980 to 2011. (Data from BP 2015.)

The most common criticism of the peak oil theory is that reserves are so large that we won't soon run out. This argument misses the main point: oil becomes scarce not when we run out, but much sooner when production peaks and then begins to decline. This is why the timing of the production peak is more important than the reserve depletion time *R/P*. However, the *R/P* ratio is useful because it can be easily calculated and used to measure how scarce a nonrenewable resource is and to identify resources that are in short supply and for which we need to find substitutes quickly (Keller 2011).

Others argue that oil production, or at least combined conventional and unconventional oil and gas production, will not rapidly decline but will plateau or slowly decline (Cheney and Hawkes 2007). While production of conventional oil and gas may decline steeply over time, substitution with unconventional oil such as **tar sands** combined with improvements in extraction technologies will slow the rate of production decline for combined conventional and unconventional oil and gas. Even in this best-case scenario where world oil production plateaus rather than peaks, oil prices will still climb considerably because demand will continue to increase exponentially as global population continues to increase exponentially and as the economies of China and India expand at an exponential rate. The United States will be competing with China, India, and every other country in the world for oil, which will drive up oil prices. Thus, while the timing of peak oil is uncertain, there is a consensus that oil and prices will increase in the long term. However, as Hopkins (2008) points out, the exact date of the peak doesn't matter; what matters is that it is near, and we haven't begun to prepare for it. Furthermore, when looked at on a per capita basis, global oil production peaked in 1979 and has remained nearly constant since 1982 (Figure 6.5). Looking at production rates on a per capita basis gives a more accurate basis for evaluating the balance of supply and demand than total resource production rates because the amount of resource available per person is what determines the resource price.

Finally, the most important question about oil is not how much remains in the ground, but how much can we mine and still maintain economic and energy profits, that is, the size of the reserve (Hall and Day 2009). We get an energy profit when we get more energy from oil than the amount we use to produce it. The **energy return on energy investment** (EROEI) is a ratio of the energy produced by extraction to the energy consumed by extraction and transportation. As EROEI decreases, the cost per unit energy increases. The EROEI of U.S. petroleum declined from roughly 100:1 in 1930, to 40:1 in 1970, to about 14:1 in 2000 (Hall and Day 2009). For the tar sands that produce a major amount of oil consumed in the United States, the ratio is much less than 10:1, perhaps even close to 1:1. As EROEI decreases, the amount of environmental damage caused by production of each barrel of oil, including emissions of greenhouse gases, continues to rise. For example, mining of oil sands in Alberta, Canada increased the concentrations of 13 high toxicity elements in the

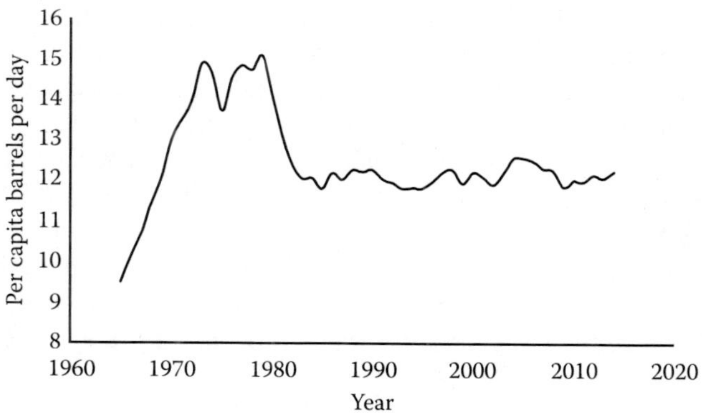

FIGURE 6.5 World oil production per capita. (Population data from UN 2012. "UN World Population Prospects: The 2012 Revision." http://esa.un.org/wpp/Excel-Data/Interpolated.htm, oil production data from BP. 2015. "Statistical Review of World Energy 2015." http://www.bp.com/content/dam/bp/pdf/Energy-economics/statistical-review-2015/bp-statistical-review-of-world-energy-2015-full-report.pdf.)

Athabasca River; for seven of those elements, concentrations exceeded the maximum acceptable level as defined by Canada or the province of Alberta (Kelly et al. 2010).

According to Hirsch et al. (2005), "The world has never faced a problem like this. Without massive mitigation more than a decade before the fact, the problem will be pervasive and will not be temporary. Previous energy transitions (wood to coal and coal to oil) were gradual and evolutionary; oil peaking will be abrupt and revolutionary." Based on the arguments and data presented, we can expect that oil will eventually become scarce, exploration and enhanced recovery are unlikely to relieve that scarcity in the long term, and oil prices will begin a long-term rise as demand increasingly outstrips supply. We will now delve into the consequences of a world that reaches peak oil.

6.1.2 Consequences of Peak Oil

Oil shortages have already had a major impact on society. High gas prices in 2007–2008 led to many public protests and riots worldwide.* Rising gas prices, and the realization that the gigantic cars manufactured by the U.S. auto industry were unsustainable, caused the **automotive industry crisis of 2008–2010**. The American auto manufacturing industry was so unsustainable that doubling the price of gas during the **2003–2008 energy crisis** caused an almost complete collapse of the industry.†

What will the post-peak world be like? It's hard for us to know, but given our dependence on oil, it seems unlikely that the change from cheap to expensive oil won't have big repercussions.

6.1.2.1 Environmental and Social Costs of Oil Use

The environmental consequences of peak oil and the costs of our oil dependence are well illustrated by the **Deepwater Horizon oil spill** in the Gulf of Mexico in 2010, the largest marine oil spill in the history of the petroleum industry. The Deepwater Horizon rig was drilling 41 miles off the

* For a sampling from 2007–2008, see "Transporters, Farmers to Protest Failure to Cut Fuel Prices in India," http://www.thaindian.com/newsportal/business/transporters-farmers-to-protest-failure-to-cut-fuel-prices_100148001.html, "Truckers Protest Fuel Prices in Mexico City," http://www.cnn.com/2009/WORLD/americas/02/24/mexico.protest/index.html, "Scores of Bikers in UK Have Caused Rush-Hour Disruption in a Protest Against Rising Fuel Prices," http://latestnews.virginmedia.com/news/uk/2008/06/05/bikers_stage_fuel_price_protest, "Truckers to Protest Fuel Costs in U.S.," http://www.usatoday.com/money/industries/energy/2008-03-30-truckers_N.htm, "Hundreds Protest Against Steep Fuel Price Rises in Burma," http://www.irrawaddy.org/multimedia.php?art_id=8391.

† "Gas Prices Put Detroit Big Three in Crisis Mode," http://www.nbcnews.com/id/24896359/, retrieved April 27, 2016.

Louisiana coast in water 5000 feet deep when it exploded on April 20, killing 11 platform workers. Before British Petroleum (BP) capped it on July 15, 5 million barrels of crude oil had gushed from the drill hole, causing widespread damage to shorelines and fisheries.* About 2 million barrels are thought to have ended up in the deep ocean, covering ~3200 km^2 and representing between 4 and 31% of all oil sequestered in the deep ocean (Valentine et al. 2014). The federal government closed nearly 36% of federally owned area in the Gulf of Mexico to fishing, costing the fishing industry billions of dollars. The U.S. Travel Industry estimates that the three-year cost to lost tourism could exceed $23 billion. Costs to BP had risen to $3 billion by July 5, 2010. Plant deaths caused by spilled oil caused increased rates of land erosion (Silliman et al. 2012), and corals were damaged over large areas.† On the positive side, the environmental impacts of the spill are expected to be relatively short-lived.‡

A clue to how the spill relates to peak oil is contained in the name: the Deepwater Horizon was in deep water because oil companies had already drilled all of the shallower, easier to drill locations. Drilling for oil is becoming riskier and more expensive as we are forced to mine more extreme environments; the easy oil is already gone.

The largest environmental cost of oil use is associated emission of greenhouse gases (Chapter 7). The social costs of oil use also deserve closer inspection. In Figure 6.6, the countries that produce more oil than they consume plot in the "sustainable field" because they can meet their own needs. These countries can choose whether to export oil, but countries in the "unsustainable" field such as the United States are forced to import oil. The dependence of importing countries like the United States on oil from exporting countries has caused many social problems, including decreased national security.

In his book, *Hot, Flat, and Crowded*, Thomas Friedman (2008) argues that the global dependence on oil has made governments of oil states powerful, and that power has prevented or even reversed political reforms. Of the 23 countries that get the majority of their income from oil and gas, none are democracies. Saudi Arabia, Iran, and Russia can treat the United States with impunity because oil income has made them powerful. Friedman's First Law of Petropolitics states, "In oil-rich petrolist states, the price of oil and the pace of freedom tend to move in opposite directions…Petrolist states (are) authoritarian states (or ones with weak state institutions) that are highly dependent on oil production for the bulk of their exports and government income." Governments of petrolist states get their money from oil sales, not taxes, and they use the money to placate their citizens through subsidies. If the price of oil plummets, petrolist countries like Iran will have economic crises and decreased social stability.

In general, the **resource curse** affects third-world countries that sell their natural resources and use the money to develop in unsustainable ways. Typically, a minority of citizens controls the resources, and they become fabulously rich while the vast majority of citizens remain destitute. The resulting concentration of power prevents the development of democracy. According to Friedman (2008), "Our addiction to oil makes global warming warmer, petrodictators stronger, clean air dirtier, poor people poorer, democratic countries weaker, and radical terrorists richer." The substitution of renewable energy sources for oil (Chapters 10 and 11) may end the resource curse affecting oil-rich countries and reduce the negative environmental and social impacts of oil production and use.

6.1.2.2 Effects on Transportation and the Economy

Rising oil prices will cause many changes in transportation (Section 11.2.1.1), including electrification, increases in efficiency and use of mass transit and car sharing, and declines in the use of

* On Scene Coordinator Report on Deepwater Horizon Oil Spill, September 2011, http://www.uscg.mil/foia/docs/dwh/fosc_dwh_report.pdf, retrieved April 27, 2016.

† http://www.nbcnews.com/science/environment/coral-damage-goes-deeper-gulf-mexicos-oil-spill-zone-n167001, retrieved April 27, 2016.

‡ "The Legacy of the Gulf Spill: What to Expect for the Future?" John Mcquaid, 2010, Yale E360, http://e360.yale.edu/content/print.msp?id=2302.

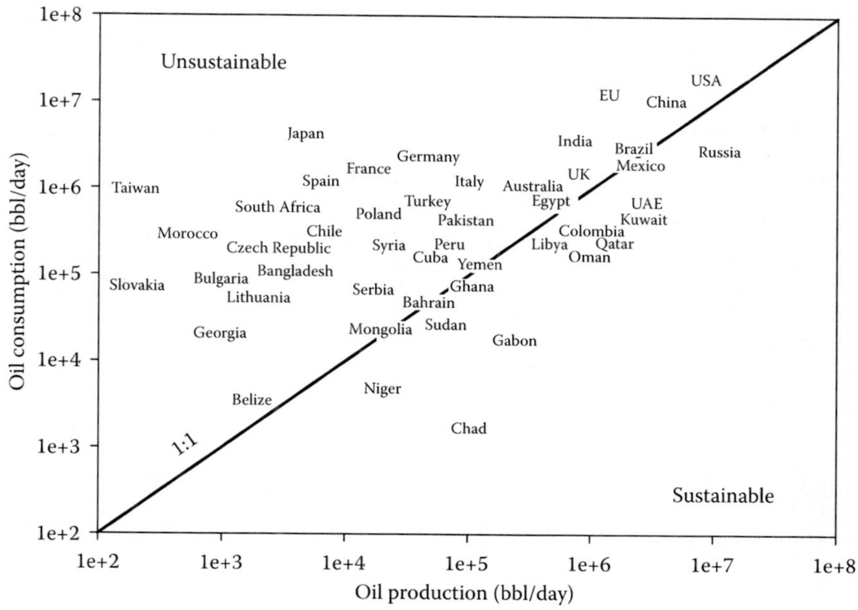

FIGURE 6.6 Oil production versus consumption in billion barrels per day for various countries. Along the "1:1" line production equals consumption. (Based on 2013–2014 data from the CIA World Factbook, https://www.cia.gov/library/publications/the-world-factbook/, retrieved April 27, 2016.)

personal transportation (Gilbert and Perl 2010; MacKay 2009). These factors will work together to reduce oil demand. In addition to changing personal transportation, higher oil prices will lead to higher food prices. Oil provides the energy to transport the food in a typical American meal an average of 1500 miles (Weber and Matthews 2008) and that allows our industrial food production system to inefficiently consume ten calories to produce only one calorie of food energy (Manning 2004). When oil becomes too expensive, consumers may switch to cheaper transportation fuels such as ethanol, and farmers will sell their crops for fuel rather than food because they will earn more money. If this happens, the resulting food shortages will drive food prices even higher. In March 2011, global food prices were at the highest inflation-adjusted level in 20 years.*

The United States is particularly vulnerable to the challenges presented by peak oil because it has a low population density, and because the United States built its cities for cars rather than people, leading to urban sprawl. However, peak oil is unlikely to precipitate a collapse. Electric cars are no longer a boutique purchase for environmentalists concerned about global warming; they are now mass-produced by major auto manufacturers for consumers concerned about the rising price of gasoline and the environmental impacts of conventional autos (see Section 11.2.1.1 for a discussion of the environmental impacts of electric cars). As peak oil kicks in and gas prices rise, electric cars will look increasingly attractive to buyers. As demand grows, manufacturing capacity will increase, and prices will decrease due to economies of scale. The market should help ensure a successful transition to electric vehicles. The ride may be bumpy, but peak oil and global climate change (GCC) together make electrification of personal transport likely, reducing the impacts of peak oil.

Peak oil will also reduce economic capital. As people adapt to preserve economic capital, the changes will become social as individuals work together as communities to adapt to an oil-free, low energy lifestyle. Long-distance transport of food currently depends on cheap liquid fuels. Oil price increases caused by a growing supply-demand gap may sharply curtail transport, in turn creating

* Alessandro Rizzo, AP, March 4, 2011, *The Tennessean*, page 3A.

a supply-demand gap for food that can only be filled by increasing local food production. This is what happened during an energy famine in Cuba in the 1990s, which serves as a useful case study.

6.1.3 Case Study: Cuba

As described in Friedrichs (2010), several countries have experienced oil shortages in the past, and have responded in different ways. In the early 1940s, Japan lacked sufficient oil to fuel its growing military. This situation became a crisis when the United States put an embargo in place. Japan's response to the resulting oil shortage was military aggression in the form of a preemptive strike at Pearl Harbor. North Korea faced an energy crisis in the early 1990s when its main source of oil, the USSR, collapsed. Rather than engage citizens in a countrywide mitigation response, leaders chose to protect their interests by becoming increasingly authoritarian. This ineffective response led to a country-wide famine causing 3 million to 6 million deaths. Like North Korea, Cuba also faced an energy crisis following the collapse of the USSR. However, they responded in a very different and more effective way.

Many ideas presented in this book are touched on in the 53-minute video "The Power of Community: How Cuba Survived the Peak Oil Crisis" (Morgan 2006). It describes the **Special Period**, an economic depression that began in 1991 after the collapse of Cuba' primary sponsor, the USSR.* The depression peaked by the mid-1990s and decreased in severity by the end of the decade. During the Special Period, Cuba experienced an energy famine because oil imports dropped from 13 million to 4 million barrels per year. Cuba's peak oil crisis resulted from the physical scarcity of oil within Cuba, but more importantly from a supply disruption caused by the collapse of its trading partner. This crisis transformed Cuba's society and economy, as exemplified by the Cuban government's change of its 30-year motto from "Socialism or Death" to "A Better World is Possible," and led to the nationwide adoption of sustainable agriculture. Cuba's successful transition from a peak- to a post-petroleum world teaches us many lessons that may be useful when our own countries are forced to make this transition.

Food shortages became the first problem to develop during the Special Period, for two reasons. First, an intensification of the U.S. embargo in the 1990s led to an 80% decrease in food imports and exacerbated food shortages. Second, Cuba's agricultural system was heavily dependent on oil, being the most heavily industrialized in Latin America. Cuba's dependence on foreign countries for oil and food made it vulnerable to energy and food shortages. The oil shortage meant that Cuban farmers couldn't use energy-hungry farm tractors or combine harvesters to produce the food, or trucks to transport the food to consumers. To compound the problem, people could no longer rely on refrigerators because Cuba produced most of its electricity by burning oil, and the oil shortage led to widespread blackouts. The only option was to eat fresh food when it was available. Thus, farmers had to transform the agricultural system completely by relocalizing it and changing farming methods from those of industrial agriculture to **permaculture** (Section 13.3.3). Society became more decentralized as many people moved from cities to farms, and people became more self-sufficient as they learned to produce their own food. This process took three to five years, during which there were constant food shortages, and adult Cubans lost an average of 10 pounds (Murphy and Morgan 2013). Government food distributions and rationing kept people from starving.

This process followed the classic adaptive cycle. Cuba experienced rapid population† and economic growth‡ between 1960 and 1991. It dealt with this problem by increasing food and energy imports, which increased efficiency but which made the country less self-reliant and less resilient. By 1990, Cuba was well into the conservation phase. In the mid-1990s, external factors led to a

* http://blogs.worldwatch.org/cubas-power-the-energy-revolution-part-1/, retrieved May 2, 2016.

† http://www.tradingeconomics.com/cuba/population, retrieved June 22, 2016.

‡ http://www.tradingeconomics.com/cuba/gdp, retrieved June 22, 2016.

collapse, followed by reorganization and a new growth phase. Since that time Cuba passed through the demographic transition, so that its population has remained relatively constant since 2002.*

Fortunately, Cuba had a high adaptive capacity that made it resilient, and it was able to avoid widespread starvation. Cuba had 2% of the population of Latin America but 11% of the scientists. Before the Special Period, scientists had conducted research on sustainable organic farming, and once the need arose, they implemented these methods nationwide. It took three to five years to make damaged soils fertile and productive again through systematic application of compost and **green manure** and use of **crop rotation**. Nationwide, farmers decreased oil-derived pesticide use from 21,000 tons to only 1000 tons per year by using crop-interplanting methods and biopesticides. Now, 80% of the food produced in Cuba is organic. The Cuban diet has changed in response: It is now more vegan-like and healthier, with greatly decreased consumption of meat, sugar, and dairy products and increased fiber content.

The urban agricultural movement was also effective. It started as a survivalist response by individuals, but grew when entire communities began to convert idle plots of land to community gardens. These communities used permaculture methods to create natural gardens on roofs and patios. Now each neighborhood has a kiosk to sell fruits and vegetables.

During the Special Period, Cuba experimented with several different agricultural systems ranging from fully state-controlled to privatized **community-supported agriculture (CSA)** systems. They found that private farmers were by far the most productive per acre. As a result, President Raul Castro took steps to privatize agriculture in Cuba, loosening state controls on commerce to let Cubans grow and sell their own fruits and vegetables in hopes of increasing domestic food production.† Urban gardens and farms became an important new source of fruits and vegetables.

During the Special Period, the impact of peak oil extended far beyond agriculture to include energy production, the economy, transportation, and health. Before becoming economically and politically independent, Cuba had to become energy independent. In 2005, the Cuban Energy Revolution legislation implemented many reforms that increased energy security.‡ First, consumption was decreased by promoting energy efficiency and conservation. The government gave homeowners free energy efficient lightbulbs, rice cookers, and pressure cookers, and they replaced many old, inefficient appliances. They also implemented a tariff on electricity use greater than 50 kWh per month, ensuring that the poor could still afford electricity but still providing an incentive for heavy users to decrease their electricity consumption. Second, they improved electricity availability and reliability. They decentralized energy production and upgraded the transmission network. They replaced old oil-fired power plants with more efficient diesel micro-generators, and installed backup generators in critical facilities such as hospitals. Third, they increased renewable energy production to 6% through installation of micro-hydro and solar-electric systems. Fourth, they increased domestic production of oil and gas to meet about half of demand (Venezuela provides the remainder). Fifth, they exported these policies to other developing countries by providing knowledge and installing solar-electric panels and energy efficient lighting. Between 1990 and 2009 Cuba decreased its CO_2 emissions 25%, from 3.2 to 2.4 tons per person (Murphy and Morgan 2013). Cubans now use one-eighth the energy that Americans use.

Oil shortages forced people to abandon their cars, transforming the transportation system by making it more energy efficient. In small towns people turned to horses for transportation. For transportation over short distances, city dwellers walked or used bicycles. For longer distances Cuba had to develop a mass transit system overnight, relying mostly on buses.

---

* http://www.tradingeconomics.com/cuba/population, retrieved June 22, 2016.

† AP, August 28, 2010. These steps were also taken to counteract the loss in agricultural capacity caused by hurricanes. Hurricane Michelle devastated Cuba's agricultural sector in 2002 (USDA 2008). "Cuba's Food & Agriculture Situation Report," http://www.fas.usda.gov/itp/cuba/CubaSituation0308.pdf. Two more hurricanes hit in 2008, causing Cuba to import as much as 80% of its food, a figure that hoarding may have inflated (see http://www2.tbo.com/content/2010/aug/12/cuba-cut-us-food-imports-28-year/news-breaking/).

‡ http://blogs.worldwatch.org/cubas-power-the-energy-revolution-part-1/, retrieved June 22, 2016.

TABLE 6.1
Comparison of Cuba and the United States

Statistic	Cuba	U.S.
Physicians per 1000 people	6.4	2.7
Hospital beds per 1000 people	5.9	3.1
Medical expenditures as % of GDP	11.8	16.2
Life expectancy in years	77.7	78.4
Infant mortality rate per 1000 live births	4.8	6.1
Adult obesity rate in %	11.8	35.7
Education expenditures as % of GDP	5.5	13.6
Average length of time in school in years	18	15
Energy use in tons of oil equivalent per person per year	1.0	7.0
CO_2 emissions in tons per person per year	2.4	16.9

Source: Statistics from Murphy, P. and F. Morgan. 2013. "Cuba: Lessons from a Forced Decline." In *State of the World 2013: Is Sustainability Still Possible?* 332–42. The Worldwatch Institute and Island Press.

Peak oil forced many of Cuba's economic sectors to relocalize to reduce transportation energy requirements. Like food production, both health care and education relocalized. Doctors and nurses began to serve patients in their own neighborhoods by paying house calls. Cubans became healthier due to increased exercise and a switch to a healthier diet of locally produced fresh vegetables, so they needed less medical care (Table 6.1). Universities relocalized by decreasing in size but increasing in number so they could serve local populations. Even now, mass transportation in cities is inadequate, so the commercial sector is also relocalizing. Rather than segregating residential and commercial housing, the current trend is to build mixed-use communities that are self-reliant because all amenities are local.

The people of Cuba displayed impressive resilience during the Special Period. They were forced to live with less and to change their way of thinking and way of life. Nevertheless, they successfully adapted, and are still happy. Cubans survived despite their government's planned economy; perhaps during "long emergencies" such as the Special Period success is determined less by the type of government than by how resilient communities are. Cuba's plan must be working because it is sustainable (EF $< B$) and has a high Human Development Index (HDI > 0.8) (Figure 3.9). Now Cuba is entering a new phase of economic development, with the United States opening its embassy and EU countries increasing trade. While its per capita energy consumption is low, the hope is that economic development will allow Cuba to decrease its reliance on fossil fuels and improve the efficiency of energy production.*

Becoming sustainable does not require the same form of government that Cuba has; it simply requires resilient communities and wise policies. While the energy transition in Cuba was painful for its people, its society is now more sustainable. The hope is that the United States and other developed countries will move toward sustainability voluntarily rather than waiting until they are forced to make the transition.

6.1.4 U.S. Solutions to Oil Scarcity

Americans consume an average of 2.5 gallons of petroleum per person per day (CIA World Factbook 2014), more than any other nation. What do we use all of this oil for besides transportation? Oil is

* http://blogs.worldwatch.org/how-will-cubas-reopening-affect-the-countrys-energy-future-part-ii-looking-ahead/, retrieved May 2, 2016.

a critically important part of our industrial agriculture system, and is the raw material for many chemical products, including pharmaceuticals, solvents, fertilizers, pesticides, and plastics; the 16% not used for energy production is converted into these other materials. Some argue that we should save our remaining oil for more valuable applications than personal transportation (Deffeyes 2001). For example, we can't make most plastics without oil. An oil shortage could cause shortages in wax, asphalt, tar, oil-based paints, resins and epoxies, synthetic fibers and rubber, and many lubricants, solvents, detergents, pharmaceuticals, inks and dyes, food additives, agrochemicals, adhesives, and materials for electronics and personal care products.

Several alternative responses to the peak oil dilemma exist. The usual economic response to a resource shortage is to find substitutes. One type of oil substitute, plant oils, is increasingly being used to produce biodiesel and bioplastics. In Chapter 9 we will show that other potential substitutes such as tar sands and liquefied coal would cause more harm than good. As mentioned previously, natural gas (Chapter 9) and electricity produced using renewable energy sources (Chapter 10) are viable oil substitutes with much fewer drawbacks. Another response that is more sustainable but hasn't received much attention is **automotive oil recycling**. All major car manufacturers have approved the use of American Petroleum Institute (API) certified re-refined oil in their vehicles.

The response of the U.S. Department of Energy to the impending peak oil crisis was to produce the **Hirsch report** in 2005 (Hirsch, Bezdek, and Wendling 2005). Unsurprisingly, this report advocated the **business as usual** approach of using technology to extract every drop of oil from the ground. It concluded that peak oil and the consequent shortage of liquid fuels will have a strong negative impact on society and the economy, and that to reduce risk the United States should take steps to decrease demand and increase supply at least 10 years in advance of the peak.

Steps proposed by others to reduce demand include replacing conventional cars with electric cars and increasing fuel efficiency of vehicles. Statistics highlight the advantages of increasing the fuel efficiency of vehicles*:

- An average U.S. car mileage increase of 5 miles per gallon would reduce oil consumption by more than the expected peak annual oil production from drilling in the Alaska National Wildlife Refuge (ANWR).
- An average U.S. car mileage increase of 10 miles per gallon (recommended as feasible by the National Academy of Sciences) would replace one quarter of the total amount of recoverable oil in ANWR each year, based on the most optimistic reserve estimate of 10 billion barrels of oil.

However, the Hirsch report concludes that steps to reduce demand will be inadequate, and so we must increase the supply of liquid fuels. To increase supply, it advocates coal liquefaction, use of enhanced oil recovery technologies, expansion of the use of liquefied natural gas, and mining of tar sands. The report estimates it will take 10 to 20 years to complete these steps, and that the sooner we start the process, the less it will cost.

Unfortunately, the Hirsch report completely ignores the problem of global climate change. Its proposed peak oil mitigation strategy of increasing supply by using coal liquefaction and mining of tar sands would greatly exacerbate the GCC problem. The steps the Hirsch report advocates for reducing peak oil risk are exactly the opposite of the steps the Stern report (2007) advocates for reducing GCC risk. Clearly to find a sustainable solution we must consider these problems simultaneously (Hopkins 2008): how can we reduce the risks of both peak oil and GCC? We will look at the GCC problem in the next chapter before attempting to answer this important question in Chapter 11.

* From http://serc.carleton.edu/introgeo/interactive/qotd/ANWR.html.

6.2 NATIONAL AND GLOBAL RESOURCE PRODUCTION PEAKS

> We are not good at recognizing distant threats even if their probability is 100%. Society ignoring [peak oil] is like the people of Pompeii ignoring the rumblings below Vesuvius.
>
> **James Schlesinger**
> *Former U.S. Energy Secretary*

We can apply Hubbert's approach of constructing resource availability curves to any nonrenewable resource on either a local or a global basis. Many countries are already post-peak for production of certain resources. For example, the United States imports 100% of the following resources that it uses: arsenic trioxide, asbestos, bauxite and alumina, columbium (niobium), fluorspar, graphite, manganese, mica, quartz crystal, strontium, thallium, thorium, and yttrium (Keller 2011). Because we have global trade, local scarcity has not resulted in a crisis. Countries that have a surplus of a resource export it, and countries erase their deficits by importing. In a globalized world, the problem occurs when *global* annual production of a nonrenewable resource peaks and then begins to decline. During the decline, resource production cannot keep pace with demand, and resource prices rise. Peak oil may cause shortages of many other resources because oil provides the energy to mine and transport those resources. If the United States doesn't have enough oil to transport all of the resources that we import, we will have more than just an energy problem.

What nonrenewable resources may become scarce in the twenty-first century? Hubbert predicted that copper, tin, lead, and zinc would reach peak production within decades (Hubbert 1987). At the current rate of consumption, reserve depletion estimates are 60, 40, 40, and 45 years, respectively, and indium, which is used in LCDs and solar cells, may run out in only 15 years (Ragnarsdottir 2008). Phosphate, which is an essential component of fertilizers, may reach peak production within the next 60 to 70 years (Oelkers and Valsami-Jones 2008), which could subsequently decrease agricultural productivity and cause widespread food shortages.

Why are we at risk of running out of these resources? Besides rising population and per capita consumption rates, other reasons people overexploit environmental resources include (Diamond 2005):

- They initially seem inexhaustibly abundant.
- Signs of incipient depletion become masked by normal fluctuations in resource levels between years or decades (e.g., rainfall).
- It's difficult to get people to agree to exercise restraint in harvesting a shared resource (Tragedy of the Commons).
- The complexity of ecosystems often makes the consequences of human-caused perturbations impossible to predict.

It's not just nonrenewable resources that we have to worry about. In some cases, the harvesting rate of a renewable resource may greatly exceed its renewal rate. For example, deep (fossil) groundwaters have been in the ground for hundreds or thousands of years, which means it would take that long to replace them at natural recharge rates. In many areas of the world, the groundwater extraction rate is much greater than the recharge rate, so the groundwater reserve is shrinking, as made visible by the **water table** (the level below which the sediment or soil is saturated, meaning all pores are filled with water) falling in unconfined aquifers. According to the second law of sustainability, the sustainable approach to renewable resource use is not to use renewable resources faster than nature can renew them. Consequently, hydrologist Luna Leopold advocated treating groundwater as a nonrenewable resource that we should use only during droughts (Keller 2011). When we use renewable resources faster than they can be replaced, they effectively become nonrenewable, and we can expect the production rate to peak and then decline. For example, in Figure 4.6 we saw that the global wild fish catch production curve appears similar to that of a nonrenewable resource, having peaked around 1990 due to overfishing, although it has not declined. Fortunately, **aquaculture** (fish farming, see Section 13.3.3.3) is expanding and serving to substitute for fish hunting, which has softened the blow. As human population and resource demand continue to increase, we can expect to see the production of more resources peak and then begin to decline. The important question is will we always find adequate substitutes as we did for marine fish?

6.3 MINERAL RESOURCES

The average American uses 5677 pounds of aluminum ore, 65,480 pounds of cement, 19,815 pounds of phosphate rock, and 19,245 pounds of clays during their lifetime (Walther 2014). We use smaller amounts of materials such as gold that are rare and therefore expensive. In this section we will learn about some of the solid, naturally occurring materials that provide us with nonrenewable resources.

Most natural materials come from minerals mined from rocks contained in the Earth's crust. The abundance of an element is measured by its average concentration in the continental crust. Generally, the abundance of elements in the Earth decreases with increasing atomic number, so heavy elements such as gold (atomic number = 79) are less abundant. Natural processes concentrate elements in small areas in the Earth's crust to form **ore deposits**. Some elements must be concentrated thousands of times in the crust before they can be economically mined.

Although mineral resources are nonrenewable, the abundance of some elements in the Earth's crust is so high that we need not be concerned about their depletion. Examples include silicon and aluminum, the second and third most abundant elements in the Earth's crust (oxygen is the most abundant). However, energy shortages could cause aluminum metal to become scarce because large amounts of energy are needed to convert naturally occurring aluminum oxide to aluminum metal.

It is important to be able to identify critically important minerals with shrinking supplies so we can anticipate future shortages and decrease consumption rates, either through conservation and recycling or by substitution. If a resource becomes increasingly scarce while demand remains steady or increases, then price will increase, leading to a decrease in demand and consumption rate. This negative feedback in the market leads to efficient use of scarce resources. The concern is that resources that are essential for human well-being will become too expensive for the poor to afford.

As humanity approaches the physical limits imposed by our environment we will experience natural resource crises with increasing regularity. For example, in 2011 increasing scarcity caused the price of some rare earths to increase as much as 750%.* The rising scarcity of rare earths is particularly troubling from a sustainability standpoint because some sustainable technologies require them. Wind turbines use the rare earth neodymium in their magnets, and electric cars use lanthanum in their batteries. China supplies 97% of the world's rare earths, but it only has 48% of the world's reserves, while the United States has 13%.† China is starting to decrease exports because its growing economy uses 60% of the supply, and this is causing prices to rise.

Gardner (2013) states that in the United States 95% of the material flowing through the economy is nonrenewable. UNEP estimates that more than half of 60 metals studied have global recycling rates less than 1%. There is concern that many nonrenewable resources may soon become scarce. First, demand is rising faster than ever before. Second, prices are rising: Between 2002 and 2010, the average prices for 86 metals rose 6.4% annually. Supply has not been able to keep pace with demand. Third, metal ore grades have been declining for over a century; this means higher environmental impacts and more energy required to mine and process the ore.

In this section we will look at a few selective mineral commodities to illustrate important principles, focusing on minerals that are critically important to our economy but that are currently used unsustainably and may soon become scarce.

---

* "A Scarcity of Rare Metals Is Hindering Green Technologies," Nicola Jones, Yale environment 360, November 18, 2013, http://e360.yale.edu/feature/a_scarcity_of_rare_metals_is_hindering_green_technologies/2711/, retrieved June 24, 2016.

† Tim Folger, *National Geographic*, June 2011, "The Secret (Chinese) Ingredients of (Almost) Everything," http://ngm.nationalgeographic.com/print/2011/06/rare-earth-elements/folger-text, retrieved June 24, 2016.

6.3.1 Phosphorus

Phosphorus is a critically important element because it may ultimately limit global food production. It was first recognized by Leibig in 1840 that phosphorous is the limiting nutrient for plant growth and food production; this is especially true now that we can manufacture nitrogen fertilizer using the Haber-Bosch process (Cordell and White 2014). Roughly 70% of phosphorus that is mined today is used for fertilizer (Walther 2014). The United States produced roughly 28 million metric tons of phosphate rock in 2011, second only to China. According to the USGS, in 2014 the world reserve of phosphate rock was 67 billion metric tons and annual production was 0.22 billion metric tons, giving an R/P = 305 years.* These reserve estimates from the USGS have been criticized as being too high. Furthermore, using R/P to estimate reserve depletion times assumes that demand remains constant, but global phosphate demand may increase 2% to 2.5% annually over the next few decades, suggesting that half of global phosphate reserves will be consumed in the next 60 to 70 years (Oelkers and Valsami-Jones 2008). Recent Hubbert peak-based estimates of peak production range from 2025 to 2084 (Cordell and White 2014). This is a concern because there is no substitute for phosphorus in food production.

Mining and agricultural use have tripled the flow of phosphorous to the oceans. Most of the phosphorous losses are from agriculture because plants only use 15% to 30% of the phosphorous applied as fertilizer (Cordell and White 2014). Globally only ~1/5 of phosphorous used for food production ends up in food and in the United States ~66% of phosphorous applied to agricultural fields ends up in surface water bodies. The 2013 EPA National River and Stream Assessment found high levels of phosphorus in 40% of rivers and streams. The phosphorous flows to lakes and the ocean where it causes eutrophication, leading to annual economic losses in the United States of $2.2 billion. In the United Kingdom, roughly 60% of phosphorus in surface waters comes from agricultural activities and 40% from water treatment facilities (Parsons and Smith 2008).

Global phosphorous consumption as fertilizer is increasing (Tilman et al. 2002), and is expected to continue increasing into the future. An increase in global population to 10 billion to 12 billion by 2100 (Section 3.4) will place an increased demand on food and therefore phosphate fertilizer. Per capita phosphorous consumption is also expected to increase due to increasing per capita consumption of meat and dairy products (Chapter 13), production of biofuel crops (Section 10.3), and use in lithium-iron-phosphate batteries for electric vehicles (Cordell and White 2014).

Several strategies should be adopted for increasing phosphorus supply. Advancements in technologies could decrease the costs of recovering phosphorus from phosphate deposits, which would increase phosphorus reserves. Policy changes are also needed to increase production in countries such as Morocco with large underutilized phosphate reserves.

The demand for phosphorus could also be decreased by increasing use efficiency, that is, reducing leakage from the terrestrial phosphorous cycle. We will see in Chapter 14 that pollution is a resource that is out of place. While 1/6 of the world's farmers cannot afford to buy phosphate fertilizer, in much of the world excess phosphate is causing eutrophication (Cordell and White 2014). That phosphate is scarce in many regions and polluting in others indicates that the problem is a result of poor global management rather than physical scarcity of this valuable resource.

Phosphorus demand would decrease if the cost of externalities such as eutrophication and resulting loss of ecosystem services were incorporated into the economic costs of phosphate fertilizers and detergents. Remember that an **externality** is a cost paid not by the producers or users of a product, but by society at large. Advancements in technology could also decrease phosphorus concentrations in effluents from wastewater treatment plants. For example, the phosphate mineral struvite can be recovered during wastewater treatment and used as fertilizer (Manning 2008). Adoption of sustainable agriculture practices such as the use of manure in organic farming could

* USGS Mineral Commodity Summaries, January 2015, http://minerals.usgs.gov/minerals/pubs/commodity/phosphate_rock/, retrieved September 30, 2015.

lower synthetic phosphorus inputs by recycling phosphorus and closing the resource loop. Another organic farming approach is to apply powdered phosphorite rock directly to agricultural fields. The lower solubility of apatite in the phosphorite slows the release of phosphorus compared with soluble phosphate salts in synthetic fertilizers. Pressure to adopt these conservation measures will increase when global phosphorous production peaks and prices begin to increase. A glimpse of the future was provided in 2008 when phosphate prices increased by 800%, leading to rapid increases in fertilizer and food prices, which sparked fertilizer and food riots globally (Cordell and White 2014).

6.3.2 Uranium

With an atomic number of 92, uranium is the heaviest naturally occurring element, and therefore is relatively scarce. Uranium's principal economic use is as fuel for nuclear fission reactors. Uranium is not a renewable resource, with global production peaking in 1980 (Vance 2006). Yet the most recent and comprehensive report estimates that more than 100 years' supply exists at 2008 rates of consumption, and that even if consumption grows rapidly, we would consume less than half the identified resources by 2035 (OECD 2010). Likewise, the MIT Energy Initiative (2011) declared that we have sufficient uranium reserves to last through this century. However, 100 years' supply from land mines does not qualify traditional nuclear power as a sustainable energy source. For example, Mackay (2009) regards 1000 years' supply of a resource as a minimum to be considered sustainable, but also argues that extraction of uranium from seawater could greatly increase the supply of fissionable uranium.

Price is a measure of the scarcity of any resource. The fact that uranium prices increased rapidly beginning in 2005 and then leveled off at a much higher level than before 2005 suggests that uranium may be becoming scarce. As an element becomes scarcer, prices increase, converting previously uneconomical resources into reserves, increasing reserve estimates, and pushing reserve depletion estimates farther into the future. In 2009, 50,600 metric tons of uranium were produced, and reserves totaled 5.5 million metric tons (Walther 2014), yielding an R/P of 109 years. However, uranium demand and prices decreased following the Fukushima nuclear disaster in 2011 (Section 9.5.3).*

6.3.3 Metals

Of the elements in the periodic table, 62 elements are metals or metalloids. These elements are used to manufacture many materials used by society. Some of these metals are relatively rare, which is reflected in their high prices. Despite this, most metals are not significantly recycled, and so the use of these nonrenewable resources in the fabrication of materials draws on raw material (primary resource) rather than recycled material (secondary resource). As a result, concern is growing that we may see shortages of some critically important metals. As these metals become scarcer, their prices will increase, which will accelerate exploration and production and increase supply. However, if supply is unable to keep pace with demand, substitute materials will need to be developed. There are a few examples of this happening with metals, including the development of substitutes for cobalt in batteries and magnets when the supply chain for cobalt was disrupted in the 1970s (Graedel et al. 2015).

Graedel et al. (2015) evaluated the potential for metal substitution, and found that none of the 62 metals has substitutes with equivalent performance across all of their applications, and some metals have no good substitutes for their major uses. This analysis makes clear that substitution of metal for metal will not always solve the problem of metal scarcity. While per capita demand of some abundant metals such as Fe and Al in the United States has leveled off, this "saturation" of the

* https://ycharts.com/indicators/uranium_spot_price, retrieved May 2, 2016.

market is not expected for scarcer metals that have only recently been used, whose use is increasing, and whose recycling rates are low. Furthermore, the development of new technologies for production of advanced materials that may serve as substitutes is lagging behind demand and will not be available in the market for some time. As a result, "often no suitable substitute can be found no matter what price is offered without performance and function being seriously compromised. It thus appears that society will need to pay more attention to the acquisition and maintenance of nonrenewable resources than has been the case in the past" (Graedel et al. 2015). In Chapter 14, we will cover the topics of metal pollution and recycling.

6.3.4 Environmental Impacts of Mining

Extracting minerals from ore deposits in the Earth's crust requires at a minimum the removal of overlying rock called the overburden, the collection of ore, transportation of the mineral to the point of use, and return of the overburden. Mining usually includes other steps, some of which have severe environmental impacts including pollution of air and water and negative impacts on ecosystems and human health (Keller 2011).

Mining is very energy intensive and therefore responsible for emission of large quantities of greenhouse gases to the atmosphere. Other environmental impacts are usually unique to the resource or rock type being mined. For example, coal and other subsurface resources are often mined by exposing them at the surface for **strip mining** (by mountaintop removal, see Chapter 9). Coal strip mining often exposes the mineral pyrite to the atmosphere, which causes it to oxidize and produce **acid mine drainage** (Raymond and Oh 2009). Although waste rock is required to be returned after removal of the coal, the land surface after completion of mining is often more susceptible to soil erosion and resulting sediment pollution (Lottermoser 2010). Solution mining, also called **in situ leaching**, can be used to mine mineral deposits that are water soluble. Rock salt from the subsurface is mined by drilling wells down to the underlying deposit and then injecting water to dissolve the salts. The resulting brine is pumped back to the surface, where the water is evaporated and the salt collected. Removal of rock salt from great depths usually causes overlying rock layers to collapse and fill the openings formed by solution mining. The resulting land subsidence can cause structural damage to buildings and roads, increase erosion, and contaminate shallow aquifers.

Because of the often high environmental impacts of mining, regulations are required to reduce the impacts. In the absence of regulations, mine operators will usually use the methods with the lowest economic costs. Imposing regulations usually causes production costs to increase so that mining becomes less profitable. As a result, many mining companies have chosen to move their operations to developing countries without environmental regulations. This has improved the state of the environment in the United States, but it has also made us less aware of the environmental impacts of our resource use. This practice is inherently inequitable and therefore unsustainable because Americans enjoy the benefits but do not pay the full costs of their resource use.

Many people protest against mining regulations in particular and environmental protections in general, claiming they are too expensive. To a certain extent they are correct. Most nations of the world cannot afford the costly environmental protections the wealthy developed countries have developed. They have not passed the peak in the environmental Kuznet curve, so as their economies grow environmental damage increases—witness the terrible state of the environment in China (Section 16.2). Many countries will never be able to afford these environmental protections. In addition, it is not enough to reuse and recycle materials; these processes are not 100% efficient, so if we maintain the same high level of resource consumption, even with widespread reuse and recycling we would still have to continue mining for new raw materials.

So what is the solution to reducing pollution caused by resource consumption? Rich and poor countries alike must shift focus from the end of the material stream (the tailpipe) to the beginning. The pollution problem primarily results from material throughputs that are too high. The more material we remove from the earth, the more waste we produce. Much of the material we mine never becomes part of a product; most of the Earth's material that we move is **overburden** and mine **tailings**. If we could decrease our need for metals and other nonrenewables by a factor of 10, we would reduce the waste produced by at least a factor of 10. This is what the developed countries must do to allow developing countries to reach the same level of prosperity they have without exceeding the Earth's carrying capacity: decrease material throughputs by a factor of 10 (Schmidt-Bleek 2007). Research suggests that by increasing resource use efficiency the developed countries can achieve this "factor of 10" decrease in resource use without decreasing their prosperity. This **dematerialization** of the economy would decouple resource use from income growth. What would a dematerialized world look like? It would shift jobs from natural resource extraction to services. This shift has already occurred in the developed countries. Unfortunately, developed countries have accomplished this primarily by outsourcing their polluting resource extraction industries to developing countries.

6.4 CONCLUSIONS

Oil is a nonrenewable resource and the most important transportation fuel. Peak oil theory postulates that global oil production will eventually peak and then decline, causing oil prices to sharply increase. This will eventually happen for all nonrenewable resources for which we find no suitable substitutes because per capita consumption rates are increasing exponentially in parallel with population growth. Peak production will occur soon for phosphorous used for food production, and has already occurred for uranium used for energy production.

Some argue that the data do not support the peak oil theory.* These views are part of the normal cycle of gradual acceptance of a repugnant idea. Bardi (2009) posits that peak oil is an idea that is working its way through four stages of acceptance:

Stage 1: Never heard of it.
Stage 2: It is wrong.
Stage 3: It is right, but irrelevant.
Stage 4: It is what I had been saying all along, or it's too late to do anything about it.

We saw a similar sequence of stages in the "Limits to Growth" debate, although that debate is stuck at either stage 2 or stage 3, depending on which group is polled. In the next chapter we will see how the theory of stratospheric ozone destruction by chlorofluorocarbons progressed through all four stages, and how the theory of anthropogenic global climate change progressed through all four stages in the scientific community but is currently stuck in stages two to three in the court of public opinion. We will also see that global climate change caused by use of fossil fuels is an even bigger problem than peak oil.

To avoid catastrophic climate change and energy shortages that could disrupt the global economy we must rapidly transition from nonrenewable fossil fuels like oil to renewable energy sources. Because burning oil releases greenhouse gases, reducing demand by quickly transitioning to a transportation system fueled only by renewable energy would avert a peak oil crisis and at the same time mitigate GCC. The first step to reducing oil dependence is to electrify transportation. Individuals can reduce their oil dependence by using electric cars, moving closer to their workplaces, and buying locally produced goods. Reducing transportation miles by relocalizing will help communities transition smoothly. These and other solutions to be discussed in coming chapters would help mitigate both the GCC and peak oil problems.

* See Michael Lynch, "'Peak Oil' Is a Waste of Energy," op-ed published in the *New York Times* on August 25, 2009, http://www.nytimes.com/2009/08/25/opinion/25lynch.html.

WEB RESOURCES

- ASPO International, The Association for the Study of Peak Oil and Gas: http://www.peak oil.net/
- http://www.ted.com/talks/lisa_margonelli_the_political_chemistry_of_oil.html
- http://www.ted.com/talks/rob_hopkins_transition_to_a_world_without_oil.html
- Oilcrash—The documentary: http://www.oilcrashmovie.com/index.html
- The Great Change Into a Post Petroleum World: http://www.thegreatchange.com/
- There's No Tomorrow: http://youtube/IipwQryubIE

HOMEWORK PROBLEMS

1. Find updated versions of Figures 6.3, 6.4, and 6.5. Have trends changed, and if so, why?
2. Using data from the most recent annual "Statistical Review of World Energy" report produced by British Petroleum (http://www.bp.com/en/global/corporate/energy-economics/statistical -review-of-world-energy.html) plot historical global oil production. Has production peaked and started to decline? If so, when did the peak occur?
3. Find data on historical prices for the following resources. What do the trends tell you about changes in the relative abundance or scarcity of the resource?
   a. Oil
   b. Phosphorous
   c. Uranium
   d. Neodymium

7 Global Climate Change

Global warming is too serious for the world any longer to ignore its danger or split into opposing factions on it.

Great Britain Prime Minister Tony Blair Speech
September 27, 2005

The most important challenge we face is climate change. Make no mistake: Climate change is no longer a threat—it's a reality.

U.S. President Barack Obama
*June 2016**

Perhaps the greatest challenge to sustainability is global climate change (GCC). Burning fossil fuels releases carbon dioxide (CO_2), a known greenhouse gas, into the atmosphere. This has led to a steady rise in the atmospheric concentration of carbon dioxide and in global surface temperatures. That humans can change the Earth on a global scale is shocking, but greenhouse gas pollution is just the type of "sink" problem that the Limits to Growth study predicted in 1972 (Section 4.6; Meadows et al. 1972). This problem results from too many people producing too much waste for the environment to absorb. Average global temperature has risen by 0.76°C (1.4°F) since 1850 (Figure 7.1) and is projected to increase another 0.5°C to 1.0°C (0.9°F to 1.8°F) due to greenhouse gases *we have already added* to the atmosphere (Dawson and Spannagle 2009). Potential effects of GCC include rising sea levels, increased species extinctions, greater weather extremes (heat waves, floods, and droughts), and the resulting effects on people (famine, destruction of coastal communities, and environmental migration) (Prugh 2015). Continued anthropogenic emissions of greenhouse gases is causing Earth's climate system and ecosystems to lose resilience, and positive feedbacks may push them past a threshold into a new climate regime that can provide fewer ecosystem services, resulting in a global decrease in human well-being.

Business as usual models that assume no major changes in climate policies, and therefore continuing increases in atmospheric concentrations of greenhouse gases, project that global mean temperature will rise an additional 4°C (7°F) by 2100 (IPCC 2014). The consequences of such rapid and dramatic global change are largely unknown, but preliminary estimates suggest that sea level will rise a little over 3 feet by 2100, weather hazards will become more severe, and that *by the year 2100 climate-related deaths will be in the hundreds of millions and economic losses will be trillions of dollars* (Stern and Treasury 2007). Recent estimates of climate change-caused slowing of economic growth at 0.25% per year, rising to 0.28% per year in the near future (Carleton and Hsiang 2016). A 3°C rise in average global temperature could also put 30% to 50% of plants and animals at risk of extinction (IPCC 2007). Passing a tipping point that leads to irreversible change in the climate system would amplify climate-related risks (Figure 4.3). High levels of uncertainty about the consequences of GCC demand that we apply the precautionary principle and reduce carbon emissions. Because the atmosphere is a shared resource and therefore is subject to the Tragedy of the Commons (Section 5.4.3), all countries must agree to reduce greenhouse gas emissions to avoid a climate catastrophe. However, uncertainty over the economic costs of climate change mitigation, opposition from powerful fossil energy interests, and the difficulty of attaining international agreements have led to an inadequate global response to the threat of GCC. An adequate response to GCC will require a combination of mitigation, regulation, and adaptation measures.

In this chapter we will review the theory behind GCC, the supporting evidence, future projections of GCC, and potential consequences. In Chapter 8 we will examine the public debate and

* "In Yosemite, Obama Presses Climate Change During Pitch for National Parks," https://www.washingtonpost.com/news/post-politics/wp/2016/06/18/in-yosemite-obama-presses-climate-change-during-pitch-for-national-parks/.

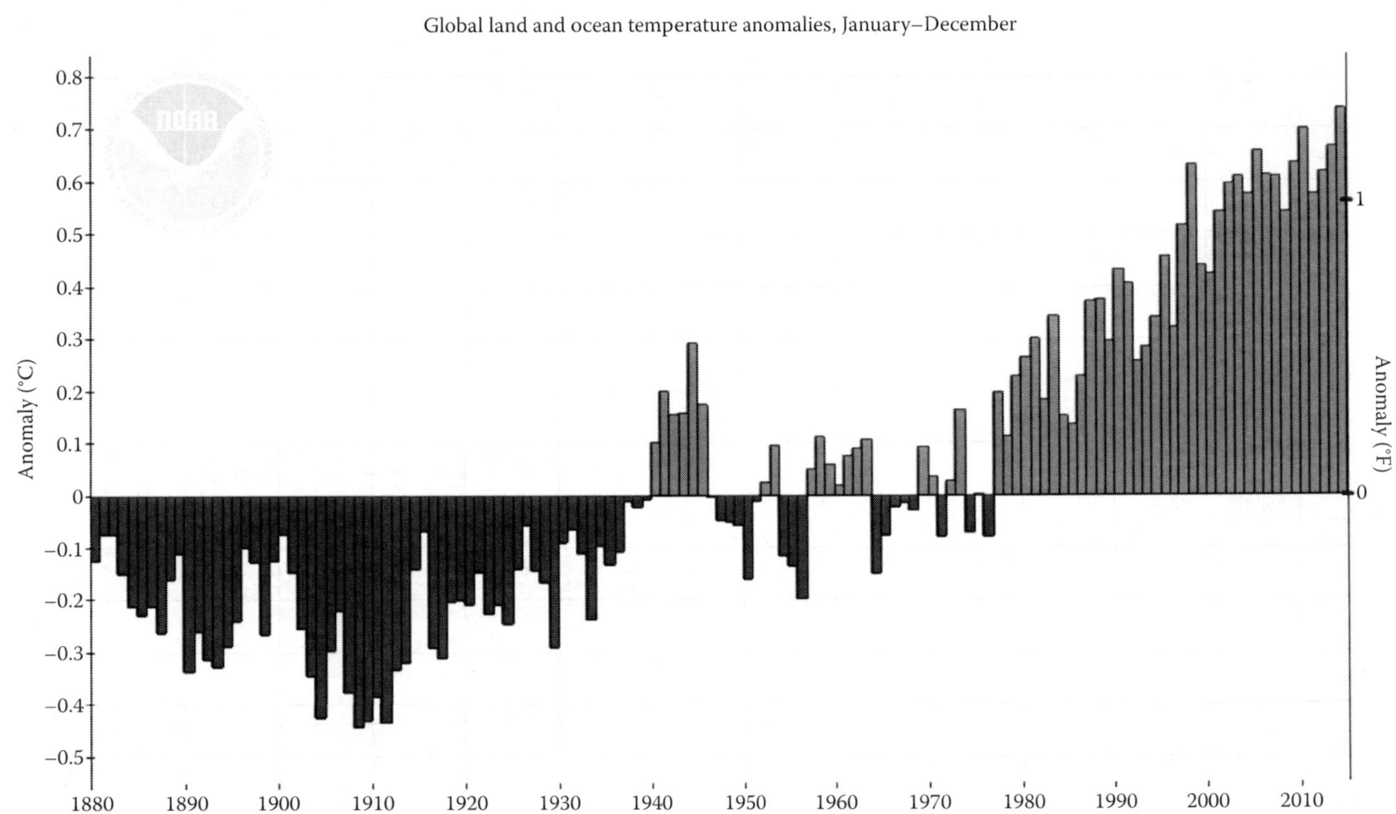

FIGURE 7.1 Annual global land and ocean temperature anomalies. (Downloaded from http://www.ncdc.noaa.gov/cag/time-series/global/globe/land_ocean/ytd/12/1880-2015 on 9/28/2015.)

evaluate proposed solutions to the GCC problem. As we will see in Chapter 8, climate change deniers cannot be persuaded by scientific evidence, and so I devote very little space in this book to presenting evidence of GCC; that evidence can be found in many other places, including the IPCC reports (IPCC 2007, 2014).

7.1 CLIMATE CHANGE IN THE ANTHROPOCENE

Ever since multicellular organisms appeared roughly 600 million years ago, Earth's climate has wobbled between inhospitable and hospitable. Now humans could push Earth's climate back into the inhospitable zone. Human activity has already increased species extinction rates and may ultimately cause a modern mass extinction event (Barnosky et al. 2011); a human-induced rapid climate transition would increase the extinction rate above the current high level, resulting in decreased ecosystem services and human well-being (more on this in Chapter 15).

To understand these changes, we must make clear the meaning of "climate." Climate is what you expect, but weather is what you get. Climate is the long-term characterization of the "average" weather. It changes over decades, while weather changes on a daily and even hourly basis. Humans often overgeneralize, in space and time, the short-term changes in weather (Pollack 2005). An example of overgeneralizing in a geographic sense is, "We had a wet summer, so everyone in the United States had a wet summer." We overgeneralize in a temporal sense by saying, "This week is the coldest I can remember; we must be entering a new Ice Age." We make both types of mistakes when we generalize short-term changes in local weather to long-term changes in global climate, for example, "This week in Nashville is the hottest I can remember; it must be global warming."

GCC has happened often during Earth's long history. Much of what we know about these changes comes from the study of rocks, sediments, and ice cores that contain information about paleoclimate. Ancient climate change resulted from natural variations in solar output, the Earth's orbit around the sun, the spatial distribution of the continents, oceanic circulation patterns, and rates of volcanic activity. In the last 5,000 to 6,000 years, human activity has also contributed to climate change.

Since roughly the beginning of the Holocene epoch 10,000 years ago, Earth's climate has been hospitable and stable (Mathez 2009). Some argue that this climate stability allowed for the invention and spread of agriculture. With agriculture humans started to influence global climate, first with deforestation beginning ~8,000 years ago that released carbon dioxide to the atmosphere. Beginning ~5,000 years ago irrigation agriculture in the form of rice cultivation and the proliferation of domestic grazing animals increased emissions of the even stronger greenhouse gas methane (CH_4) (Ruddiman, Kutzbach, and Vavrus 2011). These agricultural emissions increased global mean temperatures roughly 2°C at high latitudes, which may have staved off a glacial advance in Canada.

The spread of agriculture initiated the **Anthropocene** period, defined as the time when humans have started to change the Earth on a global scale (Zalasiewicz et al. 2010). With a warm, stable climate and the switch from a hunter-gatherer to a settled lifestyle, humans could afford to make large investments in infrastructure, knowing that essential resources like water were unlikely to disappear because of sudden climate change. Humans built cities along coastlines, confident that no rapid sea level rise would inundate them.

However, anthropogenic inputs of greenhouse gases to the atmosphere have caused Earth to shift to an unstable, rapidly changing climate. The atmospheric carbon dioxide concentration increased faster in the twentieth century (Figure 7.2) than any natural factors that affect global temperatures have changed over the last 22,000 years (Joos and Spahni 2008). Thus, we infer that new, non-natural processes, such as anthropogenic greenhouse gas emissions, are responsible for the observed changes in atmospheric greenhouse gas concentrations (Figure 7.2) and global temperature (Figure 7.1). These changes are irreversible over a timescale of hundreds of years, the amount of time required to remove anthropogenic carbon dioxide from the atmosphere (Solomon et al. 2009).

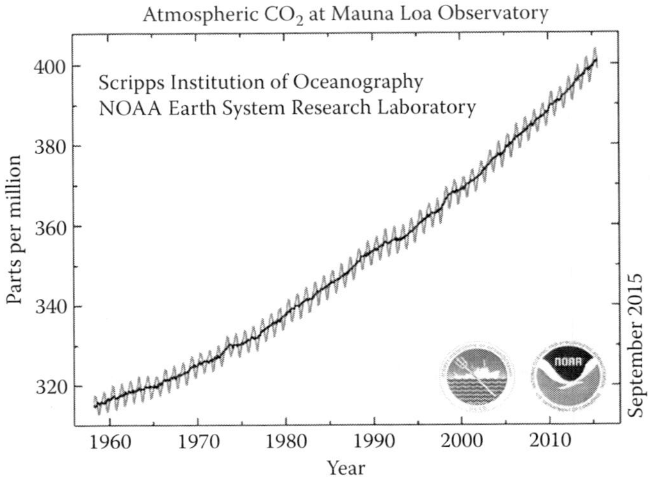

FIGURE 7.2 Measurements of atmospheric CO_2 concentration (mole fraction in dry air expressed in parts per million) at the Mauna Loa Observatory. The oscillations are caused by seasonal changes in vegetation. (From http://www.esrl.noaa.gov/gmd/ccgg/trends/index.html, downloaded September 28, 2015.)

7.2 GREENHOUSE GASES

The concentration of carbon dioxide in the atmosphere has increased much faster than scientists anticipated at the end of the nineteenth century (Richter 2010). During the twentieth century, per capita income more than doubled and population increased by roughly four times, so the economy increased 10-fold. Since energy fuels the economy and consumption, the rate of energy consumption also increased 10-fold. As shown in Chapter 3, the Kaya identity can be used to calculate the amount of CO_2 emitted by burning oil and other fossil fuels. The global carbon footprint can be calculated by summing CO_2 emissions over all sources. Doing this we find that during the twentieth century global CO_2 emissions also increased 10-fold (Figure 8.2). Recent estimates indicate that the rate of anthropogenic carbon emissions to the atmosphere is higher than at any time in the last 66 million years (Zeebe, Ridgwell, and Zachos 2016). Now we know that the rate at which the oceans absorb CO_2 is only one-tenth of what early researchers thought, making matters even worse. All of these factors (increased sources from population and economic growth causing increased burning of fossil fuels, and decreased carbon sinks) combined to reduce the estimated doubling time of CO_2 in the atmosphere (Section 3.4.1) from 1,000 years in 1896 to ~100 years today (Richter 2010). Atmospheric carbon dioxide concentration has already increased about 40%, from 270 parts per million (ppm) in pre-industrial times to ~400 ppm today. This would not be a problem if natural processes rapidly removed excess carbon dioxide from the atmosphere, but that is not so; the atmospheric residence time (the time it would take natural processes to remove a gas from the atmosphere if we stopped emitting it) of carbon dioxide is >100 years and can be as much as 1,000 years (Archer et al. 2009; Richter 2010).

It's important to know that carbon dioxide is not the only important greenhouse gas; others include methane and nitrous oxide (Figure 7.3). Together, the presence of these greenhouse gases in the atmosphere increases the average global surface temperature by 34°C (61°F) (Emanuel 2012). Scientists usually report greenhouse gas emissions as carbon dioxide equivalents CO_2e or the number of molecules of CO_2 that would have an equal warming effect of one molecule of another greenhouse gas. Although carbon dioxide is the weakest of the greenhouse gases, it has the largest effect on global warming because we emit such large volumes of carbon dioxide during fossil fuel burning. However, global anthropogenic emissions of other greenhouse gases are also increasing (Figure 7.3).

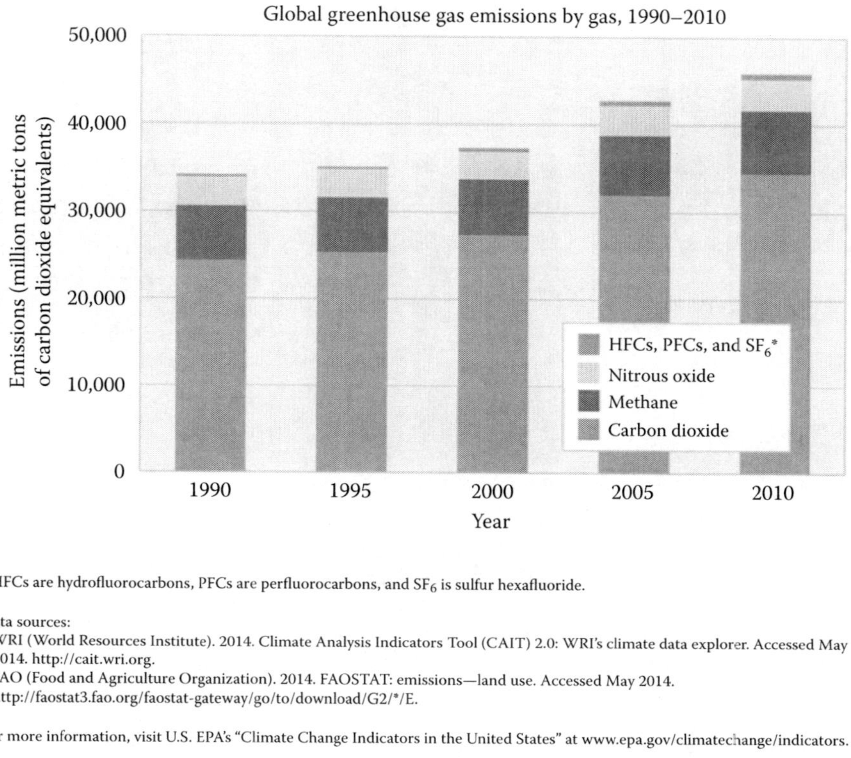

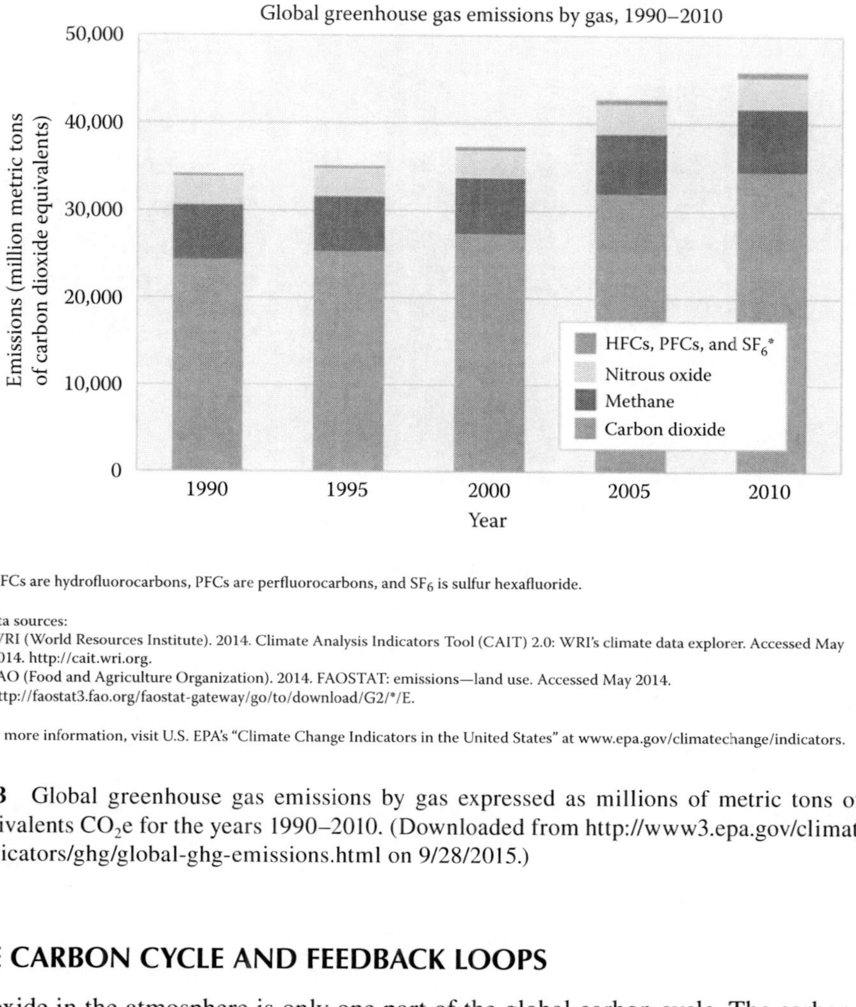

FIGURE 7.3 Global greenhouse gas emissions by gas expressed as millions of metric tons of carbon dioxide equivalents CO_2e for the years 1990–2010. (Downloaded from http://www3.epa.gov/climatechange/science/indicators/ghg/global-ghg-emissions.html on 9/28/2015.)

7.3 THE CARBON CYCLE AND FEEDBACK LOOPS

Carbon dioxide in the atmosphere is only one part of the global carbon cycle. The carbon cycle is critical to understanding the greenhouse effect and global warming. Understanding the role of feedbacks in the climate system is essential for predicting the environmental effects of anthropogenic activity. For example, photosynthesis counteracts anthropogenic carbon dioxide emissions. As we pump increasing amounts of carbon dioxide into the atmosphere and temperature rises, the Earth acts more like a greenhouse and plants grow faster. Increased plant growth removes more carbon dioxide from the atmosphere and stores it in plant tissue (Equation 2.1), acting as a sink by removing some, but not all, anthropogenic carbon dioxide emissions. Increased atmospheric carbon dioxide concentration also causes more carbon dioxide to dissolve in seawater to form carbonic acid, leading to ocean acidification (Chapter 15). Thus, photosynthesis and carbon dioxide dissolution in seawater counteract the addition of carbon dioxide to the atmosphere, according to **Le Chatlier's principle**. Another negative feedback that is poorly understood involves land (soil) acting as a net carbon dioxide sink, absorbing more carbon dioxide than it releases. These negative feedbacks slow the rate of accumulation of carbon dioxide in the atmosphere. The carbon dioxide content of the atmosphere is still increasing, but not as fast as it would without these negative feedback loops.

Because of anthropogenic activity, the global carbon cycle is in a state of imbalance, that is, it is not in a steady state (Figure 7.4). For the atmosphere stock the total inflow of carbon from land, ocean, and fossil fuels exceeds the outflow by 3 PgC/year, where Pg is petagram = 10^{15} grams. Fossil fuels add 6.5 Pg to the atmosphere each year. Negative feedbacks cause the land and ocean together to take 4–5 more Pg than they give to the atmosphere, so the net C inflow to the atmosphere is 1.5–2.5 Pg/y

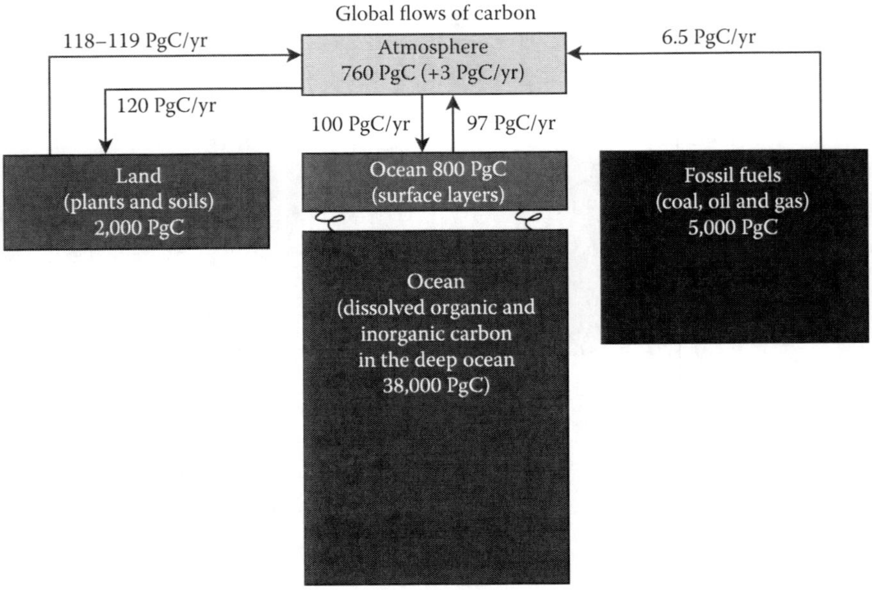

FIGURE 7.4 A simplified stock and flow diagram of the global carbon cycle. (Downloaded from http://science.nasa.gov/earth-science/oceanography/ocean-earth-system/ocean-carbon-cycle/ on September 28, 2015.)

(the figure estimates an annual net flux of 3 Pg/y). If the net flux (inflows—outflows) is 2 Pg/y, and the total amount of carbon in the atmosphere is 760 Pg, then the response time (the amount of time it takes for the concentration in the atmosphere to double) would be 760/2 = 380 years. This example is too simplistic, however, because feedbacks will change the fluxes over time.

In contrast to negative feedback, positive feedback amplifies change. As noted by Brand (2009), "The word 'positive' in the phrase positive feedback does not mean 'good.' It usually means trouble, because a small perturbation can result in big changes…The climate system is nonlinear, which means its output is not always proportional to its input; occasionally, unexpectedly, tiny changes in initial conditions provoke huge responses." One example of a positive feedback is **ice-albedo feedback**: Warming temperatures cause sea ice to melt, so that dark seawater that absorbs sunlight (low albedo) replaces ice that reflects sunlight back into space (high albedo), leading to even more rapid warming (Brand 2009). In a positive feedback loop subsequent changes reinforce the initial change, so the phenomenon is sometimes referred to as reinforcing feedback. Another example of positive feedback in the global climate system is the release of the greenhouse gas methane during melting of Arctic permafrost.

Climate is stable when positive and negative feedbacks cancel each other out. Global climate is delicately balanced between climate thresholds, or tipping points, that separate different climate regimes with different stable states. According to climatologist Jim Hansen, we have already passed the threshold of 350 ppm CO_2 in the atmosphere that could push Earth's climate system into a new regime, but the full effect of our massive addition of carbon dioxide to the atmosphere has not yet been realized. In effect, we are in climate overshoot, and will soon have irreversible and catastrophic climate change if we don't reverse the trend (Hansen et al. 2008). We can still avoid catastrophe if we act soon to bring the concentration down to 350 ppm or less.*

The Greenland and Antarctic ice cores reveal abrupt changes in climate over the last 600,000 years. Deeper back in time, catastrophic climate change during the Paleocene-Eocene Thermal Maximum (PETM) 56 million years ago caused ocean acidification and increased species extinction

* See http://www.350.org/en/about/science.

rates (Zalasiewicz et al. 2010). The PETM may have been caused by a positive feedback in which a warming atmosphere-ocean system caused the decomposition of methane ice stored in shallow marine sediments, and the released methane caused increased warming of the atmosphere.

Accumulation of anthropogenic greenhouse gases in the atmosphere may push the modern climate system toward a tipping point that, once passed, would cause catastrophic and irreversible changes. Many potential tipping points exist in the global climate system. For example, the Western Antarctic Ice Sheet could melt enough to cause it to slide into the ocean and melt, causing ice-albedo feedback and a 5 m (16-foot) rise in sea level. Another possible tipping point involves shutting off the Atlantic conveyor belt, which is part of the global **thermohaline circulation** system in the Earth's oceans (Broecker 1997). This would lead to abrupt cooling in northern Europe and other irreversible changes that are difficult to predict. If we pass a climate tipping point and enter a new climate regime, adaptation will become much more difficult and costly.

According to Lovelock (2006), the positive feedbacks in the Earth's climate system overwhelm the negative feedbacks, meaning "the Earth system is now in positive feedback and is moving ineluctably toward the stable state of one of the past hot climates," which is globally ~5°C higher than now. Once it gets there, negative feedbacks kick in and stabilize global temperature. Lovelock estimates that by 2100 this will cause Earth's carrying capacity to decrease to less than 1 billion people, suggesting that the global human population will decrease by more than 6 billion. Let's hope that Lovelock is wrong.

7.4 CHANGES IN GLOBAL TEMPERATURE OVER TIME

So, we can agree that carbon dioxide is a greenhouse gas, and that human activity has increased the carbon dioxide concentration in the atmosphere. This leads to warming of the atmosphere, which will thermally equilibrate with the land surface and oceans through heat transfer, causing them to warm also. Thus, the entire Earth warms, as shown in Figure 7.1, which plots the deviation from the long-term average global temperature as an anomaly in °C. So, for example, the average global temperature in 2014 was 0.74°C warmer than average, making it the warmest year on record (Mann et al. 2016). Average annual surface temperatures jump up and down from year to year due to random variations called system "noise." Some years are hotter than expected, and some cooler. It is the long-term trend, not the year to year variations, that characterizes climate change. Since 1980, the amount and rate of heating has been higher than over the previous 100 years (Figure 7.1). This acceleration of warming to unnaturally high rates is what has scientists concerned (Richter 2010).

Many modern global changes attest to global warming. Climatologists use globally distributed weather stations that continuously monitor temperature and other parameters; this instrumental record extends back to about 1850. They also use remotely sensed data collected by satellites to estimate temperatures of various layers of the atmosphere. These direct observations show that the Earth's surface has warmed 0.4°C to 0.8°C (~1°F) during the twentieth century. Consistent with these measured changes are observed shrinking and thinning of Arctic ice, loss of Antarctic ice shelf volume (Paolo, Fricker, and Padman 2015), receding of most Alpine glaciers globally,* lengthening of growing seasons, and migration of animals and plants to higher latitudes (Emanuel 2012).

Premodern GCC is characterized by paleoclimatologists using many temperature proxies, which are indirect methods of estimating past temperatures. For example, a tree growth ring can be dated simply by counting the number of rings that grew around it, and the average temperature during the year that a ring grew is estimated from its thickness (Richter 2010). The isotopic composition of layers in ice cores, corals, and cave deposits extend the continuous temperature record much farther into the past than the most ancient trees, as do layers of sediment in lakes and oceans.† Temperatures measured as a function of depth within boreholes can be used to estimate past surface temperatures

* See http://www.ted.com/talks/lang/eng/james_balog_time_lapse_proof_of_extreme_ice_loss.html.

† See http://palaeo.gly.bris.ac.uk/communication/Willson/isotopeevidence.html for a good explanation.

(Pollack 2005). For example, the heat absorbed by the soil in a warm year propagates like a wave very slowly into the Earth. Knowing the rate at which a heat pulse migrates downward allows us to estimate the amount of time required for it to migrate to a specific measured depth. These data show that the average global surface temperature was stable from 1000 A.D. until roughly 1800 A.D., when it began to rise rapidly because of the Industrial Revolution (Mann and Kump 2009). However, climate changes that started near the end of the last ice age ~10,000 years ago likely contributed to population declines and destruction of civilizations in Central and South America, Mesopotamia, and what is now the southwestern United States (Emanuel 2012). Studies have shown that climate changes have destabilized civilizations by decreasing food security, most commonly by drought, and through infectious disease epidemics (McMichael 2012).

Paleoclimate records from 10,000 to 1 million years ago primarily come from ice cores retrieved from thick continental ice sheets. We are currently in the Pleistocene ice age that started about 2.6 million years ago. Since then, the world has seen cycles of glaciation with ice sheets advancing and retreating on 40,000- and 100,000-year time scales. The ice core records show two dominant climate regimes—glacial and interglacial. The Earth is currently in an interglacial known as the Holocene, which started about 10,000 years ago. All that remains of the continental ice sheets are the Greenland and Antarctic ice sheets and several isolated smaller glaciers. These ice sheets have given us the ice core records that have been so useful for reconstructing the history of climate and atmospheric composition over the last 600,000 years (Petit et al. 1999).

Each ice layer in the Greenland ice cores gives us information about the climate in Greenland at the time it formed. The age of the layer can be determined by counting layers from the surface downward, or by using **radiometric dating**. Historical atmospheric temperatures are estimated using oxygen isotopes. Samples of ancient atmosphere trapped as air bubbles can be analyzed to measure the atmospheric concentration of greenhouse gases at the time the snow contained in the layer was deposited.

Plotting temperature measured from the ice cores as a function of time shows peaks and troughs corresponding to ice ages (glacials) and warm periods (interglacials). The timing of these peaks and troughs coincide with cyclical changes in Earth's orbit called **Milankovitch cycles** that change the Earth's tilt and the distribution of sunlight with latitude (Emanuel 2012). At times when the Arctic regions receive less sunlight, less ice melts, more sunlight is reflected, and temperatures decrease. The agreement of the timing of the ice core temperature peaks and troughs with Milankovitch cycles gives credence to the isotopic and dating methods used for analyzing the ice cores. Ice ages occurred roughly every 100,000 years, which is the longest-period Milankovitch cycle where the Earth's orbit shifts from circular to slightly elliptical (Richter 2010). However, Milankovitch cycles alone cannot explain the last century's increases in global temperature (Lean and Rind 2009). Changes in solar insolation likely triggered feedbacks in the climate system that amplified the changes, pushing the climate system over a threshold and causing global climate to shift between glacial and interglacial regimes (Hansen et al. 2013).

Over the last 420,000 years average global temperatures varied from +3°C (+5°F) to −8°C (−14°F) relative to today's average surface temperature, and atmospheric carbon dioxide concentrations varied between 190 and 300 ppm (Petit et al. 1999). Today's atmospheric carbon dioxide concentration of 400 ppm is higher than any time in the last 420,000 years. Since the ice core record shows that temperature closely tracks carbon dioxide, we can expect temperatures to reach levels higher than any recorded in the past 420,000 years.

Unfortunately, the high-resolution ice core records do not cover a time with carbon dioxide concentrations and surface temperatures as high as we are experiencing today. Further increases in carbon dioxide levels and temperature will bring us increasingly outside the range of well-understood climate conditions. Although we don't have as good estimates of global temperatures and atmospheric greenhouse gas concentrations prior to the oldest ice core records, which extend back roughly 800,000 years, we do know that Earth's history included long periods with carbon dioxide levels and temperatures similar to and even higher than we have today. Scientists must use the rock record to learn more about "deep time" climate episodes such as this so they can reduce uncertainty about what will happen as global warming intensifies today.

Of course, anthropogenic greenhouse gas emissions are not the only cause of GCC. Natural causes of GCC include variations in volcanic **aerosol** emissions, sunlight intensity, and **El-Nino Southern Oscillation (ENSO)** intensity. However, measurements show that these natural drivers alone cannot account for the observed increase in global temperatures (Lean and Rind 2009). Since 1980 volcanic aerosols have caused global cooling for a few years after major eruptions like Mount Pinatubo in 1991, but volcanic emissions have not caused global heating. The effect of solar intensity on global temperature is small, and is sometimes in the opposite direction of observed temperature changes (global temperature increases when solar intensity decreases because other climate effects are more important). Furthermore, if the sun were responsible for global warming, then the atmosphere would be uniformly heated from top to bottom. However, heating of the atmosphere has been concentrated in the lowermost layers, consistent with greenhouse gases trapping infrared radiation emitted from the Earth's surface. ENSO events cause temperature highs and lows but cannot explain the long-term increase in global temperature.

Since no combination of the natural climate drivers can explain the observed decades-long trend of increasing global temperature, we conclude that human activities that have increased the concentrations of greenhouse gases in the atmosphere are causing global surface temperatures to increase. According to the most recent report of the Intergovernmental Panel on Climate Change, "It is extremely likely that human influence has been the dominant cause of the observed warming since the mid-20th century" (IPCC 2014). This conclusion is supported by an excellent positive correlation between atmospheric carbon dioxide concentration and average surface temperature from 1880 to the present, consistent with the idea that increased carbon dioxide is associated with increases in temperature. Data from ice cores collected in Antarctica show that this correlation stretches back 420,000 years (Petit et al. 1999). The positive correlation between temperature and atmospheric carbon dioxide concentration suggests, but does not prove, a cause and effect relationship. However, we can say with a high level of confidence that when atmospheric carbon dioxide concentration is high, average global surface temperatures are high.

Given that burning of fossil fuels is causing the Earth to heat up, what can we expect for the future? First we will examine projections of future atmospheric carbon dioxide concentrations and surface temperatures, and then we will explore the potential consequences.

7.5 PROJECTIONS OF FUTURE ATMOSPHERIC CARBON DIOXIDE CONCENTRATIONS AND TEMPERATURES

Estimates of how much carbon dioxide we will emit in the near-future range widely. The actual amounts of greenhouse gas emitted will depend on the future strength of the global economy (greenhouse gas emission rates are higher when the economy is strong) and on the success of international climate mitigation agreements. Accurately estimating future atmospheric concentrations of greenhouse gases is even harder because the concentrations depend not only on emission rates (source terms) but also removal rates (sink terms) that are difficult to quantify. The existence of many recognized, and probably some unrecognized, feedback loops in the carbon cycle complicate the relationship between emission rates and actual atmospheric concentrations. Negative feedbacks such as increasing plant productivity will likely dampen the increases in atmospheric carbon dioxide concentration and surface temperatures resulting from an increase in carbon dioxide emission rates. On the other hand, positive feedback loops could cause the concentration of carbon dioxide in the atmosphere to rise to levels higher than predicted by increased emissions alone. For example, an increase in atmospheric carbon dioxide concentration resulting from an increase in carbon dioxide emission rate would cause global surface temperatures to rise, continental ice sheets to melt and expose permafrost to sunlight, and organic matter in the permafrost to decompose and release additional amounts of the greenhouse gases carbon dioxide and methane.

Unfortunately, the number of recognized positive feedback loops outnumbers the known negative, balancing feedback loops. If the net effect of feedback loops is positive, the climate system is unstable and the risk of a catastrophic temperature increase is high. The presence of many feedback loops in the climate system introduces significant uncertainty into climate scenarios. We need more research to reduce these uncertainties. However, all future scenarios produced using climate models, using a wide range of input parameters and various combinations of feedback loops, show global temperatures increasing over the next century (IPCC 2014).

7.6 PROBLEMS (POTENTIAL CONSEQUENCES)

We have reviewed what we know about GCC based on study of modern and ancient records. We have also explored the more uncertain future projections of carbon dioxide concentrations and temperatures. What effects will future higher temperatures have? This question is very difficult to answer because we have not done this experiment before. Furthermore, the atmosphere and oceans form a very dynamic, complex climate system with countless feedback loops and rapid, nonlinear responses, making it very difficult to predict how climate will change. Estimates of global climate change are improving, but much uncertainty about the effects of increased atmospheric greenhouse gas concentrations remains. However, even if we stopped emitting greenhouse gases now, some of the greenhouse gases we have already emitted will remain in the atmosphere for more than 1,000 years because they have long atmospheric residence times (Archer et al. 2009) and because the oceans will retain for centuries the heat they have already absorbed (Solomon et al. 2009). Thus, increases in temperature and the environmental changes they cause will be irreversible, and we will be forced to adapt to them. Here we will examine potential consequences of GCC and attempt to give some measure of associated uncertainties.

7.6.1 Sea Level Rise

One effect of GCC that scientists agree on is sea level rise. Warming causes ice on land to melt and enter the sea, increasing the amount of water in the world's oceans. This effect is well documented: nearly all of the world's glaciers are receding and thinning. The loss of ice volume on land becomes a gain in seawater volume. Warming also causes seawater to expand, further increasing the volume of seawater. These factors caused sea level to rise faster in the twentieth century than at any time in the previous 2,700 years (Kopp et al. 2016).

The range of temperature increases in the IPCC (2007) future emissions scenarios correspond to a range of sea level increases from 0.75 to 1.9 meters between the years 1990 and 2100 (Vermeer and Rahmstorf 2009). A 3°C increase in global average temperature would raise sea level ~0.8 m = 2.6 ft, causing the global loss of 2,223 km^2 (858 square miles) of land and estimated economic losses of $944 billion (Mann and Kump 2009). Thus, by the end of this century crop-producing river deltas in countries such as Bangladesh and Vietnam will likely be partially inundated (Brown 2011). Many inhabited small islands and some coastal cities like New Orleans will likely be partially or completely abandoned. In coastal New York City mean flood heights have increased more than 1.2 m (3.9 feet) in the last 1,200 years, and what were once 500-year flood events now have a recurrence interval of 24 years (Reed et al. 2015). If the Greenland ice sheet were to melt, sea level would rise roughly 6.7 m = 22 feet, affecting 11 of the 15 largest cities in the world (Emanuel 2012). The resulting environmentally caused human migration would be costly and would stress social support systems.

Inundation is not the only threat presented by sea level rise: salinization of coastal aquifers may have an even greater impact in the short term, and is already a chronic problem in low-lying countries like Bangladesh (see case study in Section 12.5). As sea level rises, salt water advances into coastal aquifers and less dense freshwater floats on the denser salt water. Pumping from wells near the coastline pulls the saltwater toward the surface, eventually contaminating the well water, surface aquifers, and soils. Salinization makes soil useless for agriculture because salt is poison to plants. Thus, **saltwater intrusion** caused by sea level rise and overpumping will displace coastal farmers, who would move inland to compete with other farmers for precious land.

7.6.2 Extreme Weather Events

Another concern is that weather events may become more intense. Higher temperature means a greater amount of stored energy. Gigantic storms can unleash this energy. Warming oceans cause larger and more frequent hurricanes, which are the costliest type of natural disaster in the United States (Emanuel 2012). Flooding, drought, and heat waves are all likely to become more frequent and intense in certain parts of the world. The 2010 extreme heat wave in Russia and record flooding in Pakistan and Australia are examples of the extreme weather events we can expect from GCC, and are signs of instability in the global climate system (Brown 2011). During the Russian heat wave in July to August 2010, the average Moscow July temperature was 8°C (14°F) above normal. The heat wave started many forest fires with economic losses estimated at $300 billion, and the resulting release of pollutants such as particulate matter combined with the high temperatures to cause more than 56,000 deaths. This event was also a global disaster because Russia, a grain exporter, lost 40% of its wheat crop, causing world wheat prices to increase 60% over two months (Brown 2011).

The western United States and California in particular are currently experiencing the worst drought on record. The drought has caused water shortages, dropping groundwater levels, and increased wildfire risk (Diffenbaugh, Swain, and Touma 2015). Droughts are found to be more likely under warm conditions, that is, they do not result solely from precipitation deficits. Furthermore, California droughts have occurred more frequently in the past two decades than in the preceding century. The wildfire season in California is now 2 1/2 months longer than it was several decades ago. Annual U.S. Forest Service expenditures on wildfire management have more than tripled since 1991. The fires burn hotter and can destroy seeds stored in the soil, or even worse destroy the soil by burning all of its stored organic carbon, making it susceptible to erosion. Without soil the forest cannot grow back, at least not until an adequate new layer of soil forms, which can take a few hundred years, or thousands of years in arid climates. Modeling shows that between 1984 and 2015, anthropogenic climate change roughly doubled the area of the western United States affected by forest fires, and this trend will continue as long as fuel (dry wood) remains available (Abatzoglou and Williams 2016).

7.6.3 Reduced Water and Food Security and Loss of Ecosystem Services

Global agricultural productivity will likely decrease in response to rising temperatures. A 1°C increase in temperature during the growing season translates into a global average decrease of 10% in cereal crop yields (Brown 2009). Since average global temperatures may increase 2°C to 3°C by the end of this century, agricultural productivity may decrease 20% to 30%, causing widespread starvation. In some regions, agricultural productivity is already decreasing and will further decrease due to longer, more intense droughts and resulting desertification (Karl, Melillo, and Peterson 2009). Current temperatures have decreased U.S. corn yields roughly 48% (Carleton and Hsiang 2016). Although the WHO and IPCC have maintained that malnutrition will be a significant impact of GCC, not enough robust studies have been conducted to assess the magnitude of this impact or to develop effective adaptation strategies (Phalkey et al. 2015).

Our understanding of the effects of GCC on ecosystems is very limited. In response to warming at all altitudes, bird species in New Guinea moved their elevation ranges up mountains by an average of 95 to 153 meters between the 1960s and 2012 so as to stay within their maximum and minimum temperature limits (Freeman and Class Freeman 2014). Insects are also moving to higher elevations, including malaria-carrying mosquitos, which will affect the millions of New Guinea Highlanders who have lived at elevations above 1,500 m in part to avoid malaria.

Because the rate of GCC is unnaturally high, species will have difficulty adapting. Some probably will not be able to migrate to higher latitudes or higher altitudes fast enough to stay within a livable range of temperatures (Schloss, Nuñez, and Lawler 2012). Others will migrate into ecosystems to which they have not adapted and will therefore fall victim to predators or starvation. Some migrating species may displace other species in ecosystems they enter; the increasingly common problem of invasive species has caused a variety of problems around the globe. GCC will undoubtedly increase species extinction rates (IPCC 2007). The resulting disruption of ecosystems would decrease the flow of ecosystem services to humanity. Lands that are currently marginally inhabitable may become uninhabitable. In those regions, the human toll will surely rise. The World Health Organization (WHO) estimates that climate change already causes more than 150,000 deaths each year (Steffen 2006). This number will increase as population increases and global warming intensifies.

In summary, GCC will affect coastal areas, agriculture, water supply, human health, and many other aspects of society and the natural environment (Karl, Melillo, and Peterson 2009). There will be winners and losers: some areas may become more hospitable, while others will become less so, but overall the negatives seem to outweigh the positives.

7.6.4 Environmental Migration

The social and economic changes caused by climate change are not well understood, but evidence indicates that they are large (Carleton and Hsiang 2016). For example, temperature increases since 1980 are estimated to increase conflict risk in Africa by 11% (Carleton and Hsiang 2016). Climate changes such as increased frequency and intensity of floods and droughts deplete household economic and social capital and reduce the resilience of communities. Communities will try to adapt to these changes through economic diversification, planting of drought-tolerant crops, purchase of crop insurance, and building of flood defense infrastructure such as levees. However, as climate change impacts become more intense, vulnerable populations, mostly in developing countries, will migrate temporarily or permanently.

Only recently has environmental degradation become recognized as a major driving force for human migration. An **environmental migrant** may move in response to sudden or long-term environmental change. Sudden changes include floods, storms, and heat waves, which usually cause temporary, short distance displacement (Renner 2013). Slow changes such as drought, sea level rise, and soil and water salination are more likely to cause permanent migration. For example, residents of several Pacific island nations are planning to abandon, or are already leaving, their disappearing islands. However, migrants rarely move in direct response to environmental changes. Rather, they migrate in response to resulting decreases in their economic, water, and food security. For example, Gray and Mueller (2012) found that crop failures are a stronger migration motivator than flooding. Scientists estimate that 1.4 million to 6.7 million adults will emigrate from Mexico to the United States because of declines in agricultural productivity in response to warming associated with GCC (Feng, Krueger, and Oppenheimer 2010).

Because multiple driving forces are often at work, the term *mixed migration* is increasingly used (Renner 2013). More research is needed to identify the causes of mixed migration so that aid can be provided to affected populations in a timely fashion. Ideally, governments and aid organizations will have enough knowledge to anticipate mass migration and head it off by providing the resources needed to stabilize the well-being of affected communities. Even then, environmental migration can quickly overwhelm the capacity of governments in the developing world to provide assistance. As climate change kicks into high gear, even governments in the developed world will struggle to provide emergency aid and to rebuild devastated communities. The U.S. government had trouble providing security to those affected by Hurricane Katrina in 2005, which caused many residents of affected areas to migrate permanently. In 2015, Europe was struggling to accommodate civil war refugees from Syria, Iraq, and Afghanistan. Multiple disasters, whether natural or human, occurring closely in time may overstress social support systems, resulting in increased migration. Climate change-induced natural disasters and resulting environmental migration may provide the trigger for the social change needed to spur society to address the climate change problem.

7.6.5 Conclusions

Global climate change will have major impacts on the sustainability of socioecological systems around the globe. In some places, the changes will be beneficial. For example, communities at high latitudes may have longer growing seasons and fewer cold-related deaths. However, most studies have concluded that on a global scale and for most countries, positive impacts will be smaller and fewer than negative impacts. Climate-related human health risks will include extreme heat, reduced outdoor air quality, flooding, reduced mental health and well-being, and infection by insects, water, and food (Crimmins et al. 2016). Some populations such as low income and immigrant groups will be more vulnerable to these negative impacts. Besides human health impacts, some of the current major impacts that the most recent IPCC assessment found to be at least 90% likely included (IPCC 2014):

- Negative impacts of climate change on crop yields have been more common than positive impacts.
- Impacts from recent climate-related extremes, such as heat waves, droughts, floods, cyclones, and wildfires, reveal significant vulnerability and exposure of some ecosystems and many human systems to current climate variability.
- Climate-related hazards exacerbate other stressors, often with negative outcomes for livelihoods, especially for people living in poverty.
- Uncertainties about future vulnerability, exposure, and responses of interlinked human and natural systems are large.

In the next chapter we discuss the social dimensions of climate change and potential solutions. We will see that as the debate about global climate change is settled in the scientific community, the focus of research has shifted from whether climate change is occurring, to what is causing climate change, and now to what are the current and anticipated future impacts, and how can we mitigate and adapt to those impacts.

WEB RESOURCES

- Carbon Dioxide Information Analysis Center: http://cdiac.ornl.gov/
- Climate change: http://nas-sites.org/americasclimatechoices/summaries-and-videos-from-the-acc-series/climate-change-lines-of-evidence-videos/
- CO_2 emissions, birth and death rates by country, simulated real-time: http://breathingearth.net/
- Global Warming—Understanding the Forecast: http://forecast.uchicago.edu/
- How to Talk to a Climate Skeptic: Responses to the most common skeptical arguments on global warming: http://www.grist.org/article/series/skeptics/
- RealClimate: Climate Science from Climate Scientists: http://www.realclimate.org/

HOMEWORK PROBLEMS

1. Think about the environment in and around your hometown. How will its three pillars of sustainability be affected by GCC?
2. Use the numbers in Figure 7.4 to calculate the average residence time of carbon in the oceans.
3. Class discussion: Why do so many people refuse to accept that humans are causing GCC?

8 Responses to Global Climate Change

The biggest challenge for humanity may not be to master the intricacies of climate science but rather to answer the much more vexing questions of how political systems operate and why they are so resistant to heeding science's alarm bells.

Michael Renner
Climate Change and Displacements, 2013

8.1 WHY ARE WE NOT ADDRESSING THE CLIMATE CRISIS?

All truth passes through three stages. First, it is ridiculed. Second, it is violently opposed. Third, it is accepted as being self-evident.

Arthur Schopenhauer
German philosopher, 1788–1860

Because climate change occurs over decades and centuries, and because it took scientists decades to develop methods and collect enough data to be confident in their conclusions, it took several decades for the scientific community to reach a consensus on climate change. The vast majority of climate scientists agreed in the late 1990s that global warming is occurring, and then in the early 2000s that warming is primarily human-induced. Polls in 2009 found that 97% to 98% of climate scientists actively engaged in scientific research were convinced of the reality of anthropogenic climate change, and "the relative climate expertise and scientific prominence of the researchers unconvinced of anthropogenic climate change are substantially below that of the convinced researchers" (Anderegg et al. 2010).* More recent studies show that number is now well over 99%.[†] So climate science experts agree that humans are causing GCC.

Despite the scientific consensus, in the United States in 2015 only 63% of Americans believe global warming is happening (Howe et al. 2015). A 2014 survey of citizens from 20 countries found that the United States had, by far, the largest percentage (32%) who denied that climate change is the result of human activity.[‡] While evidence of anthropogenic climate change accumulated, the percentage of Republicans who believed that GCC was occurring actually decreased from 50% to 30% between 2001 and 2010.[§] The Yale Project on Climate Change Communication found that in 2014 Americans could be classified into six groups based on their perception of global warming.[¶] Only 44% fall in the two groups "alarmed" and "concerned," meaning that a minority of Americans consider global warming serious enough to support government action.

So why has public disagreement about GCC persisted? Here we examine why lingering public doubts persist. It's important to understand why public consensus on controversial problems with a scientific component can take so long to develop. The long timeframes that are usually required to

* Also see http://climate.nasa.gov/scientific-consensus/ and EOS v. 90 Number 3, p. 22.
† http://www.msnbc.com/msnbc/how-climate-change-deniers-got-it-very-wrong.
‡ http://ecowatch.com/2014/07/22/americans-lead-world-climate-denial/.
§ http://nymag.com/daily/intelligencer/2015/09/sunniest-climate-change-story-ever-read.html.
¶ "Global Warming's Six Americas' Perceptions of the Health Risks," http://climatecommunication.yale.edu/publications/global-warmings-six-americas-perceptions-of-the-health-risks/, retrieved August 29, 2016.

develop a public consensus can often lead to long and expensive delays in implementing solutions. Scientists have tried for nearly three decades to convince politicians and the public that anthropogenic climate change is real and represents a potentially serious threat to human well-being, and have mostly failed. Many scientists thought that we simply needed more and better data to convince the public, but this strategy has not worked. We need to develop a better understanding of social psychology and use that knowledge to communicate science to the public more effectively, and to give them information that is both actionable and relevant.

Part of the reason there is not a public consensus is that GCC is complex. Furthermore, the culprit is fossil fuel use, and extremely powerful and wealthy business concerns have campaigned against this consensus to protect their profits. This situation closely parallels that of the tobacco companies in the 1970s, who paid lobbyists and scientists large sums of money to spread falsehoods about the link between smoking and cancer* (Oreskes and Conway 2010). Unfortunately, this has led to a politicization of the issue and a polarization of the debate over the reality of GCC. On one side of the debate are climate change deniers, a group mostly driven by opposition to any type of government regulation. The term "denier" is preferred over "skeptic" because healthy skepticism is necessary for good science, but deniers ignore all evidence except that which supports their beliefs, and that evidence usually turns out to be anecdotal (James Hansen 2006). Climate change deniers usually have a strong belief in the ability of a free market to solve all problems. However, as pointed out by Nicolas Stern in the famous "Stern Review," GCC is the largest example of market failure (Stern and Treasury 2007). The market does not place a price on carbon emissions, and does not account for externalities such as economic losses caused by climate change. Theoretically, the free market will set prices that lead to the most efficient use of fossil fuels, but even this doesn't happen because government subsidies to fossil fuel producers results in false price signals that contribute to market failure. We can't rely on the market alone to mitigate GCC.

On the other side of the global warming debate are extreme environmentalists or advocates of **deindustrialization** who often exaggerate the threat of global warming and other environmental problems. They tend to advocate solutions that are impractical from a monetary or societal perspective but that fit their prejudices against possible solutions like nuclear power. Extreme environmentalists will make claims such as "we can solve the energy problem through conservation alone," not mentioning that would require that everyone give up their cars, move into much smaller homes, and grow their own food. Such unrealistic positions on the issues make it harder to reach a public consensus on environmental problems.

People who fall into these two extreme groups usually cannot be reasoned with. No amount of evidence will make them shift their positions, which are irrational. For example, Nobel prize-winning physicist Burton Richter observed that climate change deniers "agree that the greenhouse effect is real, and that greenhouse gases in the atmosphere are the main control on the average temperature of the planet. Why they do not agree that changing the greenhouse gas concentration changes the temperature is beyond me" (Richter 2010).

The politicization of GCC has caused many people to base their opinions and decisions on emotions rather than facts and logic. Unfortunately, slandering individuals or groups who say things you don't want to hear is easier than listening to the messages carefully and building an informed opinion. This explains the popularity of talk shows, and the public reaction to climate scientists' theory of climate change: kill the messenger!

Why do so many Americans refuse to accept the scientific evidence of GCC? The reasons include the following (see Marshall 2014):

- The public's low regard of science caused by
    - Scientist's inability to communicate to the public effectively
    - The perception that scientists are politically biased
    - A shortage of transparency in the science enterprise

* See Buckley, C. (2006). *Thank You For Smoking*, Random House, for an insightful and amusing illustration of how corporations conspire to hide the truth.

- The lack of public understanding of complex issues with a scientific component
- The inability of humans to detect slow changes, which is related to their inability to distinguish climate and weather
- The view that emission limits are an infringement on personal freedom
- The "secret war" waged by fossil fuel companies that stand to lose money if America adopts emissions standards
  - Funding of a handful of climate denial scientists (Oreskes and Conway 2010)
  - Funding of climate denial lobby groups such as the **American Legislative Exchange Council (Alec)***

Even without these biases, persuading Americans to change their lifestyles to stop something they cannot detect with their own senses to prevent an undefined threat in the future is a steep, uphill battle (Marshall 2014).

The spread of misinformation by oil companies has delayed political action on global warming for decades. Biases and misinformation prevent the public from reaching the correct conclusion about GCC. An example of a tactic used by climate deniers to spread misinformation is "cherry picking," which involves selectively choosing the data that supports your argument and ignoring any evidence that doesn't. Let's use Figure 8.1 as a hypothetical example. The data points could represent sea surface temperatures, average atmospheric global temperatures, or annual ice loss.

Points 1–5 show no trend over time. Point 6 has an anomalously high value, meaning that no data points near it in time have similar values. Points 7–19 show continuously increasing values. Points 21–22 seem to fall on the same trend, but point 20 has an anomalously low value that does not fall on the trend. A scientist, statistician, or any reasonably objective person would say that in years 7–22 there is a distinct trend of increasing values over time, with only one year not falling on the trend. Someone trying to convince themselves or others that there is no trend would say that year 20 was no higher than year 6, so there is no significant trend. From year 5 to year 22, the cherry picker has thrown out 16 values and kept only two. Now if these data points were yearly averages of data collected by satellites, each data point may have cost roughly $1 million. Does it make sense to throw out $16 million worth of data without justification and keep only $2 million worth?

In contrast to oil companies, the insurance industry has accepted the reality of GCC and is actively lobbying Congress to enact legislation to fight GCC. They recognize that GCC can increase the frequency and intensity of tropical storms, cause loss of shore-front homes due to a sea level rise, increase the spread of disease, and reduce agricultural productivity, all which represent financial risk to insurers (Pollack 2005). International insurance companies are losing large amounts of money due to the effects of GCC such as increased frequencies and severity of storms (Steffen 2006).

Because insurance companies will be among the first to pay the consequences of climate change, they are less likely to let ideology compromise the accuracy of their risk assessments. Insurance is a hedge against unexpected change, so when unexpected change occurs insurance companies lose. Thus, insurance companies want to reduce uncertainties in their models of the future and prevent increases in the probabilities of catastrophic events by mitigating the risks presented by climate change.

Our federal government needs to invest in GCC mitigation and adaptation measures to reduce potential adverse impacts. We spend trillions of dollars per year for national defense as an insurance policy against external aggression, but we spend almost nothing to insure ourselves against the threats posed by GCC (Pollack 2005). Here, the old adage, "An ounce of prevention is worth a pound of cure" is appropriate: it is usually a lot cheaper to prevent a problem than to deal later with

* http://www.theguardian.com/environment/2015/mar/25/fossil-fuel-firms-are-still-bankrolling-climate-denial-lobby-groups, retrieved March 15, 2016.

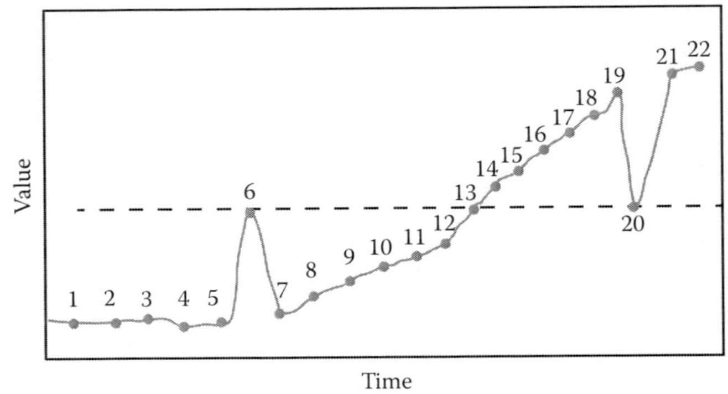

FIGURE 8.1 Hypothetical trend of a measured quantity over time.

its consequences. The sustainability perspective says that we should apply the precautionary principle when making decisions involving global risks with high levels of uncertainty.

The approach that government and many people take to climate change is similar to the approach they take when driving in a lane that is about to close. Prudent people change lanes when they see a warning sign. The probability that they can change lanes without slowing is high because they have much time and therefore opportunities to change lanes. However, some people won't change lanes until they are forced to when their lane ends. Because they didn't use the warning sign, they have only one chance to change lanes. The probability of their being able to change lanes without slowing or stopping is low. They may find it very difficult to change lanes and get back up to speed. They may even get into an accident. By ignoring warnings, they miss most of the opportunities to make change easy. Likewise, if we keep driving down the same path and don't heed the warning signs about GCC, we will miss most of the opportunities to make change easy. As a result, we may be forced to slow drastically (greatly decrease our consumption rates and quality of life) to make the necessary changes.

The lack of an adequate response to the threat of GCC is similar to the response to the early warnings of environmentalists. For example, the authors of the "Limits to Growth" series of books have warned the public every decade about the dangers of ecological overshoot and the potential for societal collapse by the middle of the twenty-first century. Critics responded that we don't need to worry, but as evidence mounted supporting the claims of environmentalists, the critics changed their reasons not to worry. It is the same story with climate change deniers, who have stated:

Late 1990s: GCC is not happening.
Early 2000s: GCC is happening, but humans are not responsible.
Late 2000s: Humans are responsible, but GCC is not a serious problem.
Early 2010s: GCC is a serious problem, but now it's too late/too expensive to do anything about it.

People who say that we don't have to worry about making changes to the Earth and its atmosphere, or that the changes we make may even prove beneficial, should think of this analogy: the Earth is a complex system that we don't understand. Making changes to it without knowing the consequences is like an untrained mechanic bashing the working engine of a flawless Ferrari with a wrench in hopes of improving its performance. The Ferrari is a complex system of working parts, and almost any change will have deleterious effects. In fact, breaking one part of the engine can lead to other parts breaking down if it is kept operating (and we can't stop the Earth system from operating to repair it). Our tweaking of the much more complex Earth system, with its many working connected parts, could lead to the failure of individual parts or complete subsystems (atmospheric or oceanic circulation patterns, ecosystems, etc.). Again, the precautionary principle states that we would be unwise to make global-scale changes such as changing the concentration of greenhouse gases in the atmosphere without having any idea of what the consequences will be. As Donald Rumsfeld said, "There are the known knowns, the known unknowns, and the unknown unknowns." For global climate, we know there are known unknowns, and almost certainly there are unknown unknowns.

Two strategies help when trying to choose the correct side from two competing schools of thought. First, which side is supported by experts in the field? In the case of GCC, this argument clearly favors the proponents of anthropogenic climate change. A few meteorologists are climate change deniers, but they are not climate scientists who keep up with the cutting edge research in climate science; rather, they are simply people who report the daily weather, and in most cases they know little about climate change. Another strategy is to select the side that has less motivation to be biased. Here climate scientists win again over climate change deniers: each climate scientist has little or nothing to gain from agreeing with the 99% of climate scientists who agree that humans are causing climate change. In fact, if they want to be in the spotlight or make money by selling books or by touring the speaking circuit, they would choose the denialist side. The popular book *Freakonomics* posits that individuals often benefit by choosing to be in the minority; for example, this has been well-documented for stock trading. Not surprisingly, the co-author of *Freakonomics* and author of *SuperFreakonomics*, Stephen Dubner, is a climate change denier: it helps him sell more books. So deniers have an economic incentive to deny that humans are causing GCC. Exxon's own researchers warned in 1978 that their products were causing atmospheric CO_2 levels to rise, which would cause global temperatures to increase 2°C to 3°C, but Exxon executives chose to suppress this information and continue to fund climate change deniers so that they could continue to reap profits.* Climate scientists could almost certainly earn more money from oil companies and conservative think tanks by denying that climate change is happening, and several scientists and their organizations have done so.†

Scientists are by nature skeptical. They demand evidence, and continuously attempt to disprove theories. Those who are successful in doing so gain the respect of the scientific community, which provides a powerful incentive. Theories that survive many attempts at falsifying them are gradually accepted by the skeptics, and a consensus emerges, which has happened in the case of anthropogenic global climate change. Theories that gain consensus in the scientific community are rarely falsified, but the desire to achieve fame in the scientific community will always motivate scientists to continue attempts at falsifying those theories. Occasionally a scientific revolution occurs when a foundational theory is falsified and a new theory is accepted in its place (Kuhn 1970). For example, the Plate Tectonic Revolution occurred in the 1960s, and it profoundly changed the geosciences. However, it's important to note that the theory that preceded plate tectonics gradually unraveled, as it did not explain a growing number of measurements. Currently a large number of different types of data support the theory of anthropogenic climate change, so it seems highly unlikely that this theory will be overturned in the scientific community, despite the strong incentive to disprove it.

Climate change deniers are also motivated by political ideology. Libertarians are almost universally climate change deniers; their view is based not on scientific data but on the fear that government will expand its scope and power in order to regulate carbon emissions. Our conclusion is clear: because climate scientists are the experts on global climate change, and because they have no financial or ideological incentives to argue for human influence on global climate, they are much more likely to be correct than climate change deniers. Put another way, increasing the amount of

* http://www.pbs.org/wgbh/frontline/article/investigation-finds-exxon-ignored-its-own-early-climate-change-warnings/, retrieved March 15, 2016.

† https://www.washingtonpost.com/news/morning-mix/wp/2015/02/23/the-favorite-scientist-of-climate-change-deniers-is-under-fire-for-taking-oil-money/, retrieved August 23, 2016. http://www.ucsusa.org/global_warming/solutions/fight-misinformation/global-warming-skeptic.html#.V7yPKSgrJN0, retrieved August 23, 2016.

information and removing biases that cloud objectivity increases the likelihood of drawing an accurate conclusion. Most climate scientists have used this objective approach to arrive at the conclusion that humans are causing GCC.

No amount of scientific evidence will convince active climate change deniers that climate change is occurring. An important study concluded that "public divisions over climate change stem not from the public's incomprehension of science but from a distinctive conflict of interest: between the personal interest individuals have in forming beliefs in line with those held by others with whom they share close ties and the collective one they all share in making use of the best available science to promote common welfare (Kahan et al. 2012)." Recent studies suggest that the only way to persuade climate change deniers to adopt climate change mitigation measures such as cap and trade is to convince them that even if climate change claims are false, mitigation measures will have a positive effect on social welfare such as greater technological and economic development (Bain et al. 2012). While this may treat the symptoms of anti-science denialism, it doesn't address the causes. What can cure the disease? Effective teaching of evidence-based science in public schools is a start, but society needs to find new ways to stem the spread of anti-science denialism, which is making our country less competitive and harming future generations.

8.2 THE SOCIAL PSYCHOLOGY OF GCC

In recent years, experiments and studies in social psychology have shown that, when it comes to complex issues such as climate change, people tend to think with their emotional brain rather than their rational brain. No amount of evidence will convince the rational brain to overrule the emotional brain. In fact, showing the evidence of climate change to skeptics only makes them more convinced they are right (Feinberg and Willer 2011). Many arguments presented below follow Marshall (2014), who provides an excellent overview.

Because GCC is such a complex topic, we tend to rely on stories to simplify the issue and shape our opinions. Effective stories have similar structures: they involve a perpetrator, an effect, and a motive. For environmentalists, the compelling story is that oil companies or right-wing oil billionaires such as the Koch brothers (perpetrators) fund efforts to spread misinformation (effect) to increase their wealth (motive). For climate deniers, the persuasive stories are that governments justify carbon taxes in order to extend their control over our lives, or that rogue scientists are conspiring to fake evidence in order to secure larger research grants.

Let's examine the latter story in detail. It was popularized in Michael Crichton's book *State of Fear*. First, the word "rogue": Is it possible that the 99% of climate scientists who accept anthropogenic climate change are all deceitful and unprincipled? If so, their stated objective is to "secure larger research grants." To do what? Manufacture more fake evidence? What do they accomplish by doing that? They don't benefit financially. And let's be clear: smart people don't choose to become scientists in order to make money; they would make far more with less effort by going into business or finance. Most scientists work very hard to learn the secrets of the natural world. Their mission is to weed out the false stories or hypotheses until only the true stories remain; they are dedicated to scientific truth, and the search for truth is the core of their identity. To throw away their integrity by manufacturing lies would be to betray themselves. Yes, in the past a small percentage of scientists have falsified data to gain fame or collect money from advocacy groups, and today some scientists who are climate change deniers are spreading lies to promote their political ideologies (Oreskes and Conway 2010), just as a handful of scientists claimed for decades that tobacco does not cause lung cancer. But the fact remains that of the scientists who actually study climate, more than 99% accept that humans are causing climate change. Stories like *State of Fear* support deniers' preconceived notions about government and government scientists; **confirmation bias** leads them to ignore contradictory facts and to cherry pick data to find evidence that they believe supports their story.

Climate change is the greatest threat to global sustainability now and in the future. How can we achieve sustainability, how can we preserve our economic, social, and environmental capital if we ignore the risks? What are our chances of survival as a human race if we cannot accept the reality of a global security threat, if we think only with our emotional brain and not our rational brain?

To be fair, global climate change may be the most difficult problem to solve that the human race has ever faced, partly because evolution has poorly prepared us to deal with it. Here are some of the problems social psychologists have identified that prevent people from recognizing GCC as a serious problem (mostly after Marshall 2014):

- **GCC is a wicked problem.** It is multifaceted in every respect. Our understanding of GCC is always incomplete and constantly changing. This requires people to constantly reevaluate the evidence and their position on the issue. Tame problems can be solved using standard methods: Collect data, analyze it, and then propose and implement a solution. This approach doesn't always work for wicked problems.
- **GCC is multivalent.** It has many meanings, and lacks a distinct set of defining qualities, because it lacks a single cause, solution, and geographic location. We can call it a problem of ethics, technology, governance, economics, land use, social justice, or simply an ideological battle between left and right worldviews (which is what it has become in the United States). Historically, each problem type has a different approach to finding a solution. Finally, there is no defined enemy we can use to frame the battle and rally against. These qualities make analysis of the problem susceptible to confirmation bias and miscategorization, leading people to choose the easiest option, "believe what they want to believe" and do nothing.
- **GCC lacks salience.** The threats presented by GCC are not concrete, immediate, and indisputable. It does not demand our immediate attention because it is abstract, distant, and disputed.
- **Cognitive biases work against acceptance of GCC.** Cognitive biases are mental shortcuts that are useful for simple decisions but not for interpreting complex problems like GCC. Our perception of the GCC problem is biased because we favor certainty over uncertainty and are more sensitive to short-term costs than long-term costs (Kahneman 2011). Environmentalists emphasize the long-term costs of inaction, while climate change deniers use the more effective approach of highlighting the short-term costs and deemphasizing the long-term costs. In general, people tend to avoid short-term costs and act in their own self-interest. Cost-benefit analyses rarely persuade people to change their minds on the issue. Marshall (2014) sums up the problem by stating, "the more important question in trying to understand why people are so unwilling to accept climate change is whether humans as a whole are innately disposed to disregard any threat that requires sustained payments in order to avoid greater, but less certain, long-term losses."
- **People's capacity for worrying about problems is limited and rationed.** Other concerns get higher priority and cause GCC to get less of our attention. This causes the media to give it less attention, which reinforces our inattention, resulting in a positive feedback loop. In general, issues of immediate personal concern like the economy and terrorism receive the highest priority; GCC is neither immediate nor personal.
- **People feel powerless to do anything about GCC.** Reading this book will hopefully persuade you that this is wrong: Through bottom-up efforts we can reduce the carbon footprints of individuals, and through social movements we can instigate top-down solutions.
- **Discussion of GCC carries a social stigma.** Because the mention of GCC invokes worry and guilt, it has become socially unacceptable to bring up GCC in social conversations. Try doing it in a group of people: you will find that many will immediately move to another group, or will try to change the topic of conversation. Very few people will actually engage in a discussion about climate change. To avoid anxiety, people use innate defense mechanisms that include ignoring, denying, or disavowing (i.e., the active choice to not notice). There is also anxiety that our views may threaten our standing in social groups of which

we are members. Conservatives in the United States feel, either consciously or subconsciously, that they will be ostracized by their social groups if they admit that GCC is real. Social pressures keep group members in line; to oppose their social group puts their self-identify at risk.

- **People underestimate the risks of rare events.** People tend to underestimate the probabilities of rare events, especially if they have never experienced one. A person who has lived through an earthquake will likely judge a future earthquake more likely than a person who has not. This is one reason why people tend to underestimate the risks posed by long-term environmental changes. As observed by Kahneman (2011), "When it comes to rare events, our mind is not designed to get things quite right. For the residents of a planet that may be exposed to events no one has yet experienced, this is not good news."

In summary, humans and human societies are poorly equipped to handle the GCC problem. To mitigate the problem effectively, we must overcome a suite of cognitive biases and a sense of powerlessness and make GCC a higher priority. Your sense of powerlessness may be partially alleviated if you consider that humanity joined together to solve the ozone hole problem (see Section 8.3.4). Furthermore, progress has been made on a host of social issues involving race, child abuse, and homosexuality. Society passed a tipping point on each of these issues; we may be on the verge of a tipping point that will lead to a public consensus on the reality of anthropogenic GCC and the critical need to address it now. In the developed world we have all of the resources we need to solve the problem: wealth, education, renewable energy and energy efficiency technologies, and international cooperation during a time when most of the world is at peace.

8.3 SOLUTIONS

We can evade reality, but we cannot evade the consequences of evading reality.

Ayn Rand

The lack of agreement on cause and effects of GCC has slowed society's response to a potentially terrible threat. Despite having knowledge of this threat for decades, we have not stopped its cause—emission of carbon dioxide during burning of fossil fuels. Instead we have increased the rate of carbon dioxide emissions (Figure 8.2). Lack of action on climate mitigation has reduced our security and the security of future generations.

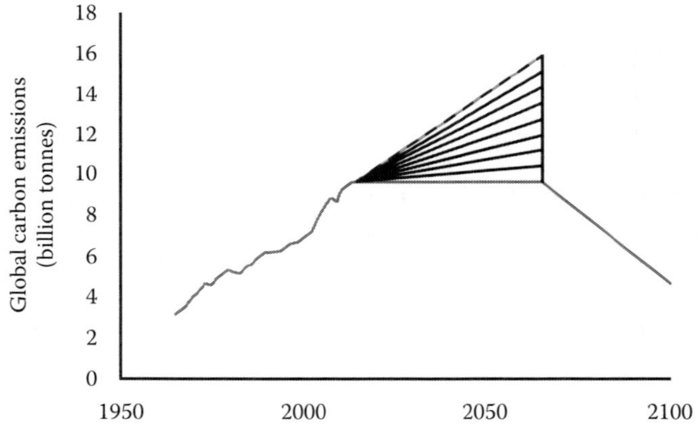

FIGURE 8.2 The carbon emission stabilization wedges of Pacala and Socolow (2004). The stabilization triangle is comprised of eight stabilization wedges. Global CO_2 emissions 1965–2014 from BP (2015). The dashed line is projected "business as usual" emissions.

Public opposition to the scientific consensus on GCC is rooted in the fear that mitigation measures will be costly and therefore harmful to the economy. Sustainability thinking is useful for putting this potential roadblock into perspective. Sustainability requires that we meet present needs without compromising the ability of future generations to meet their own needs, and that we do so by preserving economic, social, and environmental capital. Let's examine whether we currently meet these requirements on the issue of GCC.

Humanity's carbon footprint makes up about half the world's ecological footprint (Figure 3.4). Our carbon emissions in the form of carbon dioxide and methane are disrupting climate and damaging ecosystem services. Now we are approaching what scientists believe are dangerously high levels of carbon in the atmosphere. We must consider from a sustainability perspective the potential impacts to future generations before we choose to continue emitting carbon. The sooner we act to reduce carbon emissions, the less severe the consequences will be, and the less expensive the solutions will be.

Because of GCC, global temperature is changing faster than at any time in the Earth's past. If continued, changes in global temperature will outpace the ability of plants and animals to adapt by migration and evolution. GCC is altering global ecosystems and causing increasing rates of species extinctions (Chapter 15). The resulting loss of species and ecosystem services will make it harder for future generations to meet their needs. Sustainability thinking suggests that we must act to preserve environmental capital and protect ecosystem services.

GCC is putting millions of the world's most vulnerable citizens at risk. Residents of low-lying coastal communities face having to abandon their ancestral homes due to sea level rise and associated groundwater salinization. Humans could adapt through use of technology. However, most cannot afford the costs of these technologies, so the death rate in affected undeveloped countries will likely increase. Americans can afford air conditioners, imported food, bottled water, and seawalls. This is a great injustice of global warming: those most responsible for global warming (e.g., U.S. citizens) are likely to suffer the least from it. Distributional equity requires that we preserve social capital by reducing these risks and compensating those affected.

Finally, GCC impacts will be felt most strongly by future generations. Intergenerational equity compels us to combat GCC to protect the interests of our descendants. We may be making many areas of the Earth uninhabitable for our offspring. Almost certainly, life will be more difficult for the next generation. We will burden them with the consequences of GCC, an enormous financial debt (witness the exploding budget deficits of many governments), and shortages in key resources such as oil. The current generation must look for ways to soften the blow to our offspring from our actions and decisions. Most parents make sacrifices for their children's current welfare. Truly responsible parents also make sacrifices for their children's future welfare (e.g., saving money for them to go to college). We must now make other kinds of sacrifices, ones that will make our lifestyles more sustainable and therefore easier for our children to maintain in the future.

We must act to preserve social and environmental capital and protect present and future citizens of the Earth by taking drastic steps to mitigate GCC, *even if it means losing some economic capital in the short term*. Such short-term losses could preserve or even grow economic capital in the long term. Even in the short term, many steps that we can take to mitigate GCC can save money. Clearly these are the steps we must first focus on.

It doesn't matter whether humans caused global warming; we still have to deal with the impacts. To reduce future climate impacts, we will need to use three primary approaches: mitigation, regulation, and adaptation. **Climate change mitigation** approaches reduce carbon emissions by reducing use of fossil fuels through conservation, efficiency, use of low carbon energy, and **carbon capture and storage** (Figure 8.3). The regulation approach to reducing greenhouse gas emissions usually calls for market-based policy instruments such as **emissions trading** (cap and trade) and **carbon taxes**. Adaptation approaches include migration and **climate engineering**, which involves active intervention in the climate system using **greenhouse gas removal** or **solar radiation management (SRM)** to reflect sunlight into space. Let's examine these options in more detail.

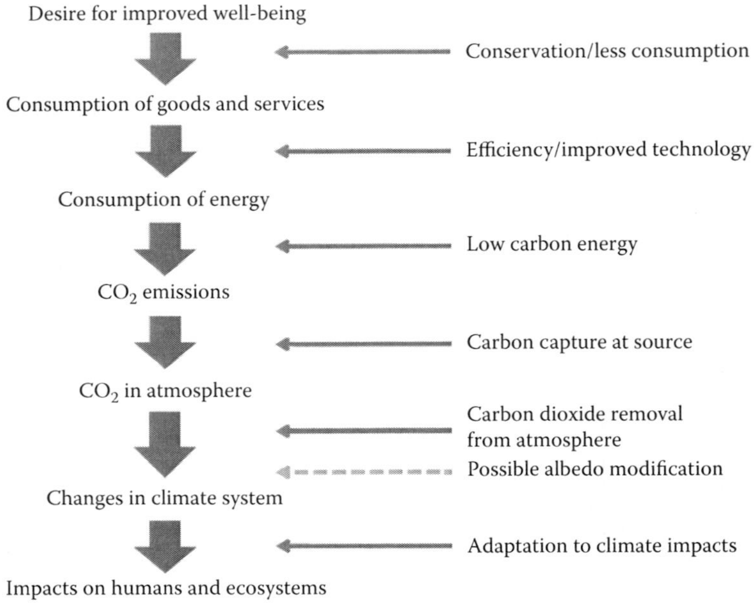

FIGURE 8.3 Solutions for mitigating and adapting to global climate change. (Reprinted with permission from the National Academies Press, Copyright 2015, National Academy of Sciences (U.S.). 2015a. "Climate Intervention: Carbon Dioxide Removal and Reliable Sequestration." doi:10.17226/18805.)

8.3.1 Mitigation

We are already actively intervening in the Earth's climate system. At first our intervention was unknowing, but now we know that we are adding greenhouse gases to the atmosphere, and that we must regulate the composition of the atmosphere so that Earth's climate will stay within the narrow range of conditions that have allowed us to prosper as a species. Throughout its 4.5-billion-year history, Earth was inhospitable to us more often than not, so we must make every effort to keep it hospitable by stopping GCC.

Scientists believe that a global atmospheric mean temperature increase of more than 2°C to 3°C above preindustrial levels may greatly increase risks to human health and the economy (IPCC 2007). To prevent it from rising above this level would require not only keeping atmospheric CO_2 from increasing above its 2016 level of 404 ppm, but decreasing it to a maximum of 350 ppm, which means we must start acting *now* (Hansen et al. 2008). Recent work has shown that to limit global warming to a 2°C rise would require that we leave much of our global fossil fuel reserves in the ground, including one-third of the oil, one-half of the natural gas, and over 80% of the coal (McGlade and Ekins 2015).

Mitigation steps taken early have a greater effect in reducing climate change than comparable reductions made later (Karl, Melillo, and Peterson 2009). Economist Nicolas Stern estimated the cost of mitigating climate change to be ~2% of Gross World Product,* corresponding to a total cost of $1.3 trillion or $210 per capita in 2008. Stern warned that this was much less than the cost of inaction. Reducing emission of greenhouse gases other than carbon dioxide will help, but carbon dioxide has the largest climate impact of the greenhouse gases (Figure 7.3), so we will focus on mitigation of carbon dioxide emissions by reducing carbon sources and increasing carbon sinks.

* Jowit and Wintour, *The Guardian*, June 26, 2008, http://www.guardian.co.uk/environment/2008/jun/26/climatechange.scienceofclimatechange.

8.3.1.1 Reducing Carbon Sources

Carbon Capture and Storage (CCS; the "S" can also stand for "Sequestration") aims to eliminate the release of the greenhouse gas carbon dioxide from the smokestacks of fossil fuel burning power plants. The United States plans to use CCS to make power plants "clean" by capturing and storing all of the carbon dioxide underground (more on this in Section 9.3.3). CCS holds promise for mitigating anthropogenic climate change. However, it will make fossil fuel-produced electricity more expensive and take decades to set up on a large scale.

CCS aims to immobilize carbon dioxide deep in the Earth, isolating it from the atmosphere. The carbon comes from fossil fuels that we extract from the Earth. CCS closes the material cycle by returning the carbon to the Earth's interior. This can be accomplished by three types of carbon storage (Oelkers and Cole 2008). Geological storage involves injecting carbon dioxide into spent petroleum reservoirs or saline aquifers where it dissolves in or displaces the brines. Ocean storage requires injecting carbon dioxide into the ocean at depths greater than 3.5 km where it can dissolve in seawater, form liquid carbon dioxide lakes on the seafloor, or form a slush of clathrates (carbon dioxide hydrates). However, carbon dioxide can react with water to form carbonic acid. If it would exacerbate the ocean acidification problem, then ocean storage would be unacceptable. Mineral storage relies on carbon dioxide reacting with unstable silicate minerals in magnesium-rich rocks to form carbonate minerals (Broecker 2008). Certain rock types can rapidly react with atmospheric carbon dioxide to form carbonate minerals for long-term storage (Kelemen and Matter 2008). Estimates of geological carbon dioxide storage capacity in the United States suggest that we can use it to maintain emissions at current levels for at least 100 years (Szulczewski et al. 2012). However, we have shown that current emission levels are already too high.

What are the most effective mitigation options for reducing carbon dioxide emissions? Energy use is responsible for 70% of greenhouse gas emissions (Richter 2010). To reduce energy-related greenhouse gas emissions we must reduce the use of fossil fuels, particularly coal, that emit large amounts of carbon dioxide per unit energy. Old, polluting coal-fired power plants should either be retrofitted for CCS or replaced with power sources that emit little carbon dioxide such as wind, solar, hydro, geothermal, and possibly nuclear (Chapters 9 and 10). Second, we must reduce overall energy use through conservation and increased efficiency, not just in homes, where energy use accounts for 17% of greenhouse gas emissions in the United States, but also in industry (30%), transportation (28%), commercial (17%), and agriculture (8%). Third, we must discourage deforestation, as it changes a carbon sink into a carbon source and is responsible for ~30% of global greenhouse gas emissions (Richter 2010).

Socolow and Pacala (2004) devised a popular graphical approach to formulating policies for carbon mitigation.* Their plan aims to keep atmospheric carbon dioxide from doubling in concentration from the preindustrial level of 280 ppm to 560 ppm (in 2016 the concentration reached 404 ppm†). It would hold carbon emissions at the current level of 7 billion ton (gigatons) of carbon per year (Gt C/y) for the next 50 years and stabilize atmospheric carbon dioxide concentration at 500 ppm. Socolow and Pacala (2004) divided the growing gap between future "business as usual" (BAU) emissions and the desired flat emissions trajectory (7 Gt C/y) into stabilization wedges (Figure 8.2). They offered 15 possible mitigation steps or "wedges" that use existing technologies to fill the gap. Examples include doubling the current global nuclear capacity to replace coal-based electricity, increasing the fuel economy for 2 billion cars from 30 to 60 mpg (increased efficiency), and decreasing the number of car miles traveled by half (conservation). Each emission reduction wedge starts at zero in 2005 and increases linearly until it reaches 1 Gt C/y of reduced carbon emissions in 2055. We need at least eight wedges to achieve the desired reduction in carbon emissions (Friedman 2008). Adding those wedges, we obtain a "stabilization triangle," located between the desired flat trajectory and BAU, which removes exactly one-third of BAU emissions between 2005

---

* See http://cmi.princeton.edu/wedges/intro.php for more information.

† You can obtain the current concentration from the web page http://www.esrl.noaa.gov/gmd/ccgg/trends/global.html.

and 2055 and one-half of BAU emissions in the year 2055. The reduction in carbon emissions must add up to at least 200 billion tons by mid-century. We would use new technologies in the second half of the twenty-first century to decrease net emissions, eventually to zero. If we delay action and continue BAU, by the year 2055 annual carbon emissions will double and atmospheric carbon dioxide concentration will be triple the preindustrial value.

The study of Socolow and Pacala (2004) did not estimate the cost of each carbon emission stabilization wedge. Studies have shown that, while some greenhouse gas abatement options are expensive, others have negative costs, meaning we can save money by adopting them. The most recent version of their **marginal abatement cost** curve shows that 35% of identified greenhouse gas abatement strategies have negative costs.* For example, switching residential lighting from incandescent or halogen to compact fluorescent light (CFL) or light emitting diode (LED) bulbs has large negative costs, meaning that by switching we can save money while reducing greenhouse gas emissions. Furthermore, it's relatively easy and inexpensive to switch bulbs, making this a good example of the "low hanging fruit" that our society must focus on first if we are serious about reducing greenhouse gas emissions (Dietz et al. 2009). A total of 75% of abatement strategies cost less than €20 ($27 USD) per ton CO_2e (Enkvist, Dinkel, and Lin 2010). This is close to the Obama administration's current (and very low) estimate of the **social cost** of carbon,† which is the **marginal cost** of emitting 1 ton of CO_2e. Over time the social cost of carbon will rise as temperatures rise, and even more abatement strategies will become profitable when considering social costs. Since 35% of abatement strategies are already economically profitable, the United States has no excuse to continue dragging its feet on climate change mitigation measures. We should adopt abatement strategies with cost savings quickly to reduce climate change risks and save money. McKinsey and Company estimate that it would take only 0.6% of the world's GDP to reduce greenhouse gas emissions so that CO_2e never rises above 450 ppm (Creyts 2007).

Some propose that GCC mitigation measures should first focus on reducing carbon dioxide emissions. However, there are several reasons for concentrating on reducing the emissions of greenhouse gases with short atmospheric removal times such as methane: stop emitting them and their atmospheric concentrations drop sharply. Furthermore, methane is a powerful greenhouse gas, and it acts as a catalyst to produce ground level ozone, a pollutant that damages people's lungs and crops. Finally, CH_4 gets converted to CO_2 in the atmosphere, which lengthens the impact of CH_4 emissions because CO_2 has a longer residence time in the atmosphere (5 to 200 years, compared with 12 years for CH_4, [IPCC 2007]). Using current technologies to prevent methane and soot emissions, we could reduce global warming by about 0.5°C by 2050 (Shindell et al. 2012).

The most effective climate change mitigation approaches that reduce greenhouse gas sources include reversing the trend of deforestation, increasing energy efficiency, and transitioning from conventional fossil fuel power plants to low greenhouse gas emission power plants that use either renewable energy sources (Chapter 10) or fossil fuels coupled with CCS (Chapter 9). The United States is currently phasing out old, inefficient coal-burning power plants. However, decreasing coal use in one country may lead to increased coal consumption in other countries because coal is a globally traded commodity. When countries choose to reduce emissions by reducing coal consumption, the reduced demand causes prices to drop, incentivizing other countries to increase their coal consumption (Steckel, Edenhofer, and Jakob 2015). When U.S. coal consumption decreased, domestic coal producers increased exports. Developing countries are capitalizing on low coal prices by building new coal-fired power plants, but this locks them into coal use for decades, and growth in coal use is the primary cause of continuing increases in global CO_2 emissions. Global coal consumption will not decrease until coal-produced electricity becomes more expensive than other forms of electricity. One way to increase the cost is to eliminate fossil fuel subsidies. Spending the savings

* http://www.mckinsey.com/business-functions/sustainability-and-resource-productivity/our-insights/impact-of-the-financial-crisis-on-carbon-economics-version-21, retrieved August 29, 2016.

† http://www.grist.org/article/2010-04-23-what-is-the-social-cost-of-carbon/.

on poverty alleviation could improve human well-being in multiple ways (Steckel, Edenhofer, and Jakob 2015).

In addressing climate change mitigation, we should focus on "win-win" choices. Why not invest in an energy efficiency technology that reduces greenhouse gas emissions if it also saves money in the long term? It's a good economic decision even if you are not concerned about greenhouse gases. And if you think that scientist's claims that global warming is occurring and that it is caused by human activity are false, you have to admit that there is a tiny chance they may be true; we cannot be 100% sure of their truth or falsehood. You can then view the investment as insurance against the small probability that GCC may cause you harm.

Individuals can take many steps to reduce their carbon emissions and mitigate GCC (more on this in Chapter 11). The following approaches reduce carbon emissions through conservation and increased efficiency:

- Use mass transit, bike, walk, or roller skate instead of driving a car.
- Tune up your furnace.
- Caulk, weatherstrip, insulate, and replace old windows in your home.
- Buy appliances with a U.S. EPA Energy Star label.

We will explore more energy solutions in Chapter 11.

8.3.2 Regulation

8.3.2.1 Domestic Regulation

In the past, governments relied on strict regulations focused on reducing point-source pollution. These regulations were often costly to enforce. The favored approach now is to use the carrot rather than the stick: Economic incentives are used to guide people and industries to desired outcomes. These market-based policies make compliance economically desirable by assigning a price to carbon emissions that accounts for the resulting negative externalities. The two principal **carbon pricing** policy instruments for mitigating greenhouse gas emissions are **carbon taxes** and **emissions trading** (cap and trade) (Dawson and Spannagle 2009). A carbon tax is a **Pigovian tax** because it adds the negative externalities of carbon fuel use by making the economic cost equal to the true cost as estimated using LCA + FCA, making the market for carbon fuels more efficient. Many countries in Europe have reduced carbon emissions by imposing carbon taxes.

Although carbon taxes would be an effective approach to reducing carbon emissions, most U.S. citizens are opposed to increasing taxes on gasoline and other fossil fuels. Emissions trading is now the approach that is favored over traditional regulations and taxes (Dawson and Spannagle 2009). It provides flexibility, allowing the market to decide the most cost-effective approaches to reducing emissions, and therefore achieves emissions reductions at lower cost than the regulatory approach (Jaffe, Newell, and Stavins 2005).

Emission trading involves setting an upper limit to emissions (the cap), and then giving emitters tradable emissions allowances. Traders buy and sell allowances as commodities, and the commodity market determines the price of an emission allowance. Those who emit more must purchase emission allowances from those who emit less, providing a financial incentive to reduce emissions. This approach was first shown to work effectively in the 1990s by the U.S. Sulfur Dioxide (SO_2) allowance trading system. The **European Union Emissions Trading Scheme** began operating in 2004, and is now part of a global carbon-trading market.

In an emissions trading scheme, the government sets the allowable emissions amount, and the market sets the emissions cost. In contrast, for an emission tax system, the government sets the emissions cost and the market determines the emissions amount. Since the objective of greenhouse gas emission regulations is to hold atmospheric carbon dioxide concentration to below 450 ppm, which sets an upper limit to carbon dioxide emissions, an emission trading scheme is preferable. It is also preferable because emission trading is more effective when regulating a few large emitters (power plants). Carbon taxes work better for many small emitters, which makes imposing high gas taxes a good approach to limiting carbon dioxide emissions from automobiles.

Another promising, commons-based approach to reducing greenhouse gas emissions is the **cap and dividend** approach of Peter Barnes (Barnes 2006; Walljasper 2010). The atmosphere is a commons, a natural resource we all share and depend on for life. Currently corporations effectively own the sky: they use it as a dumping ground for their wastes. A commons approach would give ownership to all citizens, and polluters would have to compensate citizens by paying dividends for use of their resource. Government could limit the total amount of carbon emitted to the atmosphere annually by setting a cap and requiring polluters to purchase permits. Over time, government would reduce the cap and therefore the number of permits issued. This would cause the price of permits to rise, resulting in higher dividends paid to each citizen. This in turn would protect the purchasing power of consumers if carbon emitters raise the prices of their products to offset the costs of their permits. This system is simple and transparent, and because it would result in higher carbon prices, it gives consumers a stronger incentive to reduce fossil fuel consumption. It would cost more for ~30% of Americans with the highest incomes who use the most energy, while ~70% would come out ahead, with the poor benefitting the most. A different approach called **Cap and Share** was devised by **FEASTA**, the Foundation for the Economics of Sustainability. In this system scientists set a cap, the amount of greenhouse gases that can be safely emitted. The government shares out emission permits to citizens, who can then sell their permits at post offices and banks. Fossil fuel companies buy permits to cover their emissions. Finally, inspectors enforce the cap by ensuring that emission permits match the emissions.

Another option is to tax fossil fuel producers rather than consumers. Generally, fossil fuel sytems have open material cycles, and regulations and taxes are targeted at the end of the cycle, the "tailpipe" where greenhouse gases are emitted. An alternative approach is to tax at the start of the material cycle where fossil fuel extraction occurs. This approach would reduce the profit incentive for fossil fuel extraction by making producers pay for the externalities introduced by fossil fuel use. Costs would be passed on to all energy consumers, and the cost would be proportional to the amount of energy consumed. Consumers would pay the true cost of fossil fuel consumption, not a price made artificially low through subsidies and not accounting for externalities. The incentives for reducing fossil fuel consumption would be increased across the board.

Most researchers and economists have concluded that a portfolio of policies is needed to promote the development of new technologies and reduce emissions, and the optimal portfolio mix will depend on the energy sector and country. More than one policy is required because more than one type of market failure affects greenhouse gas emissions. One failure is due to negative pollution externalities, and the other to positive knowledge externalities or **knowledge spillover**, in which a firm incurs all of the costs of developing new knowledge but does not capture all of the benefits, because knowledge is a public good (Jaffe, Newell, and Stavins 2005). For example, a wind turbine manufacturer may spend a great deal of money on R&D (research and development) before discovering a new technology that makes its wind turbines more efficient. While patents help ensure that the innovator will recoup much of its R&D costs, society at large will accrue greater benefits than the innovator. Also, there is always the risk that R&D will not be successful in developing beneficial new technologies. Some studies have tried to account for these factors when identifying an optimal mix of policies for reducing greenhouse gas emissions (Fischer and Newell 2008).

Climate change mitigation represents a huge business opportunity. Research in the public and private sectors is needed to develop new, effective technologies for reducing greenhouse gas emissions. The federal government has put the United States at a disadvantage by not creating a carbon trading system that puts a price on carbon emissions. European countries adopted the **Kyoto protocol** (a 1997 international treaty to reduce greenhouse gas emissions), created a carbon market, and are now developing and purchasing technological tools for greenhouse gas remediation in order to offset their emissions. Domestic entrepreneurs and companies are much less likely to develop and market new technologies without the incentive of a domestic market. The United States must adopt a carbon trading scheme to become a player in this new market sector. We are rapidly falling behind while members of Congress argue about a scientific issue that has already been resolved. Opponents argue that adopting a carbon trading system will hurt the economy, but in reality we are missing the opportunity to add a whole new sector to our economy.

Most economists agree that GCC (not GCC mitigation) will harm not only ecosystems but also the economy, and therefore society must make investments to prevent future harm. They also agree that the cost is paid most fairly and effectively by internalizing the external costs of activities that emit greenhouse gases (Jaffe, Newell, and Stavins 2005). What is debated is how fast we should act. Should we accept responsibility for our actions now, or leave that burden to future generations? How much should we spend now on climate change mitigation?

These questions have led to the development of two competing schools of thought led by William Nordhaus from Yale University and Sir Nicholas Stern of the London School of Economics. Both use very complicated models that produce the same results if given the same assumptions. The two groups arrive at different conclusions largely due to their subjectively using different values of the **discount rate**, which is the present value of a future cost. Both Nordhaus and Stern estimate the cost of future harm from GCC, and then use their preferred discount rate to estimate how much money we should invest now to offset that cost and break even (Stern and Treasury 2007; Nordhaus 2007). The discount rate actually consists of two parts, a wealth factor that accounts for changes in per capita income over time and can be extrapolated from historical data, and the social discount rate, which is a subjective measure of how important it is to ensure the well-being of future generations. Sustainability requires that we use a low discount rate to ensure intergenerational equity, which is the approach Stern takes. A low discount rate requires that we spend more money now to reduce harm to future generations. In contrast, Nordhaus effectively discounts future generations by using a high social discount rate. Thus, the Nordhaus school concludes that we don't have to act now because we will have so much money in the future that we will easily be able to deal with GCC. This is the conclusion that free market enthusiasts like, and it is the primary reason that they are employing every conceivable means to avoid paying the costs now, including denying the existence of global warming. Instead of a constant discount rate as in the United States, a different approach taken by France and the United Kingdom is to use a declining discount rate (Arrow et al. 2013).

Proponents of the Nordhaus school are betting that the economy will grow faster than the cost of climate change mitigation, and that we will always be able to counteract the effects of GCC. If we lose that gamble, it will be future generations that will pay the consequences. Discounting the needs of future generations in this way is irresponsible and inconsistent with the principles of sustainability. Nate Lewis, a chemistry professor at Caltech, puts the economist's approach into perspective: "I haven't talked much about economics, but I will say that it's easy to prove, thinking 100 years out, on a risk-adjusted net-present-value basis, that the earth is simply not worth saving. It's a fully depreciated, four-billion-year-old asset. Unless you have policy incentives that reflect the true cost of doing this experiment, the economically efficient thing to do is just what we are doing now."* Economic considerations aside, sustainability and intergenerational equity demand that we reduce carbon emissions today to reduce the risks faced by future generations.

If government regulation and oversight prove inadequate, legal tactics may be necessary to reduce greenhouse gas emissions. Recent work has shown that 90 companies are responsible for almost two-thirds of all industrial greenhouse gas emissions (Heede 2014). Half of those emissions occurred after 1988, when James Hansen testified to Congress about global warming (Starr 2016). Exxon Mobil scientists knew seven years before that year that the fossil fuels they produced caused global warming.† Now fossil fuel producers are starting to face lawsuits from groups like Pacific

* calteches.library.caltech.edu/700/2/Power.pdf, retrieved August 23, 2016.

† https://www.theguardian.com/environment/2015/jul/08/exxon-climate-change-1981-climate-denier-funding, retrieved September 27, 2016.

Islanders who are suffering the consequences of unrestrained use of fossil fuels. Some claim that it's unfair to target or blame fossil fuel producers, since we all benefit from their use. They argue that our lifestyle choices have caused global warming, and therefore we all share the blame. But the atmosphere is a public good subject to the Tragedy of the Commons; there will always be freeriders who will continue to consume fossil fuels after others have stopped. Targeting the 90 companies responsible for most greenhouse gas emissions is likely to be a more effective approach to protecting this public good than persuading billions of people to change their lifestyles. In the past when government regulation and enforcement failed to protect public goods and common pool resources, NGOs and private parties effectively used the courtroom to compel companies to compensate harmed parties for damages; rising legal costs then forced the companies to stop polluting. This could also be an effective strategy for reducing greenhouse gas emissions, and is justifiable since greenhouse gas polluters like ExxonMobil have known for decades that their product was causing harm, and chose to do nothing to reduce public risk.

8.3.2.2 International Regulation

Citizens of most countries in the world burn fossil fuels for energy. All countries profit economically from burning fossil fuels and emitting carbon. So how can they all agree to limit carbon emissions? The waste absorption capacity of the atmosphere is limited and therefore has high subtractability (Table 5.1). It is also an open-access regime, meaning it has low excludability. This makes the waste adsorption capacity of the atmosphere a CPR, which means it is susceptible to the free rider problem and the Tragedy of the Commons (MacKay et al. 2015). Furthermore, because access is global, it can only effectively be managed through international cooperation and regulation (Farley 2010).

Because each country earns a greater profit if they put the CPR (the waste adsorption capacity of the atmosphere) to greater use, and there is no penalty for using more than their neighbors, there is no incentive to preserve the resource. Countries do not want to sign an international agreement that limits their ability to make profits from putting carbon dioxide into the atmosphere. Yet every country knows that unregulated carbon dioxide emissions will likely harm future generations. How much are we willing to discount the needs of future generations? Humanity must use smart management of shared resources like the atmosphere to prevent individuals from consuming or destroying them in an economic "free for all."

The **United Nations Framework Convention on Climate Change** (UNFCCC) took the first step toward smart management of the atmosphere as a shared resource by adopting the Kyoto protocol in 1997. This protocol called for a reduction in emission of four greenhouse gases—carbon dioxide, methane, nitrous oxide, and sulfur hexafluoride—and two groups of gases—hydrofluorocarbons and perfluorocarbons. The 1987 Montreal protocol and subsequent stricter amendments already restricted emission of chlorofluorocarbon (CFC) compounds, another group of greenhouse gases. By July 2010, 191 countries had signed the Kyoto protocol. Although the United States was the world's largest emitter of greenhouse gases in 1997, it never signed the protocol.

The 1997 Kyoto protocol that called for individual country commitments was a failure because it lacked enforcement provisions, so some countries signed the agreement and then backed out of it. Cooperating countries felt that defector countries took advantage of them, and trust was diminished. Research suggests that common commitments combined with enforcement is the most effective approach to managing the global commons (both CPRs and public goods). MacKay et al. (2015) state, "A uniform carbon price is widely accepted as the most cost-effective way to curb emissions." They suggest that each country set its average carbon price (cost per unit of greenhouse gas emitted) at least as high as the negotiated global carbon price. Countries would have flexibility in how they set that price, through either carbon taxes, cap and trade, or a hybrid scheme. Overall taxes in a country would not have to rise if taxes were shifted from good things like employment to bad things like pollution.

The **2015 United Nations Climate Change Conference** (COP21) agreement was a step forward. Remarkably, 195 countries signed the legally binding agreement without objections. To simplify discussions and help quickly move countries to a consensus, negotiators held "indabas," in which each country stated what threshold they did not want to cross (e.g., 2°C global temperature increase) and proposed a solution.* COP21 calls for global carbon emissions to be decreased to levels that would limit the global temperature increase to 2°C above preindustrial levels and for zero net anthropogenic greenhouse gas emissions to be reached between the years 2050 and 2100.† Unfortunately, it seems doomed to failure. COP21 does not take effect until 55 countries that produce >55% of greenhouse gas emissions ratify it, and it does not call for specific emission commitments from countries. Instead, it calls for participating countries to set their own emission reduction target, but there are no enforcement provisions that would punish countries for not setting or meeting a target. Individual country commitments without enforcement will likely fail, as has happened before. The United States is already on track to miss its emissions goals for the year 2025 (Greenblatt and Wei 2016). However, progress is being made on climate mitigation. An amendment to the 1987 Montreal Protocol signed by over 170 countries in 2016 legally binds those countries to reducing emissions of hydrofluorocarbons, greenhouse gases used in air conditioners and refrigerators.‡ Also, while efforts to pass climate change mitigation legislation have stalled in many countries, investors and NGOs have pressured corporations to reduce their carbon footprints, and in response large private corporations such as Google and Microsoft have made commitments to become carbon-neutral (Vandenbergh and Gilligan 2014). Such forms of private environmental governance are proving more effective at protecting public goods and CPRs than government legislation.

8.3.3 Adaptation

While mitigation reduces probabilities of climate change disasters, adaptation reduces the impacts; both reduce risk, which equals probability × impact (Chapter 4). If mitigation and regulation are insufficient to prevent the harmful effects of GCC, we will be forced to adapt to a new environment. The further temperature increases, the harder it will become to adapt.

Adaptation is a cyclical process that starts with understanding of a problem, making plans to reduce or eliminate the problem, and then implementing and managing the solution (Figure 8.4). The nature of the problem often changes over time, so the cycle must periodically start over and progress through the stages of understanding, planning, and managing (Moser and Ekstrom 2010). This process is most likely to be effective if it involves affected stakeholders, requires a consensus on chosen actions, and considers all dimensions of the problem including economic, social, and environmental.

Complex, dynamic systems like regional climate systems pass through adaptive cycles. Because the climate and the environment will continue to change, and at rates higher than at any time in human history, adaptation will be difficult (Karl, Melillo, and Peterson 2009). The IPCC found with high confidence that "responding to climate-related risks involves decision making in a changing world, with continuing uncertainty about the severity and timing of climate-change impacts and with limits to the effectiveness of adaptation," and "adaptation and mitigation choices in the near term will affect the risks of climate change throughout the 21st century" (IPCC 2014).

Given the uncertainties in the timing and extent of climate system changes, the process of **adaptive management** is often used in the decision making process. It involves continuous monitoring of a system such as regional climate and continuous refinement or change in adaptive responses. Individuals and communities that anticipate the risks associated with climate change

---

* *Quartz*, http://qz.com/572623/this-simple-negotiation-tactic-brought-195-countries-to-consensus-in-the-paris-climate-talks/?utm_source=pocket&utm_medium=email&utm_campaign=pockethits, accessed January 4, 2016.

† "Obama: Climate Agreement 'Best Chance We Have' to Save the Planet." CNN, John D. Sutter, Joshua Berlinger and Ralph Ellis, http://www.cnn.com/2015/12/12/world/global-climate-change-conference-vote/, accessed January 4, 2016.

‡ "Nations, Fighting Powerful Refrigerant that Warms Planet, Reach Landmark Deal," http://mobile.nytimes.com/2016/10/15/world/africa/kigali-deal-hfc-air-conditioners.html?_r=0, retrieved October 21, 2016.

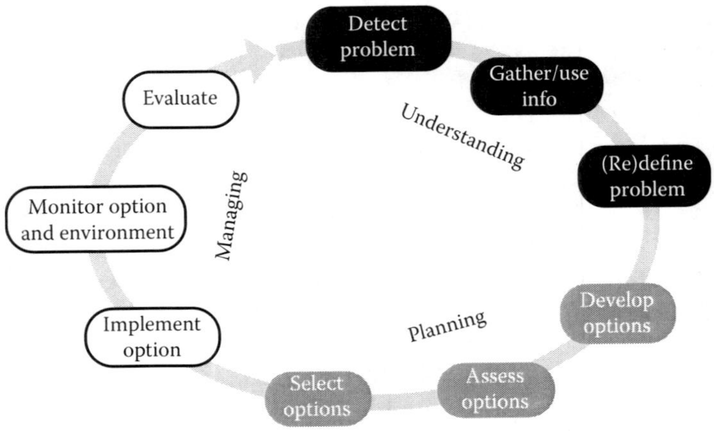

FIGURE 8.4 Phases and subprocesses throughout the adaptation process. (From Moser, Susanne C, and Julia A. Ekstrom. 2010. "A Framework to Diagnose Barriers to Climate Change Adaptation." *Proceedings of the National Academy of Sciences* 107 (51): 22026–31. doi:10.1073/pnas.1007887107.)

and respond by becoming more sustainable and resilient will be better prepared to adapt when it becomes necessary.

Anticipating all of the potential types of adaptation that humans will employ in the future is impossible. The types of adaptation will depend on how climate change is manifested in a specific region and the rate and direction of change. Thus, adaptation responses will vary by region and over time. Let's look at agricultural adaptations as an example. Currently warming is strongest at high latitudes, where agricultural yields will likely increase as growing seasons lengthen (Dawson and Spannagle 2009). However, at low latitudes temperatures will approach the limits of adaptation. Reduced crop yields may result in widespread starvation. Farmers will need to adapt by switching to heat-resistant crops, and in areas of drought they will need to adopt **rainwater harvesting**, **drip irrigation**, and drought resistant crops. New technologies may be needed to adapt successfully to the changing conditions. If adaptation is unsuccessful, then people will be forced to migrate to habitable areas.

8.3.3.1 Migration as a Climate Adaptation

Perhaps the most common form of adaptation will be migration. Rising sea levels and increasing storm strength will force seaside communities to migrate landward. People will also migrate away from regions that experience severe drought or food shortages. Environmental refugees will become commonplace across the world.

According to Gemenne (2015), migration attributes include duration, character (voluntary versus involuntary), and destination (in-country or transboundary). These factors depend on event rate (rapid or slow onset), scale/intensity, and frequency (chronic or episodic). For example, sea level rise is a slow, continuous change, which allows affected communities to plan and prepare for relocation. Hydrometeorological events can have slow onset and long durations (e.g., drought) or rapid onset and short durations (e.g., cyclones). Most environmental migration is caused by natural hazards that are weather-related, with floods and storms topping the list of causes. The intensity and frequency of hydrometeorological events and resulting environmental migration events are expected to increase as anthropogenic climate change increases in intensity.

The poor often lack the resources required to relocate. Policies should enable affected parties to choose the adaptation strategies that best suit their needs. The most vulnerable populations must be assisted by removing barriers to migration, which include financial, administrative, and informational. Policies must also ensure that adequate infrastructure is in place in destination areas (de Sherbinin et al. 2011).

8.3.3.2 Increasing Carbon Sinks and Reducing Solar Radiation: Climate Engineering

Climate engineering (also referred to as "geoengineering") is defined as "the deliberate large-scale manipulation of the planetary environment to counteract anthropogenic climate change" (Nicholson 2013). According to the U.S. National Academy of Sciences (2015a,b), society has "reached a point where the severity of the potential risks from climate change appears to outweigh the potential risks from the moral hazard" of conducting geoengineering experiments. There are two distinct approaches to countering the effects of greenhouse gas accumulation in the atmosphere—**greenhouse gas removal** and **solar radiation management** (SRM).

Greenhouse gas removal uses enhanced natural or artificial greenhouse gas sinks to extract carbon dioxide from the atmosphere. Note that CCS alone is not a form of climate engineering because it prevents carbon dioxide from entering the atmosphere rather than extracting what is already in the atmosphere. Removal can be accomplished using trees, chemical pumps, ocean fertilization, low-till agriculture for soil carbon storage (Chapter 13), direct air capture + carbon storage, biofuel production + carbon storage (Chapter 10), or enhanced weathering of rocks (National Academy of Sciences [U.S.] 2015a). Compared to SRM, greenhouse gas removal is a lower risk option but is slower to reduce atmospheric warming.

To evaluate the effectiveness and potential costs associated with direct air capture + CCS, consider a candle. While burning, a candle emits soot particles, black carbon visible as smoke. Black soot from incompletely burning organic matter such as candle wax can settle on reflective white snow and absorb more incident and reflected sunlight, causing surface warming. What is the best approach to mitigate soot emissions? The soot is concentrated near the candle, so we can capture it easily by placing a collecting cup over the candle with a filter or scrubber that captures the soot particles. Capturing the pollution at the source where it is concentrated makes physical and economic sense. Some propose capturing trace greenhouse gases like carbon dioxide and methane from the atmosphere. How would you collect the soot particles after the smoke has dissipated? It would be nearly impossible. You would be working against entropy, which requires energy. When the soot is concentrated near the candle, the entropy of the system is low; over time the soot diffuses into the air, the particulate concentration decreases, entropy increases, and more energy is required to recover the particulates from a much larger volume of air. So is capturing the particulates at the source the easiest, most cost-effective approach to reducing carbon pollution from the candle? No, putting the candle out is. In general, conservation is the most cost-effective approach to mitigating greenhouse gas emissions, followed by capture at the source; direct air capture of trace gases after they have dispersed is the worst approach.

The premise behind SRM is that decreasing incident solar radiation at the Earth's surface by 1% to 2% could counter the effects of a doubling of atmospheric CO_2 (Nicholson 2013). SRM is a more intrusive approach that can only be performed globally and has higher costs and risks than greenhouse gas removal. Many political and ethical factors need to be considered before implementing such a global plan. A country could implement such a plan without the approval of any other countries, but all countries would be affected, potentially in a very negative way (National Academy of Sciences [U.S.] 2015b). It is considered a method of last resort because it is likely to have unintended consequences, but it has the advantage that it can mitigate GCC faster than greenhouse gas removal, so it could be used in the case of a planetary emergency. The leading candidate for SRM proposed by Nobel laureate Paul Crutzen would involve injecting sulfur dioxide into the stratosphere, which would form sulfate aerosols that reflect sunlight (Flannery 2008). When Mt. Pinatubo erupted in 1991, it injected 20 million tons of SO_2 into the atmosphere, causing the Earth to cool 0.5°C (0.9°F) for a year or so. An unwanted side effect is that sulfates destroy stratospheric ozone, which increases ground levels of harmful ultraviolet radiation. Other approaches to making the Earth's surface "shinier" include pumping sea salt into clouds and using white roofs on buildings. Painting roofs white is a low-cost, low risk SRM approach, and therefore should be implemented first.

The problem with climate engineering is that we are experimenting with a complex system that we don't fully understand. Each climate engineering option is likely to have unintended consequences, and it's possible that none of them may be effective. As observed by Nobel-prize winning physicist Burton Richter (2010), "It is not smart to count on introducing new effects you don't fully understand to cancel another effect you do not fully understand. Doing two dumb things rarely gives a smart result." However, other smart people are strong proponents of climate engineering. According to David Keith, a professor of applied physics and public policy at Harvard University, "The impacts of climate change are on the order of a trillion dollars a year, as are the costs of cutting emissions." Thus, he advocates the potentially cheapest option: SRM, specifically using airplanes to dump SO_2 particles into the atmosphere to reflect sunlight back into space.* However, evidence is accumulating that the cost of cutting carbon emissions has been overestimated. For example, between 2014 and 2016 the global economy grew 6.5% without increasing carbon dioxide emissions and without expensive mitigation actions, suggesting that we may be seeing the beginning of a decoupling between economic growth and environmental impacts (Pearce 2016).

Because all climate engineering choices are likely to have unintended consequences, society should only use them if a climate catastrophe strikes, such as when the global climate system passes a tipping point. We must mitigate carbon emissions before we consider climate engineering options. However, we should be researching climate engineering options now so that we will be prepared if a climate catastrophe strikes.

8.3.4 Parallels between the Greenhouse Gas—Climate Change and CFC—Ozone Hole Problems

Ozone (O_3) is an unstable, highly reactive molecule that is a pollutant in the troposphere (photochemical smog) but occurs naturally in the upper stratosphere, where it absorbs ultraviolet (UV) radiation. Stratospheric **ozone depletion** translates into higher ground-level UV radiation levels, which can lead to elevated rates of skin cancer.

Photochemical reactions produce ozone naturally in the stratosphere. Atmospheric chemists Rowland and Molina (who later won the Nobel Prize) predicted in the 1970s that CFCs, synthetic compounds used as refrigerants, could destroy stratospheric ozone. DuPont designed CFCs to be highly stable so they would last long, and as a result, CFCs can persist until they reach the stratosphere. Rowland and Molina predicted that photochemical reactions would release chlorine, which would then act as a catalyst for the breakdown of ozone (Molina and Rowland 1974). Each chlorine atom produced by the breakdown of CFCs can destroy millions of ozone molecules (Meadows, Randers, and Meadows 2004).

Special conditions promote the destruction of ozone in the stratosphere above Antarctica during the winter, leading to the development of an ozone hole. Ironically, for many years the development of the ozone hole went undetected. Although the United States had satellites measuring stratospheric ozone concentration, scientists programmed their computers to throw out values lower than a certain concentration because they assumed that those measurements were in error. Once they started monitoring the ozone hole, atmospheric scientists realized that Rowland and Molina were correct, and that the size of the ozone hole was growing each year. Measurements showed that ozone levels were low when chlorine monoxide levels were high. This negative correlation strongly suggests cause and effect, that is, chlorine monoxide was destroying ozone. We now know that higher levels of UV radiation are reaching the ground in the winter at high latitudes in the southern hemisphere because of formation of the ozone hole.

The strong consensus that artificial chemicals were destroying the ozone layer led to the first and most successful international environmental agreement, the Montreal Protocol of 1987. This called for the phasing out of CFC production. Although CFC releases to the atmosphere have greatly

* Phil McKenna (2011) "British to Test Geoengineering Scheme: Can a Garden Hose to the Stratosphere Really Keep the Planet Cool?," *Technology Review*, http://www.technologyreview.com/printer_friendly_article.aspx?id=38564.

decreased, it will take 50 to 100 years for natural processes to remove the chlorine from the ozone layer and replace the ozone that it destroyed.

The manufacture of CFCs was an example of unsustainable industrial chemical production. Until recently, chemical companies designed synthetic chemicals to be stable and durable. However, it was the stability of CFCs that allowed them to reach the stratosphere before decomposing. Other older-generation chemicals that persist for a long time in the environment include PCBs and DDT. DDT was particularly troublesome because it has a tendency to bioaccumulate, so that it becomes concentrated in organisms at the top of the food chain. That is how the DDT problem was first recognized: bald eagles at the top of the food chain were unable to reproduce because the buildup of DDT in the bodies of females caused their eggs to have thinner shells, causing the eggs to break. The newer approach of **green chemistry** aims to produce chemicals "in a manner that is sustainable, safe, and non-polluting and that consumes minimum amounts of materials and energy while producing little or no waste material" (Manahan 2006). Green chemists design chemicals to decompose when they escape to the environment and products to biodegrade when they are disposed of. The goal is to reduce environmental damage by using sustainable practices.

The development of the greenhouse gas-GCC problem closely parallels that of the chlorofluorocarbon (CFC)-ozone hole issue (see Meadows et al. 2004 for a full account). Initially, industry and the government disputed the scientific claim that CFCs destroy ozone, saying scientists and environmentalists were being alarmists. Publication in 1987 of the smoking gun proving the link between CFCs and ozone destruction finally silenced the skeptics. Due to industry opposition, the government didn't ban CFCs in aerosols, so the public voted with their wallets and stopped buying aerosol cans, which greatly decreased the production of CFCs in 1974. Industry said it would be too expensive to find and deploy substitutes. In the end, phasing out CFCs was much less expensive and disruptive than predicted by skeptics and industry (about $40 billion globally). Only when a crisis was reached—the discovery of the ozone hole in 1985—did the government take action and sign the Montreal Protocol in 1987. They acted even though our knowledge of how CFCs created the ozone hole was imperfect. Perfect knowledge or scientific proof is not necessary for action. This crisis was scary because it was the first time that the public recognized that we could change the Earth on a global scale, and that those changes could endanger us. Developing countries such as China refused to sign the protocol unless developed countries shared technical knowledge and established an international fund to ease the transition away from CFCs. The United States initially balked, but then signed a stricter protocol in London in 1990. The developing countries China and India later signed on. Finally, when the ozone hole problem grew beyond expectations, governments revised the Montreal protocol with stricter limits in London (1990), Copenhagen (1992), Vienna (1995), and Montreal (1997). Scientists later found that the 1987 and 1990 emission limits were inadequate, that is, the stratospheric concentrations of chlorine and bromine would have continued to increase without the stricter emission limits adopted later. Scientists now expect the ozone hole to disappear by the middle of the twenty-first century, and several lines of evidence indicate that the ozone layer above Antarctica is already healing (Solomon et al. 2016).

How did we solve the CFC-ozone problem? By reducing the need for CFCs, adopting temporary substitutes such as HCFCs, and shifting to alternatives that do not harm the ozone layer (Meadows et al. 2004). In the process, governments learned some valuable lessons about atmospheric pollution and the need for international regulatory agreements. First, continuous environmental monitoring with quick and honest reporting of results is essential. Second, the ozone hole and other issues involving environmental regulations show that industry often exaggerates or overestimates the economic consequences of meeting new environmental regulations. According to Meadows et al. (2004), the reason is most likely because they "systematically underestimate the capacity for technological advance and social change." Finally, when knowledge is incomplete, stakeholders must write environmental agreements flexibly and review them regularly (Meadows et al. 2004). The successful international response to the CFC-ozone hole problem gives us hope that we can tackle the greenhouse gas-GCC problem before it is too late.

In the end, a public and political consensus on climate change may not be necessary to reduce carbon emissions. A technological revolution has caused renewable energy costs to plummet, leading to the substitution of renewable energy sources for fossil fuels and decreasing carbon emissions (following statistics from Chait*). The price for 1 kW of solar power decreased from $101 in 1975 to $0.61 in 2015. China is expected to increase its solar capacity by 18 GW in 2015 alone (for comparison, the United States had 20 GW of solar capacity in 2015). The year 2014 was the first in which the global economy grew without carbon emissions increasing. U.S. carbon emissions peaked in 2007 and have fallen ever since, mostly because coal use decreased 21% due the number of coal-burning power plants decreasing from 523 to 323.

8.4 CONCLUSIONS

The Earth is like a compost heap that provides energy to microorganisms (Bauman and Klein 2014). Activity of the microorganisms produces heat, and population growth can increase the temperature of the compost heap until it kills the micro-organisms. Similarly, the Earth provides energy to humans, who reproduce and generate heat, warming the planet. The amount of heat produced is related to population and activity levels, as reflected in the Kaya identity. Any solution to this problem must overcome the Tragedy of the Commons and reduce the factors in the Kaya identity.

Despite abundant scientific evidence of anthropogenic GCC, many factors have kept public opinions divided, and political action has been ineffective at solving this problem. This is because the debate is no longer about the science; it is about competing ideologies and values.† A three-pronged approach of mitigation, regulation, and adaptation will be needed to prevent catastrophic GCC. The COP21 international agreement signed in 2015 is a hopeful sign, but stronger commitments must be made to prevent the average global temperature from increasing more than 2°C. The 1987 Montreal protocol to eliminate CFC emissions solved the ozone hole problem, showing that international treaties can successfully protect the atmospheric commons.

In the next three chapters we will look at the energy problem that is so closely tied to the climate change problem.

HOMEWORK PROBLEMS

1. What are the three primary approaches to reduce climate impacts in the future? List and give an example of each.
2. Think about the mitigation and regulation strategies discussed in Sections 8.3.1 and 8.3.2. What combination of these do you think would be most effective, affordable, and realistic? Why? On what scale?
3. Visit the URLs in the captions of Figures 7.1 and 7.2. Do updated versions of those figures show any changes in global temperature or atmospheric CO_2 concentration over time?

* "The Sunniest Climate-Change Story You've Ever Read," Jonathan Chait, *New York Magazine*, September 7, 2015, http://nymag.com/daily/intelligencer/2015/09/sunniest-climate-change-story-ever-read.html#, retrieved March 15, 2016.

† http://ssir.org/articles/entry/climate_science_as_culture_war, retrieved August 23, 2016.

9 Nonrenewable Energy Sources

Energy may be our most important resource, and energy availability will likely be the ultimate limit on population and economic growth. Abundant energy makes it easy to be sustainable because it can be used to produce essential resources such as food and potable water where they are scarce. However, many areas of the world are experiencing energy shortages. Furthermore, 88% of the world's energy comes from burning fossil fuels, which adds the greenhouse gas carbon dioxide to the atmosphere, intensifying global climate change (Princen, Manno, and Martin 2013). The fossil fuel era began when fossil fuels surpassed wood as the dominant energy source, but it is not permanent. To become sustainable, our society must transition from nonrenewable fossil fuels to renewable energy sources. Having covered oil in Chapter 6, we will review sustainability issues associated with the remaining nonrenewable energy sources in this chapter before reviewing renewable energy sources in Chapter 10. In Chapter 11 we will compare the costs of these energy sources and review policy options.

9.1 INTRODUCTION

Energy comes in many forms, including chemical, mechanical, electrical, heat, and light. We use energy to perform work, usually by converting it from one form that stores it to another. For example, fossil fuels and electrochemical batteries store chemical energy. To propel conventional autos that have internal combustion engines we burn gasoline or diesel to convert the stored chemical energy to mechanical energy. Electric cars convert chemical energy stored in electrochemical batteries to mechanical energy, with the added advantage that they don't emit carbon dioxide.

Some forms of energy are more useful than others. Heat is considered low quality energy because it is dispersed. Electricity is a high quality form of energy because we can transport it through conductive wires, store it in batteries, and use it to make other forms of energy. We also can use it for electronic communication, which makes it indispensable to modern society. As a result, we typically convert other forms of energy to electricity. For example, mechanical energy in flowing water or steam is converted to electrical energy using a turbine. According to the second law of thermodynamics, each time we convert other forms of energy to electricity we lose some energy, but the increased utility of electricity balances this loss.

Energy availability is what most likely limited human population in the past. Humans first obtained energy from biomass (wood and dung), then coal, and finally energy-dense oil (National Science Board and Foundation 2009). The availability of coal in the United Kingdom enabled the beginning of the Industrial Revolution, and coal and other fossil fuels allowed it to spread to other parts of the world, causing rapid growth in energy consumption. Global energy consumption continued to increase during the twentieth century. Abundant energy has allowed humans to multiply and increase their standard of living to levels never previously reached. Energy from oil fueled the green revolution of the twentieth century that quadrupled agricultural productivity worldwide, allowing the global population to increase at an exponential rate. We currently use oil and the energy it provides to make fertilizers and pesticides, to run the machines that plant and harvest crops and till the soil, and to transport the harvest to market. We use energy to pump groundwater to the surface for irrigation in arid regions, transforming deserts to productive cropland. With unlimited energy, we can turn the most inhospitable environment into heaven on Earth. However, we do not have unlimited energy, and even worse, our use of fossil fuels may make Earth inhospitable.

To discuss energy and power we must first introduce their scientific definitions and units of measure. **Energy** is a measure of how much work can be performed. In this book we will usually use the kilowatt-hour (kWh) as a measure of stored or used energy. **Power** is the rate at which we use energy in kilowatt-hours per day (kWh/d). The total amount of energy used to do work depends on the rate at which you use the energy (power) and how much time you work:

$$\text{energy (kWh)} = \text{power (kWh/d)} \times \text{time (d)} \tag{9.1}$$

Since Americans use 250 kWh daily and the average human power output is 1 kWh/day, we use the equivalent of 250 "energy slaves" per day. Europeans consume half of that (MacKay 2009).

Efficiency is the ratio of the work output to the energy input; sometimes power is used in place of energy because power is the rate of energy use. If we express efficiency as useful power output divided by total power input, values range from zero (completely inefficient) to one, or as a percentage from 0% to 100%. In cases where the numerator and denominator have different units of measure, efficiency is not unitless. For example, fuel efficiency measures the distance traveled (miles) per amount of fossil fuel (per gallon of gasoline). Lighting efficiency expresses the amount of light (lumens) per quantity of power (watts). Efficient systems use most input energy to do useful work; inefficient systems use much of the input energy to produce wasted heat or noise.

Energy supply is classified as either primary energy or secondary energy. **Primary energy** is energy captured from nature that is used directly such as oil and coal. Figure 9.1 shows that 36% of primary energy consumed globally comes from oil. Burning that oil releases 40% of the carbon dioxide released by burning fuels. **Secondary energy** is produced by converting primary energy to a form such as electricity or hydrogen gas that can be easily transported, referred to as an **energy carrier**. Some energy is lost during the conversion of primary to secondary energy.

Primary energy use in the United States increased steadily from 1983 until the recession in 2008, and then began to increase again (Figure 9.1). Experts anticipate the global consumption of energy to double between now and 2050 due to increases in global population and wealth resulting from the globalization of markets (Friedman 2008). According to the IPAT formula, energy consumption is the product of population P, affluence (GDP/person), and energy intensity (E/GDP) (de Vries 2013):

$$I = E = P \times A \times T = P \times \text{GDP}/P \times E/\text{GDP} \tag{9.2}$$

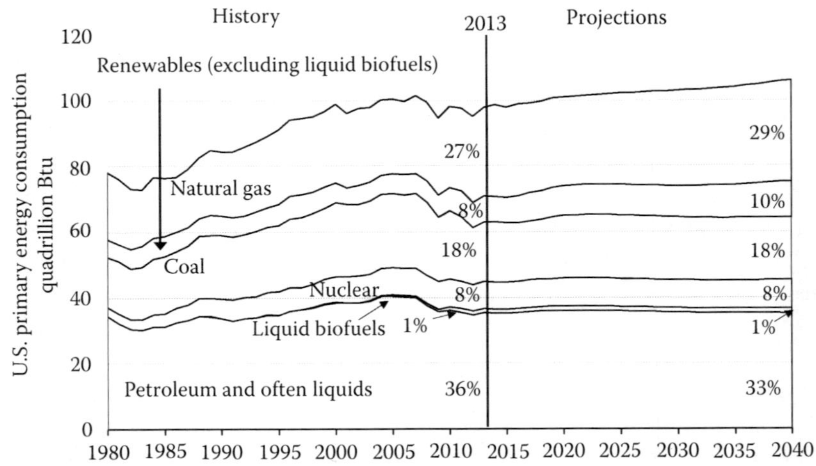

FIGURE 9.1 Primary energy use by fuel in the United States between the years 1980 and 2013, and projected from 2014 to 2040. (From EIA. 2015. "Annual Energy Outlook 2015." http://www.eia.gov/forecasts/aeo/pdf/0383(2015).pdf.)

Burning fossil fuels and biofuels for energy is responsible for 70% of greenhouse gas emissions, with the remaining 30% coming from deforestation (Brown 2009). Despite rising concerns about GCC, global CO_2 emissions continue to rise (Figure 8.2). Natural removal of excess greenhouse gas from the atmosphere will take centuries. Thus, reducing greenhouse gas emissions by reducing use of fossil fuels is essential for the well-being of future generations (Richter 2010). In this chapter, we will explore the problems posed by use of the nonrenewable energy sources coal, natural gas, and nuclear power.

9.2 LIFE CYCLE ASSESSMENT OF ENERGY COSTS

To measure the total environmental impact of energy sources, we need to measure the impact at every stage of their life cycle using Life Cycle Assessment (LCA). For example, LCA can be used to estimate the energy, water, and carbon footprints of different products or activities. It does so by summing impacts (measured as energy consumed, water consumed, or CO_2 emitted) over the four phases of the life cycle: raw materials (*R*), production (*P*), use (*U*), and disposal (*D*):

$$\text{Total impact footprint} = R + P + U + D \tag{9.3}$$

Remember RPUD. We can express the amount of energy used to manufacture a product by summing the amounts used in the *R* and *P* phases of the life cycle to estimate the **embodied energy**.

LCA requires large amounts of data. For example, to estimate the total energy cost of a car, you would need to know how much energy was consumed for production of the raw materials for every part, for production of every car part, and for use, maintenance, and disposal. You would have to add up the energy costs through the entire supply chain, a formidable task. Only recently have these types of calculators become available. A good example is the Economic Input-Output Life Cycle Analysis (EIO-LCA) calculator created by the Carnegie Mellon Green Design Institute.* It calculates embodied energy and greenhouse gas emissions in CO_2e for a wide range of products and activities. Table 9.1 shows that $1 million of cattle ranching economic activity consumes 18.8 Terajoules (TJ) energy (5.2 million kWh). Energy is consumed to produce grain feed, transport the grain and the meat, produce the fertilizers, pesticides, and antibiotics, and to raise the animals. In addition, energy is used to mine and distill the oil and to produce the electricity used at all stages of the life cycle. The EIO-LCA calculator also tells us that $1 million of cattle ranching economic activity has a carbon footprint of 8,550 metric tons of CO_2e. Per dollar of economic activity, the amounts of energy consumed and CO_2e produced are larger for cattle farming than almost any other economic activity.

During the fuel use phase power plants emit 10 billion tons of carbon dioxide per year, about one-quarter of global carbon dioxide emissions. The United States accounts for ~25% of that total.† Carbon dioxide is also emitted during the raw material, production, and disposal phases of fuel use. For example, consider the equipment used for mining, processing, and transportation of coal, and the equipment used to capture and dispose of coal ash, all powered by fossil fuels. Table 9.2 shows LCA estimates of **emission intensity**, the amount of carbon dioxide emitted per unit of electricity produced for different primary sources of energy. Lower values mean lower greenhouse gas emissions. As mentioned previously, coal emits the most carbon dioxide per unit energy, followed closely by oil. Natural gas emits 62% as much carbon dioxide per kWh as coal, a significant improvement, but still much higher than the nonfossil fuels. To mitigate GCC, we must phase out coal, oil, and natural gas as energy sources, and replace them with solar, biomass, wind, geothermal, and hydroelectric. Although nuclear has a small carbon footprint, it is a nonrenewable and therefore nonsustainable source of energy, and it has safety concerns that, although often exaggerated, make it less attractive than the renewable energy sources (discussed in detail below). Table 9.2 also shows the water intensity, the amount of water consumed to produce 1 kWh of electricity. The water intensity

* http://www.eiolca.net/.

† http://www.sciencedaily.com/releases/2007/11/071114163448.htm.

TABLE 9.1
Embodied Energy (Terajoules) per Million Dollars of Economic Activity for Each Sector of the Cattle Ranching Industry

Sector	Total Energy (TJ)
Total for all sectors	19
Cattle ranching and farming	7.3
Power generation and supply	4.4
All other crop farming	1.2
Petroleum refineries	1.1
Grain farming	0.8
Truck transportation	0.6
Oil and gas extraction	0.5
Fertilizer manufacturing	0.5
Pipeline transportation	0.2
Other basic organic chemical manufacturing	0.1

Source: EIO-LCA online tool available at *http://www.eiolca.net/index.html*. Uses data from year 2002.

TABLE 9.2
Carbon Dioxide Intensities and Water Intensities of Energy Sources

Energy Source	CO_2 Intensity (l/kWh)	H_2O Intensity (l/kWh)
Solar	26	
Biomass	21	360
Wind	6.7	0
Geothermal	8.4	5.3
Hydro	11	17
Coal	530	1.9
Natural gas	330	0.6
Oil	500	1.6
Nuclear	17	2.6
Photovoltaic		0
Thermal		3.2

Source: Data from Cho, A. 2010. "Energy's Tricky Tradeoffs." *Science* 329 (5993): 786–87. doi:10.1126/science.329.5993.786.

of biomass is two orders of magnitude higher than any other energy source, making it a poor choice for arid regions that need to conserve water. Armed with this information, let's look at other pros and cons for each energy source.

9.3 COAL

Never mind when fossil fuels are going to run out; never mind whether climate change is happening; burning fossil fuels is not sustainable anyway.

David JC MacKay
Sustainable Energy—without the Hot Air, 2009

Coal is a black or brown sedimentary rock formed from the remains of fossilized plants. Most coal formed in swamps during the Carboniferous period 359–299 million years ago. Because coal takes millions of years to form, it is considered a nonrenewable resource, meaning that coal use is unsustainable. Assuming a constant rate of consumption, *R/P* estimates indicate global depletion of coal will occur in just over 100 years (BP 2015). However, because global coal consumption has been increasing (although it may have leveled off beginning in 2013), global coal reserves would be depleted in 60 to 90 years (MacKay 2009), and peak global coal production will occur much sooner. Britain's coal production peaked in 1913, and since then its global influence has decreased. In the United States, energy production from coal peaked in the late 1990s, and global coal demand is projected to outstrip supply by the year 2020 (Heinberg and Fridley 2010).

In the United States we have abundant coal. This combined with generous federal subsidies has made coal an inexpensive energy source. Substituting coal for oil can reduce U.S. dependence on foreign oil and make us less vulnerable to Peak Oil. However, coal is the dirtiest source of energy. It has the largest carbon footprint, meaning it emits more carbon dioxide per unit energy over its life cycle than any other energy source (Table 9.2). Coal burning releases more carbon dioxide worldwide than any other human activity. It supplies 50% of electricity but releases more than 70% of the electrical sector's carbon dioxide emissions (Brown 2009). Coal burning also releases toxic metals like mercury, sulfurous and nitrous oxides that contribute to acid rain, and particulates and ozone that contribute to ground-level air pollution. In addition, coal mining and its associated activities (mountaintop removal, fly ash settling ponds, etc.) are very harmful to the environment and have serious safety concerns.

However, even with an all-out push to expand the use of non-renewable energy and nuclear power, most countries will still get most of their energy from fossil fuels for the next few decades. The United States probably cannot reduce the energy derived from fossil fuels below 80% by 2040 (Figure 9.1). Consequently the IPCC, acknowledging that coal will remain an important source of energy, advocates the development and use of carbon capture and storage (CCS, see Sections 8.3.1.1 and 9.3.3) to reduce carbon dioxide emissions from coal (IPCC 2007). So is it possible for coal to be an environmentally friendly source of energy? Here we look in detail at three topics related to that question. We will show that **mountaintop removal mining** causes serious environmental damage, and that coal-fired power plants release toxic heavy metals, greenhouse gases, and the pollutants that cause acid rain. Solutions exist to reduce the severity of these problems, but they would make coal more expensive, and it is still unknown whether CCS can be scaled up to significantly reduce greenhouse gas emissions.

9.3.1 Coal Mining: Mountaintop Removal

One problem associated with all forms of coal mining is that coal usually contains sulfide minerals such as pyrite (FeS_2) that dissolve in water when exposed at the surface. This process makes the water acidic, resulting in **acid mine drainage (AMD)**. Acidic water is very good at dissolving toxic heavy metals, so AMD can mobilize these metals and transport them to locations where people can be exposed. Heavy metals are also more bioavailable in acidic water, meaning plants more readily take them up before animals consume them. **Bioaccumulation** occurs when contaminant input to an organism is faster than output so that the concentration of the contaminant in the organism increases. **Biomagnification** causes the concentrations of contaminants such as heavy metals to increase as they move up the food chain so that species at the top of the food chain (e.g., humans) are exposed to the highest concentrations. Bioaccumulation and biomagnification make the release of heavy metals to the environment during coal mining and burning a serious health risk. For example, they can cause toxic organic compounds and heavy metals to become concentrated in the placenta of a pregnant woman, putting the health of the fetus at risk.

We mine coal in a variety of ways. Coal beds are sedimentary layers that are often flat and flat-lying. If the coal bed, or seam, is close to the surface, miners can strip off the overburden (rock layers above it), a process called strip-mining. This approach is inexpensive, but if the mining company does not reclaim the land after mining, it can cause extensive environmental damage.

Proper reclamation requires that miners cover the coal tailings and exposed bedrock with soil and contour it to approximate the original land surface. Historically, underground coal mining was the most common coal mining practice, but strip mining and mountaintop removal mining have superseded it because underground coal mining is very unsafe and expensive (Keller 2011). In the United States coal comes mostly from strip mines in the west and mountaintop removal mines in the east (Keller 2011).

In the Appalachian Mountains mountaintop removal mining is the preferred method of coal extraction. It involves using dynamite for piecemeal removal of the parts of a mountain that overlie a coal seam. Only recently has our society become aware of the extensive environmental and safety problems associated with mountaintop removal. The main problem is that miners dump the overburden into stream channels, contaminating the streams and blocking their flow. A study of the effects of mountaintop removal mining with valley fills (Palmer et al. 2010) concluded that it can cause permanent loss of ecosystems in the filled valleys; that the frequency and magnitude of floods downstream of mined areas increase; and that valley fill can contaminate water and lead to decreases in stream biodiversity, even long after mining ceases. Even streams draining reclaimed areas show continuing evidence of water quality degradation (Lindberg et al. 2011). Of particular concern is the mobilization of the heavy metal selenium, which was found at unsafe levels in 73 out of 78 surveyed streams (Palmer et al. 2010). Bioaccumulation and biomagnification of selenium have led to the publication of safety advisories that recommend limiting consumption of fish caught in affected states such as Kentucky and West Virginia. Residents of areas affected by mountaintop removal mining with valley fills have higher rates of cancer, chronic heart, lung, and kidney disease, and mortality than the general population, and postmining mitigation has not been effective at reducing environmental and health problems (Palmer et al. 2010).

Residents of Appalachia are divided over whether to allow mountaintop removal mining to continue.* Those who have jobs with coal companies are unwilling to give them up. Yet mountaintop removal coal mining and its associated lifestyle are unsustainable. Given recent trends, the mines in their towns will likely be closed within a generation. If they can't manage without coal mining, what do they expect their children to do? This is a serious social problem with no easy answers. Meanwhile, the availability of high-grade coal in Appalachia is declining, so production is shifting to the lower-grade coal of the western United States. As a result, the energy content of extracted coal in the United States has been declining since 1998 (Hughes 2010), and unemployment in Appalachian coal mining towns is on the rise.

Coal will not be a safe, environmentally friendly energy source until the government bans the use of mountaintop removal mining with valley fills and coal slurry ponds. Environmentally friendlier options exist, such as burial of fly ash, but they cost more. Coal companies have externalized the environmental and social costs of coal mining, which has kept coal a relatively inexpensive source of energy in the United States. Until Americans start to pay the true cost of coal, they will continue to inflict damage on Appalachian ecosystems and communities.

9.3.2 Coal Burning: Toxic Heavy Metals, Acid Rain, and Ash Waste

Coal contains sulfur and nitrogen, and releases them to the atmosphere as sulfur and nitrous oxides during burning. These oxides react with water in the atmosphere to form sulfuric and nitric acids, which acidify rainwater (Langmuir 1997). Acid rain has caused serious ecological damage in areas such as New England that are downwind from many coal-fired power plants. Most fish eggs do not hatch when pH decreases below 5.† Sulfur emissions have been greatly reduced by an emission trading system (cap and trade) established by Congress in 1990 as an amendment to the Clean Air

* "Appalachia Turns on Itself," Jason Howard, *The New York Times*, July 8, 2012, http://www.nytimes.com/2012/07/09/opinion/appalachia-turns-on-itself.html?_r=1.

† http://www3.epa.gov/acidrain/effects/surface_water.html.

Act. This has provided power utilities with a strong incentive to use low-sulfur coal and to install scrubbers that remove sulfur oxides from the exhaust air in smoke stacks.

Coal also contains radioactive and toxic elements. It contains trace amounts of the naturally occurring radioactive elements uranium and thorium. These elements are concentrated in the fly ash produced during burning. Most of the fly ash is recovered for disposal, but some escapes to the atmosphere, exposing people who live downwind of the plant to low levels of radiation.* In fact, it has been known for decades that coal-fired power plants expose people to higher levels of radiation than nuclear power plants (McBride et al. 1978). The toxic heavy metals mercury, arsenic, and lead are also concentrated in fly ash. Around 1300 coal-fired power plants across the United States collectively emit some 50 tons of mercury annually into the air, the largest single source of mercury emissions to the air in the United States.† Mercury is the most toxic naturally occurring element: It is a neurotoxin and, as methyl mercury, it biomagnifies in the food chain, reaching its highest levels in a mother's breast milk, which puts mothers and infants at greatest risk. In 2011, the EPA enacted new regulations to limit mercury emissions from coal-fired power plants, which they stated would save $9 in health care costs for every dollar invested in pollution prevention.‡

The radioactive and toxic elements contained in coal are concentrated in the fly ash collected from coal-fired power plants. Unfortunately, the EPA did not start regulating the disposal of fly ash until 2015.§ As a result, coal companies dispose of it using the cheapest method possible: Often they simply pump it as a water slurry into holding ponds enclosed by earthen dams. These fly ash retention ponds are disasters waiting to happen. A pond behind the Kingston coal-fired power plant in eastern Tennessee collapsed on December 22, 2008, releasing 4 million cubic meters (1.1 billion gallons) of coal fly ash slurry. This was the largest fly ash release in U.S. history, enough to cover 300 acres in a layer of toxic coal fly ash sludge several feet thick.¶ The cleanup will take years to complete and will cost the Tennessee Valley Authority (TVA) near $1 billion. These risks caused the EPA to classify 44 fly ash storage ponds in 26 communities as high hazards, meaning that failure of the dams and uncontrolled release of the slurry could cause death and significant property damage.** The health risks presented by these spills appear to be low for water, as the ash seems to retain heavy metals even when submerged for long periods in water (Ruhl et al. 2009). However, ecosystems downstream may be negatively impacted, and there is a risk that particles containing radioactive isotopes, mercury, or arsenic could become airborne and inhaled. A more sustainable alternative to storage of fly ash in retention ponds is to use it in construction materials or as a soil amendment (Ferreira, Ribeiro, and Ottosen 2003).

9.3.3 Carbon Capture and Storage

Carbon emissions from coal-fired power plants could be captured and stored in the ground. However, several obstacles to widespread adoption of carbon capture and storage (CCS) exist:

- CCS is energy-intensive, meaning that coal-fired power plants with CCS must burn 20% to 44% more coal to generate the same amount of electricity (Viebahn, Vallentin, and Fischedick 2009).
- Large-scale application of CCS technology is unlikely before 2020. If the United States expands the use of coal, it will build many coal-fired power plants within the next decade

* See https://www3.epa.gov/radtown/, retrieved August 25, 2016.
† http://www.epa.gov/hg/about.htm.
‡ http://www.cnn.com/2011/12/21/health/epa-mercury-rule/index.html.
§ https://www.epa.gov/coalash/coal-ash-rule, retrieved August 25, 2016.
¶ http://earthobservatory.nasa.gov/NaturalHazards/view.php?id=36352.
** "E.P.A. Lists 'High Hazard' Coal Ash Dumps," Shaila Dewan, *The New York Times*, June 30, 2009, http://www.nytimes.com/2009/07/01/science/earth/01ash.html?ref=us?_r=2, retrieved August 25, 2016.

before CCS technology is available. Fortunately, new coal-fired power plants are not being built in the United States because natural gas has become cheaper than coal.

- CCS is costly. Adding CCS to a coal-fired power plant would increase the cost of electricity produced by 50% (see Chapter 11).
- Carbon dioxide injected at high pressures into deep aquifers may rise through fractures, potentially contaminating shallow groundwaters that may be used for drinking water. Dissolving carbon dioxide in water lowers the pH, which may increase the concentrations of toxic elements such as heavy metals (Zhu and Schwartz 2011).
- Better alternatives to fossil fuel CCS plants exist; greenhouse gas emissions for electricity produced by solar, thermal, and wind are <10% of emissions from fossil fuel CCS plants (Table 9.2).
- Injection of carbon dioxide into brittle rocks can trigger earthquakes that would likely fracture the rocks and make them permeable to carbon dioxide flow (Zoback and Gorelick 2012).

And many unanswered questions remain: Where would we inject the carbon dioxide, given the anticipated objections? What is the probability of it escaping, perhaps explosively, after injection? Who would be liable if this happened? CCS may not be a panacea. And widespread adoption of CCS technology would reduce carbon dioxide emissions but would not solve all of the other problems associated with coal mining, burning, and waste disposal.

However, given that expansion of nonrenewable energy sources cannot keep pace with demand, it makes sense to try to make abundant coal a cleaner energy source. We are making progress: American Electric's Mountaineer Power Plant is the first plant in the world to use CCS. Scrubbers at the plant remove 98% of sulfur dioxide emissions and 90% of nitrogen oxides (Biello 2010). Although they capture only 2% of carbon dioxide emissions, the project represents "proof of concept," and they plan to expand the CCS capacity. Power utilities are already building more plants with CCS, including the resurrected FutureGen plant and plants in France, Germany, and China. Many other CCS plants are in the planning stages (Biello 2010). Although these projects cost billions of dollars and may double the cost of coal-produced electricity, companies are moving forward in anticipation of binding carbon dioxide emissions standards resulting from an international climate change agreement.

9.3.4 The True Cost of Coal Use

The true cost of coal appears to be much greater than the market cost. To calculate the true cost of coal use we must include all externalities, from mining through burning, waste disposal, climate effects, and health effects from pollution. The U.S. National Research Council published a report in which it estimated the external costs of energy sources (USNRC 2010). It used LCA to evaluate external costs and benefits related to health, environment, security, and infrastructure for each type of power production over its entire life cycle. Because it had been previously identified as a large external cost, the study focused on air pollution, especially the effects of emissions of particulate matter (PM), sulfur dioxide (SO_2), and oxides of nitrogen (NOx). By determining the monetary value of energy-related damages to health and the environment, it found that most external costs were related to health, particularly premature mortality. In part this finding resulted from a lack of data to assess the external costs resulting from loss of ecosystem services or nongrain agricultural crops.

Most external costs are incurred during fuel use, that is, during electricity production. The study found that of all the energy sources used to produce electricity, coal was responsible for by far the largest damages. Damages resulting from emissions of PM, SO_2, and NOx from coal-fired power plants in 2005 amounted to roughly $62 billion, or $156 million on average per plant (USNRC 2010). These external costs amount to 3.2 cents/kWh, more than 20 times higher per kWh than for electricity produced from natural gas. Considering that in 2005 the economic cost of coal-produced electricity was ~3.0 cents/kWh, adding external costs would more than double the true cost of coal. These external cost estimates do not even include the effects on global warming, which the study treated separately due to large uncertainties.

Making the hidden costs of energy production public may spur policy changes that correct the failure of the market to produce accurate price signals. Plants with the highest external costs per unit of power generated should be targeted for early closing or conversion to natural gas, biomass- or waste-burning facilities. Americans will save large amounts of money and be safer and healthier if the United States phases out these coal-burning power plants as quickly as possible. And we are making good progress on that front. Between 2009 and 2015 more than 200 coal-fired power plants closed in the United States (Chait 2015).

Expanding the use of coal in the United States without CCS and mining reform would be a huge mistake. Conventional coal use leads to mountaintop removal, pollution, failure of coal slurry retention ponds (e.g., Martin County, KY 2000) and fly ash retention ponds (e.g., Kingston, TN 2008), and maximum carbon dioxide emissions and climate change. Because federal regulation of the coal industry has been inadequate, some states are taking action. For example, in 2016 Oregon voted to eliminate coal from its power supply by 2030.* Coal production in the United States is now at its lowest level in 30 years.† Until banning mountaintop removal mining and the construction of power plants without CCS makes coal truly clean, the United States must make every effort to decrease conventional coal use and to replace coal with sustainable, renewable, nonpolluting energy sources.

Unfortunately, global coal consumption is rapidly rising, driven primarily by low cost and the resulting expansion of coal-fired power plants in China and developing countries (Steckel, Edenhofer, and Jakob 2015). This has caused the carbon intensity of global energy production to increase. In 2012 China was responsible for 47% of global coal consumption and 82% of the growth in coal consumption since the year 2000.‡ Building new coal-fired power plants locks a country into coal use for the expected lifetime of each plant, typically about 30 years, which will make it more difficult to reduce future carbon emissions. However, the good news is that between 2008 and the first quarter of 2015 China's annual coal production growth fell from 12.7% to –3.5%, caused by rapid increases in renewable energy production and a slowdown of China's economic growth.§

9.4 NATURAL GAS

Natural gas consists primarily of methane plus minor amounts of other light hydrocarbons. Conventional natural gas is associated with oil because it forms from similar material through similar processes. Oil and gas rise through permeable rocks in the Earth's crust until they become trapped by an impermeable layer. The less dense gas often rises above the oil, so when a drill penetrates the overlying impermeable layer the gas can be released explosively if proper techniques are not used to contain the flow of gas. This is what caused the **Deepwater Horizon** disaster. Methane gas that escapes to the atmosphere is considered a pollutant and is a potent greenhouse gas. It has a half-life of only seven years in the atmosphere because it reacts with atmospheric oxygen to form carbon dioxide and water.

Natural gas is considered a "clean" fossil fuel because burning it emits less carbon dioxide and sulfur per unit energy than coal or oil. It is the fuel most commonly used for heating. Because it is a gas, natural gas takes up more volume than liquid oil and gasoline, making it more expensive to transport. Currently natural gas is compressed and transported under high pressure as a liquid. This combined with it being highly flammable means that natural gas can cause dangerous explosions and fires. The difficulty of transporting natural gas means that it is rarely exported, so unlike

* https://e360.yale.edu/digest/oregon_to_eliminate_coal_from_state_energy_mix_by_2030/4668/, retrieved March 15, 2016.
† https://e360.yale.edu/digest/us_coal_production_drops_to_30-year_low_in_2015/4629/, retrieved March 15, 2016.
‡ http://www.eia.gov/todayinenergy/detail.cfm?id=9751.
§ https://www.aei.org/publication/the-great-chinese-coal-collapse/.

the globally traded commodities oil and coal, natural gas is continentally traded (Hughes 2010), although global exports of liquefied natural gas are expected to expand.*

In 2013 in the United States natural gas provided 27% of primary energy (Figure 9.1). Natural gas is an efficient and safe fuel for automobiles, with an octane rating of 135 (Deffeyes 2001). Worldwide there are about 15 million **natural gas vehicles** fueled by compressed natural gas in countries like Iran, Pakistan, and Argentina (Deffeyes 2001). Natural gas can also be used to produce hydrogen for use in autos with hydrogen fuel cells. It is also used to synthesize ammonia for use in fertilizer production. For these reasons, energy experts expect natural gas to fill the gap during the transition from coal and oil to sustainable renewable energy sources (Meadows, Randers, and Meadows 2004).

Hubbert (1956) used his empirical method to predict future conventional natural gas production in the United States. Later, Roger Naill (1973) used a system dynamics model to do the same. Although the approaches were very different, the results were similar: a plot of production rate versus time yielded a bell-shaped curve that peaked in the early- to mid-1970s, consistent with the observed peak in 1973. Numerous system dynamics studies of fossil fuel production have confirmed the accuracy of the Hubbert approach for estimating peak production.† However, exploration in the United States and worldwide has since uncovered large unconventional natural gas reservoirs, which combined with improved drilling and recovery technologies have roughly doubled natural gas reserves since Hubbert made his first prediction. Development of unconventional gas sources caused natural gas production in 2011 to exceed the previous peak in 1973 (BP 2012). In the United States, the natural gas production curve has two peaks, much like that for oil (Figure 6.3). So while the Hubbert curve approach may work for conventional gas alone, it is not suitable for making projections for combined conventional and unconventional production. Globally, depletion of natural gas reserves would occur in ~50 years at current rates of consumption (BP 2015).

Unconventional natural gas is obtained from sources other than oil fields. Unlike oil, natural gas is not restricted in depth to a "window," so deeper drilling can uncover new reserves of unconventional natural gas. Unconventional natural gas can also be produced by biofuel production methods that transform biomass to natural gas through bacteria-mediated anaerobic decay. The same processes occur in swamps and landfills, where bacteria obtain energy by catalyzing the breakdown of heavy hydrocarbons to form methane (swamp gas). Decomposition of organic material in landfills and sewage also produces methane. Landfills used to burn off produced methane to prevent explosions, but it is becoming more common for landfills to recover it for use as a fuel. Similarly, sewer treatment plants in large cities such as Los Angeles and New York City are starting to recover methane and use it to produce electricity. Another unconventional natural gas source is **methane clathrates** or "methane ice" stored in sediments on the continental shelf. Methane clathrates may store more energy than all of the other fossil fuels combined (Lavelle 2012). However, we have yet to develop a safe method for extracting natural gas from these deposits.

9.4.1 Hydraulic Fracturing

In the United States, the most important unconventional natural gas source is shale, which now provides more than one-third of our natural gas (Lavelle 2012). Production of shale gas in the United States increased dramatically between 2008 and 2015. The shale formation with the largest natural gas reserves is the Marcellus, which stretches from West Virginia to New York through Ohio and Pennsylvania. Natural gas reserve estimates for the Marcellus Shale in trillion cubic feet increased from only 2 in 2002 to between 43 and 144 in 2011.‡ This attracted the attention of drillers, who have greatly expanded the combined use of high-volume **hydraulic fracturing** (HVHF, also known

---

* http://www.eia.gov/forecasts/ieo/nat_gas.cfm, retrieved August 25, 2016.

† http://www.systemdynamics.org/DL-IntroSysDyn/energy.htm.

‡ Kevin Begos, "Marcellus Shale Gas Reserves Top Earlier Estimates," AP, August 24, 2011.

as "fracking") and **directional drilling** to recover gas from shale at depths from 1 to 6 kilometers (3,000 to 20,000 feet). Shale is impermeable, which is why the gas remains trapped in the Marcellus Shale. To recover the gas, drillers inject large volumes of water and chemicals under very high pressure to fracture the underlying impermeable rocks, making them permeable so that natural gas can be extracted. The HVHF fluid contains ~99.4% water and 0.5% sand, and <0.5% of proprietary chemicals that are toxic to wildlife and humans such as benzene, ethylbenzene, and toluene (Walther 2014). Although the concentrations of the chemicals may be low, the amounts pumped into the ground can be large because each well requires as much as 7.5 million to 15 million liters of HVHF fluid per well. Only 9% to 53% of HVHF fluid flows back up the well and is recovered as wastewater, meaning that usually more than half of the fluid remains in the ground (Vidic et al. 2013). Wastewaters from HVHF are usually highly saline and can contain low levels of radioactive radium, radon, uranium, and thorium (Kharaka et al. 2013). Wastewaters are generally reused for HVHF or transported to a treatment plant to remove the chemicals.

HVHF can have impacts on water quantity, water quality, earthquake risks, air pollution, human health, the economy, and global greenhouse gas emissions. Because smaller volumes of rock are sampled by a HVHF well than a conventional well, more wells are required to retrieve the same volume of natural gas (Walther 2013). Over 3,000 wells have been drilled into the Marcellus Shale, mostly in Pennsylvania and Ohio. This combined with the large volumes of fluid that must be transported to and from the site means that HVHF operations can make life unpleasant, especially in populated areas. While there is concern in water-stressed areas that HVHF uses large amounts of valuable water, LCA has shown that in Texas the water footprint of natural gas production by HVHF is smaller than that of conventional thermoelectric generation (Grubert, Beach, and Webber 2012). Industry is starting to use saline water and recycled wastewater to reduce the impact of HVHF on freshwater supplies (King 2012; Vidic et al. 2013). Ironically, HVHF is exempt from regulation under the Clean Water Act, and oil and gas companies want it to remain unregulated so they can continue to externalize their costs and make greater profits.

Much of the public debate on HVHF centers on the use of proprietary chemicals that mining companies inject into the subsurface to lubricate and prevent corrosion during the natural gas extraction process. These chemicals can potentially contaminate groundwater supplies by ascending through vertical fractures. However, to date evidence of groundwater contamination by HVHF fluids is rare, but not unheard of (Llewellyn et al. 2015). In contrast, elevated methane concentrations have been found in groundwater samples near shale gas wells in Pennsylvania, and evidence indicates that the methane is from deep source rocks, further implicating the unconventional gas drilling operations (Jackson et al. 2013). This is believed to be caused by improper well construction, which has affected ~1% to 3% of gas wells in Pennsylvania (Vidic et al. 2013).

Because it affects a relatively small volume of rock, experience in the United States has shown that the risk of HVHF inducing earthquakes that can be felt is small (National Research Council 2013), although it appears to be more common in western Canada (Atkinson et al. 2016). Disposal of HVHF wastewater by injection into wells has a greater potential to cause small earthquakes.* A dozen earthquakes in northeastern Ohio associated with injection of gas drilling wastewater led the state to temporarily ban wastewater injection in January 2012. Drilling resumed in March 2012 but with much stricter regulations.† A two-year study of the Barnett Shale found that disposal of drilling wastewater in injection wells induces small magnitude earthquakes (Frohlich 2012).

Unconventional oil and gas operations can cause environmental damage, but the impacts can be reduced by using **best practices** (Northrup and Wittemyer 2013). In areas near these operations human health may be negatively impacted by increased concentrations of volatile organic compounds, but while little research has been done in this area, estimates suggest that levels are below

* http://earthquake.usgs.gov/research/induced/, retrieved August 25, 2016.

† http://usatoday30.usatoday.com/money/story/2012-03-09/fracking-gas-drilling-earthquakes/53435232/1, retrieved February 21, 2013.

the level of concern (Bunch et al. 2013). Social impacts on communities near large unconventional oil and gas operations include negative impacts such as increased noise and traffic congestion, road damage, increased prices for housing, and increased crime and mental health issues caused by stress, while positive impacts include local job creation and economic growth. These contrasting impacts can lead to polarization of attitudes toward oil and gas companies in affected communities (Schafft, Borlu, and Glenna 2013).

The glut of natural gas from hydraulic fracturing the Marcellus Shale caused natural gas prices in the United States to drop precipitously in 2008. These low prices have persisted through 2015, making natural gas a cheaper source of energy than coal. This has caused many power utilities to convert their power plants from coal to natural gas, which emits less carbon dioxide and other pollutants. For the first time in the United States, natural gas produced roughly the same amount of electricity as coal in 2015.* Together with a sluggish economy and high oil prices, the switch from coal to natural gas caused U.S. carbon dioxide emissions in billion metric tons to decrease from roughly 6 in 2006 to 5.3 in 2013 (EIA 2015). However, there is controversy over whether substituting unconventional natural gas for coal is actually reducing greenhouse gas emissions, as the previously quoted emission estimates do not account for leakage of greenhouse gases during production. An early LCA analysis suggested that methane leakage at well sites or during transport made the carbon footprint of natural gas worse than coal (Howarth, Santoro, and Ingraffea 2011). More recent direct measurements show that emissions from well operations are lower than previously thought and similar to EPA estimates of ~0.5% of gross production (Allen et al. 2013). Studies based on larger-scale atmospheric measurements indicate greater losses, and in 2016 the EPA increased its estimates of CH_4 emissions from the oil and gas industry; it now accounts for one-third of total U.S. methane emission.† Another problem is that the recent very low prices for natural gas may be slowing the adoption of renewable energy sources that have much smaller carbon footprints (Schrag 2012). The good news is that substitution of natural gas for coal leads to reduced emissions of sulfur, nitrogen mercury, and particulates (Jackson et al. 2014), and the external costs per kWh for natural gas are much lower than for coal. We conclude that because it is a relatively clean-burning fossil fuel, use of natural gas is preferable to coal. However, use of natural gas as an energy source should eventually be phased out to mitigate climate change unless it employs CCS to eliminate emission of carbon dioxide produced during burning.

9.5 NUCLEAR POWER

Nuclear power is an important energy source. The United States has the most nuclear reactors in operation of any country (104), producing about 20% of electricity in the United States (Walther 2014). However, safety concerns and high costs have slowed the rate of nuclear power plant construction, and as a result global nuclear power production peaked in 2006 (Brown et al. 2015).

Importantly, nuclear power provides ~70% of the zero-carbon electricity in the United States (MIT Energy Initiative 2011). However, in the United States, nuclear power has always been a controversial energy source because it uses radioactive fuel and produces radioactive waste, and radiation is known to cause cancer. This has raised safety concerns, which resulted in high levels of oversight and regulation, which in turn have made nuclear power plant construction and operation very expensive. Furthermore, the United States does not have a site for long-term geologic storage of radioactive waste from nuclear reactors. As a result, nuclear power plant construction in the United States came to a standstill in the 1980s that continued through the 1990s. However, recent concerns about GCC have renewed interest in nuclear power. Nuclear power plants do not emit carbon dioxide, making them an attractive option for mitigating GCC.

* http://www.eia.gov/electricity/monthly/epm_table_grapher.cfm?t=epmt_1_1, retrieved September 9, 2016.

† https://www.washingtonpost.com/news/energy-environment/wp/2016/04/15/epa-issues-large-upward-revision-to-u-s-methane-emissions/.

Historical data and modeling show that nuclear (along with wind) is the safest form of energy, with the lowest death rate per unit energy of all forms of energy (MacKay 2009; OECD and NEA 2010). Planned third generation plants are smaller and safer. However, concerns about lack of storage facilities for nuclear waste continue. Also, the 2011 **Fukushima Daiichi nuclear disaster** in Japan has called the safety of nuclear power plants and the practice of storing nuclear waste in pools on-site into question. We will examine the pros and cons of nuclear power in detail.

9.5.1 Nuclear Fission

In atoms the nucleus stores nuclear energy and the electrons store chemical energy. The nucleus contains more than one million times more energy than the electrons, so the mass of fuel and waste for nuclear energy is about one million times smaller than for sources of chemical energy such as fossil fuels (MacKay 2009). For example, 2 g of natural uranium in a fission reactor produces the same amount of energy as 16,000 g of fossil fuels, and the proportions of waste produced are similar. This gives nuclear energy a distinct advantage; MacKay (2009) believes that the volume of waste is so small that "nuclear waste is only a minor worry, compared with all the other forms of waste we are inflicting on future generations."

Atoms with medium-sized nuclei are the most stable, that is, they have the lowest energy per unit mass. Thus, splitting a large nucleus, such as that in uranium, into several medium-sized nuclei releases large amounts of energy. The products of a nuclear fission weigh less than the reactants, and the difference is the mass converted to energy according to Einstein's famous equation $E = mc^2$, where c, the speed of light, is a very large number.

Most of the energy produced in nuclear power plants comes from the fission of an isotope of uranium $^{235}_{92}U$, where 235 is the atomic mass (the number of protons plus neutrons in the nucleus) and 92 is the atomic number (the number of protons in the nucleus). This isotope makes up 0.7% of natural uranium, but we must enrich it to 3% before we can use it as fuel in a burner reactor. MacKay (2009) calculates that seawater extraction of uranium used in inefficient "once-through" fission reactors could globally produce 7 kWh per day per person for 1,600 years (the overturn time of oceanic circulation). Using fast breeder reactors that are 60 times more efficient than once-through reactors would increase that to 420 kWh/d per person, more than even Americans consume (MacKay 2009). Thus, nuclear fission could be a significant long-term source of energy.

9.5.2 Nuclear Reactors

A **nuclear reactor** includes a core, control rods, coolant, and reactor vessel.* In the reactor neutrons smash into ^{235}U atoms, splitting them apart and releasing energy. This also releases neutrons that can split other atoms. Control rods absorb neutrons and moderate the rate of reaction, allowing the nuclear "chain reaction" to continue at a constant rate (Walther 2014). **Nuclear fission** releases energy primarily as heat, which power plants use to boil water. The resulting steam flows upward and pushes the blades of a turbine to generate electricity. In the famous **Chernobyl disaster**, operators pulled the control rods too far out of the reactor, causing ^{235}U nuclei to fission too fast. This increased the temperature until the fuel became so hot that it melted through the bottom of the containment vessel, causing a **nuclear meltdown**. In contrast to a nuclear power plant, a **nuclear bomb** initiates an uncontrolled, runaway fission process that suddenly releases a huge amount of energy. Fortunately, nuclear explosions are not possible in nuclear power plants.

Designs for U.S. nuclear power plants are not optimal or standardized, making power plants expensive to build and maintain. In contrast, France has only one nuclear power plant design, which decreases the cost per unit through economies of scale. Every power plant in the United States is different, meaning that each plant has been custom-built at great expense, and whenever a part

* http://www.youtube.com/watch?v=fjgdgAhOzXQ&feature=related and http://www.youtube.com/watch?v=u0VjHg0juz4.

breaks a replacement part must be custom-fabricated at great cost. If the United States chooses to start building new nuclear power plants, it should settle on a single, optimal design to reduce costs and increase safety and efficiency.

Use of **small modular reactors** that are manufactured in a factory and transported to a site fully constructed would partially or completely solve problems associated with nuclear power production in the United States, as would third-generation "mini" nuclear reactors.* Decentralizing nuclear power production by building many small reactors rather than a few large reactors more evenly and fairly distributes the risk, making it more equitable. Also, it reduces line transmission loss and the need to build high-capacity transmission lines. One manufacturer, Babcock and Wilcox Company, claims that "each BWXT mPower reactor brought online will contribute to the reduction of approximately 61 million metric tons of carbon dioxide emissions…over the life of the reactor.†" Furthermore, throughout its projected 60-year lifespan, it would store the **spent nuclear fuel** (SNF) underground, at least temporarily negating the need for transportation and storage at a centralized location. Older plants thermally polluted rivers by releasing their cooling water downstream, but modern reactors use cooling towers, and future reactors will likely be air-cooled. Thus, the new generation of nuclear power plants may solve many problems that have plagued the industry in the past. The TVA hopes to finish construction of the first of these plants by 2018.

9.5.3 Obstacles to the Expansion of Nuclear Power

Several obstacles prevent growth of nuclear power in the United States. First, a large part of the public resists expansion of nuclear power because they fear all things nuclear. Nuclear power will always be associated in people's minds with the use of nuclear bombs in WWII and the fear associated with proliferation of nuclear warheads during the Cold War and the accidents at Three Mile Island, Chernobyl, and Fukushima (Weart 1987). Furthermore, radioactivity is particularly frightening to people because it is invisible and outside their normal experience. People generally most fear events that are unpredictable and have high costs such as nuclear meltdowns.

Psychologists have shown that people are insensitive to variations in the level of risk when probabilities are low. They will judge a probability of 0.0001% the same as a probability of 0.01%, so that their level of worry is not proportional to the probability of the threat. Worry will only be eliminated if the probability is reduced to zero by an outright ban on nuclear power. When people focus on the potential for a nuclear disaster, fearful thoughts cause them to overweight its probability. This effect is even more pronounced if an accident occurred recently, a cognitive bias called the **availability heuristic** (Kahneman 2011).

Despite the fear it invokes, nuclear power has a remarkable safety record in the United States. The only significant nuclear power plant accident ever in the United States was the **Three Mile Island accident** in central Pennsylvania in 1979, which released a small amount of radioactivity into the environment (Walther 2014). Producing electricity with a nuclear reactor is one of the safest activities in which our society engages. Individuals face far less risk living near a nuclear power plant than driving a car, walking along a street, smoking, bicycling, or swimming. Historical data (MacKay 2009) and probabilistic safety assessments considering both immediate and delayed fatalities (OECD and NEA 2010) show that nuclear power is much safer than coal power. Even the levels of radiation exposure near power plants are higher for coal-fired than nuclear (McBride et al. 1978). Furthermore, replacing coal-fired power plants with nuclear power plants is estimated to have prevented emissions of 64 Gigatonnes CO_2e and 1.8 million air pollution-caused deaths, and could save millions more lives by mid-century (Kharecha and Hansen 2013).

Besides cost, the most important obstacle to growth of nuclear power in the United States is that we have no site to store the radioactive SNF from fission reactors. However, the recent recognition of

* *The Tennessean*, June 11, 2009, "TVA Plans Mini Nuke Reactor for Tenn.," Bill Theobald and Dave Flessner.

† http://www.bwxt.com/nuclear-energy/utility-solutions/smr/bwxt-mpower, retrieved August 25, 2016.

the need to reduce carbon dioxide emissions has reopened the debate: should we expand the use of nuclear power in the United States? Though they vary widely, LCA estimates of the carbon footprint of nuclear power are far lower than all fossil fuels (Table 9.2). With the renewed interest in nuclear energy, the U.S. Nuclear Regulatory Commission issued licenses to seven new plants between 2012 and 2016.* However, several questions remain unanswered. Can the United States choose a site and build a facility for storage of SNF? If not, is the current industry practice of storing SNF on-site in pools safe? And would nuclear energy still be cost-effective if we include the cost of waste disposal?

One possible solution to the SNF problem is to use breeder reactors that reprocess the fuel, which is 60 times more efficient than the once-through reactors we currently use. Recycling the waste sounds like a good choice from an environmental standpoint, as it would reduce how much SNF must be disposed of and the required amount of environmentally harmful uranium mining. However, breeder reactors are very expensive and difficult to operate, and since none of the handful of breeder reactors provided affordable power, most have been shut down (Daniel 2012). Also, breeder reactors produce large amounts of plutonium, which raises risks associated with waste disposal and proliferation of material that can be used to make atomic bombs.

If in the future we build a centralized storage facility, the waste will be so radioactive that by law we will have to monitor it for thousands of years. According to Kellogg and Pettigrew (2008), "The energy needed to run even so much as a light bulb, let alone a full security operation guarding spent fuel, for the duration of nuclear waste's radioactivity would rival that of all the energy ever produced by all of the world's nuclear power plants combined." Thus, they argue, uranium fuel production and waste disposal would consume far more energy than fission reactors produce, making the EROEI of nuclear power less than one, which would mean that nuclear power is an unsustainable form of energy. However, this argument fails to take into account the economic and technological benefits of producing power today, which is an investment that spurs economic growth and may lead to the development of new energy producing technologies. Like other investments, the benefits are compounded over time, so over a million years the economic returns would dwarf the cost of operating a light bulb for one million years. Furthermore, the argument shifts the timeframe for decision-making to a million years, while our sustainability-driven decision making operates on a timescale of hundreds to thousands of years. So the energy produced by fission reactors today is almost certainly worth more than the energy used to ensure safe storage of waste in the future.

A problem associated with nuclear power in the past has been that it is highly centralized, with a few large power plants. Plans for waste disposal have been even more centralized, with the most recent plan to store all of the nation's waste in one site, Yucca Mountain, a plan that was abandoned in 2011.† This would have led to an unequal geographic distribution of risk, and the potential lack of equity has led to large-scale opposition to nuclear power. A more equitable solution would be to decentralize nuclear power production and waste disposal, although it's unclear if that would reduce overall risk.

---

* http://www.nrc.gov/reactors/new-reactors/col-holder.html, retrieved August 25, 2016.

† After the federal government spent $13.5 billion developing a high-level nuclear waste disposal site at Yucca Mountain, about 100 miles northwest of Las Vegas, Nevada, funding ended in 2011. The United States spent more than 30 years developing the Yucca Mountain site before, according to the Government Accounting Office, it fell victim to politics: Senate majority leader Harry Reid represented the southern part of Nevada that includes Yucca Mountain, where resistance to the project has always been strong, and he followed through on his vow to kill the project.

Even if the United States had followed through, the Yucca Mountain facility would not have been large enough to accept all of the waste we would have produced by the time it opened. The United States currently has 103 operating nuclear power plants (Wallace 2005). By law, the capacity of the Yucca Mountain facility was limited to 70,000 tons, of which 63,000 tons were designated for SNF and 7,000 tons for defense waste. However, estimates are that by 2050 the United States will have 84,000 tons of SNF (Carter and Pigford 1998). The United States now has SNF at more than 100 sites in 42 states (J.C.S. Long and Ewing 2004), and we have now eliminated our only option for safely disposing of it. The federal government now pays large annual fines to the utility companies for breach of contract: they had promised to take the SNF off the hands of the utility companies by now, but utilities still store SNF at every nuclear plant.

Nuclear power suffered a setback when an M 9.0 earthquake struck northern Japan March 11, 2011. The earthquake and the subsequent 40 m high tsunami that it generated left nearly 16,000 dead and 2,500 missing, and damaged several nuclear power plants in Fukushima (Normile 2016b). The nuclear reactors lost cooling water due to damaged pumps or ruptured pipes, causing a complete meltdown in Unit 1 and partial meltdowns in Units 2 and 3, causing releases of large amounts of radiation to the environment. The backup plan of using seawater to cool the reactors led to large releases of radioactivity into the ocean. One reactor had two hydrogen explosions that occurred when the zirconium casings on the fuel rods heated to temperatures >1,200°C and reacted with water to produce H_2 gas, which exploded when it came into contact with atmospheric oxygen. The Fukushima disaster ranked at the top of the seven-step International Nuclear and Radiological Event Scale, equaling the 1986 Chernobyl reactor explosion.* The nuclear fuel has still not been located, but must be removed before the site can be remediated, which is expected to cost more than $9 billion. Fortunately, because evacuations were swift, winds blew most radiation quickly out to sea, and the reactors released only one-tenth the radiation released in the Chernobyl disaster. Radiation exposures were low enough that scientists do not expect dramatic increases in cancer rates (Normile 2016a).

The nuclear disaster in Japan has highlighted the dangers associated with on-site storage of SNF in pools. SNF recently removed from reactors is even more radioactive than the fuel inside the reactor but has no protective container surrounding it like the reactor fuel, which is encased in 6-inch-thick steel walls. Experts in the United States have been calling for a safer form of SNF storage called dry-cask storage, but even when dry-cask storage is used, SNF must first cool off in pools for a few years after being removed from the reactor before they can be transferred to casks.†

These problems will likely increase resistance to the construction of new nuclear power plants in the United States, especially in areas like the West Coast that are vulnerable to major earthquakes and tsunamis. Nuclear power plants are very safe under normal operating conditions, but natural hazards can make them become very dangerous. Once again, the precautionary principle applies: we did not anticipate an earthquake and tsunami of the magnitude seen in Japan in March 2011. Now we realize that nature can destroy any safeguard system we devise. It is safer to rely on renewable sources of energy like solar and wind because they do not use hazardous materials that can endanger people during extreme events, and they are generally smaller in scale because renewable energy production is more distributed than nuclear.

Many studies have shown that the United States can meet its energy needs with renewable energy alone (see Chapters 10 and 11). In the United States renewable energy sources are limited not by thermodynamics, but by a lack of political will.

9.5.4 Summary

So what are the advantages of nuclear power plants? They have near-zero carbon dioxide and pollutant emissions, the supply can last more than 1,000 years if we use breeder reactors or extract uranium from seawater, and nuclear energy is safe compared with other forms of energy production. What are the disadvantages? The nuclear fuel cycle releases small amounts of radiation to the environment at every stage. Nuclear reactor accidents pose a very small but real risk. Terrorists or hostile countries could steal enriched uranium destined for fission reactors or plutonium from breeder reactors to make nuclear bombs. The United States has no permanent SNF disposal facilities, and won't have any for at least 20 more years. Finally, nuclear power is not cost-effective: electricity generated using nuclear fission reactors is projected to be more expensive in 2020 than electricity produced using natural gas, wind, or hydroelectric and about the same as coal (EIA 2015, Chapter 11).

---

* http://www.cnn.com/2011/WORLD/asiapcf/05/06/japan.nuclear/index.html?eref=mrss_igoogle_cnn.

† http://www.nrc.gov/waste/spent-fuel-storage/dry-cask-storage.html, retrieved 8/25/2016.

However, as noted previously, including externalized costs like healthcare makes coal power more expensive than nuclear power.

Nuclear power is a very complicated, expensive, centralized form of energy production that requires much government involvement (regulation and oversight), has a very vocal opposition, and large potential problems. Decentralized, renewable energy sources pose fewer risks, and some are currently more cost-effective (Chapter 11). Distributed energy systems that are modular and have diverse, renewable sources are more resilient than centralized, resource-intensive nuclear power systems.

Because it can theoretically obtain sufficient energy from safer renewable sources alone, the United States should start building new nuclear power plants only if an all-out effort to rapidly build new renewable power plants nationwide and to decrease demand through conservation and efficiency measures fails to satisfy our energy needs. However, we will show below that in many countries and regions, primarily those with high population density and insufficient available land such as Japan and Europe, low-energy density renewable energy sources cannot provide enough energy to meet current energy needs. Countries without enough available space to produce wind or solar energy or biofuels will likely have to rely on nuclear energy. And in light of the fact that, since 2011, Germany and Japan, two countries with high population densities, have dramatically decreased the share of their electricity produced by nuclear power without negatively impacting their economies,* it's fair to ask why nuclear power would be necessary anywhere.

9.6 CONCLUSIONS

The many problems caused by coal mining and burning earn it a spot in the ABCs of unsustainability: autos and airplanes, beef, and coal. Coal is the least preferred energy source, followed by oil and natural gas. Nuclear power should be used only in cases where safer renewable energy sources are inadequate. We will take a detailed look at sustainable energy production in Chapter 10, and sustainable energy consumption and policy in Chapter 11.

HOMEWORK PROBLEMS

1. Should clean coal technology be developed, given our energy needs, supply of coal, and economic stability?
2. What is the dominant preferred method of coal mining in the Appalachians?
3. Name one desirable and one undesirable trait of nuclear power.
4. How do natural gas and nuclear energy production contribute to global climate change mitigation?
5. What is the biggest obstacle to the expansion of nuclear power?
6. Use the U.S. Energy Information Administration website (http://www.eia.gov) to find historical data on consumption of coal and natural gas in the last decade. Plot the data in a spreadsheet and compare the trends. Is increased natural gas production causing a decrease in coal production, that is, is displacement or substitution occurring rather than simply increased fossil fuel consumption?

---

* http://www.earth-policy.org/plan_b_updates/2012/update103.

10 Renewable Energy Sources

> In a sense, the fossil fuels are a one-time gift that lifted us up from subsistence agriculture and should eventually lead us to a future based on renewable resources.
>
> **Kenneth Deffeyes**
> *2001*

We cannot create or destroy energy, we can only capture it. The sun provides, either directly or indirectly, nearly all of the energy available to us. Plants capture solar energy directly through photosynthesis. Fossil fuels contain the energy of sunlight captured directly by plants millions of years ago. Photovoltaic (PV) cells also capture sunlight energy directly. Other energy sources capture the energy of sunlight indirectly. Energy from the sun powers the flow of air and water. We usually capture the kinetic energy of wind and water by using a **turbine** that transfers the energy to an **alternator**, an electrical generator that produces alternating current.

In this chapter we will see that renewable wind, water, and sun energy sources are sustainable because they are renewable, clean, safe, and nearly carbon-free. Although they have low energy densities, meaning that they require larger areas of land or water than nonrenewable energy sources to produce a fixed amount of energy, they are sufficient to meet the energy needs of most countries. Perhaps the greatest challenge facing humanity is to transition to renewable energy as rapidly as possible to mitigate global climate change.

Sunlight and wind are public goods because they have low excludability and low subtractability, meaning they are freely available to everyone (Section 5.4.3). This is fortunate, as other sources of energy such as fossil fuels have often been monopolized by the wealthy and powerful to further their wealth and power. Even the poor can afford to use solar cooking, passive solar heating, and biofuels, and as prices for PV panels continue to decrease they are becoming increasingly common in poor villages.

As we saw in Chapter 9, nonrenewable energy sources are stock-limited. In contrast, renewable energy sources are flow-limited: we can't capture energy faster than flowing wind or water provide it. Fortunately, wind and water flow continuously, driven by energy from the sun, so the supplies of renewable energy sources are continuously replenished. Because renewable energy sources are flow-limited, they cannot support an indefinitely growing population, but they can provide an energy base for a sustainable society (Meadows, Randers, and Meadows 2004).

However, we have a long way to go before renewables become our dominant energy source. It takes decades to transform the energy infrastructure. Coal burning power plants will have to be replaced by new wind farms, solar panel arrays, biofuel production facilities, hydroelectric installations and, in select areas, geothermal plants. Our current electrical grid will have to be replaced with a smart grid that can transport electricity great distances from wind and solar sources with minimal power loss and that can deal with power generation intermittency (see Chapter 11). Currently the growth of renewable energy production is slightly greater than the growth of the Total Primary Energy Supply (TPES), which consists of all energy sources we get energy directly from and excludes secondary energy sources such as electricity and hydrogen gas. The percentage of TPES from renewables (including hydroelectricity) increased from 6.4% in 2011 to 9.3% in 2014 (BP 2015), or ~1% per year.

Why hasn't the percentage of electricity from renewables increased faster than 1% per year? Because electricity consumption is also growing exponentially. This makes it harder for renewables to displace nonrenewables; the growth in electricity consumption eats up most of the growth in renewable energy. Also, the installed base of renewable energy production is so small that it will take several decades for it to catch up with increasing energy demand and make a dent in the amount of fossil fuel consumed. The International Energy Agency (IEA) estimates that we must spend roughly 1% of global GDP between now and 2050 to wean the world off fossil fuels and cut carbon dioxide emissions in half (Sawin and Moomaw 2009). Some believe that we need nuclear power or coal + CCS to address climate change and rising energy demand (Deffeyes 2001; MacKay 2009). However, Sawin and Moomaw (2009) and Jacobsen and Delucchi (2011) claim that renewables plus increased energy efficiency are enough, and that they are the only technologies available now that can do the job.

Renewable sources of energy can potentially provide far more energy than we consume. As a point of reference, the global TPES is currently about 14 million MW. The Earth receives 8,000 times that amount from the sun, with about half reaching the ground (Richter 2010). However, in practice we can never capture all that energy. Renewable energy sources are just too diffuse to provide enough energy for people in all locations on Earth. In certain areas, with abundant sunlight and wind and low population densities, it will be possible, but many countries such as the United Kingdom cannot currently rely solely on renewable energy without reducing their energy consumption, as shown convincingly by MacKay (2009). Which renewable energy source is potentially sustainable for a community will depend not only on local availability but also whether ecologically sensitive areas will be affected, what the resource demands are (e.g., don't use water-intensive technologies in areas with water scarcity), and other economic and social constraints.

Renewable energy will not increase market penetration unless it is cheaper than fossil fuels. To compare the cost of electricity, a secondary energy source, produced by different primary energy sources, we compare the levelized cost of electricity (LCOE) produced using different primary energy sources in Figure 10.1. The LCOE is the break-even cost of electricity over the lifetime of use, a useful measure for comparing the cost effectiveness of different energy sources. It includes all of the different factors that affect the total cost, including the initial capital, discount rate, operation and maintenance, and fuel. It's like an economic Life Cycle Assessment (LCA) that accounts for all phases of product use (Section 9.2). Comparing LCOE estimates for the United States in the year 2020 we see that renewable energy sources like geothermal and wind are cheaper than any other primary energy sources, followed by natural gas, hydroelectric, coal, and nuclear. Natural gas + CCS and biomass are slightly more expensive, and solar PV is significantly more expensive, although the price of solar PV continues to drop dramatically. Because geothermal and wind

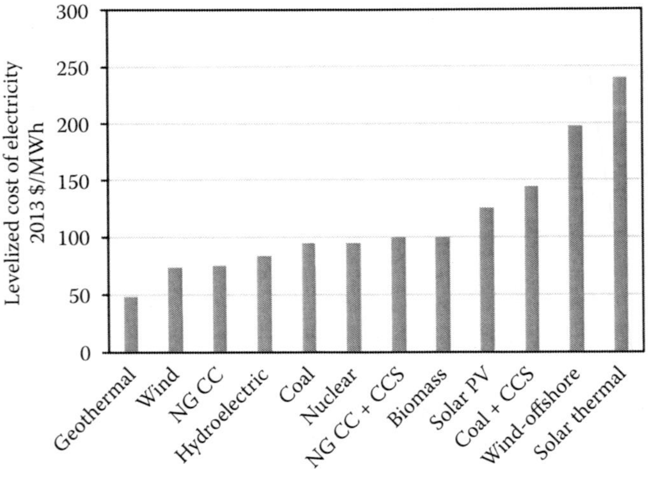

FIGURE 10.1 Levelized electricity costs for new power plants in the United States in the year 2020 (2013 dollars per megawatthour). "CC" is combined cycle, "CCS" is carbon capture and storage. (Data source: EIA. 2015. "Annual Energy Outlook 2015." http://www.eia.gov/forecasts/aeo/pdf/0383(2015).pdf.)

are the least expensive sources of electricity and also happen to have the lowest carbon footprints (Table 9.2), areas with high generation potential for these energy sources (high surface heat flow or continuously high wind velocity) should use them first to increase local energy production. In the following sections we will look at the promises and shortcomings of each type of renewable energy.

10.1 WIND

> First, there is the power of the Wind, constantly exerted over the globe....Here is an almost incalculable power at our disposal, yet how trifling the use we make of it! It only serves to turn a few mills, blow a few vessels across the ocean, and a few trivial ends besides. What a poor compliment do we pay to our indefatigable and energetic servant!
>
> **Henry David Thoreau**
> *From Paradise (To Be) Regained, 1843*

Wind power has great potential as an energy source. It is renewable, has small carbon and water footprints (Table 9.2), and produces no waste, giving it a small environmental impact throughout its life cycle (Steffen 2006). Wind turbines may be large, but they don't take up much space on the ground, making them compatible with other land uses.* Also, the technology is simple and scalable, meaning that autonomous communities can use it as a distributed energy source (Kellogg and Pettigrew 2008). Wind alone holds much more energy than we use. Harnessing the wind in only three states—North Dakota, Kansas, and Texas—could provide enough energy for the entire United States, while wind farms placed up to 50 miles offshore could provide 70% of national electricity needs (Brown 2009). Wind-produced electricity is projected to be cheaper in the year 2020 than electricity produced by any other primary energy source except geothermal (Figure 10.1).

Wind is stronger at higher altitudes and in areas without obstructions such as buildings or trees. So, for example, hilltops in rural areas are good sites for wind turbines, but ground-based turbines in cities are not efficient. Wind strength also varies strongly by time of day, by season, and by region, which means that it is not a sufficient source of energy at all times in all areas. When wind strength is low, we must supplement it with at least one other source of energy.

Wind turbines produce energy by using blades shaped as air foils and mounted on a rotating axis that is usually horizontal. The larger the blades, the more wind energy they can capture. We can use the captured energy to do mechanical work such as drive a water pump, or convert it to electricity using an alternator. We can use the electricity immediately, or store it in batteries for use when wind strength is low.

The number one cause of climate change agent in the United States is electricity production, and fossil fuels produce three-quarters of it. We could replace that three-quarters with 400,000 windmills rated at 2.5 MW (Komanoff 2006). Each windmill requires 60 acres, but the footprint is only 380 square feet, leaving the remaining 99% for other purposes such as farming. In fact, farmers can earn $3,000 to $10,000 per year by simply leasing the 380 square feet to the windmill owner, who will still profit because the windmill produces an average of $300,000 worth of electricity per year (Brown 2009).

Between the years 2000 and 2014 global wind energy production in TWh increased from 29.5 to 706.2 (BP 2015), and in the United States wind energy production is growing at an exponential rate and is projected to continue increasing dramatically through 2040 (Figure 10.2). Most wind installations have been on land, but now offshore wind installations are rapidly expanding, though not yet in the United States.† Sustaining the exponential growth rate of wind power could make wind a major source of electricity in the United States, and even help to reduce fossil fuel use.

Opponents have put forward few arguments against wind. Early wind turbines had smaller blades that turned rapidly and often killed birds, but the slowly-rotating blades of new, large wind turbines kill fewer birds; in fact, skyscrapers, cars, and cats kill far more birds (Brown 2009). Some have complained

* http://www.earth-policy.org/plan_b_updates/2012/update108.
† http://www.earth-policy.org/plan_b_updates/2012/update106.

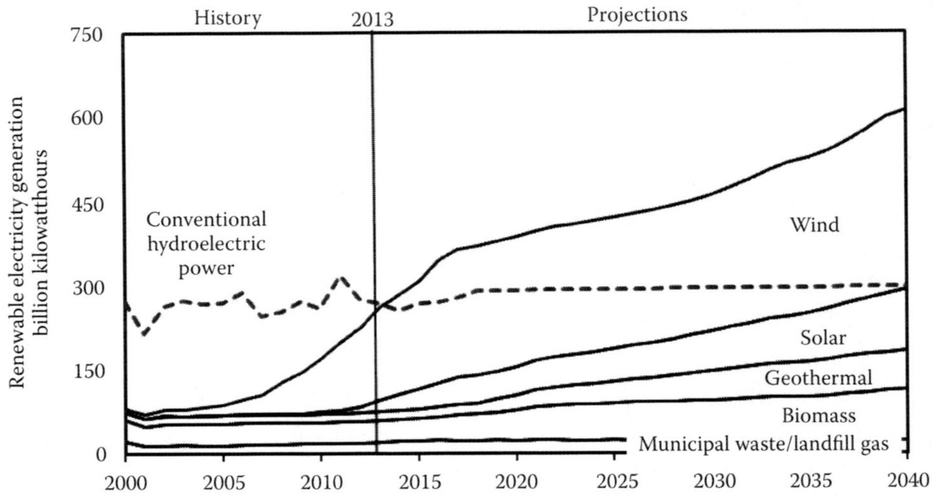

FIGURE 10.2 U.S. renewable electricity generation by fuel type in billion kilowatt hours. (EIA Annual Energy Outlook reference case, 2015, http://www.eia.gov/forecasts/aeo/pdf/0383(2015).pdf.)

about noise, but the new models are very quiet. Others complain that they ruin the landscape because they are unsightly. However, the urgency of the climate change problem should trump any arguments based on aesthetics. Replacing coal-fired power plants with wind turbines will save lives.

One problem is that the electricity generation capacity of wind turbines is variable, changing with the weather. Solar energy is also an **intermittent energy source** (we can only capture solar energy during the day). Coupling wind with solar power can partially solve the intermittency problem because solar power is often complementary to wind power, that is, the intermittencies of wind and solar power partially cancel each other because the wind is generally stronger when the sun is not shining, and vice versa. However, our current electrical grid lacks the capacity to transport electricity from remote wind farms and to balance supply and demand, and this has limited the growth of wind energy.

We could potentially bypass the grid capacity problem by taking a decentralized approach and placing many small wind turbines where we need electricity. The government could offer tax incentives for people to purchase their own wind turbines. In rural areas zoning regulations may permit installation of smaller wind turbines. However, because wind power production scales with the square of the blade radius, power production drops dramatically as the blades decrease in size, so wind turbines have to be very large to be effective. Decentralized wind power production makes the most sense for homes not connected to the electrical grid (due to choice or geographic isolation) and in windy areas.

U.S. energy officials want 20% of the nation's energy to come from wind by 2030.* The United States needs new interstate power transmission lines and power stations, but the high cost and political squabbling (the NIMBY syndrome) have delayed construction of this critical part of the nation's infrastructure. Despite these problems, U.S. installed wind power capacity continues to rapidly increase, and in 2014 the United States remains the world's largest producer of wind energy, accounting for 26% of global wind energy production (BP 2015).

It's clear that the United States should expand its wind power production capacity. Deploying wind and solar on a partially decentralized basis may be the most effective and expedient approach for the United States to rapidly transition from fossil fuels to renewable energy sources and become energy independent. However, a potentially limiting factor will be the availability of the rare earth

* *The Tennessean*, June 18, 2009.

element neodymium, which is used to make the magnets in generators. To produce 50% of global energy demand in 2030 would require at least a five time increase in neodymium production rates, which could only be sustained for about 100 years given proven reserves; neodymium recycling might be required thereafter (Makhijani and Ochs 2013).

10.2 SOLAR

> I'd put my money on the sun and solar energy. What a source of power! I hope we don't have to wait until oil and coal run out before we tackle that.
>
> **Thomas Edison**
> *1931*

The sun is the ultimate source of energy. We are fortunate that plants can harness some of that energy through photosynthesis, which forms the basis for the entire food chain and produces all of the oxygen in the atmosphere. However, we need new technologies that can efficiently and safely convert sunlight into usable electricity that can be available on demand (even when the sun is not shining).

We can actively harness energy from the sun to produce electricity using **photovoltaic** (PV) and **solar thermal energy** technologies. These renewable energy sources are most cost-effective in areas with abundant sunlight and land, where they are expected to generate electricity at a cost of 6–8¢ per kWh in the short term (Makhijani and Ochs 2013). **Passive solar** technologies use sunlight directly to heat water or air.

Most people are familiar with solar panels composed of PV cells that make use of the photoelectric effect discovered by Albert Einstein. Sunlight energy knocks electrons in a semiconductor out of their orbits, which generates a flow of electricity. Currently the efficiency of PV panels measured as the percentage of incident sunlight energy that is converted to electrical energy ranges between 9 and 14% (MacKay 2009), but technological advances continue to increase the efficiency and lower the cost of PV-produced electricity.

At first glance, PV panels seem like an environmentally friendly solution because they use a renewable energy source (the sun), and they do not emit pollution during operation. However, to evaluate their true environmental impact we must examine their entire life cycle. PV panels use semiconductors made from heavy metals. The mining of these metals causes much environmental damage. Manufacturing the PV cells is energy intensive and emits much pollution. These factors make PV panels very expensive, which has limited the market growth of this technology. Finally, PV cells have finite lifetimes (20–25 years), are rarely recycled, and have a high potential to pollute the environment with toxic metals after disposal, especially if disposed of improperly. Furthermore, we can only manufacture PV cells using a high-tech factory, and we cannot repair them, so they are not a sustainable source of energy for autonomous communities or for developing countries striving for self-sufficiency or having insufficient funds for purchasing them. However, they are still a more sustainable source of energy than coal, and LCA analysis suggests that PV cells can be made more sustainable by developing, using, and recycling thinner solar cells (Fthenakis 2009).

Because of their high cost in the past, solar PV systems have been most useful for people in remote areas that are off the electrical grid. However, PV cells are becoming increasingly affordable, efficient and eco-friendly, and people who live in sunny areas should seriously consider adding them to their home designs. The most promising technology is a solar panel made from a thin film of cadmium (Cd) and tellurium (Te). The latest life cycle analysis suggests that these PV cells recoup their embodied energy (the energy required for the raw materials and production) within 1.1 years in places like sunny Spain.* Many areas in the United States would take less than 2.7 years to break-even.

* http://www.worldchanging.com/archives/009864.html.

An advantage of PV panels is that the developing world can deploy them in areas not connected to the electric grid. After paying for the equipment and installation, any energy collected is free. Local solar does not need electric meters, electric bills, or high-voltage transmission lines, required elements of inefficient centralized electrical systems. As a result, installing solar cells on every rooftop is now often cheaper than building a central power plant and a grid (Brown 2009). PV can potentially produce 340,000 GW globally, and use of rooftops alone in the United States could produce 600 GW of electricity, more than 20% of current demand (Makhijani and Ochs 2013).

Some PV cell designs do not seem sustainable because they use exotic metals such as indium, selenium, gallium, or tellurium whose supplies may become depleted if the cells are manufactured in high volumes, driving up costs or creating production bottlenecks. However, the price of PV cells has been dropping dramatically due to innovations and economies of scale. As prices drop, demand increases, production increases, and prices drop more, a positive feedback loop that has a positive outcome. PV cell owners can sell their surplus electricity and add it to the grid, a process called **net metering** that creates a strong incentive for homeowners to purchase rooftop PV systems.

Solar thermal energy, also referred to as **concentrated solar power**, focuses the sun's heat on a liquid that then drives a turbine to generate electricity. Solar thermal energy systems typically use mirrors to focus sunlight. A common design uses a long, curved (parabolic trough) mirror that tracks the sun across the sky and concentrates sunlight on a pipe to heat water or oil. The hot liquid is stored in a tank, and then passes through a heat exchanger where steam is generated to drive a steam turbine and power an electric generator. In a closed-loop system, the liquid is reused for many heating-cooling cycles. Another type of solar thermal energy focuses sunlight on a cylinder that contains an expandable gas and a piston. The gas heats and expands, driving a crank that pushes a piston. It then cools and contracts, withdrawing the piston. The movement of the piston generates electricity. This "Stirling" engine is an external combustion engine because solar heating outside the cylinder (not an explosion within the cylinder) supplies the energy (Steffen 2006, p. 174).* Stirling Energy Systems is installing 10,000 "Suncatcher" units to form a 500 MW solar power plant 70 miles southeast of Los Angeles. The Suncatchers are two to three times more efficient than conventional PV cells, but unlike solar PV, solar thermal energy systems are only practical when deployed on a large scale by utilities. Like solar PV, solar thermal energy has the potential to generate huge amounts of power, about 240 TW globally (Makhijani and Ochs 2013). An advantage of solar thermal energy is that its parts are made from common materials such as glass, steel, and concrete. One disadvantage of solar thermal energy is that it is water intensive, requiring about 1.9–3.0 L per kWh, although the technology for air-cooled solar thermal energy plants that consume 90% less water is mature (Makhijani and Ochs 2013).

All forms of solar power have the advantage of providing peak electricity during the day when demand is high. The disadvantage of solar power is intermittency. A national smart grid could partially solve the intermittency problem by moving electricity from areas with excess renewable energy (sunny and windy) to areas with a deficit. Complementary renewable energy sources such as wind can partially compensate for the intermittency of solar power production. However, this approach is efficient only when the intermittent sources contribute less than 20% of total electric power. This means that excess renewable energy must be stored for later use during periods of peak demand. Options for **energy storage** include **rechargeable batteries**, **pumped storage hydroelectricity** (discussed below), and **compressed air energy storage** (MacKay 2009).

The use of distributed batteries stored in plug-in electric vehicles is a particularly promising solution to the wind intermittency problem. Electric vehicle owners could plug their cars in at night to charge during off-peak hours. At work during the day they could plug their cars in so that they give electricity back to the grid during peak demand (Brown 2009). One problem is that off-peak charging has higher emission intensity because the utility produces more solar power during the day but must rely on fossil fuels like coal for baseload electricity generation at night.

* See http://www.youtube.com/watch?v=fi0Y0Kr-_KIU.

PV and solar thermal energy are **active solar technologies** because they actively and directly use the sun's energy to produce electricity. In contrast, **passive solar** technologies passively absorb heat from the sun's rays and have no electronic components and no moving parts. Examples of passive solar technologies include solar collectors such as solar ovens, parabolic solar cookers, solar water heaters, and passive solar architecture (Kellogg and Pettigrew 2008). Passive solar technologies have all of the benefits of active solar technologies like PV panels (a renewable energy source, no pollution during operation, decentralized) but none of the drawbacks. They are technologically simple, easily constructed (often from recycled parts), and are nonpolluting throughout their life cycle, meaning that they represent a truly sustainable option for autonomous communities.

Cooking food requires much energy, mainly because water has a high heat capacity. In the developing world, cooking is often very inefficient and environmentally damaging: people chop down and burn trees in open fires that emit harmful smoke and waste most of the heat. Chopping down trees for fuel at a rate faster than they can grow back is unsustainable. A sustainable alternative is to use passive solar technologies for outdoor cooking. Solar ovens have glass covers that permit sunlight to enter and dark interior glazes that absorb the sunlight and heat up. The glass acts as a thermal insulator that traps the heat, allowing the interior to heat to the boiling temperature of water (100°C or 212°F), sufficient to cook most vegetables or thinly sliced meat in a matter of hours. Higher temperatures require a parabolic solar cooker, which you can build by lining a recycled satellite dish with mirror shards, aluminum sheeting, or Mylar plastic (Kellogg and Pettigrew 2008). To heat food or water, place it at the focal point of the mirror.

Another example of a sustainable passive solar technology is rooftop solar water heaters, which provide hot water for roughly 160 million people in China (Brown 2009) and harness as much energy as produced by 54 coal-fired power plants. More than 30 million homes in China have solar water heaters, and the government plans to triple the number by 2020.* The simple devices consist of angled rows of dark glass tubes filled with cold water. As the sun heats the water, it expands and rises into an insulated tank, where it can remain hot for days. New models add an electrical heater to the tank for supplemental heating on cold days. In Europe where the cost of energy is high, rooftop solar collectors are cost-effective. Many provide not only hot water but also space heating. Estimates suggest that solar energy can meet most of Europe's low-temperature heating needs (Brown 2009).

Solar power has taken longer to develop and implement than it should have because wealthy and powerful oil and coal companies and coal-burning utilities have fought hard against it. For this reason and the resulting lack of political support only one of the ten leading PV manufacturers is in the United States, even though the PV cell was invented here (Friedman 2008). However, great potential remains for expanded use of solar thermal energy, passive solar collectors, and PV cells in the sun-rich southwestern United States. Although the United States lags behind the EU, Israel, China, and other countries in the development of solar energy, solar power is rapidly growing in capacity. Federal tax incentives will only increase the rate of deployment. Currently the cost of solar electricity is too high for widespread adoption. As shown in Figure 10.1, in 2020 the levelized cost of solar energy is projected to be higher for PV and solar thermal energy than most other forms of primary energy production. However, the cost of PV-generated electricity continues to drop dramatically and is expected to continue to do so until at least 2020.† Declining costs are expected to eventually lead to widespread adoption of solar PV.

10.3 BIOFUELS

Like modern solar thermal energy and photovoltaics, the ancient process of photosynthesis captures sunlight energy, but stores the energy in biomass. **Biofuels** are produced through treatment of

* David Pierson, *L.A. Times*, September 13, 2009.

† http://www.mckinsey.com/client_service/sustainability/latest_thinking/solar_powers_next_shining, published May 2012, retrieved February 20, 2013.

biomass, including plants and animal waste. Biofuels are appealing because they can provide reliable, renewable baseload power with a small carbon footprint (Makhijani and Ochs 2013). However, biofuel feedstock production often competes with food crop production for land, which has caused global food security to decrease. The amount of energy we can obtain from biofuels is limited by the low efficiency of photosynthesis and land availability. However, the technology of biofuel production is rapidly advancing. Its potential may be realized if we can develop feedstocks with high yields that can be grown on marginal lands so that they don't compete with food crops, and if they are grown using sustainable agriculture techniques.

Through photosynthesis, plants capture energy from the sun and carbon dioxide from the atmosphere and temporarily store them in organic molecules that make up plant tissue. When we burn biomass, we release the stored solar energy as heat. Burning biomass also returns the carbon dioxide stored in biomass to the atmosphere. The net change in atmospheric carbon dioxide concentration in the biomass lifecycle is small because the amount removed by photosynthesis equals the amount returned by burning. However, biofuels do not have a zero carbon footprint because fossil fuel energy is used in biomass production. Despite this, the average carbon footprint of biofuels is far less than for fossil fuels (Table 9.2), which makes biofuels attractive energy sources. Biofuels have a much smaller carbon footprint than fossil fuels because burning biofuels only releases the carbon dioxide that the plant temporarily stored when it grew a few years ago, while burning fossil fuels releases carbon dioxide stored in the Earth for millions of years. If we combine the burning of biofuels with CCS, we can even make the carbon footprint negative, which is an attractive option for climate change mitigation.

Humans learned thousands of years ago how to use the solar energy stored in plants. They would burn wood to cook food and provide space heating. Until recently, biofuels could provide enough energy for the small human population. However, biofuels alone cannot provide enough energy for the developed world. New technologies may make biofuels an important energy source, but to this point politicians and the media have overhyped them. The problem is that the low efficiency (1% to 2%, MacKay 2009) of solar energy captured by photosynthesis severely limits the rate of biomass energy production. Furthermore, the low energy density of biomass production means that biofuel production requires large areas of land. For example, to meet its current energy needs the city of San Jose, which occupies 460 km^2, would need the following areas in km^2 for each fuel type: natural gas 2.9, hydroelectric 13, coal 38, nuclear 42, solar 75, wind 530, and biomass 2700, almost six times the area of the city consuming the energy (Cho 2010). Thus, cities can rely solely on wind or biomass only if they have abundant open land for energy production.

In the developing world the biofuel wood is still used as an energy source for cooking. This leads to problems of deforestation and respiratory diseases. **Solar cookers** can eliminate the need for biofuels for cooking. Solar ovens and efficient solar stoves reduce the economic, health, and environmental costs of cooking by reducing or eliminating biofuel consumption (Steffen 2006).

Most biofuel electricity generation in the United States comes from burning wood chips (Brown 2009). However, the United States currently underutilizes organic waste for energy production. For example, **concentrated animal feeding operations** (CAFOs) produce vast quantities of manure that we can use as fuel and fertilizer. Decomposing the waste in an **anaerobic digestor** produces methane (natural gas) for fuel and a nutrient-rich solid waste for fertilization. Using the solid waste as agricultural fertilizer closes the resource loop. Likewise, we can recover energy from organic wastes in landfills. **Anaerobic decomposition** of this waste naturally produces methane. In the past, we wasted landfill methane by burning it off to reduce the risk of an explosion, but now we capture it at many landfills and use it as a source of energy. Other types of biomass we can use as biofuels include yard waste, livestock waste, and sewage, which all contain energy-rich organic compounds. Closing resource loops in this way reduces waste and environmental impacts.

We also use liquid biofuels such as ethanol and biodiesel, primarily for transportation. Energy utilities produce ethanol by using enzymes to break down plant biomass into sugars, adding yeast to produce alcohol, and then distilling the alcohol. The rising cost of oil and gasoline has led to a rapid increase in the use of ethanol as a gasoline substitute in the United States. However, the United States primarily uses the food crop corn for ethanol production. The competing demands for food and fuel caused corn prices to more than double between 2005 and 2007. This reduced food security for people with low incomes who use corn as a staple crop. In 2007, the United States used 20% of its corn harvests to produce only 4% of its automotive fuel (Brown 2009). Thus, corn-based ethanol can substitute for only a small portion of the gasoline demand. A recent report concludes that corn "is among the least efficient, most polluting, and overall least sustainable biofuel feedstocks" (Food & Water Watch and Network for New Energy Choices 2007). Recognizing this, scientists are now conducting research to develop new nonfood sources of ethanol. Cellulosic plant-based materials such as **switchgrass**, wheat straw, and corn stalks and husks have higher EROEI and lower carbon footprints than food crops (Brown 2009), but are still are very inefficient sources of energy for transportation (Jacobson 2009). Converting waste such as corn stalks and husks into energy is highly sustainable.

In summary, the jury is still out on how much energy we can obtain from biofuels, and what the best approach is for biofuel energy production. Biofuels will never become our primary source of energy because photosynthesis is an inefficient form of energy production. However, the United States clearly underutilizes biofuels. We must try to expand the use of this renewable and potentially sustainable energy source, keeping in mind that biofuels are only as sustainable as the agriculture or forestry practices that produce them (Meadows, Randers, and Meadows 2004). We must also find ways to increase biofuel production without decreasing food production.

10.4 WATER

> We have a lot of ways to meet our energy needs. These salmon only have one river forever. If we do not support them, they will go extinct.
>
> **Todd True**
> *Quoted in "Agency Sued Over Putting Hydropower Ahead of Fish," Seattle Post-Intelligencer, May 4, 2001*

10.4.1 Hydroelectric Power from Dams

Moving water is a more attractive source of energy than moving air because water is 1,000 times denser than air and therefore has 1,000 times the power density of wind (MacKay 2009). **Hydroelectricity** was the first renewable energy source deployed on a large scale and still produces more electricity than all of the other renewable energy sources combined, about 7% of global TPES (BP 2015). Global hydropower potential is roughly 1.6 TW, but deployment is sometimes blocked or scaled back due to environmental risks including damage to river ecosystems, flooding of terrestrial ecosystems and human developments, eutrophication and methane emissions from decomposing plankton (especially in the tropics), and changes to sediment transport and water quality (Makhijani and Ochs 2013). The carbon dioxide intensity of hydroelectricity is smaller than all energy sources except wind and geothermal (Table 9.2). Until recently hydropower was only deployed on large scales by utilities, but **small hydro** is now being deployed where large dams are not practical, providing distributed power to small communities or industrial plants.

Hydroelectric technology is simple and reliable.* All that hydroelectric plants need to produce electricity is a rapid flow of water, usually associated with a sudden drop in elevation. Power generation potential increases with increasing elevation drop and increasing water velocity. Waterfalls are natural elevation drops, but few waterfalls large enough to produce significant amounts of electricity exist, so we build dams to serve as artificial elevation drops for water. The dams hold back the flow of a river, impounding the water upstream of the dam to form a reservoir. As water accumulates

* See http://www.youtube.com/watch?v=cEL7yc8R42k&feature=related.

behind the dam, the water level rises, producing a large drop in water elevation across the dam. **Hydroelectric dam** operators control the rate of water flow and therefore the rate of electricity production. When they release water, it falls and pushes the blades of a turbine, which moves copper coils between magnets in a generator, inducing a flow of electricity. **Tidal barrages** work in a similar way: water flows in with the tide, and at peak high tides a dam rises out of the stream channel and traps the water on the upstream side. As the tide lets out, the elevation drop increases. At low tides operators release the water to produce electricity.

Today dams produce most hydroelectric power. During the use phase of their life cycle hydroelectric dams have long lifetimes, zero fuel costs, zero direct greenhouse gas emissions, small carbon footprints (Table 9.2), low external costs, and low operating costs. In areas with strongly seasonal river flow, dams regulate water flow and make it more constant. Dams also have more consistent power generation than wind farms. However, like wind and solar, population centers are often far from hydroelectric power supplies, requiring long-distance high voltage transmission lines.

Because we can easily regulate hydroelectric power generation, it is effective for balancing loads; we can match power output to demand. **Pumped storage** is a form of hydroelectric power used for load balancing, that is, it can supply energy when intermittency problems cause reduced renewable energy production. Pumps push water up to a reservoir at high elevation during periods of low demand energy, and operators release water to generate electricity during peak demand (MacKay 2009). The efficiency and carbon footprint of pumped storage depends on the efficiency and life cycle-wide greenhouse gas emissions of the method used for pumping.

Although hydroelectric power has many advantages over fossil fuels, large dams often cause more problems than they solve, including loss of fertile land, displacement of people, and the potential for catastrophic dam collapses. It's possible that in some areas more power could be produced by draining a reservoir and covering the exposed land with PV panels or wind turbines. Most importantly, dams act as a barrier to the flow of sediment and organisms along the river, two problems we will now examine more closely.

Water flowing in rivers carries large quantities of sediment derived from upstream erosion. Unobstructed rivers deposit most sediment in their deltas, which form when fast-moving river water enters a larger body of water. Because water velocity decreases as river water enters a reservoir, suspended sediments settle out, a process called **siltation**. Over time reservoirs fill with sediment and lose their storage capacity. Energy is required to dredge the sediment out of the reservoir and dump it downstream.

Another sediment transport problem caused by dams is **downstream erosion**. Water released from dams has little sediment, so it effectively scours the stream channel downstream of the dam, causing erosion and loss of species' habitats. By reducing the flow of sediment downstream, dams deprive downstream areas of fresh sediment, decreasing the fertility of agricultural lands and altering river delta ecosystems. Since natural rivers supply most of the sediment (e.g., sand) to coastlines, beaches and deltas can disappear when a dam steals their sediment. This is because deltas are always sinking due to the great weight of sediment deposited on them. Fresh sediment deposition keeps river deltas at near-constant elevation. However, once a dam removes the sediment source, the delta continues to sink.* Without continuous sediment deposition, the delta's elevation decreases, making it more susceptible to erosion and hurricane damage, or causing it to sink below sea level completely (Syvitski et al. 2009).

Dams cause another major problem by blocking the migration of river species. Dams can make it impossible for fish and eels to reach their spawning grounds. This has led to the loss of extremely valuable salmon runs in the Pacific Northwest. It has also caused a dramatic drop in the global population of eels, which migrate along rivers. A hydroelectric dam on the St. Lawrence impedes the migrations of eels to and from what was the single largest nursery for the American eel—Lake Ontario and its tributaries. The population of eels migrating along the St. Lawrence dropped from

* Isostatic adjustments and sediment compaction are slow processes that have long relaxation times.

nearly a million in the 1980s, to 100,000 in the early 1990s, to less than 10,000 in the late 1990s, and virtually to 0 in 2000 (Prosek 2010).

Hydroelectric dams have similarly hurt eel populations worldwide. International trade in eels is a multibillion dollar industry. Eel fisheries employ about 25,000 fishermen in the EU to feed the voracious appetite for eels in Japan, and those jobs are now at risk. Some European countries like Ireland have temporarily banned eel fishing to prevent a complete collapse of the eel population (Prosek 2010). However, populations of eel and fish that migrate along rivers are unlikely to recover unless we remove the dams that prevent their migrations or provide an alternative route for them to migrate.

Small dams cause less environmental damage than large dams, though it's not clear if they cause less damage per unit of energy. If they do, it would make sense to expand the use of small hydro projects that produce less than 30 MW.* In 2006, the DOE identified 5400 potential small hydro sites in the United States that could collectively produce 18,000 MW (Richter 2010). Small dams are useful for areas not connected to the grid because they complement PV energy by producing more energy in the winter when PV power output is lowest.

Hydropower production in the United States has not increased significantly since 2000 (Figure 10.2), and future increases in the generating capacity of conventional hydroelectric power are unlikely because hydropower sources are now almost fully utilized. In fact, with all of the problems associated with large dams, the trend is to remove, rather than construct, large dams, so the output of conventional hydroelectric power in the United States is currently decreasing. Between 2013 and 2014 hydropower production in TWh in the United States decreased 3.7% to 261, but world production increased 2.0% to 3885 (BP 2015).

Since 2000, dam construction in South America and Asia has greatly increased global hydropower production. Brazil and other countries that share Amazon tributaries have proposed 151 dams. They are starting near the headwaters of the Amazon and working their way down. The dams are supposed to be environmentally friendly: Low head design means that water upstream does not need to be as deep, so the submerged area can be smaller. This is important because in the tropics the surface water in lakes and reservoirs is always warm, so there is no turnover in the fall and spring as occurs in lakes in temperate zones. The lack of turnover causes **eutrophication** (see Chapter 14) in all reservoirs with appreciable nutrient levels, which can result in release of large volumes of methane produced by anaerobic decomposition of organic matter deep in the reservoir. Methane emissions from reservoirs in the tropics may offset the climate change mitigation benefits of substituting hydroelectricity for fossil fuel electricity production (Fearnside and Pueyo 2012). The problem is exacerbated by the trapping of nutrients in dam reservoirs. While agricultural fertilizer runoff has increased the global river fluxes of nutrients, dams are projected to decrease the global river flux by 17% due to storage in reservoir sediments (Maavara et al. 2015). Changes in runoff and river fluxes greatly affect food webs downstream.

Hydroelectric dam construction and operation along the Amazon has caused numerous problems, including deforestation, downstream fish kills from production and release of anaerobic waters in the reservoir, downstream bank erosion, loss of river connectivity for migratory aquatic species, and displacement of indigenous peoples. A recent study of potential environmental impacts found that 47% of the proposed dams would have large negative impacts, while only 20% are low impact (Finer and Jenkins 2012). Worst of all, the people living near the dams who are affected by these problems will receive none of the benefit. The electricity produced by the dams will be transported roughly 1500 km to large cities like Sao Paolo. Despite these problems, Brazil is moving forward with construction of large hydroelectric dams along the Amazon, which they hope will allow ocean-going vessels to travel over 1500 km to reach the foothills of the Andes through a series of locks.

The following case study further illustrates the problems caused by large hydroelectric dam projects.

---

* Since hydroelectric power is highly variable, quoted values of power output are averaged over a year.

10.4.1.1 Case Study: The Three Gorges Dam

An examination of the world's largest dam will illustrate why many countries are no longer building large dams, and some like the United States are decommissioning them. The **Three Gorges Dam** in central China is the largest dam ever built, and may be the largest engineered structure ever built (Rogers and Feiss 1998). The dam, completed in 2006, is nearly 200 m high, 2092 m long, and has a total generating capacity of 22 GW. The reservoir behind the dam is 600 km long (Gleick 2009). The cost of the dam is somewhere in the $50 billion range, although it's impossible to determine the exact amount. The largest benefit of the dam is the carbon dioxide emissions savings. Most of China's electricity comes from coal-burning power plants, which if used instead of the Three Gorges Dam to supply 22 GW would burn 50 million tons of coal and emit 100 million tons of carbon dioxide annually (Gleick 2009).

The dam is on the Yangtze River. Between one and six million people who lived along the fertile banks of the river were forced to abandon their homes, which are now underwater. Also, the dam flooded priceless archaeological sites. Before the dam construction, many citizens voiced concerns about the effects of the dam on safety and the environment. For example, without sewage treatment plants, all of the sewage produced by the million-plus residents around the reservoir washes into the reservoir, contaminating the water and causing algal blooms and eutrophication. Because the dam was built in a mountainous region with high erosion rates, siltation is quickly filling the reservoir. Agricultural fields downstream have lost the supply of fresh sediment that kept their fields fertile. The downstream loss of sediment is also leading to coastal erosion near the Yangtze River delta in the East China Sea. By blocking migration of fish and other species to spawning grounds, dams along the Yangtze River are causing species to go extinct, and reduced harvest size is hurting downstream river fisheries. Since the dam was completed, the frequency of earthquakes and landslides along the steep slopes of the land surrounding the reservoir increased (Gleick 2009). This could be explained by water seeping into and lubricating faults and cracks, which act as slip surfaces. Also, water seeping into soil and porous sediments and rock can add weight to a slope, helping to destabilize it, and sediment is filing the reservoir faster than anticipated.

Chinese officials admitted in 2007 that problems associated with the dam, including landslides and pollution, are more serious than expected (Gleick 2009). Benefits are easier to identify and quantify than costs, but time is showing that the costs are quite high. It remains to be seen whether the Three Gorges Dam will cause more problems than it solved.

10.4.2 Ocean Power

Tidal power and **wave power** are relatively new methods for harnessing the energy of flowing water (Brown 2009). The total energy contained in tides and waves is about four times the TPES (Richter 2010), but we can capture only a small fraction of that. Many techniques exist for capturing ocean power,* but current global electricity production is only 10 MW, about 0.001% of the world's total electricity production of two million MW (Richter 2010). All forms of ocean power must contend with potential damage from storms and corrosive saltwater and incorporate plans to minimize damage to marine ecosystems.

Lunar tidal power captures the energy of water flowing in and out with the tides. It is a reliable and essentially continuous source of energy, unlike wind and sun that depend on weather and time of day. For example, a **tidal barrage** is a dam-like structure that can be raised to trap water in a tidal lagoon at high tide, and then lowered to release the water at low tide to produce hydroelectricity. Approaches like this can reduce the intermittency problems of wind and solar. The technology is simple and inexpensive, and installations don't take up valuable land. However, even if deployed on a countrywide scale in an island country like the United Kingdom, tidal power would supply only 11 kWh/d per person, <6% of power currently consumed (MacKay 2009).

* See http://science.howstuffworks.com/environmental/green-tech/energy-production/ocean-power.htm.

What about the energy stored in waves? If you have body-surfed or watched large waves crashing onto rocky shores, you know that waves can pack a punch. However, as stated concisely by Mackay (2009), "Sun makes wind and wind makes waves…Wind is second-hand solar energy…Waves are thus third-hand solar energy." Since energy is lost at each conversion step (sun to wind, wind to wave), waves hold much less energy than wind or sunlight.

Currently companies are developing interesting technologies to harness wave energy. However, wave machines are expensive, and if deployed around the United Kingdom they would produce only 4 kWh/d per person, only 3% of the average UK power consumption of 125 kWh/d per person. In the United States, the maximum amount produced per person would be <4 kWh/d, much less than 2% of the average American power consumption of 250 kWh/d. Ocean power installations must also compete for space with shipping lanes, **marine protected areas**, and pipeline infrastructures, and technologies that use permanent magnet generators also face potential shortages of the rare earth neodymium (Makhijani and Ochs 2013). We conclude that ocean power is currently an insignificant source of renewable energy, and even when fully developed it will provide only a small fraction of TPES in most countries.

10.5 GEOTHERMAL

The Earth stores an enormous amount of heat energy. Radioactive decay in the Earth's interior produced most of this heat. Because rock is a poor thermal conductor, it has trapped much of the heat produced during the 4.5 billion years of Earth's history. Heat is constantly leaking out at the Earth's surface, but its rate of escape is highest in areas of volcanic activity such as the Pacific **Ring of Fire**, which includes nearly two billion people in the United States, Japan, China, and Indonesia. These areas with high surface heat flow have great potential for production of **geothermal energy**, a renewable source of energy powered by the flow of heat from the Earth's interior to the surface. For example, geothermal heats nearly 90% of the homes in Iceland and produces 25% of electricity in the Philippines (Brown 2009).

Two different methods are used to extract thermal energy by moving heat. A **geothermal heat pump** is effective for decentralized space heating and cooling almost everywhere. In contrast, a **geothermal power plant** is used to produce steam to drive turbines and generate electricity, but is feasible only in areas with temperatures exceeding the boiling point of water at shallow depths (usually volcanically active areas). Unlike solar and wind, which are intermittent, geothermal energy can provide reliable baseload electricity generation, although it has a high water use intensity of 1–3 L per kWh (Makhijani and Ochs 2013). Geothermal currently produces only 0.3% of the world's electricity, but much higher percentages in volcanically active countries like Iceland, Nicaragua, and New Zealand.

In the past, only areas with high surface heat flow and permeable rocks could be used to capture geothermal energy. Natural water convection in the subsurface would transport the heat to drilled wells. New **enhanced geothermal systems** can capture geothermal energy even from hot dry rock that is impermeable by pumping high pressure cold water into wells. The use of enhanced geothermal systems has greatly increased the potential for centralized geothermal energy production, although it is still in the demonstration phase (Makhijani and Ochs 2013). Areas with high surface heat flow should invest in the construction of geothermal power plants to make their energy supply sustainable.

As a low-grade form of energy, less useful than electricity, geothermal heat is most efficiently used for space heating, and this application is rapidly rising globally. Decentralized approaches to using the Earth as a heat source or sink include passive and active approaches. Geothermal energy passively heats greenhouses and ponds used for aquaculture and natural spring waters used in spas and public bathhouses. The active approach is to use geothermal heat pumps for space heating and cooling. Because the temperature below ground does not fluctuate seasonally, geothermal heat pumps can heat homes in the winter and cool them in the summer. The heat pump continually transfers heat from the hot to the cold region. This approach is efficient because it produces primary energy, with none of the energy loss associated with conversion to secondary energy like electricity. Especially when they use electricity from renewable sources, heat pumps are the most energy efficient, clean, and cost-effective option for space heating and cooling, using 25% to 50% less electricity than conventional heating and cooling systems (Brown et al. 2015). Thus, a truly sustainable home should incorporate a geothermal heat pump in its design.

10.6 SECONDARY ENERGY: ELECTRICITY AND HYDROGEN

We use primary energy sources such as fossil fuels and renewables to produce forms of secondary energy such as electricity and hydrogen gas that are used as **energy carriers**. These forms of energy are useful because we can easily store and transport them, so it is worth losing some primary energy to produce them. However, both need a transportation infrastructure to make them widely available. In the United States, we have an outdated, inefficient national electrical grid that does not extend to some remote areas with high renewable energy production potential. The hope is that it will be replaced in the near future with efficient, high voltage transmission lines and a **smart grid** (see Chapter 11). We don't yet have a hydrogen transmission and distribution network.

Hydrogen gas can be produced by electrolysis of water or by reacting natural gas with steam (Makhijani and Ochs 2013). Of these, only generation by electrolysis is sustainable because it doesn't require the nonrenewable resource natural gas. Both require high energy inputs, but this can be provided by excess renewable generation, in which case hydrogen acts as a battery to store energy until it can be used during peak demand periods. The high costs of production, storage (it has low energy density), and of building a distribution network combined with safety concerns has delayed widespread adoption of hydrogen gas as an energy carrier.

10.7 SUMMARY: ARE RENEWABLES ENOUGH, AND CAN WE AFFORD THEM?

Lester Brown claims in his book *Plan B 4.0* (2009) that by 2020 the world could increase the amount of renewably generated electricity fivefold and decrease fossil fuel-based energy use by 90%. Combined with increases in energy conservation, *Plan B 4.0* would reduce global carbon dioxide emissions 80% and stabilize atmospheric carbon dioxide below 400 ppm (the carbon dioxide concentration in 2011 was 391 ppm). Now in 2016 it's clear that we will not come close to reaching those goals, but we are making progress: in 2014 the global energy consumption in TWh for wind was 706, solar 186, and all other renewable sources 509 (BP 2015). However, these numbers are small compared with the TPES of 155,141 TWh.

Even Brown's optimistic plan does not completely phase out fossil fuels. A recent analysis by MacKay (2009) of the United Kingdom's energy needs concludes that, even ignoring economic constraints and public opposition to many renewable projects, domestically produced renewables cannot meet current energy demand. Put another way, in the ideal case of no economic and political constraints, if the United Kingdom covered all land with wind farms and PV panels and the entire coastline with offshore wind farms and various forms of ocean power production, the maximum amount of energy that they could produce using renewable sources would fall just short of current demand. Considering that energy demand keeps increasing over time, the renewable energy supply seems unlikely to meet demand fully in the United Kingdom. MacKay's analysis is convincing because he clearly presents all of the calculations, states his assumptions, and uses reliable data. His less detailed analysis of North America finds that solar energy alone can meet current and future energy needs because, unlike the United Kingdom, North America has large areas of virtually uninhabited land with abundant sunlight for solar energy collection (MacKay 2009).

MacKay concludes that the United Kingdom and other developed countries with high population densities can eliminate fossil fuels only by decreasing demand (discussed in the next chapter) and supplementing domestic renewable energy with either clean coal, nuclear fission, or imported renewable energy (MacKay 2009). The latter approach is not a globally sustainable approach, and raises many thorny political issues, so we will focus on examining the first two options. Many respected scientists who are experts on energy supply (Deffeyes 2001; Lovelock 2006) think that until we can develop sufficient renewable energy capacity, nuclear energy will be a necessary short-term baseload generation component of the energy portfolio in developed countries. Nuclear is preferable to coal, which through multiple health impacts including respiratory diseases kills far more people per unit of energy produced than nuclear. Furthermore, expanded use of coal would intensify global climate change. Global climate change is a much greater danger than radioactive waste from nuclear fission reactors, so nuclear is definitely the lesser of two evils. Thus, we may have to expand the use of nuclear power in some developed countries to phase out the use of coal. An attractive compromise would be to tie the phase-in of new nuclear power plants to the phase-out of coal-burning power plants. Policy makers should keep all energy options on the table except the expansion of coal-burning power plants.

The scale of change required to avert dangerous climate change is enormous. Climate scientists say that to avoid dangerous anthropogenic climate change we must decrease total annual carbon dioxide emissions below 5 gigatons and reduce the carbon footprint to less than 1 ton per person, lower than the current average in India (Jackson 2008). This can only be accomplished by rapid changes, including decreased total energy consumption, increased renewable energy production, and a transition to a fossil fuel-free energy economy (Brown 2009).

Without the development of a new centralized energy source, energy production may slowly transition to small, decentralized, renewable sources of energy. That may lead to the development of sustainable, autonomous communities such as **transition towns** (Kellogg and Pettigrew 2008; Hopkins 2008). For example, Eigg Island in Scotland gets up to 90% of its power from renewable energy sources.* Decentralization of power (both energy and political/social) would be beneficial to society; it would reduce our dependence on foreign countries and energy monopolies, and increase our security and self-sufficiency.

The transition to a more sustainable energy portfolio is possible using existing technologies. For example, MacKay (2009) presents six energy plans for the United Kingdom that incorporate energy savings from increased efficiency and eliminate fossil fuels. Three of these plans rely on solar energy collected in the Sahara and transported by special high voltage direct current power lines. The cost of installing the solar collectors and power lines would be tremendous, and it would transform the landscape.

In the United States, solar energy is currently too expensive to become a major source of electricity (Figure 10.1). As a result, the EIA projects that up to the year 2040 most growth in renewable energy will be in wind (Figure 10.2). However, our environmental problems will not be solved only by finding ways to produce more and cheaper energy, which is what caused many of our environmental problems in the first place. Without cheap energy we would not have been able to mine many of the metals we use today at great cost to the environment, or expand our population and transform the landscape, displacing species and causing extinctions. Cheap energy has been a double-edged sword; we have used it without being forced to think about its true costs. So in the next chapter we will focus on ways to decrease demand and make energy production more sustainable.

HOMEWORK PROBLEMS

1. Choose a country other than the United States or the United Kingdom. Acting on behalf of that country, what form of renewable energy would you most invest in? Make your case.

---

* http://inhabitat.com/scotland-will-soon-be-home-to-the-worlds-first-self-sufficient-island/, retrieved October 3, 2016.

11 Sustainable Energy Plans

Let's step back and put our current energy situation into perspective. In the United States 85% of energy comes from fossil fuels (National Research Council [U.S.] 2009). Together the United States and China cause 40% of the world's greenhouse gas emissions. Light from an incandescent bulb contains only 2% of the energy content of the coal used to produce it. To become more sustainable our society must reduce these inefficiencies and focus on energy security, sustainability, and environmental responsibility.

The required transition from fossil fuels to sustainable, renewable energy sources will be this generation's largest and most important undertaking. This energy transition will be driven by the threat of global climate change and possibly by rising oil prices caused by Peak Oil. The transition to renewable energy sources may be facilitated by the restructuring of taxes to incorporate the indirect costs of fossil fuel use (Brown 2009). We already have most of the needed technologies in place to meet the world's energy demand with only wind, water, and solar (Jacobson and Delucchi 2011). What remains is to choose the policies that will accomplish this transition most quickly and at the lowest cost. In Chapter 8 we discussed policy tools for reducing greenhouse gas emissions, and concluded that the most effective tools are emissions trading systems for power plants and carbon taxes for consumers. In this chapter, we review strategies and devise a plan for reducing energy use and for choosing the optimal mix of energy sources.

11.1 INTRODUCTION

We've learned that most of our energy comes from nonrenewable sources, and that some sources, such as oil, could become scarce within a few decades. Buying oil from foreign countries decreases political freedom in those countries, fosters conditions that promote terrorism, and decreases our national security (Friedman 2008). Burning nonrenewable fossil fuels, particularly conventional coal, releases large amounts of the greenhouse gas carbon dioxide to the atmosphere, contributing to global climate change. Conventional coal uses mining methods that pollute the environment, power plants that release a wide range of pollutants including toxic mercury, and unsafe ash and slurry ponds that store poisonous wastes. Technologies to reduce the carbon footprint of fossil fuel use such as CCS are not yet fully developed and may ultimately prove to be unfeasible. Thus, a sustainable energy plan would reduce the use of nonrenewable fossil fuels, especially conventional coal.

The challenge presented to developed countries is how to become more sustainable by decreasing their dependence on nonrenewable energy sources while maintaining high standards of living. Developed countries like the United States can't decrease their dependence on nonrenewable energy sources by abruptly decreasing their energy consumption because it would cause economic hardship. As noted by Wallace (2005), sales of electricity primarily produced by nonrenewable fossil fuels accounts for only 3% to 4% of U.S. GDP, but the other 96% to 97% of our economy depends on that 3% to 4% (e.g., see a Life Cycle Assessment of energy use by the beef industry in Section 9.2). So to maintain a high standard of living, the United States and other developed countries must substitute renewable energy sources for fossil fuels. *The transition to sustainable energy will increase national security by making the United States more energy independent, protect valuable environmental services, and keep the United States economically competitive by creating green jobs and new green industries* (National Science Board and Foundation 2009).

Two contrasting approaches to environmental resource use and development exist. Advocates of the *hard path* are usually anthropocentric exemptionalists (belief that humans are exempt from the laws of nature). They prefer the "conquer nature" approach that relies on technology for development and to fix environmental problems. Nuclear energy is an example of the hard path to energy production. The *soft path* is advocated by ecocentric naturalists who prefer to work with rather than against nature. This approach uses natural processes or designs to provide resources and to fix environmental problems. It places an emphasis on energy alternatives such as wind and solar that are renewable, flexible, decentralized, and environmentally more benign than those of the hard path. The soft path has a smaller environmental impact than the hard path because its main design principle is to modify nature as little as possible. Many examples throughout this book illustrate that the soft path is usually more effective and sustainable (less resource intensive, less polluting, fewer unintended consequences, etc.), so we favor use of the soft path when designing a sustainable energy plan.

The primary objective of a sustainable energy plan must be to reduce the environmental impacts of energy use, most importantly by reducing greenhouse gas emissions to mitigate anthropogenic climate change. This can be accomplished by reducing demand through energy conservation and increased use efficiency, and by switching from fossil fuels to clean, renewable energy sources. We also want to maintain or increase our energy security. Appropriate use of the right balance of energy sources, technologies, and policies can help achieve these objectives.

To decrease risk, improve national security by decreasing our reliance on foreign energy sources, and smooth the transition to a twenty-first-century energy base, rapidly ramp up the capacity of renewable energy sources. Use fuels that are cost-effective and safe, that is, do not present unacceptable risks or generate hazardous waste. Most importantly, use fuels that have the lowest environmental impacts, that is, fuels that do not pollute or emit greenhouse gases. Use full cost accounting to estimate the true costs of each energy source, and set policies that ensure that prices incorporate externalities.

Daly's second rule of sustainability states that renewable resources must be substituted for nonrenewable resources faster than the rate of decline of the nonrenewable resource stock in order to avoid resource shortages (Section 2.2). To meet this requirement, the combination of decreased energy demand (2) and increased renewable energy supply (3) should more than offset the decrease in fossil fuel energy production. We will examine approaches to decreasing demand and improving the energy supply by diversifying and replacing fossil fuels with alternatives that are more sustainable.

11.2 SOLUTIONS

11.2.1 Objective 1: Reduce Greenhouse Gas Emissions

We need a plan to reduce greenhouse gas emissions, specifically carbon dioxide. But how can we determine how much carbon dioxide we, our communities, or our countries emit by using energy? We can use the Kaya identity to calculate the carbon footprint of energy use by a population. To calculate the CO_2 footprint F_{CO_2}:

$$\begin{aligned} F_{CO_2} &= P \times A \times C \times T = \text{population} \times \text{per capita GDP} \\ &\quad \times \text{energy intensity} \times CO_2 \text{ emission intensity} \\ F_{CO_2} &= \#\text{ persons} \times (\text{dollars/person}) \\ &\quad \times (\text{energy consumed/dollar}) \times (\text{mass } CO_2/\text{energy consumed}) \end{aligned} \tag{11.1}$$

We use this equation because measuring the four factors on the right-hand side of the equation is easier than measuring the one factor on the left-hand side. We have historical data for these four factors, and we can project those historical trends into the future with some confidence to estimate future carbon dioxide emissions. For example, Figure 11.1 shows how energy intensity has changed in the United States since 1980, and how it is projected to change through the year 2040 (labeled "energy use per 2009 dollar of GDP").

We can reduce carbon dioxide emissions by decreasing any combination of the four factors in Equation 11.1. Since we project global population *P* to continue increasing to at least 2050, and because we must increase global per capita GDP *A* to lift the world's poor out of poverty, to decrease global CO_2 emissions we must reduce the product of energy intensity *C* (kWh of energy per unit of GDP) and carbon dioxide emission intensity *T* (mass CO_2/kWh of energy) to reduce the environmental ImPACT *I*, in this case represented by the CO_2 footprint F_{CO_2}. Multiplying *C* and *T* we obtain carbon dioxide emissions per 2009 dollar GDP, which in the United States decreased dramatically between 1980 and 2013, and is expected to continue to decrease through 2040 (dotted line in Figure 11.1).

In Figure 11.2 we show how the components of ImPACT that contribute to carbon dioxide emission changed between 1950–1990 and 1990–1997, with PACT shown in lowercase to denote annual percentage changes in these quantities (Waggoner and Ausubel 2002). Again, because government policies promote continuous economic growth (increasing per capita income *a*) and don't discourage

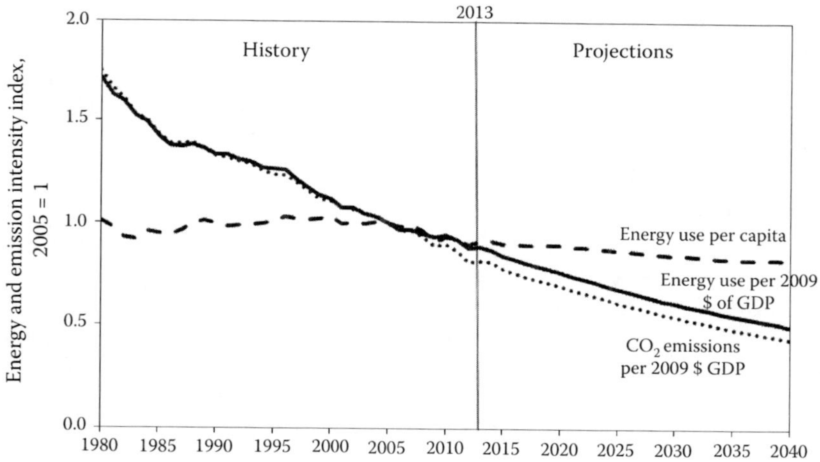

FIGURE 11.1 U.S. energy use per capita and per 2009 dollar of GDP, and carbon dioxide emissions per dollar of GDP, 1980–2040. EIA (2015) Annual Energy Outlook reference case.

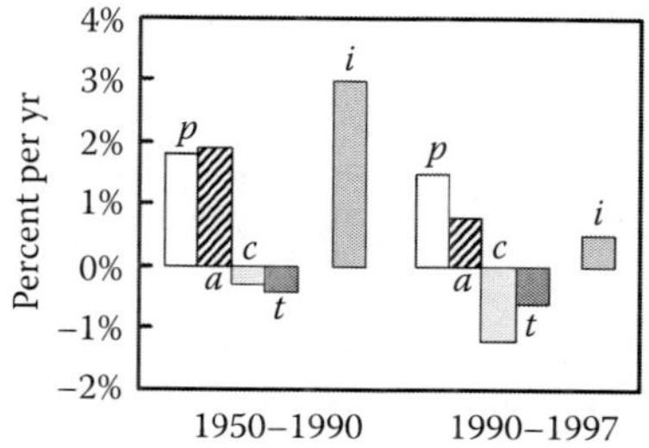

FIGURE 11.2 The changes of population (*p*), income (*a*), intensity of use of energy (*c*), and carbon emission per energy (*t*) altering global carbon emission (impact, *i*). (From Waggoner, PE, and JH Ausubel. 2002. "A Framework for Sustainability Science: A Renovated IPAT Identity." *Proceedings of the National Academy of Sciences of the United States of America* 99 (12): 7860–65. doi:10.1073/pnas.122235999. Copyright (2002) National Academy of Sciences, U.S.A.)

population growth p, emission reductions must come from reductions in energy intensity c and carbon dioxide emissions intensity t to reduce global carbon dioxide emissions i and mitigate global climate change. Between 1950 and 1990 large increases in population and per capita income offset small improvements in energy intensity and carbon dioxide emissions intensity, causing the environmental impact of carbon dioxide emissions to increase. The situation improved between 1990 and 1997, with smaller increases in p and a and larger reductions in c and t causing a much smaller annual increase in i of ~0.5%. To avoid dangerous global climate change, we need to accelerate these trends of slowing growth of population and per capita income and decreasing energy intensity and carbon dioxide emissions intensity. However, projected average annual changes between 2012 and 2040 are p 0.9%, a 2.4%, c –1.9%, and t –0.4%, which yield an average increase in impact i of 1% per year, greater than for the period 1990–1997 (EIA 2016b).

11.2.1.1 Sustainable Transportation

One area in which we could greatly reduce per capita greenhouse gas emissions without reducing quality of life is transportation. Transportation accounts for roughly one-third of our daily energy consumption (MacKay 2009) and 35% of total CO_2 emissions from energy consumption,* so there is tremendous pressure to reduce the carbon footprint of our transportation. But transportation has other negative environmental impacts. Consider the life cycle (RPUD) environmental impact of a single automobile that travels an average of 100,000 miles in its lifetime. There is the damage that results from the mining and the processing of the raw materials R (including the drilling, transporting, and refining of the oil and gas the car uses) and production of the car P; from the emission of greenhouse gases, nitrous oxides that contribute to acid rain, and ozone that causes photochemical smog during use of the car U, and the waste and pollution associated with disposal D. In addition, driving a car is one of the riskiest activities we engage in, and cars make walking and bicycling much more dangerous on shared roads. Much of our country has been paved over by roads and parking lots, which has increased flooding risks and degraded the land. Life cycle analysis shows that motor vehicles produced $56 billion in health and other nonclimate-related damages in 2005, with damages per vehicle mile traveled ranging between 1.2 and 1.7 cents (USNRC 2010).

Climate impacts can be reduced by changing the mode of transportation. For example, the climate impact of freight transportation is greatest for light truck and air transport and lowest for shipping and rail transport (Borken-Kleefeld, Berntsen, and Fuglestvedt 2010). For passenger travel, the near-term climate impact per passenger-kilometer is lowest for rail, coach (bus), two-wheel (mopeds and motorcycles) or three-wheel vehicles and highest for cars and air travel. In Japan in 1999 the energy consumption in kWh per 100 person-kilometers was rail 6, bus 19, air 51, sea 57, and car 68 (MacKay 2009).

While mass transit options such as buses and rail are the most energy efficient options, they are only cost-effective in cities with high population densities. Even when they are cost-effective, mass transit projects are often opposed in the United States because they require government coordination and funding. Free market advocates prefer to rely on the market to find the optimal (most inexpensive) transportation solution. However, individuals don't have the option to build their own train system, and they make choices that benefit only themselves (the free rider problem and the Tragedy of the Commons). Finally, individuals often don't make the best choices, choosing more expensive options than mass transit for a variety of reasons. As Kahneman (2011) observed, "Although humans are not (usually) irrational, they often need help to make more accurate judgements and better decisions, and in some cases policies and institutions can provide that help." Finally, the free market only considers financial costs and not externalities such as pollution. So, government must play a role in building and incentivizing mass transit.

* EIA, June 2015 Monthly Energy Review, http://www.energyglobal.com/downstream/the-environment/26062015/EIA-June-2015-Monthly-Energy-Review-1002/.

We also need to make cars less harmful to the environment. Hybrid cars like the Toyota Prius have already raised the bar for energy efficiency. They are more efficient than conventional cars because they produce electricity through regenerative braking, and they automatically shut off the engine when idling. In addition, the third generation Prius (model year 2009) uses solar panels to power the air conditioning. Hybrids are essentially conventional cars with the addition of an electric motor and a battery that cannot be charged from an external source. In contrast, the newer plug-in hybrid electric vehicles are electric cars with a secondary internal combustion engine to extend their travel range, and have an even smaller carbon footprint than hybrids (Frank 2007).

Can we reduce the carbon footprint of cars even further? Yes, by using the most efficient type of car, the electric car, which does not emit greenhouse gases or other pollutants during use. While production of the electricity used by electric cars produces greenhouse gases, this environmental impact will decrease over time as fossil fuel power plants are replaced by carbon-free renewable sources of energy such as wind and solar. Electric cars are becoming an important transportation option in the United States. They can be affordable (the 2016 Nissan Leaf sells in the United States for $29,000*) or luxurious and high-performance (the wildly popular Tesla Motor's "S" model can drive itself, has a base cost of $70,000 and received the highest auto rating ever awarded by Consumer Reports†).

LCA shows that for both electric and gas-powered vehicles, most of the environmental impact occurs in the use phase (Notter, Gauch et al. 2010), not in the raw materials, production, or disposal phases (remember **RPUD**). Electric cars have lower environmental impact during the use phase because they are more energy-efficient than conventional cars, and therefore have lower environmental impact over the entire life cycle. For example, Jacobson (2009) evaluated the impacts of 12 combinations of energy source and vehicle type on global warming, air pollution mortality, energy security, water supply, land use, wildlife, resource availability, thermal pollution, water chemical pollution, nuclear proliferation, and undernutrition. They found that wind power combined with battery powered electric vehicles is the most effective solution, ranking first in 7 out of the 11 categories, including the two most important, mortality and climate damage reduction. One would expect the advantages of electric cars to become even greater over time as the associated technologies mature. Electrification of personal transportation (including use of renewable energy fuels) could decrease energy consumption to one-fifth of its current value (MacKay 2009). Thus, electrifying personal transport should be a high national priority.

Other technology trends are rapidly changing personal transportation in the United States. Most auto manufacturers and **Google** are now designing and prototyping autonomous cars. These will likely be used in fleets in urban settings, where users will use their smartphones to hail a car. The number of autos required to support an urban population could decrease to less than half, which would free up much parking space that could be converted to green space. The dangers presented by drivers who are unskilled or whose abilities are impaired will be eliminated. Pollutant emissions will decrease dramatically. Finally, users will save money by not having to purchase and maintain personal vehicles.

These trends suggest that the most significant lifestyle change in the United States in the next two decades may be abandonment of the car culture, specifically the personal ownership of cars with internal combustion engines. That lifestyle won't disappear completely, but it will become less prevalent as the price of fuel increases (due to Peak Oil and carbon taxes), and better alternatives become available. The change may be traumatic, as 88% of workers in the United States travel to work by car (Brown 2009). Some people will switch to electric cars, while others will opt for alternative modes of transportation such as mopeds, bicycles, mass transit, and ridesharing options such as **Uber** and **Lyft** that make use of smartphone apps. As fuel costs rise, the trend will be for people to take fewer trips and to move closer to their jobs and schools to decrease their transportation costs.

---

* http://www.nissanusa.com/electric-cars/leaf/?dcp=ppn.63023881.&dcc=0.240189299, retrieved February 8, 2016.

† www.consumerreports.org.

Telecommuting will become even more widespread, and in many cases, videoconferencing will make travel to meetings unnecessary. The higher population densities resulting from urbanization will make mass transportation more cost-effective and popular. All of these changes will reduce traffic congestion and pollution, increase our national security by decreasing our dependence on foreign oil, reduce CO_2 emissions contributing to global warming, and increase our health (more walking) and quality of life (less time wasted in traffic, better scenery).

11.2.2 Objective 2: Reduce Per Capita Energy Demand

Reducing energy demand is equivalent to reducing the environmental impact $I = P \times A \times C \times T$, preferably by reducing energy intensity C through conservation, or reducing energy inefficiency T through technology (MacKay 2009). For example, we can reduce transportation energy consumption through conservation (e.g., driving fewer miles) and increased efficiency (e.g., getting more meters per liter). As an example, let's compare high- and low-impact city drivers. The high-impact city driver uses an energy inefficient SUV like the gas-guzzling Chevrolet Suburban that gets 4 km per liter or 0.25 L per km, while the low-impact driver uses a fuel-efficient Toyota Prius that gets 20 km per liter = 1/20 or 0.05 L per km.* The Prius owner also conserves energy by combining errands and living in a central location, reducing his or her annual kilometers traveled from 16,000 to 8,000. To calculate the environmental impacts, we ignore the term P because we are comparing two individuals and not looking at the impact of the entire population. We also ignore the term A because we are only measuring the impact of one activity rather than all of the activities that a person's income funds. The environmental impact in one year is then:

$$I_{\text{PriusOwner}} = C \times T = (\text{miles driven/year}) \times (\text{liters gas/km}) = 8{,}000 \times 0.05 = 400 \text{ L/year}$$

$$I_{\text{SuburbanOwner}} = 16{,}000 \times 0.25 = 4{,}000 \text{ L/year} \tag{11.2}$$

The Prius owner decreased fuel consumption by 90% by using technology to increase energy efficiency and by conserving energy. If gas costs $1 per liter, the economic impact on the Prius owner would be only $400 compared with $4,000 for the Suburban owner. The carbon footprint of the Prius owner is also much lower, emitting 930 kg CO_2/year while the Suburban owner emits 9,300 kg CO_2/year.† Combining conservation with efficiency can greatly reduce the economic and environmental impacts of our activities.

11.2.2.1 Increasing Energy Efficiency

The U.S. energy system has increased its efficiency, as measured by the percentage of potential energy in fuel converted to useful work, from 3% to 13%, but we still waste 87% of the energy (Gore 2009). After the OPEC oil embargo and resulting energy crisis the United States decreased oil use by 17% while increasing economic output by 27% through increased efficiency between 1977 and 1985. Without this decrease in energy intensity (amount of energy per dollar of GDP) since 1975, the U.S. economy would be consuming 80% more energy than today.‡ However, **diminishing returns** has slowed the rate of efficiency improvement since 1985. We will see this theme of diminishing returns many times; in technology development and more generally in economics it results from engineers eliminating the largest sources of inefficiency first and then focusing on progressively smaller inefficiencies.

---

* Fuel efficiencies from http://www.fueleconomy.gov/feg/bestworst.shtml.
† Calculated in WolframAlpha using EPA (2005) formula.
‡ American Council for an Energy Efficient Economy (ACEEE), http://www.aceee.org/.

Given the choice between conservation and increased efficiency, people prefer to save energy through increased efficiency because that does not require lifestyle changes. Many opportunities exist for energy savings by the end user. The cost of energy efficiency upgrades for older residential and commercial buildings is usually recovered through energy bill savings within three years. The federal government also offers tax incentives for improving the energy efficiency of homes, which has helped to create thousands of green jobs that cannot be outsourced.

How much could we reduce energy demand by increasing efficiency? According to Granade et al. (2009), a holistic approach to increasing energy efficiency nationwide could reduce total energy demand in 2020 by 2.7 trillion kWh, roughly 23% of projected demand, and reduce greenhouse gas emissions by roughly the same percentage.* A report by the National Research Council (2009) also found that by improving energy efficiency the United States could cut total energy use by 20% or more within 10 years. McKinsey & Co. concluded that energy efficiency savings in the United States could total $130 billion annually (Creyts 2007).

Increasing energy efficiency should be the top priority of a U.S. energy plan. However, many businesses and individuals resist taking simple actions to reduce energy use because of a large upfront cost, though government programs like "Energy Star" labeling make it easy to save money (Luoma 2011). We have to understand the behavioral barriers to adoption of energy efficiency improvements so we can overcome them. Individual barriers to adoption include ignorance, inability to pay upfront costs, skepticism, apathy, and inconvenience. Energy utilities can reduce ignorance through advertising and educational outreach programs, and offer loans to consumers who cannot afford upfront costs.

Finding ways to educate consumers effectively is critically important. For example, the greenhouse gas abatement cost curve makes it easy for homeowners and business owners to identify cost-saving steps that also reduce greenhouse gas emissions.† In the United States almost half of energy consumed and greenhouse gases emitted come from buildings.‡ Architects should show the greenhouse gas abatement cost curve to all clients paying for the construction of new buildings so they can use it to choose the most effective efficiency upgrades for their buildings. Stores that sell energy efficiency products should post the greenhouse gas abatement cost curve to inform consumers and persuade them to purchase efficiency upgrades that will save them money and reduce greenhouse gas emissions from their buildings. The cost abatement curve shows that residential owners can reduce energy use and save money by switching from incandescent to LED lights, buying energy efficient appliances and HVAC systems, and installing insulation retrofits. For example, one very cost-effective and significant energy efficiency upgrade is to install exterior storm windows. Storm windows are inexpensive, easy to install, and greatly improve energy efficiency.§ They also reduce noise, improve home security, preserve existing windows rather than throwing them into landfills, and keep homes cleaner because the airtight seal prevents particulates and air pollutants from entering a home through cracks or joints in existing windows.

Other barriers to adopting cost-saving energy efficiency measures include a reluctance or an inability to invest in efficiency upgrades that pay for themselves over time. Architects and engineers do not always use "systems thinking" that would help them design and engineer large, complex systems to be efficient. Also, our economic system has structural problems that prevent attainment of optimal (most efficient) economic solutions. For example, decision makers often lack a financial incentive to purchase cost-saving efficiency upgrades. The **principal agent problem** occurs when an "agent" makes a decision for another person or "principal," and the decision benefits the agent

* In 2004, the United States emitted 7 billion tonnes CO_2e, Dawson, B. and M. Spannagle (2009). *The Complete Guide to Climate Change*, Routledge. 1 tonne = 1 metric ton = 1,000 kg; in the United States a "ton" usually means a short ton of 2,000 pounds (907 kg).

† http://www.mckinsey.com/business-functions/sustainability-and-resource-productivity/our-insights/impact-of-the-financial-crisis-on-carbon-economics-version-21, retrieved August 29, 2016.

‡ http://architecture2030.org/buildings_problem_why/, retrieved July 18, 2016.

§ http://www.windowreplacementic.com/windows/stormwindows/.

but not the principal. For example, when building a new home, construction firms don't invest in efficiency upgrades because they will not benefit from the reduced operating costs, and they want to keep the purchase price low so the house will sell. The economic incentives of homebuilders and buyers are at odds, and this misalignment of interests slows the rate of adoption of energy efficiency measures.

Another example of the principal agent problem is when the landowner and the energy bill-payer are different: efficiency upgrade incentives disappear because the tenants have less incentive to invest in efficiency upgrades to homes they do not own, while landowners lack economic incentive because they do not pay the energy bills. From a system dynamics perspective, social policies must change to eliminate these structural problems in the economic system. Perhaps we can devise a system that rewards builders who construct energy efficient buildings and penalizes those who do not. For example, we could make builders responsible for paying part of future energy bills for buildings they construct. This would help encourage long-term planning, since efficiency gains continue for the life of the building. Effective regulations could eliminate the principal agent problem as it relates to energy efficiency upgrades for buildings.

Another structural barrier to adoption of energy efficiency upgrades was that energy utilities had no economic incentive to encourage efficiency: The only way they could make more money was by selling more energy. Most states penalize utility companies when energy consumption decreases, causing their utility companies to prevent clients from adopting energy efficiency measures. This is a structural problem in the energy system that needs to be corrected through policy change. During the energy crisis of the late 1970s and early 1980s, California changed policy to allow utilities to charge more for energy if they improved efficiency, so that they share in energy efficiency cost savings (Luoma 2011). This policy decoupled utility profits from energy consumption, removing the economic barrier to utilities encouraging efficiency. By offering consumers and businesses incentives for adopting energy efficiency measures, utilities could reduce energy demand and save money by delaying the construction of expensive new power plants. This innovative policy has been very effective, and is the primary reason that Californians consume about half the energy of the average American. Adopting policies that don't penalize utilities for improving energy efficiency creates a strong incentive to increase efficiency because the cost of efficiency upgrades is usually much less than the cost of building the equivalent amount of electric generating capacity (Gore 2009).

One approach for utilities to increase energy efficiency is to build, or allow their clients to build, **cogeneration** (combined heat and power) plants that are twice as efficient as conventional power plants because they use energy twice (Gore 2009). Cogeneration produces electricity, and then uses the byproduct heat to produce more electricity (by producing steam to drive a turbine) or for space heating. This approach makes businesses more profitable and greatly reduces carbon dioxide emissions.

Cogeneration systems offer other benefits: They represent a decentralized form of energy production. Cogeneration systems can be small, and only nearby buildings can use the heat they produce. This reduces transmission line loss, and reduces the required capacity for transmission and distribution wires. Utilities should build cogeneration plants near customers that could use the by-product heat. Replacing old, inefficient, polluting, greenhouse gas-belching coal-fired power plants with clean, efficient, distributed cogeneration plants could increase efficiency to 80%, greatly reducing electricity costs and carbon dioxide emissions (Gore 2009). In spite of these benefits, governments rarely offer tax credits for deploying efficient cogeneration plants.

Increasing energy efficiency is essential to becoming sustainable. However, energy efficiency improvements alone cannot solve our energy problem. Adopting all of the cost-savings efficiency upgrades listed in the cost abatement curve would reduce energy less than 20% (Section 8.3.1). Even radical improvements would decrease energy consumption less than 30%. These savings are erased as population increases. Another problem is that consumers often spend the savings on more and larger devices that consume more energy or by taking more trips. For example, over time cars have become more fuel-efficient, but people have increased the number of miles that they drive. Thus, increasing efficiency acts to decrease conservation, and this negative feedback reduces the energy savings from increased efficiency (Section 3.6.1). Jevons paradox, also known as the "take-back" or "rebound" effect, occurs when increased energy efficiency results in decreased price, which in turn causes increased consumption, thus eroding the gains from increased efficiency (Greening, Greene, and Difiglio 2000). For example, the take-back for increases in space heating efficiency is between 10% and 30%, meaning that 70% to 90% of the efficiency gains translate into reduced energy consumption. Lighting efficiency improvements are 80% to 95% effective at reducing energy consumption. For vehicle use, technological improvements in efficiency are 50% to 80% effective in reducing energy consumption (Greening, Greene, and Difiglio 2000).

As an example of the rebound effect, consider that the electrical efficiency of personal computers measured as computations per kWh doubled every 1.5 years, corresponding to an increase of over one million times between 1975 and 2010 (Koomey et al. 2011). However, personal computers consume roughly the same amount of energy as they did in 1975. Why? Consumers desired larger monitors, higher storage capacity, and software and operating systems that were more user-friendly but which required much higher computing speed. All energy efficiency improvements were erased by increased performance. The same is true for TVs: The energy savings from the switch to LED displays is being offset by increasingly large displays. Home energy savings are erased by the shift to increasingly larger homes. The average size of a new American home increased from 983 square feet in 1950 to 2,480 square feet in 2011. Furthermore, the average number of people in each home decreased from 3.4 to 2.6, meaning that Americans took up more than three times more space in 2011 than in 1950.* The energy required to heat, cool, and maintain these homes has increased proportionally. These observations make clear that reducing energy consumption is as much a behavioral problem as a technological problem. The more money we have, the more energy we consume (see the chart at www.bit.ly/1Mtxu3h).

One positive trend is that in recent years personal computing devices have been shrinking. Smaller devices use less energy. Notebook computers use less energy than desktop computers and workstations. Tablets use less energy than notebook computers, and smart phones use even less energy, yet they have more computing power than a 10-year-old desktop computer. Consumers can only use one computing device at a time, so the more they use smaller devices, the less energy they consume. Although they may have more than one device powered at a time, new devices quickly turn off when they are not being used. The trend toward increasing adoption of mobile computing devices may significantly decrease the average energy consumption of personal electronics.

11.2.2.2 Energy Conservation

Because improvements in efficiency require development and deployment of new technologies, which can be expensive, the most cost-effective approach to reducing energy use and associated environmental impacts is through conservation. Also, while there are physical limits to improvements in efficiency, energy savings from conservation can approach 100% of energy use. However, although the potential savings of conservation are greater than adoption of efficiency steps, experts believe that promoting efficiency is more effective than conservation because the latter requires lifestyle changes and behavioral plasticity is low (Dietz et al. 2009). Conservation alone will not be enough to reduce net greenhouse gas emissions to zero during the second half of the twenty-first century, as called for by the 2015 COP21 agreement. For example, much has been made of the **standby power** problem in which electronics and appliances consume electricity while turned off, yet this accounts for ~8% of residential energy demand, which with much effort might be reduced to 2% (MacKay 2009). Since residential accounts for only 21% of total U.S. energy demand, that would decrease energy demand in the United States by only 6% of 21% or 1.3%. This energy savings

* http://www.nytimes.com/2013/03/10/opinion/sunday/living-with-less-a-lot-less.html?pagewanted=1&_r=0&smid=fb-share&pagewanted=all.

would be wiped out by only one to two years of growth in energy consumption due to rising population and affluence.

Even with the economic incentive of saving money, the combined effects of increased efficiency and energy conservation have not significantly reduced per capita energy consumption in the United States, which was roughly the same in 2007 as it was in 1980, although by 2013 it had decreased slightly (Figure 11.1). The evidence suggests that we cannot meet our future energy needs through energy conservation alone without major lifestyle changes such as living in much smaller homes or abandoning cars for bicycles or mass transportation. Many people are not yet ready or are unwilling to make those choices.

On the other hand, Dietz et al. (2009) found that adoption of 17 energy conservation and efficiency approaches could reduce household emissions by 20% or 7.4% of U.S. national emissions, without government regulation or reduction in well-being. Measures with the highest potential carbon emissions reduction were, in decreasing order, use of fuel-efficient vehicles, carpooling and trip-chaining, home weatherization, efficient driving behavior, Energy-Star appliances, efficient HVAC equipment, and thermostat setbacks. However, when they took plasticity (the proportion of current nonadopters that could be induced to take action) into account, they found that the most effective behavioral changes (the actions with the highest reasonably achievable emissions reductions) in decreasing order were use of fuel-efficient vehicles, **weatherization** (protecting a home from precipitation and wind by sealing openings, and in this study by installing insulation), Energy-Star appliances, efficient HVAC equipment, efficient driving behavior, car-pooling and trip-chaining, and efficient water heating. By changing behavior and grabbing the low-hanging fruit of energy conservation and efficiency measures, Americans can mitigate global climate change and save money with little effort. If the government uses a combination of policy tools and strong social marketing to persuade multiple groups (individuals, communities, and businesses), consumers will more likely adopt these measures (Dietz et al. 2009).

Because conservation requires behavioral changes, it's important to know what behavioral interventions are effective. Research on health interventions started earlier and has progressed farther than research on environmental interventions; much work still needs to be done on the social psychology of environmental behaviors. There are several different approaches. First, information/advertising campaigns have been shown to reduce electricity use by an average of 7%, with some approaches like energy audits and consulting being more effective than others (Delmas, Fischlein, and Asensio 2013). Targeted messaging places prompts in key locations, such as a sign in a school's car pickup line saying, "Do not idle: children present." Research shows that tailoring messages to specific audiences can increase adoption rate in the targeted audience, but may have the opposite effect on others, causing the effects to cancel out. For example, labeling light bulbs "environmentally friendly" makes purchases more likely for liberals but less likely for conservatives.

A behavioral intervention approach that is become increasingly popular is feedback messaging. For example, the display on a Toyota Camry hybrid displays "Eco mode" when driving is fuel-efficient. Feedback approaches can significantly reduce home electricity use (Fischer 2008). Situational influences can also affect the likelihood of a desired environmental behavior. For example, people are more likely to recycle if they have curbside pickup and if they do not have to sort their recyclables. Related to this is the **status quo bias** where users are more likely to choose an environmental behavior if it is the default option. For example, the energy efficient mode is the default option on some HDTVs.

Pro-environmental behavior can be influenced by messages about **social norms**. Injunctive norms are messages about what you should do, and descriptive norms say what people actually do. The behavior of an individual can be influenced by comparing their behavior to the behaviors of other individuals. For example, customers that use the cloud-based software service OPower cut their energy use between 0.3% and 6.3% (Allcott 2011). The approach is to combine feedback messages with comparisons to other users. Although the overall average energy use decreased with OPower, the energy consumption of some users increased. Users that were told their energy use was lower than that of their neighbors then increased their energy use, an example of **moral licensing** where people allow themselves to indulge after doing something positive. A summary of studies of factors that influence personal energy conservation found that providing consumers information did not always reduce energy consumption; that offering rewards for reducing consumption only temporarily reduces consumption; and that feedback messaging such as that provided by OPower is effective, especially when messages are frequent (Abrahamse et al. 2005). Overall, different interventions work for different people, but more research needs to been done before we can understand the complex decision making processes that influence environmental behaviors and lead to energy conservation.

11.2.2.3 A Personal Plan to Reduce Home Energy Use

If you are a homeowner, you can substantially reduce your carbon and ecological footprints and save money by reducing your energy use through conservation (moving to a smaller home or reducing heating and cooling energy demand by adjusting the thermostat) or by increasing home energy efficiency. Many energy efficiency upgrades such as adding attic insulation pay for themselves within two to three years, and continue saving money over the lifetime of the home. They also increase the home resale value.

Before making a plan to decrease your home energy consumption, you need to identify the main energy consumers in your home. In the average U.S. home in 2008, space heating and cooling consumed 49% of the energy, with water heater (13%), lighting (10%), electronics (7%), clothes washer and dryer (6%), refrigerator (5%), and dishwasher (2%) falling well behind (Hill and O'Neill 2008). Since space heating and cooling typically dominate a home's energy budget, you can greatly decrease your carbon footprint and save money by reducing their energy demand. You can conserve energy by adjusting the thermostat to reduce the temperature difference between indoors and outdoors. You can probably tolerate home temperatures that are higher in the summer and lower in the winter than your current settings (Figure 11.3). Programmable thermostats are useful for automatically turning off an HVAC system when you leave the house and turning it back on just before you return so you don't heat or cool your home during the day when no one is home. New WiFi-enabled thermostats allow for remote control of temperature settings, while other systems such as the Nest

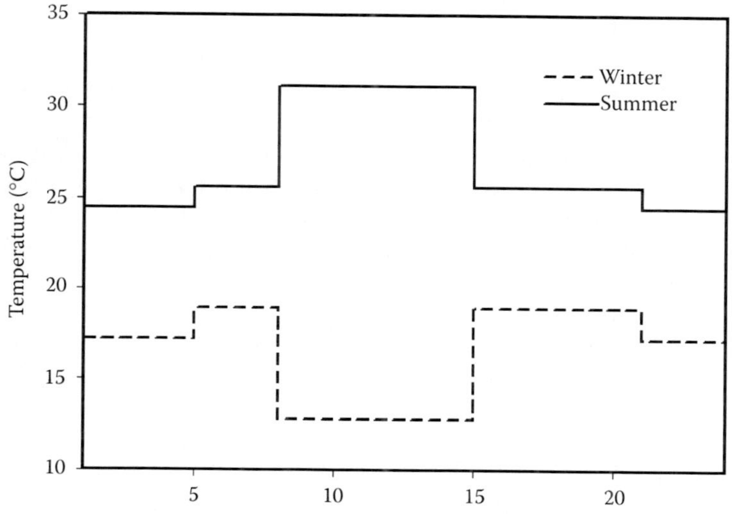

FIGURE 11.3 Thermostat settings over the course of one day, aimed at reducing home energy heating and cooling costs.

Learning Thermostat use machine learning algorithms to automatically optimize temperature settings based on user preferences.

Many people think that turning off an HVAC system when a home is unoccupied uses more energy than leaving it on because the HVAC must heat or cool a greater amount upon return of the occupant. This is not true. The power (rate of energy consumption) required to maintain a constant temperature is proportional to the temperature difference between the inside and outside. This is because the greater the temperature difference, the faster heat moves from the hot to the cold body. If the occupant turns the power off before leaving the house for an 8-hour workday, the temperature difference decreases during the day, so the house loses less heat during winter and gains less heat during summer. The energy savings from the smaller temperature difference are compounded throughout the day. The energy used to heat or cool the home to the preferred temperature just before the occupant arrives is much less than if the home were kept at the preferred temperature throughout the day. Imagine you were heating your house with a wood stove. Would you use more wood if you kept the fire roaring all night or if you let it burn out or burn slowly and then stoked it back up in the morning? Put in these terms it is easy to see that maintaining a lower heating temperature over time uses less fuel.

How else can you save energy and money at home? The first and most obvious choice is to buy a smaller home, which, all else being equal, will have lower heating and cooling costs. Second, for an older home get a home energy audit, or complete one online.* You may need to add insulation to the attic or floors. Many homes have crawlspaces beneath them, with vents that allow outside air to circulate under the house.

Water heaters are the second biggest consumers of energy in the typical American home (Hill and O'Neill 2008). Conventional water heaters store hot water in a large tank that loses heat. To compensate for heat loss, the heater must almost continuously burn fuel to maintain a constant temperature. An alternative that is already widespread in Europe and is now becoming popular in the United States is the tankless water heater, which heats water on demand and therefore only consumes energy when hot water is used. A newer technology that has a higher return on investment is the hybrid electric hot water heater, which combines a conventional water heater with an efficient heat pump. With federal tax rebates, these devices can pay for themselves in four to five years, compared with 10 to 20 for tankless and solar water heaters.†

Next, an easy way to increase the energy efficiency of your home is to switch from old fashioned incandescent bulbs to Compact Fluorescent Lights (CFLs) or Light-Emitting Diodes (LEDs). Artificial lighting accounts for 10% of the energy consumed in the average home, and improving lighting efficiency can reduce that energy by 60% to 70%.‡ CFLs are ~5 times and LEDs are ~6 times more efficient and last much longer than incandescent bulbs do.§ Substituting a standard 13 watt CFL for an incandescent bulb will reduce electricity bills by roughly $30 over the bulb lifetime, which easily repays the higher cost of the bulb (Brown 2009).¶ Shifting from incandescent bulbs to CFLs in homes, advanced fluorescent bulbs in commercial buildings, and LEDs in traffic lights** would "cut the world share of electricity used for lighting from 19% to 7%... and save enough electricity to avoid building 705 coal-fired power plants" (Brown 2009).

LEDs may represent the lighting option of the future because they are ~20% more efficient and last three to five times longer than CFLs (Brown 2009). In spring 2009, many LED lighting options appeared on store shelves. LED nightlights have light sensors that automatically turn the light on at

* http://www.energysavers.gov/your_home/energy_audits/index.cfm/mytopic=11160.

† Consumer Reports, Sept. 2010.

‡ http://www.energysavers.gov/your_home/lighting_daylighting/index.cfm/mytopic=11980.

§ Overall lighting efficiency in lumens/watt ranges from 10–17 for incandescent, 50–70 for CFL, and 60–92 for cool white LEDs: http://www.energysavers.gov/your_home/lighting_daylighting/index.cfm/mytopic=12030.

¶ Note that CFLs must be properly disposed of because they contain toxic mercury.

** Don't use LEDs for traffic lights in regions with heavy or frequent snowfall, as LED lights do not emit enough heat to melt snow, and they become buried and invisible.

night and off in the day. Outdoor solar-powered LED lights have PV cells on top that recharge batteries during the day so they can glow all night. LED lights are now commonly used for Christmas decorating not only because they are energy-efficient, but also because they are shatterproof, shock resistant, and produce almost no heat, which reduces the risk for fires.

A cost-effective alternative to artificial lighting is natural (passive solar) lighting. Exterior rooms should always have windows, but make sure they are energy-efficient. Rooms on the top floor should have skylights, but make sure they are insulated to prevent heat from escaping during the winter.

You can conserve more energy by replacing inefficient appliances with Energy Star rated appliances. The Energy Star program created by the U.S. EPA has become an international standard for rating the energy efficiency of consumer products. It is one of the reasons that the energy efficiency of appliances such as refrigerators, air conditioners, clothes washers and dryers, and dishwashers has greatly increased in recent years (Brown 2009). You can also save energy by properly maintaining your house. Plug air leaks in windows and doors. Caulk around your windows and anywhere else where heat can be lost from your home, or replace your windows with new energy-efficient windows. Clean your HVAC and change the filters regularly so it can heat and cool efficiently. Table 11.1 lists some important actions that individuals can take to conserve energy, with estimates of the energy each action can save.

TABLE 11.1
Energy Savings from Simple Individual Actions

Simple Action	Possible Savings
During winter put on a warm sweater and turn down your heating's thermostat (to 15°C to 17°C [60°F to 63°F], say). Put individual thermostats on all radiators. Make sure the heating is off when no one is at home. Do the same at work.	20 kWh/d
Read all your meters (gas, electricity, water) every week, and identify easy changes to reduce consumption (e.g., switching things off). Compare competitively with a friend. Read the meters at your place of work, too, creating a perpetual live energy audit.	4 kWh/d
Stop flying.	35 kWh/d
Drive less, drive more slowly, drive more gently, carpool, use an electric car, join a car club, cycle, walk, use trains and buses.	20 kWh/d
Keep using old gadgets (e.g., computers); don't replace them early.	4 kWh/d
Change lights to fluorescent or LED.	4 kWh/d
Don't buy clutter. Avoid packaging.	20 kWh/d
Eat vegetarian, six days out of seven.	10 kWh/d
Wash laundry in cold water.	0.5 kWh/d
Stop using a tumble-dryer; use a clothes-line.	0.5 kWh/d
Total	118 kWh/d
Major Action	**Possible Savings**
Eliminate draughts.	5 kWh/d
Insulated glazing (convert single- to double-pane windows).	10 kWh/d
Improve wall, roof, and floor insulation.	10 kWh/d
Solar hot water panels.	8 kWh/d
Photovoltaic panels.	5 kWh/d
Knock down old building and replace by new.	35 kWh/d
Replace fossil-fuel heating by ground-source or air-source heat pumps.	10 kWh/d
Total	83 kWh/d
Total for all	201 kWh/d

Source: DJC MacKay, *Sustainable Energy—Without the Hot Air*, UIT, www.uit.co.uk/sustainable. Also available free to download for personal noncommercial use from www.withouthotair.com.

According to the Annual Energy Outlook (EIA 2009), U.S. energy consumption in 2007 was 21% residential, 18% commercial, 32% industrial, and 28% transportation. If residential homes only consume 21% of power in the United States, how can consumers influence power use in the other nearly 80% of the power economy? By voting with their wallets. The choices consumers make may give them more power than their vote. Recognizing this can empower individuals. What if you only bought products made using renewable energy? You, and others like you, could become a powerful force for change; companies would have a strong incentive to switch to renewable energy sources. For example, if you have the option, enroll in a program like Green Power Switch,* a Tennessee Valley Authority (TVA) program that encourages the use of renewable energy sources. According to the Green-e database,† the TVA Green Power Switch program gets its energy from wind (63%), solar (1%), and biomass (37%).

Some readers may decide that it's just too difficult to remember all of the statistics about energy conservation, and just throw up their hands and not practice energy conservation at all. If you are leaning toward that approach, consider another course of action. If the obstacle is memorizing the relative merits of different conservation measures, then practice all of the measures, or at least all of the ones that you are willing or able to do. Motivate yourself to take energy-saving actions by picturing the money you will save. Imagine that you are picking up a free quarter from the ground when you unplug an appliance. Accept the fact that some of your actions will have smaller impacts and savings than others, but that what matters is the overall impact. Alternatively, focus only on the actions that will save the most energy (Table 11.1).

Another approach is to apply a simple rule of thumb to all actions like the "rule of halves": halve the size of your house and car, halve the time you shower, halve the number of flights you take per year, halve your commuting distance, and so on. Using this rule consistently can cut your energy and material use, your EF, and your costs in half and make your lifestyle more sustainable.

The U.S. population is still growing, mostly through immigration, and this increases total energy demand per the ImPACT identity (Figure 9.1). Furthermore, increases in energy efficiency of housing, appliances, and consumer electronics, and tax incentives to improve home energy efficiency have only slightly reduced per capita energy consumption (Figure 11.1). Thus, we cannot depend only on energy conservation and improvements in energy efficiency to meet our future energy needs. We must also grow energy supply by increasing energy production from renewable energy sources faster than we reduce energy production from fossil fuels.

11.2.3 Objective 3: Increase Renewable Energy Supply

Since per capita energy consumption is unlikely to decrease significantly and population continues to increase, power production will need to expand to meet increasing demand. We need a procedure for formulating the optimal plan for increasing energy supply. When formulating a plan, we should not take any energy options off the table a priori. Every energy choice has consequences; we must objectively weigh the positive and negative consequences of every potential energy choice, and eliminate those for which the negatives outweigh the positives.

Now that we have a plan that addresses energy objective (2) to reduce energy consumption through increased energy efficiency and conservation, we need a plan to choose the best mix of energy sources that can meet objectives (1) and (3): (1) reduce greenhouse gas emissions; (3) diversify the energy portfolio and increase the supply of safe, sustainable, renewable energy while minimizing the true cost. In order of execution the steps needed to meet these objectives are

1. Identify potential sources of energy.
2. Determine the maximum amount of energy each source can reasonably produce, taking into account physical limits and political realities (e.g., MacKay 2009).

* http://www.tva.gov/greenpowerswitch/.
† http://www.green-e.org/base/re_products?cust=.

3. Determine the true cost of each energy source and the use of other resources by using LCA and FCA (USNRC 2010). This accounts for externalities such as environmental and social costs of production and use of each energy source. These can include costs from pollution and resulting health effects (including premature mortality), loss of ecosystem services, and costs to national security.
4. Make accurate projections of energy demand, taking into account the anticipated reduction from energy conservation measures.
5. Choose the optimal mix of energy sources that can meet projected energy demand at the lowest true cost. Since the true cost includes the costs of global climate change and the use of other resources, reducing the total true cost will also reduce the total carbon, water, and land footprints of our energy use.

Let's look at examples for each step. MacKay (2009) explored steps 1 and 2 to decide whether the United Kingdom could meet its current energy needs using only renewable sources of energy. He found that current energy consumption exceeds the maximum amount of energy that all renewable energy sources could produce. He based his calculated production rates only on physical limits, however; including political constraints (what is the public willing to accept or pay for) greatly decreases the energy that can reasonably be produced sustainably. No combination of renewable energy sources can meet the current energy needs of the United Kingdom, which means that if they are to phase out fossil fuels, they must import renewable energy or expand the use of nuclear power.

The average American uses 250 kWh/d, so we need to produce an even greater amount of energy than in the United Kingdom. The best estimates for maximum per capita energy production in kWh/d for four renewable energy sources in the United States are wind 42, offshore wind 5, geothermal 8, and hydroelectric 7, for a total of 62 kWh/d, which doesn't even come close to supplying our current energy demand of 250 kWh/d per person (MacKay 2009). What could make up the difference? Currently fossil fuels and nuclear power do (Figure 9.1). Natural gas will become increasingly important, and the use of nuclear power may expand. These will be only temporary solutions. Ignoring the possible development of new technologies such as nuclear fusion, which we can't count on because it may never come to fruition, the United States can only sustainably meet its energy needs using solar power. Solar could eventually meet the current energy needs of America plus 200 million more people if we dedicate 3,600 km^2 of land (a little bigger than the state of Arizona) in the sunny desert southwest to Concentrated Solar Power farms (MacKay 2009). Solar is the only renewable energy source that can meet all of our energy needs, but first it must become cost-competitive (Figure 10.1) and then scaled up to a national scale.

Too often the critical step 2 of estimating the maximum energy production of a potential energy source is skipped when formulating energy plans. For example, in the 2000s corn ethanol production was promoted as a renewable energy source in the United States without evaluating how much energy it could produce. Biofuels can only supply a few percent of U.S. TPES because they have very low efficiency and low energy density, so it made no scientific sense to make them the centerpiece of a new energy plan (perhaps it made political sense).

Moving on to step 3 where we evaluate costs, we find that the most cost-effective fuel sources in the year 2020 are projected to be geothermal, wind, natural gas combined cycle, and hydroelectric (Figure 10.1). Coal comes in fifth place, with nuclear and biomass not far behind. However, if we use FCA, we find that coal roughly doubles in cost (USNRC 2010). If we included the costs associated with carbon dioxide emissions and resulting global climate change, or the cost of CCS to mitigate global climate change, coal would be even more expensive. The external costs of wind, solar, and even nuclear power are negligible compared with coal, although we cannot evaluate the cost of an as-yet unplanned repository for storage of high-level nuclear waste from reactors.

LCA also shows that all energy sources have a non-zero carbon dioxide intensity. The fossil fuels have the highest carbon dioxide intensity followed by nuclear, geothermal, solar PV, and concentrated solar power, with wind having the lowest carbon dioxide intensity (Table 9.2). The fossil fuels oil and natural gas are preferable to coal because we can more easily transport them, they have higher energy densities, and they have smaller carbon dioxide intensities. Thus, we remove conventional coal from our list of energy sources, recognizing that some form of clean coal + CCS could play a role in the future.

Step 3 also accounts for the use of other resources to produce a given energy type. For example, corn ethanol production requires large amounts of electrical and fossil fuel energy, making its EROEI low. It also requires large amounts of land and water. Including the externalized economic costs of converting land from, say, food production to dedicated energy production, which increases food prices, helps us come closer to estimating the true cost of biofuel energy production. As another example, production of petroleum from tar sand requires huge amounts of energy and water, which increases the carbon dioxide and water intensities of oil derived from tar sands. Thus, we should avoid energy sources with high embodied energy and water contents.

Once energy demand is determined in step 4, in step 5 we choose the optimal mix of energy sources that can meet that demand at the lowest true cost. The optimal energy source mix will depend on the location. Sun-rich regions may rely more heavily on solar power, while windy areas might rely more on wind. Decentralized renewable energy production would reduce the need for costly transmission line construction and reduce electrical energy transmission line loss. Those areas that cannot meet their energy needs using renewable energy sources alone should build high efficiency cogeneration plants powered by natural gas or small nuclear reactors.

According to the EIA projections of energy use by type (Figure 9.1), which assume that the growth rate of energy use will continue to increase, the percentage of generated electricity from renewable sources in the United States (including liquid biofuels) solely will increase from 9% in 2013 to 11% in 2040. Most growth in the percentage of U.S. TPES between 2013 and 2040 will be in wind because it is cheaper than solar power (Figure 10.2). Hydropower contributed 7% of U.S. TPES in 1980 but that share is not expected to increase. We conclude that it will take decades to replace fossil fuels with alternative renewable energy sources in the United States.

Looking at the global picture, Jacobsen and Delucchi (2011) estimate that by 2050 the entire world could be powered solely by renewables using existing technologies. This would require by 2030 the construction of ~3,800,000 5 MW wind turbines, ~49,000 300 MW concentrated solar plants, ~40,000 300 MW solar PV power plants, ~1.7 billion 3 kW rooftop PV systems, ~5,350 100 MW geothermal power plants, ~270 new 1,300 MW hydroelectric power plants, ~720,000 0.75 MW wave devices, and ~490,000 1 MW tidal turbines. Adoption of this plan would reduce energy consumption by 30%. It would use only 0.6% more land that we use for energy production today. An update of this plan provides a solution to the intermittency problem (Jacobson et al. 2015). These papers argue that we have the technological and economic ability to transition completely to renewable energy sources by 2050; the barriers are primarily social and political.

11.2.4 The Need for a Smart Electric Grid

The transition to a renewable energy system in the United States will require construction of a new, national electric grid. Although the current U.S. system was a technological marvel in the twentieth century, it is now outdated by several decades, resulting in high transmission line loss and blackouts. Blackouts alone cost the United States ~$80 billion per year from businesses and industry, and cost society ~$206 billion per year in 2007 (Gore 2009). Our nation needs a smart grid to transmit electricity from distant, intermittent wind and solar sources. According to Gore (2009), a national smart grid will consist of four interconnected elements:

1. High-voltage direct current (HVDC) long-distance, high efficiency transmission lines
2. "Smart" distribution networks connected by the Internet to smart meters at homes and other elements of the transmission and distribution grid

3. Modern, dynamic, and efficient electric-energy storage units placed throughout the transmission and distribution networks
4. Distributed intelligence with robust, information-rich, two-way communication throughout the grid

Imagine every electrical appliance in your house having a smart meter. Utilities will start using time-of-day pricing, charging more per kWh during peak demand. Smart devices will turn on when electricity is cheap and off when it is expensive, saving you money. Such a system would level the daily fluctuations in electricity demand, which would allow power plants to operate more efficiently by running near peak capacity all day. By decreasing peak energy demand, utilities would require fewer power plants to meet peak demand, meaning lower costs and fewer emissions. Computer-controlled load balancing would alleviate the intermittency problems of wind and solar. The required HVDC lines could be buried underground to alleviate public health concerns about power lines and reduce their vulnerability to extreme weather events and other natural disasters. Simultaneously burying fiber optic communication cables could greatly increase the capacity of the Internet at relatively low cost. Finally, a well-designed smart grid combined with distributed power generation would be more reliable, less costly, less harmful to the environment, and less vulnerable to terrorist attacks because its decentralized nature would make it more resilient. Although a national smart grid would be expensive, with a net investment of $338 billion to $476 billion, the savings are estimated to be three to six times greater than the costs, and it could decrease CO_2 emissions from the electric sector by 58% relative to 2005 emissions (EPRI 2011).

11.3 GOVERNMENT POLICIES

To move toward energy sustainability and security, the United States must adopt policies that promote energy independence and keep America competitive. The objective of these policies must be to reduce per capita energy consumption and replace fossil fuels with renewable energy sources as quickly as possible. Government should use the **carrot and stick** approach by including both rewards and penalties in policies to decrease energy consumption and reduce greenhouse gas emissions. Government policies should discourage energy waste by using carbon pricing (carbon taxes or emissions trading) to eliminate market failure, and encourage adoption of energy efficient technologies by offering tax incentives for their purchase and installation. Balancing revenue increases from carbon taxes with revenue losses from energy efficiency tax breaks would amount to a **tax shift**. A tax shift keeps overall revenue the same, but encourages desirable changes by providing incentives and discourages undesirable behavior (carbon emissions) by levying penalties as taxes. For example, many states have shifted taxes from necessities like groceries to harmful nonessential items like tobacco. Elimination of tax breaks for fossil fuel producers, which amount to **perverse subsidies**, should also be part of this tax shift.

The U.S. government could also increase the supply of renewable energy sources by providing incentives for industry to invest in renewable energy production. Until the United States adopts policies that promote the development of green technologies, it will continue to fall behind in the global competition for the rapidly expanding green market. China has become the leading supplier of clean energy technology such as solar PV panels. Recognizing that we are losing a competitive edge in the global economy, even politicians and industry executives who are climate change deniers are beginning to push for expanding investments in green technologies that will reduce greenhouse gas emissions. Likewise, stricter environmental regulations such as lower greenhouse gas emission limits are drivers of industry innovation. Friedman (2008) makes a convincing case that stricter environmental regulations make American businesses more competitive internationally.* Without

* See study "Innovation and Regulation in the Automobile Sector: Lessons Learned and Implications for California's CO_2 Standards," Roland Hwang and Matt Peak, April 2006.

a public consensus on the dangers of anthropogenic global warming, environmentalists are starting to emphasize the economic benefits of switching to clean, renewable energy sources.

11.4 PROBLEMS AND SOLUTIONS

Most energy options are inadequate either because the technology is not fully developed (e.g., clean coal, nuclear fusion) or because they can provide adequate energy only in specific areas (e.g., geothermal), and therefore are not global solutions. Peak oil may *force* us to reduce the amount of oil we use; we must also *choose* to phase out the use of conventional coal as quickly as possible to mitigate anthropogenic global warming. The United States must resist the temptation to use its abundant coal until it develops technologies to use coal cleanly.

As argued above, to phase out oil and coal as rapidly as possible while maintaining energy security, the United States should adopt policies to reduce per capita energy demand through conservation and efficiency, increase energy supply for a growing population, and build a new energy production infrastructure that is clean, reliable, and is primarily renewable with a low carbon footprint. We can make our energy supply more sustainable by shifting from the hard path to the soft path, from centralized to decentralized, and from nonrenewable to renewable. Combined wind and solar powered systems meet these criteria, and therefore are the most sustainable solution to the new energy crisis. Here we address the pros and cons of combined wind and solar powered systems.

Solar energy is the only renewable energy source that can fully meet U.S. energy demand, but it will be decades before we can build enough solar energy facilities to supply 100% of demand. Furthermore, solar and wind energy have several problems that may prevent them from being cost-effective. First, areas with little sun or wind or high population densities cannot produce adequate power locally from sun and wind. Thus, we must build a national grid that can transport electricity from the often remote areas with abundant sun and wind to areas that have renewable energy supplies that are insufficient to meet demand. Second, the power densities of solar and wind energy are low. This prevents them from becoming the primary energy sources in areas that have little available land because they have high population densities, which describes most of Europe. However, the United States and many other countries have enough land, incident solar energy, and wind energy to meet energy demand. In the future these countries may export electricity produced from sun and wind to other countries (e.g., the gigantic **Desertec** project in North Africa*).

A third problem is that both solar and wind energy are intermittent, meaning that their energy production rate is highly time-dependent. However, the intermittencies of solar and wind energy partially cancel each other (i.e., it is often windy when cloudy, and the wind still blows at night). Also, a national grid would compensate for local intermittency. For example, sunny areas can supply electricity to cloudy areas. This makes combined solar and wind power coupled with a national electric smart grid an attractive option. However, this system may not produce enough energy always to meet demand at all locations, which brings us to our fourth problem, insufficient and inefficient energy storage. Until we develop better batteries and other energy-storage technologies, we will be unable to rely completely on solar + wind energy, although pumped storage could currently solve the problem in many areas. In the meantime, we will need a reliable baseload energy source, which could be nuclear or natural gas.

A fifth problem is that solar and wind may not be cost-effective because building the facilities requires large amounts of money and raw materials. Deffeyes (2001) calls this the "energy-material paradox," and uses it to argue that nuclear power is cheaper than wind power. However, projections of levelized costs suggest that wind will be more cost-effective for electricity production than nuclear in 2020 (Figure 10.1).

Because of the five solar and wind power problems, some argue that nuclear power is preferable (Deffeyes 2001; Lovelock 2006). Others argue for nuclear power as an interim energy source, as

* http://www.desertec.org.

a bridge between fossil-fuel energy production and renewable energy production. Nuclear power can substitute for coal power, and is greener than coal at all stages of its life cycle except waste disposal. Because anthropogenic global warming is a much more urgent problem than the problem of disposal of small volumes of radioactive waste produced in nuclear power plants, nuclear power is preferable to coal power. In regions that have inadequate hydroelectric and geothermal energy sources and not enough land for solar and wind facilities, supplementing with nuclear power may be the greenest energy choice.

Another problem is that building renewable energy electricity results in only a small displacement of fossil fuel electricity sources (York 2012). Specifically, holding all other factors such as economic growth constant, reducing fossil fuel electricity production by 1 kWh would require nearly 13 kWh of non-fossil electricity production. Many factors affect this number, including the rebound effect, the influence of the fossil fuel lobby, and lock-in to using existing fossil fuel power plants as the base energy source. Technology fixes alone, specifically expanding non-fossil-fuel electricity production, will not significantly reduce fossil fuel use and associated carbon emissions. Providing disincentives to fossil fuel energy use such as a carbon tax would likely be more effective at reducing carbon emissions. To reduce carbon emissions most effectively, policies should aim to expand renewable energy production, but also take into account human behavior.

Princen et al. (2013) argue that GCC is only one of many problems caused by fossil fuel use, so we should focus on reducing fossil fuel extraction activities rather than carbon emissions. Several countries in the global South have restricted extraction activities. Costa Rica enacted a moratorium on oil extraction in 2002, choosing instead to develop the ecotourism industry, which would be damaged by extraction activities. Because whether continue fossil fuel extraction is a moral question, the most effective approach to rapidly phasing it out is to delegitimize fossil fuels, much as slavery and tobacco use were delegitimized. Rather than condemn the fossil fuel industry (an approach that was ineffective against slavery and tobacco), we must persuade people that fossil fuels are a poison to society because they threaten the "good life." Framing the issue in this way rather than using the polarizing strategy of placing blame is more likely to persuade skeptics that phasing out fossil fuels is in our best interest.

11.5 CONCLUSIONS AND THE FUTURE

We have shown that it will take decades for renewable energy sources to supplant fossil fuels in the United States. So what should we do meanwhile? Most energy experts expect that the United States will continue to expand its use of natural gas, which is abundant in North America and burns more cleanly than coal, but still emits significant amounts of carbon dioxide.* Other major energy trends in the United States in the next one to two decades likely will be the replacement of gas-powered cars with more efficient electric cars, and the expanded use but not domination of the renewable energy source wind. In the long term new, more cost-effective solar technologies may become dominant in the United States.

Many argue that the United States cannot reduce carbon dioxide emissions by reducing fossil fuel use without greatly harming the economy. Denmark provides an instructive counterexample. After the OPEC oil embargo in 1973–1974, Denmark decided to make changes. In 1985 it chose to use taxes to discourage the use of fossil fuels (gasoline cost about $9 per gallon in 2008), to increase efficiency, and to expand the use of renewable sources of energy. In 1973, Denmark got 99% of its energy from the Middle East, but that dropped to 0% in 2008. Did those steps kill the economy? Between 1981 and 2008 the Danish economy grew 70%, energy consumption rose very little, carbon dioxide emissions decreased, and unemployment dropped to less than 2% in 2008

* Hopefully we will not expand natural gas use by using environmentally unsafe recovery technologies; the debate over whether currently-used "fracking" methods are safe is now raging.

(Friedman 2008). Denmark has shown that countries can reduce fossil fuel use and carbon emissions without weakening the economy.

Some argue that countries like the Netherlands, Switzerland, and Sweden have made much more progress in greening their economies and switching to renewable energy sources than the United States because they are smaller countries. Yet the United States transformed itself completely in just a few years at the beginning of WWII, and today China is increasing the proportion of energy produced by renewable sources faster than almost any other country. So it's not size that is the problem; it's our political system, which takes too long to reach a consensus. The U.S. political system has too much inertia, and therefore responds too slowly to global geopolitical changes to stay competitive. Even in the midst of a climate crisis, even for an administration voted into office on a platform of "change," the pace of change in federal policy has been glacial. There are so many competing interests, and the technologies of yesterday have such a stranglehold on the entire system, that rapid change seems unlikely.

Despite these obstacles, the United States has made some progress in reducing carbon emissions. Between 2007 and 2013 greenhouse gas emissions dropped 11%, primarily due to substitution of natural gas for coal in power plants.* Now that we are putting real effort into building a national renewable energy base, we can decrease carbon dioxide emissions by much more than that without making sacrifices. As noted by economist Paul Krugman, "Once you filter out the noise generated by special-interest groups, you discover that there is widespread agreement among environmental economists that a market-based program to deal with the threat of climate change—one that limits carbon emissions by putting a price on them—can achieve large results at modest, though not trivial, cost."†

WEB RESOURCES

- A Guide to Energy Efficient Home Lighting: http://www.mortgagecalculator.org/helpful-advice/energy-efficient-lighting.php
- All About Energy and How to Use It More Efficiently: http://roscoebrown.com/how-to-use-energy-more-efficiently.html
- Energy Efficiency and Renewable Resources: www.energysavers.gov
- Energy Efficient Appliances: http://www.appliancepartspros.com/a-guidebook-to-energy-efficient-appliances.aspx
- Lighting Recycling: http://www.bellacor.com/viewArticle.cfm?articleid=54
- Switch Energy Project Documentary Film and Energy Expert Video Series: http://switchenergyproject.com/
- U.S. DOE Energy Efficiency and Renewable Energy Program: http://www.eere.energy.gov/

HOMEWORK PROBLEMS

1. What do you think a country like the United Kingdom should do to make up for the discrepancy between consumption rate and maximum sustainable energy production rate?
2. Energy intensity is often defined as the amount of energy used to produce one unit of Gross Domestic Product (GDP). Obtain historical data for the last decade on GDP from the U.S. Bureau of Economic Analysis website (http://www.bea.gov) and on total energy consumption from the U.S. Energy Information Administration website (http://www.eia.gov). In a spreadsheet for each year, calculate energy intensity by dividing total energy consumption

* http://www3.epa.gov/climatechange/ghgemissions/inventoryexplorer/#allsectors/allgas/gas/all.

† New York Times, April 11, 2010, http://www.nytimes.com/2010/04/11/magazine/11Economy-t.html?ref=magazine&pagewanted=print.

by GDP. Make a plot of year on the x-axis and energy intensity on the y-axis. How has U.S. energy intensity changed over time? Is that trend good or bad for the environment?

3. Use the "2050 Global Calculator" at http://tool.globalcalculator.org/ to find a combination of variables that limits the global temperature increase to 2°C by 2050.
4. Describe a scheme to diversify your energy portfolio to reduce risk.
5. Use the free software BEopt (Building Energy Optimization, available at https://beopt.nrel.gov/) to design an energy efficient home.
6. Use the Kaya identity to calculate the carbon footprint of the United States associated with:
   a. Air travel
   b. Auto travel
   c. Coal burning

12 Water

If there is magic on this planet, it is in water.

Loren Eiseley
1957

Water is our most essential resource, and it has no substitutes (Postel 1999). It is a prerequisite for life; the Earth has a biosphere only because it has a hydrosphere. Humans can survive only five to seven days without water, and two-thirds of our body weight is water (Klaver 2009). Besides drinking, the essential uses of water are to grow food and to remove pollutants. In 2010, the U.N. General Assembly declared access to clean water and sanitation a human right.* People agree that we all have a right to live, and since we need drinkable water to live, the corollary is that we have a right to drinkable water.

Viewed from space, the Earth seems to have plenty of water for us. However, less than 1% of water near the Earth's surface is available freshwater. As the human population continues to rise exponentially, global water demand is approaching the maximum available supply. Although water is a renewable resource, unequal geographic distribution of water and pollution that makes water unusable can cause freshwater shortages. In arid regions and during droughts, reduced water quantity and quality can decrease crop yields and cause people to consume contaminated water, leading to the spread of water-borne diseases. Climate change can reduce the amount of locally available water, resulting in falling water tables and lake and stream levels and loss of ecosystem services. Communities must adapt to changes in maximum sustainable water consumption rates by using water more efficiently.

In this chapter, we will show that, in many areas, water use is unsustainable. However, the availability of usable freshwater can be increased through the use of water markets, education, conservation and efficiency measures, recycling, and reuse.

12.1 LIFE CYCLE ANALYSIS, VIRTUAL WATER, WATER INTENSITY, AND WATER FOOTPRINTS

Since water is a renewable resource, we will never run out of it on a global scale, yet water shortages occur frequently on a local scale. Water is like paper money in a steady state economy: There is a fixed amount of it at any given time; it is circulated; it can be reused indefinitely; and we can have local shortages or surpluses of it. Unlike money, which is stored in banks, water is stored for extended periods in glaciers or groundwater aquifers, but like money, it always eventually returns to circulation.

Consumptive water use temporarily removes water from an available supply, usually by evaporation but also by incorporation into crops and livestock. Most water use in the agricultural sector is consumptive, since more than 50% of the water withdrawn from freshwater sources is lost to the atmosphere by plant transpiration and by evaporation from reservoirs and irrigated fields, leaving little for other uses. In contrast, most water use in the residential and industrial sectors is nonconsumptive, meaning the water we use, while doing the dishes, for example, is returned to surface water bodies where it can be reused (Oelkers, Hering, and Zhu 2011).

Water is required for the production of all agricultural products and most commercial products. The amount of water required for each stage of a product's life cycle, including raw materials,

* Resolution A/RES/64/292. United Nations General Assembly, July 2010, General Comment No. 15. The right to water. UN Committee on Economic, Social and Cultural Rights, November 2002, http://www.un.org/waterforlifedecade/human_right_to_water.shtml, retrieved July 26, 2016.

production, use, and disposal, can be calculated using Life Cycle Assessment. Similar to the concept of embodied energy, **virtual water** is the amount of water required to produce a product in the raw materials and production stages of its life cycle. For example, production of 1 ton of grain requires 1,000 tons of water, while production of 1 ton of beef requires about 6 tons of grain (Vaclav Smil 2008). Exporting 1 ton of beef is equivalent to exporting 6,000 tons of water, meaning beef production has a very high **water intensity** (amount of water required per unit of economic value). Countries and regions where freshwater is in short supply should refrain from exporting goods with high virtual water contents and should encourage the growth of industries with low water intensities.

Water demand is measured by the **water footprint**. According to The Living Planet Report (Hails 2008), the water footprint of a country is the total volume of water used to produce all of the goods and services consumed by its inhabitants. The water footprint of a single water-consuming activity can be calculated using the ImPACT formula. For example, the annual water footprint (WF) of sugar cane production is calculated as follows:

$$\text{WF} = P \times A \times C \times T = \text{Population} \times \text{Affluence} \times \text{Intensity of Resource Use} \times \text{Inefficiency}$$

$$\text{WF} = P \times A \times C \times T = \text{persons} \times (\text{dollars/person}) \times (\text{pounds sugar cane/dollar}) \times (\text{gallons water/pound of sugar cane}) \quad (12.1)$$

The calculated water footprint of 1 kg of pure cane sugar is 1,500 L. For comparison, the water footprint of a cotton shirt is 2,900 L, and 1 kg of beef is 15,500 L (Hails 2008). The total water footprint is obtained by summing the WF values for all water-consuming activities.

From 1997 to 2001 Americans had an average annual water footprint of 2,483,000 L (656,000 U.S. gallons), twice the average global water footprint of 1,243,000 L (328,000 gallons) per person per year (Hails 2008). In most countries, the agricultural sector is by far the largest water consumer. The virtual water in the average diet is 2,000 L (528 U.S. gallons) per day, 500 times as much as we drink (Brown 2009).

In the United States, the largest use of freshwater is for energy production (Figure 12.1). Thermoelectric energy production requires large amounts of water for cooling (Karl, Melillo, and

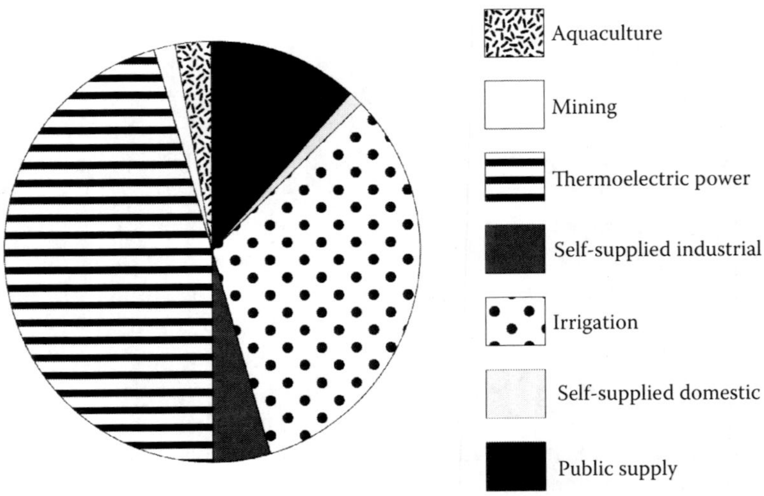

FIGURE 12.1 Water use by category in the United States in 2010. (Data from Maupin, M.A., J.F. Kenny, S.S. Hutson, J.K. Lovelace, N.L. Barber, and K.S. Linsey. 2014. "Estimated Use of Water in the United States in 2010, U.S. Geological Survey Circular 1405." http://pubs.usgs.gov/circ/1405/.)

Peterson 2009). However, much of the water use in energy production is nonconsumptive. A hydroelectric plant simply uses water to push turbine blades, leaving nearly all of that water available for other uses. An exception is biofuel production, which consumes water so it cannot be reused. Biofuels have by far the highest water intensity of all energy sources, requiring 360 L (95 U.S. gallons) of water to produce 1 kWh of energy (see Table 9.2). This makes biofuel production a poor choice for arid regions, such as the U.S. southwest. The high solar intensity of arid regions resulting from low cloud cover may compensate for this shortcoming, but solar PV has a low water intensity and also benefits from the high solar intensities, making it the best energy choice for the U.S. southwest.

While water is required to produce most forms of energy (Holland et al. 2015), energy is required to transport and treat water. This interdependence of water and energy production is called the **water–energy nexus**. Transporting water is energy-intensive because water is dense. Pumping and treating drinking water and wastewater consumes about 4% of the energy supply in the United States (Karl et al. 2009). For this reason, and to reduce evaporative losses, water transport over great distances is usually gravity driven, for example, where water starts at relatively high elevation, allowing us to direct its flow, for example, from melting glaciers to fertile valley soils.* Because we cannot easily transport water, its local abundance has limited the extent of human settlements throughout history.

12.2 THE WATER CYCLE AND WATER SYSTEMS

To understand what limits the availability of fresh water, we must start with the water cycle (Figure 4.8). The sun provides energy to purify our water continuously through distillation. For example, water evaporates from nonpotable sources of water such as the ocean, precipitates, and then collects to form **potable** sources of water (rivers and lakes). This process is fast: The average residence time of water in the atmosphere is roughly 10 days (Langmuir 1997). Gravity delivers precipitated water to the Earth's surface and then transports it on land from high to low elevations.

The availability of fresh water is primarily determined by the amount of precipitation (FAO 2005) and temperature (Chevalier and Chase 2016). Areas with high rainfall such as the Amazon River basin have more water than could be used by local populations, while areas with low rainfall such as the Sahara Desert have almost none. Freshwater availability is also changing over time due to climate change, which is altering the global distribution of precipitation. In many cases, wet regions are getting wetter and dry areas drier, making both floods and droughts more common (Emanuel 2012).

Most of the world's 1.4 billion cubic kilometers of water, 96.5%, is unusable saltwater in the oceans, 1% is unusable saline groundwater, and 1.7% is locked up in glaciers, leaving only 0.8% available as freshwater for us to use (Table 12.1). On the global scale, we have no shortage of freshwater, with supply being about 10 times the current demand (Oelkers et al. 2011). However, freshwater is not distributed equally in space or time, causing some areas to have water surpluses while others have water shortages. Much freshwater remains unused because it is not accessible for local use (Gleick and Palaniappan 2010). Thus, it is more useful to look at regional stocks and flows of water.

Until recently, water stocks such as the oceans were in a steady state, so that they did not change in volume. However, humans are now perturbing the hydrologic cycle on a global scale. Anthropogenic inputs or withdrawals become significant when they approach the size of the primary flows. Humans have an impact on the hydrologic cycle through withdrawals from surface waters (about 80% of total use) and ground waters (20%). This has caused rivers and lakes to shrink and dry up in some locations, leading to **desertification**. In 1989, humans used about 45% of the

* An exception is the Colorado River Aqueduct; see "Water's energy—and expense" http://articles.latimes.com/2011/nov/14/local/la-me-water-power-20111114 , retrieved 9/15/2016.

TABLE 12.1
Distribution of the World's Water

	Distribution Area (10^3 km^2)	Volume (10^3 km^3)	Of Total Water (%)	Of Fresh Water (%)
Total water	510,000	1,386,000	100	
Total freshwater	149,000	35,000	2.53	100
World oceans	361,300	1,340,000	96.5	
Saline groundwater		13,000	1	
Fresh groundwater		10,500	0.76	30
Antarctic glaciers	13,980	21,600	1.56	61.7
Greenland glaciers	1,800	2,340	0.17	6.7
Arctic islands	226	84	0.006	0.24
Mountain glaciers	224	40.6	0.003	0.12
Ground ice/permafrost	21,000	300	0.022	0.86
Saline lakes	822	85.4	0.006	
Freshwater lakes	1,240	91	0.007	0.26
Wetlands	2,680	11.5	0.0008	0.03
Rivers (as flows on average)		2.12	0.0002	0.006
In biological matter		1.12	0.0001	0.0003
In the atmosphere (on average)		12.9	0.0001	0.04

Source: Peter H. Gleick, *The World's Water 2008–2009*. Copyright © 2009 Pacific Institute for Studies in Development, Environment, and Security. Reproduced by permission of Island Press, Washington, DC.

total useable supply of fresh water.* By the year 2000, the amount increased to just over half of the total sustainable freshwater runoff (Meadows, Randers, and Meadows 2004). In some developed areas, the amount of water that returns to the oceans has decreased because of irrigation and evaporation from reservoirs created by dams (Oelkers et al. 2011). However, there is evidence that global discharge to the oceans increased between 1994 and 2006 due to increased ocean evaporation, which may be an early indication of intensification of the global hydrologic cycle resulting from global climate change (Syed et al. 2010).

Freshwater stocks are highly interconnected; the flow of water can transmit pollution in one stock to another. The interactions between stocks are complex and involve many feedback loops. For instance, decreasing precipitation rates lower the **water table**, the level that water rises to in an unconfined aquifer. This decreases groundwater output to surface waters by exfiltration and to the atmosphere through **evapotranspiration**, further decreasing precipitation, a positive feedback. Positive feedback loops can amplify the initial decrease in precipitation, which may lead to desertification.

Groundwater is an attractive resource because there is much more water stored in aquifers than in surface water bodies (Table 12.1), and because groundwater is available in many places where surface water is not. It is important to distinguish surface and groundwater sources when evaluating water resources, since they have significantly different recharge rates (Hornberger et al. 2014). Groundwater generally has lower recharge rates and longer residence times than surface water, and therefore provides a steadier supply of water than rivers, whose flow can have strong seasonal variations and which may completely dry up during droughts (Schwartz and Ibaraki 2011). Aquifers are also less susceptible to contamination from surface pollution than surface water bodies, particularly pathogenic microbes (Johnston et al. 2011), although some shallow aquifers have been contaminated

* Maurits la Riviére: "Threats to the World's Water" in Sept. 1989 *Scientific American*.

(Schwartz and Ibaraki 2011). In many areas, shallow subsurface aquifers are the most important source of **drinking water** (Zhu and Schwartz 2011). Water availability and quality in these aquifers is affected by climate, microbiologic activity, dissolution of soluble minerals, atmospheric inputs, physical transport, and anthropogenic contamination (Zhu and Schwartz 2011). Continuous monitoring of water availability and quality is essential for stakeholders, who may have no alternate water supplies.

In the United States, groundwater provides about 40% of the public water supply and ~20% of all water withdrawals (Maupin et al. 2014). Shallow groundwater aquifers are usually unconfined, while low permeability rock layers usually confine water in deep aquifers. Shallow aquifers can be reliable water sources when demand is lower than supply, because removed water is replaced by recharge and flow within the aquifer so that a relatively constant water table elevation is maintained (or in the case of confined aquifers, constant water pressure). However, groundwater use in many areas is unsustainable because communities pump groundwater much faster than normal recharge rates, resulting in a lowering of the water table, a sure sign of reduction of water availability (Schwartz and Ibaraki 2011).

The unsustainable practice of overpumping occurs when the pumping rate of aquifers (outflow) exceeds the natural recharge rate (inflow). This violates Daly's first law of sustainability that the sustainable rate of use of a renewable resource can be no greater than the rate of regeneration. Currently unsustainable groundwater use occurs on every continent except Antarctica (Meadows et al. 2004), and the overdraft (outflow-inflow) is greatest in Asia and North America (Gleeson et al. 2012).

The use of **fossil water** is especially unsustainable. Fossil water is very old water contained in deep aquifers. It is considered a nonrenewable source of water because the recharge rate is so low that it would take hundreds or thousands of years to replace water that is withdrawn. If we harvest aquifers too quickly, aquifer water becomes a nonrenewable resource, and over time water production follows a Hubbert curve, peaking and then declining (Figure 12.2). This is particularly worrisome in arid regions where food is unsustainably produced by harvesting fossil water. The permanent loss of groundwater supplies could seriously jeopardize food security. As noted by Charles Bowden, "Humans build their societies around consumption of fossil water long buried in the earth, and these societies, being based on temporary resources, face the problem of being temporary themselves.*"

Table 12.2 compares the concepts of global Peak Water and Peak Oil. Peak Oil will occur globally only once, and we will eventually consume all available oil, but we have renewable energy substitutes. In contrast, Peak Water can occur repeatedly, causing water scarcity. Since water has no substitute, exceeding Peak Water should be avoided by using the available water supply sustainably.

In areas where people overharvest unconfined groundwater, the depth to the water table increases each year. The water level in the High Plains Aquifer, by far the largest and most heavily used aquifer in the United States, has decreased more than 35 m (115 feet) in parts of Kansas, Oklahoma, and Texas (Schwartz and Ibaraki 2011). Unsustainable use of groundwater aquifers can also cause land subsidence, as observed in Santa Clara Valley in northern California, San Joaquin Valley in southern California, and Las Vegas, Nevada (USGS 2000). Subsidence can result in catastrophic collapse of the land surface to form sinkholes, or more gradual settling that causes structural damage to buildings and roads.

So what countries currently have sustainable freshwater supplies? In Figure 12.3, countries that use freshwater slower than it is renewed plot in the sustainable area, while countries that consume water faster than it is renewed plot in the unsustainable area. Countries that have large areas of desert like Kuwait, Saudi Arabia, and Mexico use water unsustainably. They will or already have severe water shortages because they are using their water principal in addition to the renewable portion that represents interest. The United States plots in the sustainable area because it has 6% of

* Killing the Hidden Waters, 1977.

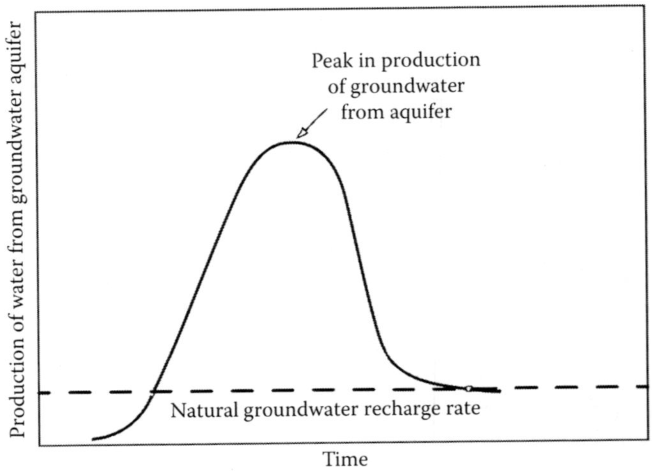

FIGURE 12.2 Unsustainable use of groundwater results in production peaking and then declining.

TABLE 12.2
Comparison of Global Peak Water and Global Peak Oil

Characteristic	Oil	Water
Quantity of resource	Finite	Literally finite, but practically unlimited at a cost
Renewable or nonrenewable	Nonrenewable resource	Renewable overall, but with locally nonrenewable stocks
Flow	Only as withdrawals from fixed stocks	Water cycle renews natural flows
Transportability	Long-distance transport is economically viable	Long-distance transport is not economically viable
Consumptive versus nonconsumptive use	Almost all use of petroleum is consumptive, converting high-quality fuel into lower quality heat	Some uses of water are consumptive, but many are not. Overall, water is not "consumed" from the hydrologic cycle
Substitutability	The energy provided by the combustion of oil can be provided by a wide range of alternatives	Water has no substitute for a wide range of functions and purposes
Prospects	Limited availability; substitution inevitable by a backstop renewable source	Locally limited, but globally unlimited after backstop source (e.g., desalination of oceans) is economically and environmentally developed

Source: Peter H. Gleick, *The World's Water 2008–2009*. Copyright © 2009 Pacific Institute for Studies in Development, Environment, and Security. Reproduced by permission of Island Press, Washington, DC.

the world's renewable freshwater for only 4.5% of the world's population (Gleick and Ajami 2011). However, if treated separately, the arid southwest regions of the United States would plot in the unsustainable area of Figure 12.3. The United States has the economic resources to make these areas livable by building a freshwater supply infrastructure that can transport water great distances and purify polluted water. However, the environmental impacts and the high costs of these activities are making these large infrastructure projects unaffordable.*

China has a similar, but more extreme, problem than the United States. It is similar because its water supply is unevenly distributed, with abundant supplies in the south but severe water shortages

* Drever, Tim, 2011, "Water Resources – Science and Values," editorial, *Elements*, 7(3), 147.

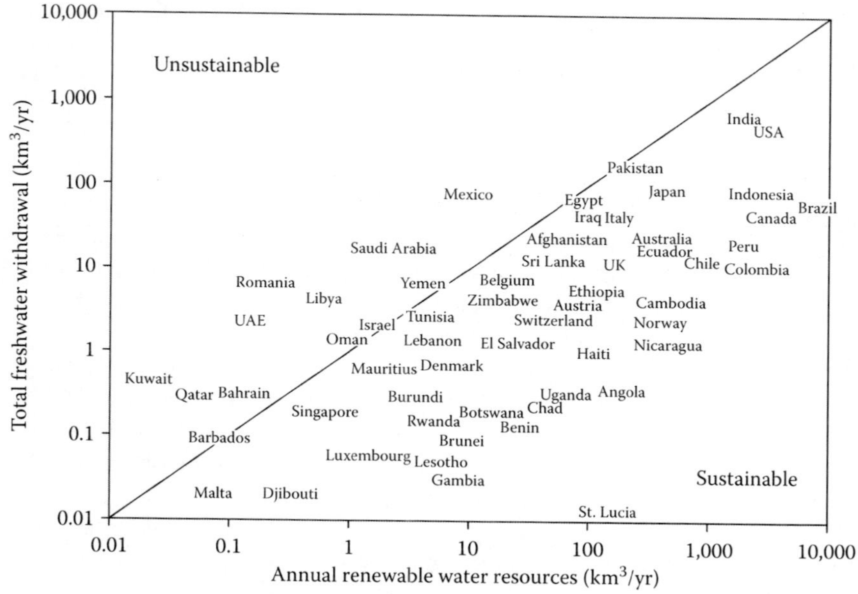

FIGURE 12.3 Water supply and demand by country. (Data from "The World's Water," Pacific Institute, 2008, Tables 1 and 2. Retrieved from http://www.worldwater.org/data.html, January 29, 2010. Data from their revised 2006 tables; not all measurements made in the same year, but range from 1999 to 2005.)

in the arid north. China's water problem is more extreme because it has 20% of the world's population but only 7% of its available freshwater (Larson 2010). Demands on the water supply are rapidly increasing as China builds new urban housing for the 350 million people expected to migrate from the countryside to cities within the next 20 years. The growing economy demands ever more energy, and power production in China has a very large water footprint. China can import all of the resources required for these new urban developments except water. To make matters worse, as demand grows water supply is shrinking due to climate change. China has no choice but to make better use of the water it has. For example, the Beijing city government recently increased residential water prices 8% to discourage practices that waste water (Larson 2010).

Cities need enormous amounts of water, but many of the world's cities are in arid regions. To supply these growing cities both water and food must be transported increasingly greater distances, requiring more energy. Los Angeles uses water transported from hundreds of kilometers away, and Beijing uses water from the Yangtze River 1,000 km to its south (Schwartz and Ibaraki 2011). A large percentage of transported water is lost to evaporation. This practice is unsustainable, and eventually the energy cost, and therefore the economic cost, of water will become so great that some of these cities may become unlivable, and collapse. Much of the population growth between now and 2050 is expected to occur in arid and semi-arid regions where water tables are already rapidly falling (Brown 2009). These regions are likely to sink into water poverty and become the source of water refugees.

12.3 PROBLEMS

12.3.1 Problems of Water Quantity: Water Scarcity

You don't miss your water until your well runs dry.

An old country proverb

Water is and has been a limiting resource in many areas of the world. Water shortages are more serious than energy shortages because all life requires water, and we need water to grow food.

The earliest civilizations in Mesopotamia such as **Sumer** most likely crumbled due to water shortages, specifically salinization of irrigated fields that caused food shortages, and the armed conflicts that followed (Postel 1999). The **Garamantian Empire** that occupied present-day Libya collapsed around 400 C.E. (**Common Era**) due to water shortages (Schwartz and Ibaraki 2011).

Because the amount of renewable freshwater available globally is fixed, while human population is increasing exponentially, the amount of water available for each person is decreasing exponentially (UNEP 2007). Between 1950 and 2000, world water use tripled (Brown 2009). Physical **water scarcity** is measured in cubic meters (m^3) of water per person per year, where regions that are water scarce have less than 1,000, while regions experiencing **water stress** have 1,000–1,700 (Falkenmark 1989). Chronic water scarcity affected 2% of the world's population in 1900, 9% in 1960, and 35% in 2005 (Oelkers et al. 2011). If present trends continue, by 2025 1.8 billion people will be living in countries or regions with water scarcity and two-thirds of the people in the world could be subject to water stress (UNEP 2007). Clearly these trends are unsustainable.

As demands on our fixed freshwater resource base have increased, **water security** has declined. Water security is defined as "the capacity of a population to safeguard sustainable access to adequate quantities of acceptable quality water for sustaining livelihoods, human well-being, and socio-economic development, for ensuring protection against water-borne pollution and water-related disasters, and for preserving ecosystems in a climate of peace and political stability" (UN Water 2013). Many different water security indices have been developed to guide water policy development aimed at reducing water scarcity (Gunda, Benneyworth, and Burchfield 2014).

The geographic availability of water has always been highly variable but has fallen within boundaries defined by the historical record, and that has allowed us to plan infrastructure to safely deal with maximum flood levels or extended droughts (Milly et al. 2008). However, climate change now makes it much more difficult to predict local water availability. The past is no longer a reliable guide to the future, and we have to live with reduced water security. This has profound implications for risk management, food security, and public health and safety (Postel 2010). Regions that historically had abundant freshwater may soon face water shortages, which will require shifts to less water-intensive industries and agricultural practices. For example, crops in parts of Kansas have dried up as the High Plains Aquifer decreased in volume in response to decades of unsustainable groundwater use. This aquifer, which supplies 30% of U.S. irrigation water, is being harvested at a rate almost seven times higher than the recharge rate (Steward et al. 2013). This has caused wells and streams in west central Kansas to run dry. It would take 500–1,300 years to refill the aquifer completely. The groundwater is a common pool resource, and its depletion is one more example of the Tragedy of the Commons. Eventually pumping rates will be forced to decline to rates no higher than the recharge rate, which is only 15% of the current pumping rate.

The results of our overuse of freshwater supplies are clear to see. Many groundwater wells have dried up, while in coastal areas salt water has replaced the freshwater that has been removed from wells. Rivers like the Colorado River in the southwestern United States no longer reach the sea: water withdrawals now consume all of the flow. The Hoover Dam on the Colorado River formed Lake Meade, which is the only significant source of water for the city of Las Vegas, Nevada. Currently the water level in Lake Meade is dropping rapidly, and may soon stop producing electricity before going completely dry (Barnett and Pierce 2008). Lake Meade reached record low water levels in 2016.*

Many lakes like Mono Lake in California have mostly or completely dried up because the rivers that supplied them were diverted for agriculture (Brown 2009). The Aral Sea on the border of Kazakhstan was once the fourth largest lake in the world, but the former Soviet Union diverted its tributary rivers for cotton production. In 2007 it was only 10% of its original size, and had split up into four small water bodies (Keller 2011). Likewise, Lake Chad was the fourth largest lake in

* "Water Levels in Lake Mead Reach Record Lows," http://abcnews.go.com/US/water-levels-lake-mead-reach-record-lows/story?id=39235749, retrieved September 14, 2016.

Africa, with a surface area of 26,000 km^2 in the 1960s that decreased to only 1,500 km^2 in 2000 (Oelkers et al. 2011). The salts from dried lakes like Mono Lake and inland seas like the Aral Sea become airborne in high winds and are deposited in agricultural fields, poisoning the soil so that it becomes unfit for agriculture. The disappearance of large bodies of water is an environmental disaster, wiping out entire ecosystems.

12.3.2 Problems of Water Quality: Pollution

While in theory the amount of renewable, accessible freshwater is constant, in reality it is decreasing in many parts of the world because of pollution. Water pollution is increasing worldwide as population expands and demands on freshwater supplies intensify. The problem arises from our use of water bodies as both sources and sinks. Only by separating these uses by designating water bodies solely for supply or for waste can we avoid poisoning our bodies with our own wastes.

Contaminated water remains the single greatest environmental cause of human sickness and death. Every year about three million people in developing countries die from water-borne diseases such as cholera, dysentery, and typhoid fever (UNEP 2007). Roughly 1.8 million children under the age of five die from diarrhea each year (Johnston et al. 2011). Since water-borne diseases are the leading killer of children less than five years old in the developing world, efforts to provide clean water in these countries should be a top priority. Inexpensive sanitation, hygiene, and safe water supply practices can reduce the number of deaths by 88% (Johnston et al. 2011). The goal of all governments and non-governmental organizations (NGOs) should be to build safe, reliable, and cost-effective water supply systems as quickly as possible to save as many lives as possible, and guarantee access to freshwater for all. International aid organizations have made much progress in reducing water-borne disease fatalities in developing countries by installing groundwater pumps and inexpensive water purification systems, and by teaching personal hygiene to residents. These are some of the most cost-effective steps to reducing premature deaths in developing countries.

In Chapter 14 we will discuss pollution in more detail. For now, suffice it to say that in developing countries the most problematic water pollutant is animal and human feces, which contain water-borne diseases. Developed countries have different pollution problems. They eliminate the transmission of water-borne diseases by using sanitation systems with separate subsystems for water purification and supply and wastewater treatment and disposal. However, developed countries use many natural and synthetic chemicals that are poisonous and can contaminate water supplies and ecosystems. For example, in 2015 it was discovered that many people in urban, poor neighborhoods in Flint, Michigan had elevated levels of lead in their blood, which can cause cognitive impairment (Hanna-Attisha et al. 2015). This mass poisoning resulted from a cost-saving measure by the government that introduced corrosive water into lead pipes, and the subsequent indifference of government officials to the plight of the urban poor, an example of the lack of **environmental justice** in many U.S. communities. Only when scientists, activists, and the media publicized the problem did the government act and declare a state of emergency.* Planned and inadvertent releases of harmful chemicals have in some cases made freshwater bodies unusable and caused premature deaths (Eby 2004; Manahan 2009). Also, if water pollution becomes more severe or widespread, it becomes costlier to chemically treat the water to make it drinkable.

Pollution is classified by source type as **point source pollution** (smokestack or wastewater pipe discharge) and nonpoint source pollution. Point source pollution is easier to identify, and environmental regulations in developed countries such as the United States have decreased emissions of pollutants from point sources, especially hazardous industrial chemicals (Keller 2011). However, nonpoint source pollution is increasing, with the most common types being fertilizers and pesticides transported to streams in runoff.

* https://www.washingtonpost.com/news/morning-mix/wp/2015/12/15/toxic-water-soaring-lead-levels-in-childrens-blood-create-state-of-emergency-in-flint-mich/, retrieved July 27, 2016.

One form of water pollution that is becoming increasingly common in coastal regions is **saltwater intrusion** into water wells (Keller 2011). In these regions, less dense freshwater floats as a lens on denser saltwater. If water is pumped from a well faster than the water can be replaced by groundwater flow, the water level in the well drops. If water is pumped too fast for too long, deep saltwater gets pulled up into the well, contaminating the well water. Saltwater intrusion is becoming more common for two reasons: Coastal regions worldwide are becoming increasingly densely populated, increasing the demand for freshwater that is met by increasing the number of wells and by pumping faster and longer on each well; and by rising sea level, which causes saline groundwater to rise to higher levels in coastal aquifers.

Saltwater contamination in coastal regions can be accelerated by unsustainable agriculture and aquaculture practices. For example, in some coastal regions farmers are converting rice farms to shrimp farms by pumping saltwater into surface ponds to produce the brackish water preferred by certain shrimp species (Chowdhury et al. 2011). The saline water may seep down and contaminate shallow freshwater aquifers, which when used for irrigation adds salts to irrigated soils and reduces crop yields because most crops are not salt-tolerant. In this situation, aquaculture sacrifices the long-term agricultural potential of a region for short-term profits. Here we have both kinds of equity problems: distributional equity is reduced when wealthy citizens of developed countries buy shrimp that is farmed unsustainably in developing countries, and intergenerational equity is reduced because future residents in areas with shrimp farming may not be able to grow enough food in their salt-contaminated soil to survive.

Another growing pollution problem is eutrophication. Addition of limiting nutrients like phosphorous and nitrogen to surface water bodies causes algae blooms. When the algae die, they decompose. This consumes the oxygen dissolved in the lake water, causing eutrophication. In temperate regions, lakes have two layers: a shallow, warm, buoyant layer and a deep, cold, dense layer. In a eutrophic lake, the shallow layer in contact with the atmosphere is oxygen-rich, but the deep layer becomes oxygen-depleted because the dead algae sink to the bottom of the lake and decompose (Langmuir 1997). In the fall and spring, the density difference between the two layers disappears and they mix. The problem is that, especially in the fall, the deep water has no oxygen, so when it mixes with the shallow water the resulting mixture does not have enough oxygen for fish to breathe, and they die in large numbers. This is still a widespread problem in many areas of the United States. For example, a huge "dead zone" has formed near the Mississippi delta in the Gulf of Mexico because fertilizer-derived nutrients caused algae blooms and eutrophication, and many other dead zones have recently been identified (Diaz and Rosenberg 2008).

The good news is that a solution to eutrophication exists. Studies have shown that phosphorous inputs cause algae blooms, so efforts to reduce eutrophication must focus on reducing phosphorous fluxes to water bodies (Schindler et al. 2008). Removing phosphorous from detergents in areas surrounding Lake Erie led to a temporary decline in algae blooms, but increasing agricultural runoff of phosphate fertilizers has led to increasing eutrophication since the 1990s, which, combined with abnormally warm waters, led to a record-setting algae bloom in 2011 (Michalak et al. 2013).

12.3.3 Use of Bottled Water

Americans purchased over 34 billion liters (nearly 9 billion gallons) of bottled water in 2008 (Gleick 2010). We pay prices that are several thousand times higher for bottled water than tap water, even though numerous studies have shown that bottled water has no taste or quality advantages over tap water (Sipes 2010). This is a glaring contradiction of the common assumption of economists that consumers behave rationally and always choose the product with the lowest price per unit of utility. There are also claims that tap water and water from public drinking fountains is less healthy than bottled water. While this may be true in many developing countries, there is no scientific evidence to support this claim in the United States. The opposite seems to be true: bottled water is sometimes contaminated with bacteria, which has led to public health outbreaks.*

In the United States the expansion of the disposable culture has led to the replacement of public water fountains with private bottled water dispensers. However, bottled water comes at an environmental cost, and sometimes a social cost. Most plastic bottles are made of polyethylene tetraphthalate, known as PET. Americans use 30 billion plastic water bottles per year, and roughly three-fourths end up in the trash, which is a problem because PET is not biodegradable (Gleick 2010). So while bottled water producers advertise that their bottles are 100% recyclable, less than 25% of them are actually recycled, and the rest take up valuable space in landfills.

Goodguide ratings for bottled water on a scale of 10 are 10 for health, 3.5–7.2 for environment, and 3.4–6.1 for society, for an average of 5.6–7.8.† Problems include manufacture of PET bottles from petroleum; the use of 3 L to 4 L of water to produce a 1-L PET bottle; the drawdown of aquifers by bottling plants; the energy used to pump the water and produce, fill, package, and transport the bottled water; the energy used by the consumer to transport the water to the place of consumption (water is very heavy); and the cost to taxpayers to dispose of the plastic bottles. The embodied energy and transportation energy of bottled water is much higher than for tap water, and therefore the carbon footprint of bottled water is much higher than for tap water. Global production and use of bottled water consumed roughly 100 million to 160 million barrels of oil in 2007 (Gleick 2010).

In the end, we have to accept that use of all disposable products that are not biodegradable is unsustainable. Taxpayers end up paying the disposal costs even when they do not use disposable products. In Chapter 14 we will examine policies that governments can adopt that shift the cost to the people using disposable products (e.g., waste pickup fees based on volume or weight) or to the companies that produce them.

12.4 SOLUTIONS

> I understood when I was just a child that without water, everything dies. I didn't understand until much later that no one "owns" water. It might rise on your property, but it just passes through. You can use it, and abuse it, but it is not yours to own. It is part of the global commons, not "property" but part of our life support system.
>
> **Marq de Villiers**
> *Water, 2000*

We are decreasing freshwater quantity in our global and local water commons through groundwater overharvesting, virtual water exports, pipeline diversions, and climate change (Walljasper 2010). Our society is also reducing freshwater quality by allowing industrial agriculture and mining and manufacturing industries to pollute water bodies. When faced with scarcity of a critical natural resource such as usable freshwater, our society turns to engineers and planners, who usually respond by trying to increase supply by adopting the hard water path rather than the **soft water path** (Gleick 2003). The hard water path involves using vast quantities of fossil fuel energy to build dams, or to drill deeper wells and pump faster. Building dams causes many environmental problems and large water losses by evaporation from reservoirs (Section 10.4.1). Dams also take water away from people downstream, decreasing distributional equity. Pumping faster from deeper wells takes water away from the next generation, decreasing intragenerational equity. It is also a short-term solution, not only because groundwater is removed faster than it can be replaced, but because groundwater salinity typically increases with depth, and eventually the well will become so deep that the water becomes unusable (Langmuir 1997).

After decades of the hard water path to water development, the amount of water that remains undeveloped has shrunk. Water sources that were easy and therefore inexpensive to develop have

* http://www.cdc.gov/healthywater/drinking/bottled/index.html, retrieved March 20, 2013.
† From http://www.goodguide.com/products?filter=bottled+water, retrieved March 20, 2013.

all but disappeared, leaving only costlier options, and infrastructure construction costs have risen faster than inflation, compounding the cost problem. Many water development projects like dams have environmental impacts that can further increase costs substantially (Vaux 2011). Finally, because global climate change often causes today's water shortages, using fossil fuel energy to increase freshwater supply exacerbates the problem. In contrast to the hard water path of increasing water supply, the soft water path uses the tools of **demand management** (recycling and reuse, conservation and efficiency measures, and price incentives) to reduce water demand. The soft water path also reduces costs by working with nature, using nature's ecosystem services that are powered by the sun rather than fossil fuels to purify and store water (Postel 2010). We will examine the tools of demand management more closely in the following sections.

12.4.1 Increase Water Supply

Because we need water to survive, it is important to increase water security and resilience by adding redundancy to water supply systems (Kellogg and Pettigrew 2008). During short- and long-term emergencies, the most critical resource is potable water, especially in arid regions. **Rainwater harvesting** is a cost-effective approach to increasing water supply. Rainwater that falls on a roof is collected at a downspout into a **rainwater tank**. Because evaporation purifies water, rainwater is usually the purest water in the hydrologic cycle, but **water purification** may still be required to make harvested rainwater potable (see Manahan [2013] for a description of methods). In areas where it is legal, rainwater harvesting is one of the easiest ways to move toward self-sufficiency.

Rainwater harvesting can be a sustainable, decentralized, cost-effective, low technology, soft path approach to increasing the supply of freshwater in developing countries. It requires little training for construction and operation. The required construction materials are widely available and inexpensive. Expanded use of rainwater harvesting in developing countries could greatly reduce deaths from water-borne diseases at low cost, but its use is not consistently promoted by NGOs and governments in those countries (Cain 2010).

Cities in arid regions are now exploring rainwater harvesting as a tool for increasing water supply, decreasing flood risk, and decreasing water treatment costs. In the past, these cities treated rainwater as waste. They designed city streets to collect rainwater and channel it into wastewater systems to help prevent flooding. Now water flows often exceed the capacity of wastewater systems, which leads to flooding and the need to bypass the wastewater treatment plant, resulting in the discharge of untreated wastewater. Furthermore, collected rainwater is sent to a wastewater treatment plant for purification, even though it is already very pure. Now city officials recognize that rainwater is a valuable resource, so they are passing laws that require new developments to harvest rainwater. For example, Tucson, AZ now requires that developers design commercial projects so that half the water used for landscaping comes from harvested rainwater.*

An example of the hard water path to increasing water supply is **desalination**. Today roughly 21,000 desalination plants operate in more than 120 countries worldwide. It currently takes 2 kWh of electricity to produce 1 cubic meter (about 264 gallons) of freshwater from saltwater, and that number will not decrease significantly because we are near the thermodynamic limit for efficiency of the **reverse osmosis** process (Postel 2010). About half of the world's desalination plants are in the Middle East, where they essentially turn oil into water (Postel 2010; Oelkers et al. 2011). Desalination plants produce concentrated brine as a waste product, which if disposed of improperly can contaminate surface and groundwater bodies (Lattemann and Höpner 2008). Although desalination capacity increased from 6.3 to 13 km^3/year between 1994 and 2004, desalination currently supplies less than 0.3% of global water needs (Oelkers et al. 2011). Desalination is only viable for wealthy countries that can afford the costs of building and operating desalination plants.†

* Botstein, Arthur H., *The Tennessean*, July 6, 2009, p. 2A.
† Drever, Tim, "Water Resources – Science and Values," editorial, *Elements*, 7(3), 147, 2011.

Furthermore, most of the energy for desalination currently comes from burning fossil fuels, and the release of greenhouse gases and resulting climate change may cause water shortages to become even greater in the future, so it is a self-defeating pursuit. Areas that still choose to use desalination should run their desalination plants at night when energy demand is low to avoid the need for additional power plant construction, and because power plants are most efficient when they operate continuously near full capacity.

12.4.2 Reduce Water Demand

Society uses water in many ways. The good news is that we can use water more efficiently. The United States has already made much progress, with water use declining 5% between 1980 and 2005, and another 13% by 2010, despite a 37% population increase over that 30-year period (Maupin et al. 2014). Increasing efficiency of irrigation methods, residential appliances, and technologies at power plants has greatly decreased water use intensity (liters per $ GDP).

According to the ImPACT identity, we can reduce demand measured by the water footprint by conserving water or by increasing water use efficiency. Conserving water means engaging in water-consuming activities less frequently. This means reducing the $A \times C$ term in the ImPACT identity by, for example, washing clothes less frequently. Conserving water also conserves energy, since energy is used to provide the water and often to power the water-consuming activity (the clothes washer), and this will reduce greenhouse gas emissions if fossil fuels are the source of that energy (Karl et al. 2009). Increasing water use efficiency using technology T means decreasing the amount of water per unit work or yield, for example, by using less water per pound of laundry. Efficiency measures are usually more popular than conservation measures because they do not involve making sacrifices. Since our goal is to decrease water demand, we must increase water use efficiency to obtain the same level of benefit from reduced water use.

Water demand can be reduced through reuse, where water discharged by one user is used by another, and recycling, where water is used multiple times before it is discharged (Manahan 2013). For example, recycling industrial and domestic wastewater could increase supply in the southwestern United States by 5% to 10% (Vaux 2011). Large-scale water recycling in municipal systems is only economical if alternative supplies are costlier or unavailable (Vaux 2011). As water shortages become more widespread, more recycling facilities will need to be built.

The potential for reuse of wastewater depends on the water quality, that is, the amounts and toxicities of pollutants. Used water that has no toxic chemicals and small amounts of suspended material that make it turbid is called **graywater**. Graywater from sources like dishwashers can safely be used for watering plants, cleaning the car, and so on (Kellogg and Pettigrew 2008). It can also be used to recharge aquifers by simply allowing it to seep into the ground. Flowing through a clean aquifer usually purifies the water to the point that it becomes drinkable without treatment. In contrast, **black water** such as sewage must be confined to septic systems and treated at wastewater treatment plants before it can be discharged; this prevents the spread of water-borne diseases. In Orange County, California on the U.S. Pacific coast treated wastewater is used to recharge aquifers for future municipal water use and to prevent saltwater intrusion (Manahan 2013).

Currently homes have two separate water systems, one for incoming fresh water and one for outgoing black water. In the future, homes will likely have a third system for graywater that can be recycled and reused. Using recycled graywater appropriately not only reduces water demand, it reduces the amount of water that must be treated. Reducing demand on municipal water supplies reduces the amount of chemicals used to treat water and the energy required to produce the chemicals and to treat and pump the water. Thus, matching water quality with the application is essential to achieving sustainability.

In general, people underestimate the amount of water they use, and the discrepancy grows larger with increasing water use demand (Attari 2014). Activities with the largest water use include swimming pools and Jacuzzis, followed by garden hoses and carwashes. Decreasing the water footprint of our homes can be accomplished using many different **point-of-use technologies** such as high efficiency toilets and showerheads. In the average American home in 2016 toilets accounted for roughly 24% of indoor household water use, showers 20%, clothes washers 17%, faucets 19%, leaks 12%, and others uses 8%, and the median annual total water use was 83,000 gallons per household (DeOreo et al. 2016). New high-efficiency toilets can reduce water consumption by 4,000 gallons per year per person.* To reduce home water use, fix leaking faucets and install faucet aerators on kitchen and bathroom faucets (Jeffery, Barclay, and Grosvenor 2008). Improvements in water use efficiency have decreased average annual household indoor water consumption in the United States by 22% between 1999 and 2016 (DeOreo et al. 2016). For more ideas on reducing your water footprint, customized for the area of the United States you live in, go to http://wateruseitwisely.com/ or http://www.h2ouse.org/.

Countries can conserve water on a larger scale. Countries with water shortages should make sure they import more water than they export. The balance sheets must reflect not only the physical amounts of liquid water but also virtual water, the water used in the production of a good or service. Countries like Israel use water efficiently because they suffer from water scarcity (Figure 12.3). Israel is the world leader in drip irrigation, which has greater than 90% efficiency (the percentage of water soaked up by the desired crop), compared with 55% for flood irrigation and 70% for sprinkler systems.† Besides decreasing demand, Israel has also worked to increase water supply through rainwater harvesting, wastewater treatment and reuse for irrigation, and seawater desalination (Tal 2006). Israel has been able to grow and prosper despite being located primarily in a desert, which gives hope that the rest of the world can live well even in a world of water scarcity.

Other countries like Australia have more abundant water (Figure 12.3), but contain subregions with high water scarcity. In the arid northern part of Victoria province in southeast Australia, farmers use an automated system to regulate water distribution in agricultural irrigation systems (Mareels et al. 2005). In the past, water flow was regulated manually by raising and lowering flumegates. Farmers placed water orders based on weather predictions and received the water three days later, and the distribution efficiency was only 70% (i.e., 30% of the water was lost en-route through evaporation, seepage, etc.). Now sensors measure the water stress of plants and send the information to computers programmed to place orders. More sensors measure the flow rate of water through the flumegates and raise or lower them accordingly. The automated system now fills water orders within two hours, and the distribution efficiency has increased to 90%. More efficient water use means higher crop yields and higher quality crops that command higher prices on the market. Often farmers have increased their profits while simultaneously decreasing their water use. In a sample farm, the payback time on the system was only 1.5 years. Systems like these, which increase water use efficiency and productivity without decreasing water quality, hold great promise for reducing water stress.

12.4.3 The Economics of Water: Public versus Private

Because freshwater is a necessity and is scarce in many developing countries, the U.N. adopted the Dublin Statement on Water and Sustainable Development in 1992. The Dublin Principles state:‡

1. Fresh water is a finite and vulnerable resource, essential to sustain life, development, and the environment.
2. Water development and management should be based on a participatory approach, involving users, planners, and policy makers at all levels.

---

* http://www.epa.gov/watersense/pubs/toilets.htm.

† http://greeninc.blogs.nytimes.com/2009/09/17/business-booming-for-drip-irrigation-firm/.

‡ http://www.un-documents.net/h2o-dub.htm, retrieved July 26, 2016.

3. Women play a central part in the provision, management, and safeguarding of water.
4. Water has an economic value in all its competing uses and should be recognized as an economic good, taking into account affordability and equity criteria.

Here we focus on principle (4) concerning the economics of water in developing countries and the United States.

12.4.3.1 Developing Countries

A major sustainability challenge is how the **water industry** can increase water security by providing drinking water to people in the developing world at an affordable price. The question is whether private companies are more effective than public utilities at doing so. The World Bank and the International Monetary Fund (IMF) favor **water privatization**, and have made privatization of water utilities a prerequisite for loans and debt restructuring for developing countries.* They argue that putting a price on water encourages conservation, and that privatization can improve services and benefit the poor. To ensure that the poor have access to water, they advocate progressive fee schedules, with minimal rates for users with low water consumption (the poor, and those who conserve water) and higher rates for high water usage.

Opponents of water privatization argue that water is part of the commons and belongs to all of us, that privatizing water changes it from a public good to a private good, and that access to fresh water is a fundamental human right that private corporations cannot take away (Bakker 2010). Furthermore, private corporations have short-term profit incentives, and without adequate regulation and oversight they may use water resources unsustainably, for example, by pumping aquifers faster than they can be recharged in order to maximize short-term profits. Some argue that water privatization is simply a way for multinational corporations in developed countries to exploit residents of developing countries by stealing their water rights (Barlow and Clarke 2002).

A point that critics of privatization often overlook is that pricing water does not require the privatization of water. The difference between private and public water utilities is that private utilities can make a profit; public utilities can only recover their costs. Both public and private water utilities can potentially charge residential users according to the amount of water used, which would encourage water economization.

In the 1990s, privatization of water in several South American countries like Bolivia was largely unsuccessful at meeting the MDG goal of halving the number of people without access to water and sanitation by 2015 (Budds and McGranahan 2003). Other case studies have shown that privatized water systems do not seem to perform better than public water systems (Prasad 2006). This is not surprising, since a water utility is a **natural monopoly**: Each water district has only one water system, so even if privatized, there is no market competition that would lead to decreased cost and increased efficiency. Furthermore, the claim of privatization advocates that public water utilities cannot provide incentives for water conservation are false, since public water utilities can adopt progressive pricing schedules with average fees that are closer to the true value of water. On balance, public utilities seem more likely than private utilities to meet the needs of the poor and to use water resources sustainably in the developing world, but assistance from the World Bank (without the requirement of water privatization) and NGOs is usually necessary to build effective water infrastructure.

12.4.3.2 The United States

In the United States, water is usually treated as a public good, not as a commodity, meaning that it has no market value (Sipes 2010). The fees we pay to water utilities only cover the costs of capturing, purifying, and transporting the water, and sometimes the cost of treating the post-use wastewater. The argument is that because water is a necessity, it should be available to everyone, no matter how

---

* World Bank, IMF: Privatize Water Utilities, http://www.npr.org/templates/story/story.php?storyId=906163.

poor (Sipes 2010). The problem is that the fees charged by water utilities are much lower than the societal value of water, which discourages water conservation and improvements in use efficiency (Oelkers, Hering, and Zhu 2011). Treating water as a commodity and trading it in an open market would ensure that water is allocated to the most beneficial (highest-value) use. Charging for water use induces economizing behavior, making water available for other users and uses. Studies have shown that as the cost of water increases, the amount of water used decreases (Vaux 2011). The higher the cost of water, the greater the incentive for efficient water use. It is difficult for all stakeholders to agree on a price for water, especially when considering that as a necessity for life, water is effectively "priceless." However, food is also a necessity, and assigning economic value to food commodities in an open market is an effective way to make food affordable. Food assistance programs guarantee that even the poorest can afford the food they need to survive. As water becomes scarcer, communities will likely adopt similar systems where users pay for water but government subsidies reduce costs for the poor, perhaps through the use of water stamps.

In the United States, there is a continuum from completely public systems to completely private systems (Sipes 2010). Each municipality usually owns its water, but it may grant a **water right** to and contract with a private company to manage or operate its water supply system. In this case, the company would operate as a private water utility that can earn a profit. The government usually only transfers water rights, not water ownership, and can reclaim the water rights.

About 10% of water sales and revenues in the United States come from private water utilities (Sipes 2010). Private utilities include California Water Services, Philadelphia Suburban, Florida Water Services, San Jose Water, and St. Louis County (Beecher 2003). All water utilities in the United States are economically regulated because they have a local monopoly on water supply, and they must follow environmental and public health regulations. Water utilities must allocate water to the residential, industrial, and agricultural sectors. There is competition for water within and between these sectors. The objective of water pricing should be to shift water demand to its highest value uses *in each sector*. The latter part is emphasized to make clear that wealthy industrial users should not be able to purchase water from other sectors such as residential. Each sector (residential, industrial, agriculture) has a different price and receives a different amount according to its needs.

Water rights can be sold or leased in markets. For example, the city of Los Angeles has purchased water rights from farmers over large areas. Water can also be traded in **commodities markets**. One problem with a water market is that there is no buyer representing the environment; without environmental protections, all of the water in the environment can be siphoned off by the other three sectors, undercutting the ability of the environment to serve as a public good by providing environmental services. This problem can be solved either by public purchase of water for environmental uses or by enforcement of environmental regulations that prevent the sale of water from protected areas for consumptive uses (Vaux 2011).

12.4.4 Sustainable Water Use

Water policies and water management practices are unsustainable in most areas of the world (Oelkers et al. 2011). Sustainable water use requires holistic approaches to ensure water demand does not exceed supply. **Integrated Water Resources Management** is "a process which promotes the coordinated development and management of water, land and related resources, in order to maximize the resultant economic and social welfare in an equitable manner without compromising the sustainability of vital ecosystems."* This is a holistic approach to water management that recognizes that many water uses are interdependent. Another approach to increasing water security is **demand management**, which can be used to economize water use through water markets, education, conservation and efficiency measures, and recycling and reuse. During temporary water

* Global Water Partnership, http://www.gwp.org/The-Challenge/What-is-IWRM/, retrieved July 27, 2016.

shortages, such as droughts, **water rationing** is also an effective water economizing approach, while education is often the most cost-effective approach to reducing consumer demand (Vaux 2011).

The water-starved city of Bolinas, California, serves as an example of the effective use of regulation for preserving water supply.* It adopted a regulation that limited the number of homes in the city to the maximum number that could be sustained by the most limiting resource, water. In contrast, the city of Los Angeles has 18 million people, but only has enough water of its own for 3 million. To be sustainable, communities should limit development to the amount that the most limiting critical resource can sustain long term. For example, a city can preserve an area of undeveloped land surrounding it that would supply necessary resources and ecosystem services. If the city population increases, then the amount of preserved land should also increase. Otherwise, ecological overshoot develops and eventually leads to collapse.

Little (2009) gives an excellent example of adaptation to water limits. When farmers in Garden City, Kansas, learned from state and federal geologists in the late 1960s that the water they were pumping was fossil water and would soon run out, they responded in two distinct ways. Most purchased more pumps and began harvesting the fossil water at a greater rate, an excellent example of the Tragedy of the Commons. Others chose to stop relying on nonrenewable water and changed to farming methods that could be sustained using only renewable water, rainfall. They started using methods like no-till agriculture and planting crops like wheat and grain sorghum that require less water than corn. By using only renewable water, these farms will remain viable even when the groundwater runs out.

So how can we manage groundwater use to keep it sustainable? A system must be put in place to ensure that this common pool resource is used sustainably. Water should be withdrawn from deep aquifers only during droughts when surface water and shallow aquifer supplies are insufficient. The system must include usage monitoring and penalties for individuals who overpump. Ostrom (1990) gives examples of how social and legal arrangements made in water-scarce southern California have successfully protected groundwater resources.

The sustainable use of water also relies on ecosystem services to help provide fresh water. Investments in preserving ecosystems that provide fresh water pay for themselves. For example, New York City relies on the water that flows down from the Catskill Mountains to the north. Preserving that ecosystem by spending $1.3 billion on upstate sewage treatment plants was cheaper than building a filtration plant in the city for $6 billion to $8 billion and operating it for $350 million to $400 million per year (Morrison 2005). Thus, the value of the water that the Catskills provide is easily hundreds of millions, if not billions of dollars per year. The Catskills provide other ecosystem services such as flood control, food, and shelter, besides their scenic beauty and the recreational activities they provide such as trout fishing, which both attract tourism dollars to the area. New York City's investment in the Catskills ecosystem made sense from both an environmental and economic standpoint. Another excellent example of the economic benefits of preserving the ecosystem services provided by natural water systems was the decision of the city of Napa, California, to restore the Napa River to its original floodplain to reduce flooding (Morrison 2005). This project cost only $250 million, but it saved an estimated $1.6 billion in flood damage repair costs over the next century. Within one year of restoration, flood insurance rates dropped 20% and real estate prices increased 20%.

Urban development can decrease the amount of water stored in the local hydrologic cycle. Replacement of water-retentive landscapes with impermeable pavement prevents water from returning to fields, meadows, wetlands, and streams and from recharging aquifers. Governments and city planners must preserve and restore green spaces that can store water. **Low impact development** makes use of permeable pavements, tree plantings and tree preservation, rain gardens, rainwater harvesting, green (vegetated) roofs and rooftop gardens, and vegetated swales to reduce flooding and pollution and to help recharge groundwater aquifers (Sipes 2010).

---

* Kelly Zitos, 2009, *The San Francisco Chronicle*, http://sfgate.com/cgi-bin/article.cgi?f=/c/a/2009/02/04/MNV415MGLA.DTL, retrieved January 25, 2016.

To move toward sustainability, the agricultural, industrial, and residential sectors should take steps to reduce their water footprints. Agriculture can reduce water consumption through the use of drip irrigation and by planting crops appropriate to the local climate. Industrial and residential users should match water quality to use, recycle, and reuse water when possible, and fix leaks. Residential users can increase water supply by harvesting rainwater, conserving water, and using water-efficient appliances.

There are also easy ways for individuals to improve water quality. Do not use pesticides. Avoid broadcast herbicides, and instead locally apply small amounts of biodegradable herbicide. Stop using antibacterial soap, which contains triclosan, an endocrine disruptor that can react with chlorine to form chloroform, a carcinogen, and can promote the growth of antibiotic-resistant bacteria. Use phosphate-free dishwashing detergent. Finally, minimize the use of dry cleaning because it usually uses the carcinogenic compound tetrachloroethylene (TCE).

Local governments and water utilities must stop the use of **perverse incentives** that encourage the wasting of water and start pricing water to reflect its value. They should charge for water based on the rate of use. Governments can offer tax incentives to all three sectors for instituting water conservation and efficiency measures and for building green infrastructure such as green roofs and permeable pavement that promote water infiltration rather than runoff and flooding. Wetlands should be set aside to purify water, recharge groundwater aquifers, and provide ecosystem services. Governments should include all stakeholders in the decision-making process to ensure that it is transparent and accountable to the public (Postel 2010). Finally, we must remember that water sustainability is not possible without climate sustainability, which also means energy sustainability (Meadows et al. 2004).*

12.5 CASE STUDY: BANGLADESH

Bangladesh provides a glimpse into both the past and the future of human civilization and water security. It is one of the poorest countries in the world, with an estimated per capita GDP at PPP of $3,390 (international dollars) in 2014, compared with the U.S. value of $54,360.† It is also one of the most densely populated. Much of the country lies within a few meters of sea level, making it very vulnerable to flooding and resulting famine. In 1970, a cyclone killed roughly half a million people, and in 1974 Bangladesh experienced a severe famine that claimed as many as 1.5 million lives (Schendel 2009).

Although political instability is common in Bangladesh, it is anything but a failed state. More NGOs operate in Bangladesh than in any other country. It was in Bangladesh that the concept of microcredit to help women buy property and start businesses was first put in practice. This combined with family planning education has decreased the fertility rate from over 6 to 2.6 children per woman. According to a report by the Gates Foundation,‡ Bangladesh has recently seen dramatic decreases in infant and child mortality rates and increased immunization coverage, which together have increased average longevity to 60 (United States is 78).

However, Bangladesh faces some serious problems. Its low elevation make it susceptible to more severe flooding as sea level rises, which can also cause saltwater intrusion in coastal aquifers. Its high population density means that increasing numbers of people are living on marginal lands that are vulnerable to flooding or salination.§ Also, government corruption is slowing reforms (in 2013, corruption perception ranked 136 out of 177 countries, United States is 19 out of 177¶).

---

* Resources: The World's Water: http://www.worldwater.org/.

† International Monetary Fund, World Economic Outlook Database, April 2016, https://www.imf.org/external/pubs/ft/weo/2016/01/weodata/index.aspx, retrieved July 26, 2016.

‡ http://www.gatesfoundation.org/learning/Pages/mdg-millennium-development-goals-report-card-case-studies.aspx.

§ In Low-Lying Bangladesh, The Sea Takes a Human Toll: http://e360.yale.edu/content/feature.msp?id=2234.

¶ http://cpi.transparency.org/cpi2013/results/.

Water security in southwest Bangladesh is threatened not only by freshwater scarcity during the dry season, but also by water pollution. Three types of drinking water pollution are common in Bangladesh: contamination by pathogens, arsenic, and salts. Pathogens that cause water-borne diseases such as diarrhea and cholera are spread by flooding, improper sanitation, and mixed use with livestock. To reduce exposure to pathogens, international aid organizations in the 1970s began installing tubewells, with millions of household wells installed by the late 1990s. These tubewells allow for hand pumping of water from shallow aquifers that are mostly pathogen-free (Ayers et al. 2016).

The second type of water pollution was recognized when water testing showed that groundwater from 6 million to 10 million tubewells in Bangladesh had arsenic concentrations higher than the World Health Organization (WHO) guideline for drinking water of 10 μg/L (World Health Organization 2011). This meant that more than 57 million people were exposed to unsafe levels of arsenic (British Geological Survey and Bangladesh Department of Public Health and Engineering 2001). Arsenic is a carcinogen to humans and exposure from drinking contaminated water can increase the risk of skin, lung, bladder and kidney cancers, hypertension, diabetes, peripheral vascular disease, and skin lesions (Hopenhayn 2006). Arsenic present in soil and irrigation (rice paddy) water can be incorporated into rice, presenting another exposure risk (Mitchell 2014a).

The third type of water pollution is salinization of surface water (including water in freshwater ponds and rice paddies) and groundwater in the coastal area. Long-term exposure to saline drinking water can cause hypertension (Environmental Protection Agency 2003). In southwest Bangladesh, high drinking water salinity has also been linked to relatively high rates of preeclampsia and gestational hypertension, with the latter occurring at higher rates in the dry season than in the wet season (Khan, Ireson, and Kovats 2011). High salinity in irrigation water and soil also decreases crop yields (Ali 2006). These three types of water pollution result in low water security in much of Bangladesh (Benneyworth et al. 2016).

After it became clear that tubewells could not be relied on to provide safe drinking water, NGOs started installing **slow sand filters**. In these simple devices, water is pumped from a surface pond up to the top of the filter, and then slowly percolates through a thick bed of sand that purifies the water. Unfortunately roughly three-fourths of the filters are nonoperational despite being only a few years old. This is an example of aid money not being wisely invested: no reliable system was put in place to maintain the filters and keep them operational. Everyone wants to be a freerider: let someone else work to maintain the filter, and I will do no work but claim the benefit of clean water. The slow sand filter problem is an excellent example of mismanagement of common pool resources. Only when the slow sand filter is managed as a **cooperative**, only when everyone makes an investment, does it continue to operate.

Now NGOs are exploring other approaches to providing safe drinking water in Bangladesh. One solution is to harvest rainwater and collect it in large cisterns. However, only wealthier people can afford these systems, and they often do not store enough water to get through the dry season. Some NGOs dig very large pits to collect rainwater during the wet season. However, if the area is inundated by a cyclone-generated storm surge the water will become contaminated with seawater. The newest approach is **aquifer storage and recovery**, where rainwater is collected during the wet season, stored in a shallow aquifer, and then recovered in the dry season using a hand pump. However, these systems require more technical expertise and maintenance, so their success rate is low.

12.6 CONCLUSIONS

Access to water is now regarded as a fundamental human right, and water availability is often the limiting factor for human development, agriculture, and energy production. Provision of safe drinking water and secure supplies of water for agriculture and energy production should be top priorities for governments of developing countries, as they are essential to improving and maintaining high levels of human well-being. Water infrastructure improvements in developing countries are often the most cost-effective public health investments, as drinking contaminated water kills millions of people annually. Because water is a common pool resource, effective management requires cooperation of stakeholders. Governments must create policies that guarantee access to drinking water, and water supply systems that charge fair rates for water use to discourage wasting water.

WEB RESOURCES

- The World's Water: http://www.worldwater.org/
- Water Conservation Tips, Facts and Resources Water—Use It Wisely: http://wateruseitwisely.com/
- Water conservation: http://www.h2ouse.org/ and http://www.homeintelligence.ca/resources/at-home-water-conservation-guide/
- American Water Works Association: http://www.drinktap.org/consumerdnn/Default.aspx?tabid=85
- Safe Drinking Water is Essential: http://www.drinking-water.org/html/en/index.html
- Water footprint and virtual water: http://www.waterfootprint.org/?page=files/home
- Water Encyclopedia Science and Issues: http://www.waterencyclopedia.com/

HOMEWORK PROBLEMS

1. Describe a scheme to diversify your water supply portfolio to reduce risk.
2. Calculate the total and average per capita water footprint for the United States.

13 Food

13.1 INTRODUCTION

> I believe that the great Creator has put ores and oil on this earth to give us a breathing spell. As we exhaust them, we must be prepared to fall back on our farms, which is God's true storehouse and can never be exhausted. We can learn to synthesize material for every human need from things that grow.
>
> **George Washington Carver**

Agriculture requires water, land, fertile soil, and a favorable and stable climate. We discussed water in Chapter 12 and climate in Chapter 7. In this chapter, we examine the environmental impacts of agriculture and the factors that influence **food security**, which exists "when all people, at all times, have physical and economic access to sufficient, safe and nutritious food to meet their dietary needs and food preferences for an active and healthy life."* We review problems that reduce food security, such as water scarcity, competition from biofuel production, overfishing, inadequate food distribution, overuse of fertilizers, and land grabbing. Solutions that can decrease food demand and its environmental impact include decreasing food waste, buying local, and moving down the food chain. Food supply can be increased without increasing the environmental impact of production by practicing sustainable farming and aquaculture, using genetically modified organisms (GMOs) responsibly, and preserving and creating soil.

Lack of a sustainable food supply has caused the collapse of many societies including the Anasazi and the Mayan civilization (Diamond 2005). In the twentieth century food security was enhanced by the **Green Revolution**, which greatly increased global food production and reduced the frequency and severity of **famines**. Despite this, in 2011 more than one billion people in the world were undernourished and two billion had nutrient deficiencies (Nierenberg 2013). Today, starvation and nutritional deficiencies are usually caused by inadequate food distribution and preservation rather than inadequate supply, resulting in roughly nine million deaths every year of causes related to hunger (Meadows, Randers, and Meadows 2004). Growing populations place ever-increasing demands on agricultural systems, and compound the problem by diverting water and land from agriculture. Industrial agriculture's response is to continuously increase **crop yields** (amount harvested per hectare of cultivated land), but growth in crop yields and agricultural research and development have slowed (Alston, Beddow, and Pardey 2009).

Two sides debate whether our food production system is sustainable: the establishment, and Malthusians. The establishment represented by the U.N. and World Bank maintain that increasing yields will keep pace with population growth. Up to the present they have been right: although pockets of starvation exist around the world, it is not due to a global shortage of food. Supply has always been close to demand globally, but food does not always make it from areas with surpluses to areas with deficits. The Malthusians led by Lester Brown and The Worldwatch Institute have maintained for decades that we are facing a global food shortage due to supply threats and growing demand (Brown 2009). According to the Malthusians, threats to food supply include a per capita and a total decrease in availability of arable land (Figure 13.1); rising water scarcity in many regions exacerbated by climate change; and slowing growth in crop yields.

* Rome Declaration on World Food Security: http://www.fao.org/docrep/003/w3613e/w3613e00.htm, retrieved November 9, 2015.

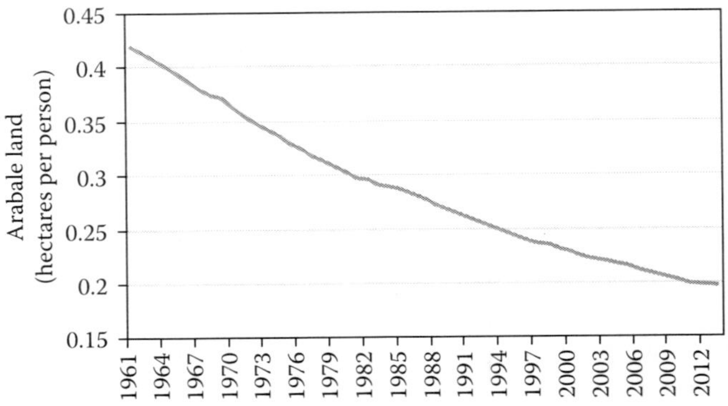

FIGURE 13.1 Historical decrease in global per capita arable land. (Arabale land data from UN FAOSTAT, downloaded from http://faostat3.fao.org/browse/R/RL/E on February 17, 2016. Population data from UN World Population Prospects: The 2012 Revision, http://esa.un.org/wpp/Excel-Data/Interpolated.htm, retrieved March 28, 2015.)

Roughly 20% of the land surface is now under cultivation (Figure 13.2). Of the world's total cropland in 2000, 62% was used for food crop production, 35% for animal feed, and 3% for biofuels and other uses. Expansion of agriculture, which now occurs primarily in the tropics through deforestation (Gibbs et al. 2010), has not significantly increased global food production, but has decreased biodiversity and ecosystem services and has increased global greenhouse gas emissions (Foley et al. 2011). Yet food demand continues to increase as population expands, biofuel production consumes food crops, and people grow their food foodprint by eating more resource-intensive foods such as beef.

Agriculture has the largest total environmental impact of all human activities. It is responsible for one-third of global greenhouse gas emissions (Nierenberg 2013), and irrigation agriculture consumes 70% of global freshwater withdrawals (Foley et al. 2011). It is nearly impossible for us to predict the effect of climate change on food supply, so setting that aside, it seems that the greatest potential threat to food security is shortage of fresh water. China and India, the two most populous countries in the world, are growing most of their crops using an unsustainable water supply.

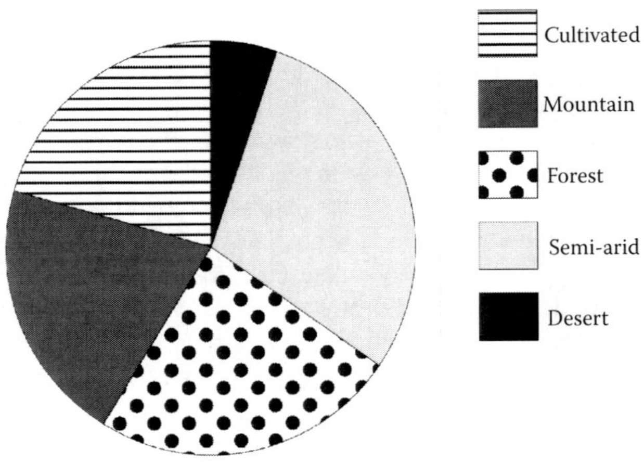

FIGURE 13.2 Percentage of land coverage as of the year 2000. (Data from Millennium Ecosystem Assessment. 2005. "Ecosystems and Human Well-Being: Synthesis." Washington, DC: Island Press. http://www.maweb.org/documents/document.356.aspx.pdf.)

Irrigated farmland produces four-fifths of China's and three-fifths of India's grain harvests (Brown 2009), and much of the water used for irrigation is nonrenewable fossil water. In contrast, in the United States, one-fifth of the grain is produced using irrigation (Brown 2009).

In this chapter we will focus on the unsustainable aspects of food production and consumption, and propose some solutions.

13.2 PROBLEMS

13.2.1 Environmental Impacts of Industrial Food Production

In one century, the United States changed from an agrarian society to an industrial society: Between 1890 and 1990 the proportion of Americans living in rural areas decreased from 60% to 25%, and the proportion of the U.S. labor force employed in agriculture dropped from nearly 50% to less than 1% (Steffen 2006). The United States is the world's largest food producer and exporter, and it is where modern industrial agriculture originated, so in this section we focus on some of the impacts of industrial food production in the United States.

Before the Green Revolution of the twentieth century, farmers practiced ancient methods of crop rotation and use of manure and **crop residue** as **organic fertilizer**. Food production was limited by nutrient availability and soil moisture. As population and the demand for food grew exponentially, farmers began to rely on outside sources such as guano (seabird dung) for the essential nutrients phosphorus and nitrogen. Depletion of guano deposits led to a food crisis that was solved by the invention of the **Haber process** in 1910 to make nitrogen **inorganic fertilizer** (Smil 2004). Every year we use more inorganic fertilizer containing nitrogen and phosphorous in order to feed the world's growing population (Tilman et al. 2002). However, resource use efficiency decreased when farmers switched from using animal waste to fertilize crops, an efficient closed loop system (Figure 13.3), to fossil fuel-based fertilizers. Manure has changed from a resource to a pollutant.

The invention of inorganic fertilizers was followed in the 1940s by the development of synthetic **pesticides** to reduce crop losses to insects. Soon Rachel Carson's book *Silent Spring* published in 1962 made the public aware that synthetic pesticides like DDT caused environmental problems. Many of these synthetic organic molecules are toxic, carcinogenic, and persistent in the environment (Manahan 2009). They also were too effective, wiping out all insects that were the primary source of food for species higher in the food chain. Another problem is that the concentrations of some pesticides increased up the food chain, a process called **bioaccumulation**. Pesticide bioaccumulation would harm species at high **trophic levels**, especially **apex predators**. For example, DDT caused the thinning of bald eagle eggs, leading to the near extinction of the national bird of the United States (Rogers and Feiss 1998). Farmers also used increasing amounts of herbicides such as **atrazine**, an **endocrine disruptor** that has been shown to feminize male frogs, implicating it as one of the causes of the global decline in amphibians (Hayes et al. 2010). Continent-scale

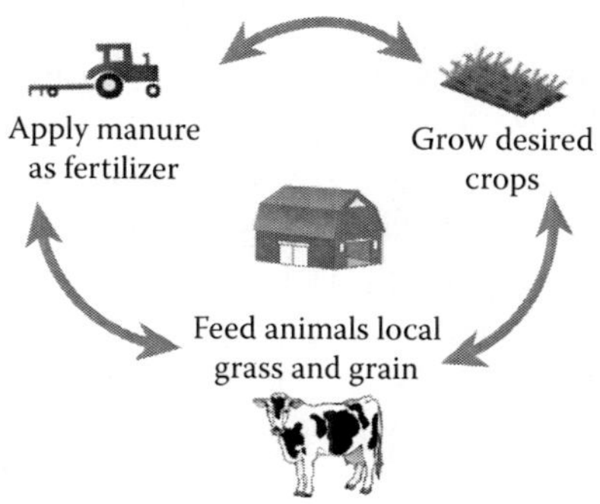

FIGURE 13.3 Sustainable agriculture: closing the resource loop makes farming more efficient.

monitoring of freshwater ecosystems has shown that the risk of chronic or lethal effects on fish, invertebrates, and algae species increased as the number of synthetic organic chemicals they were exposed to increased (Malaj et al. 2014). Besides negatively impacting nontarget species, pesticides often became ineffective because target species developed **pesticide resistance**. Awareness of the problems caused by use of synthetic chemicals in agriculture led to the banning of many types of pesticides, and to the development of the **organic farming** movement, which relies only on natural processes and compounds to produce food.

The American agricultural system experienced further changes in the 1970s. Prior to 1973, the government paid farmers to practice **crop rotation**, generally by planting **legumes** every fourth year to replenish critical nutrients such as nitrogen in the soil. For example, farmers commonly rotate soybeans with corn so that the nitrogen-fixing bacteria in soybean roots can replace the nitrogen extracted from the soil by corn. By controlling the amount of food produced, the government stabilized food prices and kept grain prices high enough to keep agriculture profitable. When the government stopped subsidizing crop rotation, farmers no longer allowed their fields to lie fallow. Instead, they replenished soil nutrients by application of increasing quantities of synthetic fertilizers (Tilman et al. 2002).

To increase efficiency, farms grew in size, and most family farms went bankrupt or were purchased by large corporate farms. To increase profits, new corn hybrids were bred to withstand higher planting densities and tolerate the application of **herbicides** to kill weeds. Soon the corn surpluses became so large that industrial agriculture had to find new markets. Cattle ranchers found they could make beef more cheaply by force-feeding corn to cattle in **concentrated animal feeding operations** (CAFOs) rather than letting cattle graze on grass. Food companies cut costs by replacing cane sugar with **high fructose corn syrup**. Corn is now the most abundant **staple food** in the American diet (Pollan 2007).

After saturating the food market, corn started to be used for biofuel production. The problem with using corn to produce biofuel is that another fuel, oil, is required to grow the corn on an industrial farm (Section 10.3). Also, biofuel production takes land away from food production, causing increased food prices. Using corn-based ethanol essentially takes food away from poor families: the amount of corn required to fill an SUV's 25-gallon tank with ethanol could feed a person for an entire year (Brown 2009). Finally, corn is a poor choice as a feedstock for ethanol production; there are far better feedstocks such as **switchgrass** that are not food. The only way to keep biofuel production from competing with food production is if biofuel feedstocks are not food crops and are grown only on land that is not productive for growing food crops.

Industrial agriculture has caused other environmental problems. Decreased natural and agricultural biodiversity has resulted from the standardization of crop strains and the replacement of polycultures with **monocultures** (single-crop). Modern grain crops are **annual plants** rather than **perennial plants**, and they did not develop slowly through natural selection (Jackson 2010). Instead, hybrids were developed quickly through cross-breeding to grow fast. These hybrids are not as well adapted to the local environment as the weeds, which is why weeds displace crops unless they are removed through use of herbicides and the crops are well fed by application of fertilizers. The less hardy and well adapted a crop, the more water, energy, and fertilizer it requires to grow and the greater the greenhouse gas emissions and nutrient runoff that causes eutrophication. Agricultural intensification has therefore increased crop yields by increasing the water and carbon footprints and the overall environmental impacts of agriculture. Also, with less biodiversity resulting from the switch to monocultures, the food supply system is at a greater risk to pathogens and is less resilient.

Working with rather than against nature can reduce the amount of oil-derived energy and nutrients used to produce crops. Examples include using perennials as crops rather than annuals, and choosing natural varieties well adapted to the local environment (known as **heirloom plants**), even if they have lower yields (Jackson 2010). The resulting increase in genetic diversity can also increase food security. Decreasing our reliance on oil in agricultural production now will mitigate climate change and better prepare farmers and our society for the post-oil world.

In conclusion, the industrial food production system in the United States is flawed because it degrades the environment. It is broken because the federal government's subsidy system rewards the overproduction of corn (Pollan 2007). These subsidies make processed foods made from corn inexpensive, leading to the expansion of fast food companies. We now live in the "age of plenty," eating more calories than in 1970 but spending a smaller proportion of our salaries on food. On the plus side, industrial agriculture requires fewer people to produce food, freeing people to do other things, and very few people in the United States are starving. Still, we must ask ourselves, is industrial agriculture good for us and the environment? In Section 13.3.3 we will examine whether a sustainable agriculture system can affordably feed everyone with reduced environmental impacts.

13.2.2 Threats to Food Security

In many areas, high-intensity agriculture is unsustainable because it relies on the nonrenewable resources oil and fossil water (Brown 2011). This may cause food production to decrease in the near future. Since global food demand continues to increase, due to the annual addition of 70 million people and the expanding use of grains as biofuels, there is concern that the number of undernourished people may increase. Fortunately, between 1990 and 2015 that number decreased from 1 billion to 793 million, and from 23% to 13% of the global population.* This was primarily due to rising development aid to meet Millennium Development Goal #1 to eradicate extreme poverty and hunger (Sachs and McArthur 2005). The questions is whether food production can keep pace with food demand, which is expected to increase 100% by mid-century (Tilman et al. 2011). Increased food demand will partly be caused by population increases, but also by increasing per capita food footprints: As people become wealthier, their diet moves up the food chain (greater meat consumption) and becomes more resource-intensive (Brown 2009). For example, an increasing proportion of crops is used to feed livestock rather than people. This combined with food waste causes food production in kcal per day per person to be 8,000–9,000 in developed countries compared with 2,000–3,000 in the poorest countries (Godfray 2011).

Threats to food security include water scarcity, climate change, decreased water quality due to salinization and pollution, increasing costs and shortages of fossil fuels, soil erosion, dietary change, the conversion of land from agricultural to urban use, and unsustainable growth in populations of humans and livestock (Brown 2009). Food security is tightly linked to water security. In 2007, agriculture accounted for ~70% of global water withdrawals.† Irrigation agriculture is particularly water-intensive, and highly unsustainable if it relies on fossil water. For example, a rapid rise and then fall of fossil water production in Saudi Arabia resulted in a similar rise and fall in wheat production, which peaked in 2004 (Brown 2011). In the United States one of the greatest threats to food security is the drying up of the High Plains (Ogallala) Aquifer (Section 12.3.1). In places like northern Texas agricultural fields are being abandoned as wells run dry.‡ The U.S. Midwest, the nation's breadbasket, may once again become the **Dust Bowl**.

Climate change is also greatly decreasing food security in some regions, most importantly by decreasing water availability. Not only are droughts occurring with increasing frequency, duration, and intensity, but also increasing temperatures are causing the loss of soil moisture in many regions. According to a report from the International Food Policy Research Institute (IFPRI) in 2009, global warming will cause global irrigated wheat yields to decrease at least 20% by 2050 (Nelson et al. 2009). According to the IFPRI models, between the years 2000 and 2050 population growth alone

* The State of Food Insecurity in the World 2015, UN FAO, http://www.fao.org/hunger/key-messages/en/, retrieved August 4, 2016.

† UN FAO Aquastat database, http://www.fao.org/nr/water/aquastat/tables/WorldData-Withdrawal_eng.pdf, retrieved August 4, 2016.

‡ Charles Laurence, "US Farmers Fear the Return of the Dust Bowl," March 7, 2011, *The Telegraph*, retrieved July 14, 2011 from http://www.telegraph.co.uk/earth/8359076/US-farmers-fear-the-return-of-the-Dust-Bowl.html.

may cause wheat prices to rise 39%, but that number increases to at least 170% when they include the effects of climate change in their models.

Although a higher atmospheric concentration of carbon dioxide may aid photosynthesis and crop growth, this effect is more than offset by the increased temperatures and decreased soil moisture it will cause* (Long et al. 2006). For example, crop yields decrease sharply when temperatures exceed ~30°C (86°F) (Schlenker and Roberts 2009). Furthermore, extreme weather events such as the drought in Australia and the 2010 heat wave in Russia caused many of the crop failures of the past decade, and scientists believe at least some of those events were caused or worsened by climate change,† and that extreme weather events will increase in frequency as climate change intensifies. In the longer term, climate change will cause profound changes in water availability in many geographic regions. For example, regional warming will cause alpine glaciers to decrease in volume and to melt earlier in the year, reducing the amount of melt-water supplied to lowlands in the spring and summer when it is most needed for agriculture (Barnett, Adam, and Lettenmaier 2005). This is already occurring in the U.S. west where farmland fed by glacial meltwaters from the Rocky Mountains is drying up, hydropower production is falling, and the intensity of regional droughts and forest fires is increasing (Struzik 2014).

Evidence suggests that global food production is struggling to keep up with growing global demand. This can be seen in an increase in food prices and price volatility (Brown 2011). An analysis of the global food system found that it is becoming less resilient due to increasing demand from a growing population and the increasing reliance of many countries on food imports (Suweis et al. 2015). The effects of temporary declines in regional food production are now felt globally, making the global food network more sensitive to external perturbations and increasingly unstable. For example, global food crises caused by extreme environmental conditions occurred in 2007 and 2010, when food supply lagged behind demand and food prices skyrocketed (Brown 2012). Some countries took steps to increase their food security by reducing food exports, which led to decreased food security in countries that relied on imports. Between 1986 and 2010, the number of countries sensitive to global food trade perturbations increased, indicating that global food security decreased (Suweis et al. 2015). Countries that rely on food imports are most vulnerable to food crises. Continued food price increases would decrease food security for the poorest and decrease global political stability (Godfray 2011).

Malthusians point to these trends and claim that we are facing a global food crisis, but their predictions in the past have frequently proved inaccurate. However, we have to admit that the current trends are troubling, and that we must come up with new solutions to prevent a global food crisis. We must solve the problem of inadequate food supply and learn how to maintain a high level of food security for the 10 billion people predicted to be on the planet in 2050. Solutions lie in proven sustainable food production and consumption approaches. The hope is that the demographic transition combined with sustainable intensification of food production will prevent a Malthusian catastrophe.

13.2.3 The Collapse of Marine Fisheries

Food security is threatened not only by potential grain shortages resulting from decreasing yields of terrestrial crops, but also by the decline in global per capita fish production due to the recent collapse of some marine fisheries. People worldwide receive ~15% of the calories and a third of the protein in their diet from seafood (Halweil and Nierenberg 2011). The poor who live in coastal

* David Biello, "Farmed Out: How Will Climate Change Impact World Food Supplies?," *Scientific American*, September 30, 2009, http://www.scientificamerican.com/article.cfm?id=how-will-climate-change-impact-world-food-supplies&print=true.

† Justin Gillis, "Food Supply Under Strain on a Warming Planet," *The New York Times*, June 4, 2011, retrieved June 5, 2011 from http://www.nytimes.com/2011/06/05/science/earth/05harvest.html?_r=2&hp=&pagewanted=print.

communities in the developing world are heavily dependent on fish for sustenance. Like nonrenewable resources, the production of a renewable resource like fish can peak and then decline if it is continuously overharvested (see discussion of peak groundwater in Section 12.2). Global "peak wild fish" production occurred in the early 1990s (Figure 4.6). Fortunately, in recent years the total catch has stabilized at around 90 million tons per year (Sohns and Crowder 2013).

Fish are the main source of animal protein for roughly 3 billion people, and the amount of fish consumed has increased at an average annual rate of 3.2%, roughly double the rate of human population growth (Auth 2015). However, overfishing of the oceans has decimated fish populations, with total marine populations decreasing ~50% between 1970 and 2012 (WWF International 2015). As of 2011 the percentage of marine stocks that were being harvested sustainably fell to 70%, and of that 70% nearly 90% were harvested at maximum sustainable rates (Auth 2015). Stocks of reproductive age fish declined 45% to 99% (Hutchings 2000). Efforts to reverse these trends have only been successful for early maturing species such as herring. Thus, the marine wild catch cannot be increased without overfishing and causing the collapse of more fisheries. Some fisheries have been overfished to the extent that they are unlikely to recover. For example, the cod population in Maine is only 3% to 4% of the target population required for maximum sustainable yield.*

A fishery is a common pool resource (CPR) because its subtractability is high and its excludability is low (Section 5.4.3). By high subtractability we mean that any taking from the CPR reduces the amount remaining for others, a characteristic that makes CPRs susceptible to overuse, resulting in the Tragedy of the Commons (Hardin 1968). Elinor Ostrom (1990) showed that the management of CPRs does not have to be tragic. She showed that many CPRs such as groundwater aquifers have been sustainably managed by communities for centuries. However, whenever money can be made by taking larger amounts from a CPR, there is always the temptation to overharvest. As observed by Meadows et al. (2004), "An unregulated market system governing a common resource with a slow regeneration rate inevitably leads to overshoot and the destruction of the commons." That is what happened with marine fisheries.

For thousands of years, people living in marine coastal areas caught seafood sustainably. The number of fishermen was too low and their fishing methods too inefficient to significantly decrease the size of marine stocks. However, in the twentieth century the increased demand caused by rising human population and the spread of seafood consumption to noncoastal regions, combined with increasing efficiency of fish hunting methods such as the use of **fishing trawlers**, resulted in the depletion of many seafood stocks. Over time the number of fishermen increased, and the technology they used to catch fish improved. Fishermen reinvested their income to buy bigger trawlers to increase their catch until the system was in overshoot, and they were catching smaller, younger fish. Eventually fish were being harvested before they had a chance to reproduce, and the annual fish harvest plummeted. Inadequate governance led to the population of large predatory fishes typically declining ~80% within 15 years of industrialized harvesting (Dietz, Ostrom, and Stern 2003).

One example of industrial overexploitation occurred in the Newfoundland cod fishery. Early warnings by independent scientists were ignored by Canadian politicians, who didn't want to impose quotas that would reduce industry profits and jobs. The fishery collapsed in 1989, and despite a complete ban on cod fishing beginning in 1992, the fishery never recovered, and tens of thousands of jobs were lost (Brand 2009). The failure of Nova Scotian inshore fisheries was also a result of federal government authorities not giving local fishing communities sufficient autonomy. Federal authorities were not aware of cooperative management schemes for CPRs, and wrongly believed that the only fishery management options were complete privatization or government regulation. They concluded that privatization was impractical, so they applied one set of rules to the entire eastern coast, which led to the collapse of many of the fisheries.

* Statement Regarding New Information Showing Continued Decline of Gulf of Maine Cod Stock, http://www.nefsc.noaa.gov/press_release/pr2014/other/MA1402/, retrieved September 1, 2016.

Because fisheries are highly variable (different species regeneration rates, migration and circulation patterns, geographies, climates, etc.), the most effective management strategy is one adapted to local conditions, which is why local community rule development and enforcement is often the best approach for management of fisheries and other CPRs. One approach is the use of **individual fishing quotas**, which was used to save a halibut fishery in Alaska (Brand 2009). Also referred to as catch shares, individual fishing quotas have been successfully used to combat overfishing. Catch shares can be sold, but rules must be in place to prevent the development of monopolies, since one of the objectives is to protect fishermen's livelihoods. An effective system of monitoring, enforcement, and use of penalties or fines for those who exceed their quotas must be in place to ensure that the maximum sustainable harvest is not exceeded. Individual fishing quota systems are usually implemented by the government as a form of regulation, but local community input is needed for setting optimal quotas, enforcement, and for adaptive management. Other sustainable approaches to protecting marine fisheries and their ecosystems are **Marine Protected Areas** with "no-take" zones (Section 15.3.1), and carefully managed **mariculture**, in which marine organisms are cultivated in the open ocean for food (Brand 2009).

Climate change can endanger fisheries just as much as overfishing. Despite implementation of a quota-based management system in 2010, the stock size of cod in the Gulf of Maine has continued to decline (Pershing et al. 2015). Quotas were set too high because policy makers did not account for the effect of increasing sea surface temperatures. Cod in the Gulf of Maine are near their southern limit, meaning they are especially sensitive to temperature increases, and Gulf of Maine surface temperatures have increased faster than 99% of the global ocean. Evidence indicates that the cod mortality rate increases with increasing mean temperature and ignoring this temperature dependence in population dynamics models results in >100% overestimates of future cod stocks. How fast the cod fisheries recover now depends as much on temperature as on fishing (Pershing et al. 2015).

Fisheries can also collapse if catches of nontarget species (**bycatch** and species with lower economic value) reduce food availability for the target species. Often bycatch is killed and simply thrown back into the ocean. To reduce waste, communities can create markets for bycatch. Also, new fishing methods can be developed and adopted that reduce bycatch. For example, using circle hooks instead of J-hooks on long fishing lines can reduce leatherback turtle bycatch by up to 90% (Sohns and Crowder 2013). Because monitoring marine species can be expensive, generally only target species are inventoried to make sure numbers are not declining. Research has shown that inventories of target species can be used to project future risk to bycatch species, which should help in setting sustainable harvest yields (Burgess, Polasky, and Tilman 2013).

13.2.4 The Food–Energy–Water Nexus

Because energy, water, and food production are interconnected, a shortage in one leads to shortages in the others, which we refer to as the food–energy–water nexus. For example, in India water shortages are causing energy shortages, and will likely lead to food shortages. Fossil water is used to produce 15% of the food in India (Brown 2009). As the water table falls, farmers must drill wells to greater depths every year. Use of electricity to pump water from depths as great as 1 km is causing electricity blackouts to become common in India (Brown 2009). Water shortages also cause energy shortages because hydroelectric and biofuel energy production is so water-intensive.

These effects can be magnified by a series of positive feedbacks that may lead to famine. For example, our energy use is causing global climate change, which causes regional water shortages. This is compensated for by using more fossil fuel energy to pump water from greater depths and to import water as grain. This increases greenhouse gas emissions, causing even greater climate shifts and regional water shortages, creating a vicious cycle that can lead to food shortages.

Another example of reinforcing feedback in the water–food–energy nexus is growing urban populations causing increased water demand that can only be supplied by diverting water from agriculture. Many large cities such as Los Angeles buy water rights from farmers and expropriate the water, resulting in lower food production, and increasing pressure on remaining farmland to increase yields by using more water. They also use more water to produce energy to transport water from greater distances. These examples show that many "solutions" to water shortages are unsustainable because they involve tradeoffs (e.g., sacrificing agricultural land for water, see Perrone and Hornberger [2016]) and cause other problems that in turn can lead to greater water and food shortages.

Because it is more economical to transport the finished product (food) than the raw material needed to produce it (water), countries like China and Saudi Arabia are compensating for water shortages by importing grains, which have high virtual water contents (Suweis et al. 2013). Another approach countries take to address water-food shortages is **land grabbing**, in which wealthy developed countries purchase farmland in regions with abundant water and fertile soil, an outgrowth of globalization that in some cases leads to exploitation of developed countries (Rulli, Saviori, and D'Odorico 2013). Both approaches require large amounts of energy for transportation, so this practice will become more expensive as energy prices rise. When this happens, food shortages will mostly occur in regions with water shortages and insufficient money to import grain.

13.3 SOLUTIONS

What changes can we make in the food supply system to ensure it can sustainably feed 10 billion people by 2050? Foley et al. (2011) consider four proposed strategies for making the world's food system more sustainable: stop expanding agriculture, especially in tropical areas where the costs of deforestation outweigh the benefits; increase yields in developing countries; increase agricultural resource efficiency, especially of water; and increase food delivery by shifting diets and reducing waste. Perhaps all of these strategies will be needed to feed 10 billion people.

Is it even possible to feed 10 billion people? It depends on what they consume (Smil 2000). If everyone on Earth becomes a vegetarian, then it may be possible. You can eat more sustainably by choosing food from lower trophic levels in the food chain (Figure 13.4). We also need to change our buying and eating habits to reduce food waste and energy consumption.

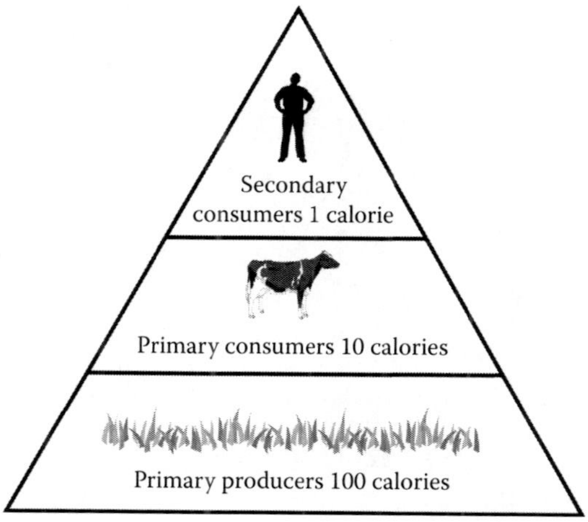

FIGURE 13.4 Stored energy decreases by a factor of ten for each step up the food chain, meaning 90% of energy is lost at each step, so it is more efficient and environmentally friendly to eat food from lower trophic levels. (Symbols courtesy of the Integration and Application Network, University of Maryland Center for Environmental Science,ian.umces.edu/symbols/.)

Governments and international aid organizations also must be part of the solution. At the apex of the Green Revolution in the 1980s, the world's food supply seemed secure, and funding for agricultural research and aid for agriculture in the developing world began to shrink. For example, the percentage of global development aid spent on agriculture dropped from more than 16% in 1980 to 4% in 2011 (Halweil and Nierenberg 2011). Now that the world's food supply is becoming less secure, the lack of agricultural innovations in the last three decades has put us at a disadvantage.

Transitioning to a sustainable agriculture system will require change on the part of farmers, food distributors, and consumers. First we will examine how to quantify the impacts of the food system using life cycle assessment, and then review approaches to reducing demand and increasing supply.

13.3.1 Life Cycle Assessment of Food Products

To identify the unsustainable parts of the food cycle we need to use life cycle assessment (LCA). Food producers can use LCA to identify the phases of the lifecycle of each food product that have the greatest environmental impacts, so they can come up with plans for reducing those impacts. Food consumers can use LCA to compare total impacts of food types summed over their life cycles; environmentally conscious consumers will choose foods with the lowest environmental impacts for purchase.

The Barilla Center for Food Nutrition did an extensive analysis of the environmental impacts of various types of food (Barilla Center for Food and Nutrition 2012). Their LCA included all phases of the food cycle: cultivation, transformation, packing, transportation, and cooking, and expressed their findings in terms of carbon, ecological, and water footprints. Beef has the highest footprint in all three categories, followed by cheese (Figure 13.5). Vegetables, fruit, and grains have much lower impacts than meat and dairy products because plants are at lower trophic levels (Figure 13.4). Consumers can decrease the ecological footprint of their diets by avoiding foods at the top of the environmental pyramid such as red meat.

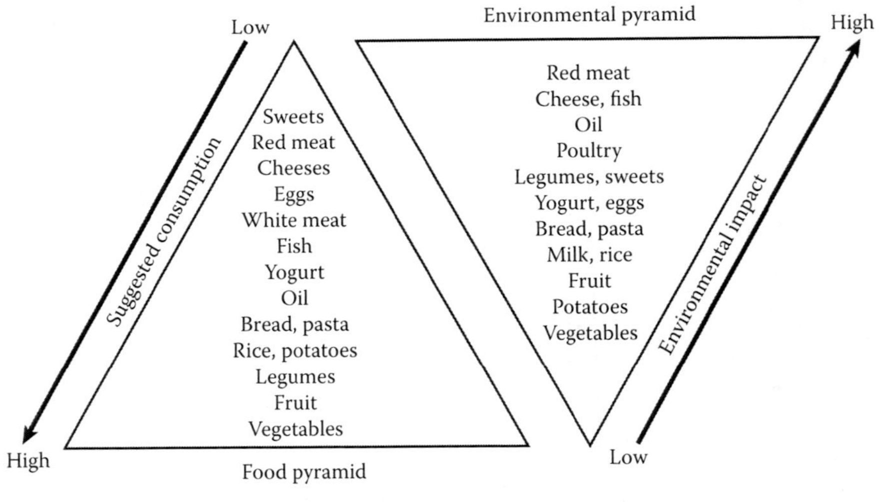

FIGURE 13.5 The food and environmental pyramids. In the food pyramid, the width of the triangle is proportional to the recommended consumption amount, with the healthiest foods like vegetables at the bottom having the highest recommended consumption levels. The environmental pyramid ranks the ecological footprint of foods over the full life cycle from highest at the top of the inverted triangle to lowest at the bottom. Note that the healthiest foods (those at the bottom of the food pyramid) have the lowest environmental impacts. (After Barilla Center for Food and Nutrition. 2012. *Eating Planet.* Edizioni Ambiente. www.barillacfn.com.)

13.3.2 Decrease Food Demand and the Environmental Impact

> The story of barbecue is the story of America: Settlers arrive on great unspoiled continent, discover wondrous riches, set them on fire and eat them.
>
> **Vince Staten**

> In general, mankind, since the improvement in cookery, eats twice as much as nature requires.
>
> **Benjamin Franklin**

Like other resources such as energy and water, food security can be increased by decreasing demand and increasing supply. Decreasing demand and adopting sustainable agriculture techniques can also reduce the environmental impacts of the food system. Demand can be decreased by reducing food waste and shifting to less resource-intensive diets, which are also healthier diets (Figure 13.5).

13.3.2.1 Conserve Food

The primary cause of hunger is not insufficient food production but inadequate food distribution. Between 25 and 50% of food never reaches the dinner table because it becomes spoiled or contaminated due to inefficient storage and distribution systems (Halweil and Nierenberg 2011). More progress has been made in improving productivity than reducing waste because there is a monetary incentive for corporations to develop new products that increase agricultural productivity. Improving food distribution systems is more difficult and less profitable because it requires solutions unique to each culture and region, and requires the cooperation of government, sometimes in the form of policy changes. Improving distribution systems often requires changes in infrastructure and therefore takes much more time than improving productivity. Also, focusing solely on the food distribution problem would ignore the driving forces of decreasing food security such as continued population growth, rising water scarcity, climate change, economic and social inequality, and ineffective government policies (Ehrlich and Harte 2015). Developed countries can best help farmers in the developing world by providing new agricultural technologies, skills, and tools.

In the United States, almost 40% of food goes to waste (Hall et al. 2009). This represents an increase of ~50% in United States per capita food waste since 1974. Wasting 40% of food nearly doubles the economic cost of food to the average consumer. Because food production consumes water and fossil fuels, food waste increases both the water and carbon footprints of food. For example, in 1995, ~27% of edible food was wasted in the United States, and the embodied energy in this wasted food was ~2% of the annual energy consumption in the United States (Cuéllar and Webber 2010). To become more sustainable Americans must conserve food and use the saved resources to provide food to the needy.

The fact that Americans waste more than one-third of their food when many Americans do not have enough food for a healthy diet is tragic (Smil 2008). However, many food conservation programs now divert edible food from the waste stream to food banks. In the past, restaurants would just throw away unused food, but now programs collect that food and distribute it to the poor and homeless, or use food waste as compost to produce soil that is distributed to farmers.* Starbucks and Panera Bread donate unsold food to food banks.† Grocery stores used to dispose of produce that was still edible but past its expiration date, but some stores have changed that policy and now distribute this food to local homeless shelters. The supermarket chain Giant Eagle repackages "ugly" produce and sells it at 20 stores, while a nonprofit called Daily Table collects expired, healthy food and sells it and food prepared from it at discounted prices in low-income neighborhoods.‡ Food Cowboy is a mobile app used to send food rejected at loading docks to nearby charities and food

* http://www.athensservices.com/recycling2/food-wasterestaurants.html, retrieved August 4, 2016.

† http://www.forbes.com/sites/geoffwilliams/2016/04/28/starbucks-finally-starts-to-donate-all-of-its-unsold-food-but-donating-isnt-as-easy-as-it-seems/#16d86aa26923, retrieved September 1, 2016.

‡ http://dailytable.org/, retrieved August 4, 2016.

banks (http://www.foodcowboy.com/); Move for Hunger (https://moveforhunger.org/) sends non-perishable food items left behind by homeowners to local food banks; and Compost Cab (http://compostcab.com/) picks up food scraps and delivers them to urban farms for composting. In 2014, California passed the Mandatory Commercial Organics Recycling law that requires businesses to recycle their organic wastes.* Programs and policies like these greatly increase sustainability by reducing waste and increasing the well-being of the needy.

13.3.2.2 Buy Locally Produced Organic Food

In the United States, food is certified by the USDA as organic if production methods promote resource cycling and soil fertility, ecological balance, and biodiversity.[†] USDA organic standards prohibit the use of synthetic fertilizers, sewage sludge, irradiation, and genetic engineering, and discourage the use of synthetic pesticides and herbicides. These standards make organic farming better for the environment than conventional farming. Although organic foods generally cost more than conventional foods, the full cost determined using full cost accounting is lower. Recent studies suggest that organic produce can be economically competitive with conventional agriculture, even when not taking externalities into account (Crowder and Reganold 2015). Increasing demand for organic foods will lead to decreasing costs through economies of scale.

Although many believe that organic produce is more nutritious than conventional produce, there is no strong evidence to support this,[‡] although substitution of organic for conventional produce can reduce exposure to pesticide residues and antibiotic-resistant bacteria (Smith-Spangler et al. 2012). Organic produce is less likely to be contaminated with bacteria than conventional produce, a problem that has caused hundreds of food-poisoning deaths in the United States within the last five years. The CDC concluded that leafy green vegetables are one of the most common sources of food poisoning.[§] However, even small organic farms can have problems with biological contamination, as shown by an outbreak of Salmonella from eggs produced at a small organic farm in Minnesota in October 2011.[¶]

You can reduce your carbon footprint by eating organic foods, but you can reduce it further by purchasing locally produced organic foods. Choosing locally produced food over industrially produced food reduces **food miles**, the distance food is transported from producer to consumer, and therefore reduces the carbon footprint of purchased food (Weber and Matthews 2008). Farmer's Markets reduce costs for locally produced foods by cutting out the middle-men such as food distributors and grocery stores, so that farmers sell direct to the consumer. At Farmer's Markets consumers often pick up boxes of fresh produce from **community-supported agriculture** groups. Consumers pay monthly or annual subscription fees, which provide a guaranteed cash flow to farmers in the community-supported agriculture group, reducing their economic risk and therefore increasing food security.

The growing demand for locally produced food is reflected by the rising number of Farmer's Markets nationally, from 340 in 1970, to 1,800 in 1994, to 5,000 by 2008 (Brand 2009). This trend, driven by **locavores**, has increased regional food production capacity and food security. It is another way for consumers to keep money in their communities and support small businesses rather than sending it to multinational corporations.

---

* http://www.calrecycle.ca.gov/recycle/commercial/organics/, retrieved August 4, 2016.

† https://www.ams.usda.gov/publications/content/organic-production-handling-standards, retrieved 8/4/2016.

‡ "Organic Food No More Nutritious Than Conventionally Grown Food," Stephanie Watson, September 5, 2012, http://www.health.harvard.edu/blog/organic-food-no-more-nutritious-than-conventionally-grown-food-201209055264.

§ "Leafy Green Vegetables Rank Among Top Causes Of Food Poisoning," Kelly Fitzgerald, January 31, 2013, http://www.medicalnewstoday.com/articles/255645.php, retrieved August 4, 2016.

¶ "Local, Organic Foods Not Always Safer," http://yourlife.usatoday.com/fitness-food/safety/story/2011-10-25/Local-organic-foods-not-always-safer/50916392/1.

13.3.2.3 Eat Healthy

Remember that sustainability means maintaining high levels of human well-being, which requires that people be healthy. In 2011 the USDA published new dietary guidelines to improve the health of Americans (USDA 2011). Recommendations include:

- **Maintain calorie balance over time to achieve and sustain a healthy weight**: many Americans must decrease the calories they consume and increase the calories they expend through physical activity.
- **Focus on consuming nutrient-dense foods and beverages**: A healthy eating pattern limits its intake of sodium, solid fats, added sugars, and refined grains and emphasizes nutrient-dense foods and beverages—vegetables, fruits, whole grains, fat-free or low-fat milk and milk products, seafood, lean meats and poultry, eggs, beans and peas, and nuts and seeds.

Many Americans find it difficult to maintain a healthy weight. Our country is in the midst of an obesity epidemic, with 72% of men and 64% of women either overweight or obese (USDA 2011). This has caused the proportion of Americans with diet-related chronic diseases to increase to 37% with cardiovascular disease, 16% with high total blood cholesterol, 11% with diabetes, and 34% with hypertension (USDA 2011). Part of the problem is that we eat too much, and another part is that we eat unhealthy food. Americans eat too much because we often have a surplus of food, and because like most animals we are genetically predisposed to eat food when it is available so that we won't starve when it's not. We eat unhealthy food because it is often more convenient and less expensive due to perverse subsidies, but also because evolution wired our brains to tell us to eat calorie-rich food whenever it is available because food was scarce (Pollan 2007). Today we often have the choice to eat calorie-rich foods, and our genes make us prefer them, so we eat too much of them. For example, since 1970 the average number of calories from sweeteners in the American diet has increased 30% (Woolf et al. 2007). Most Americans need to eat more fruits, vegetables, and whole grains and less meat, processed foods, and sugar (USDA 2011).

The **food pyramid** has commonly been used to suggest healthy proportions of different food types.* We are supposed to consume the largest proportion of foods at the base of the pyramid and the smallest proportion of foods at the top of the pyramid. Comparing the food pyramid with the environmental impacts (ecological footprint) pyramid (Figure 13.5) makes clear that the foods that are the least healthy for us are also the least healthy for the environment. Healthier foods usually come from food sources that are at lower trophic levels (Figure 13.4). A healthy diet should contain lots of fruits, vegetables, and grains, lesser amounts of oil, dairy, white meat and fish, and small amounts of cheeses, sweets, and red meat. In general, Americans eat too much food from the top of the pyramid, which causes our diets to be less healthy and to have a larger negative environmental impact. The **Mediterranean diet** is closer to the ideal, which explains why longevity is greater in Mediterranean countries (Willett 2011).

In summary, foods with the lowest environmental impacts are also the healthiest (Figure 13.5). Eating healthy foods low on the food chain increases food security and decreases your ecological footprint and carbon footprint.

13.3.2.4 Become a Vegetarian

Animals are my friends... and I don't eat my friends.

George Bernard Shaw

If slaughterhouses had glass walls, everyone would be a vegetarian.

Paul McCartney

Vegetarianism is better for the environment, better for your health, and more sustainable than the **omnivore** diet. Beyond these virtues, many people cite moral reasons for becoming

* Note that the USDA food pyramid was replaced by the MyPlate nutrition guide in 2011; see https://www.choosemyplate.gov/.

vegetarians. Eating meat, especially red meat, is less sustainable than eating vegetables and fruit (Figure 13.5).

Vegetarians do not eat meat, fish, or poultry. Instead, they eat greater amounts of fruits, vegetables, grains, and legumes. A Roper poll in 1997 found that 1% of the U.S. adult population is vegetarian (Havala 2001). Most practice **Ovo-lacto vegetarianism**, which allows the eating of milk and egg products; stricter **lacto vegetarianism** allows milk products but not eggs, while the strictest form of vegetarianism, **veganism**, does not allow the use of any animal products. **Pescetarianism** allows consumption of fish and seafood.

Research has shown that vegetarians have greater longevity and lower rates of ischemic heart disease, circulatory and cerebrovascular diseases, cancer, and type 2 diabetes than omnivores (Li 2014). In general, moving your diet down the food chain leads to better health and greater longevity (Brown 2009). People in countries like Italy that consume a Mediterranean-type diet that includes meat, cheese, and seafood, all in moderation, are healthier and live longer than most Americans who consume more meat and sweets, even though Americans on average spend more than twice as much on health care as residents in most other developed countries.*

Remember that any time we convert energy from one form to another we lose some energy. We could eat plants directly and appropriate 100% of the energy stored in plant biomass, or we could shift our diet to a higher trophic level and feed those plants to livestock, losing 90% of the energy in the process (Figure 13.4). Raising cattle for beef uses ~60% of all agricultural land yet provides <5% of the world's protein and <2% of its calories.† Thus, meat production is not resource- or energy-efficient. To produce a small amount of energy stored in meat requires a huge amount of fossil fuel energy. Thus, vegans have the smallest food energy footprint of 3 kWh/day (numbers from MacKay 2009). Adding milk and cheese to your diet to become a lactovegetarian adds 1.5 kWh/day of embodied energy, and adding two eggs a day to become a lacto-ovo vegetarian adds one more for a total of 5.5 kWh/day. LCA supports this conclusion: for every $1 million of production, milk releases 3,880 tons CO_2e and eggs 1990 tons CO_2e.‡ Much of the greenhouse gas warming potential for milk production comes from methane emissions from cows.

Adding meat to a lacto-ovo vegetarian diet raises the power requirement by 8 for a total of 12 kWh/day.§ Thus, the energy required to grow the food for a typical American diet is four times higher than that required for a vegan. Measured another way, the average American consumes 800 kg of grains per year, most of it indirectly as meat, milk, and eggs, which is four times higher than the average consumption rate in India of 200 kg of grains per year (Brown 2009).

Moreover, raising cattle for beef causes many environmental problems, including deforestation for pastureland (which reduces biodiversity) and overgrazing, which can lead to soil erosion and finally **desertification**. Deforestation and resulting desertification is a global problem that reduces biodiversity and releases carbon dioxide to the atmosphere.¶ Not only does burning down the trees create a new source of carbon dioxide, but also it causes the loss of a major sink, since the trees will no longer remove carbon dioxide from the atmosphere through photosynthesis.

In 2000 we already exceeded the proposed sustainable planetary boundary for reactive nitrogen mobilization (Pelletier and Tyedmers 2010). By the year 2050 we will be approaching the global sustainability boundaries for greenhouse gas emissions and appropriation of plant biomass. Switching to a vegetarian diet would reverse these trends, decreasing sustainability levels in all

---

* "Health Costs: How the U.S. Compares With Other Countries," Jason Kane, October 22, 2012, http://www.pbs.org/newshour/rundown/health-costs-how-the-us-compares-with-other-countries/, retrieved August 4, 2016.

† UCS Catalyst, Fall 2012, p. 11.

‡ Carnegie Mellon Economic Input-Output Life Cycle Analysis calculator, http://www.eiolca.net/.

§ Because the additions of milk, cheese, eggs, and meat reduce the amount of vegetables consumed, the required power is less than the overall sum of 13.5, closer to a value of 12 kWh/d.

¶ "Cattle Ranching Is Encroaching on Forests in Latin America," FAO Newsroom, June 8, 2005, http://www.fao.org/newsroom/en/news/2005/102924/, retrieved August 4, 2016.

three categories to well below these critical thresholds and thereby making the global food production system more resilient.

One argument against veganism is that vegans can't use grasslands that are too arid or too steep to plow as a source of food. These lands can be used for grazing livestock, however, so it follows that the world can feed more people if some eat eggs, dairy products, and meat (Brown 2009; MacKay 2009). Also, though our arguments based on thermodynamics make it seem counterintuitive, producing protein by growing corn and feeding it to poultry or catfish is more efficient than growing soybeans and eating them directly. Corn yields are four times greater than soybean yields, and poultry and catfish convert corn to protein at a 2:1 ratio. Thus, one acre could produce twice as much poultry or catfish protein as soybean protein (Brown 2009). However, as mentioned previously, a diet that includes milk and eggs has larger energy and greenhouse gas footprints than a vegan diet, so from a climate change perspective a vegan diet is more sustainable than an ovo-lacto vegetarian diet.

Although eating locally can reduce a person's carbon footprint, a much more effective approach is to reduce meat consumption (Weber and Matthews 2008). Over the entire food life cycle, 83% of greenhouse gas emissions come from production and only 11% from transportation. However, greenhouse gas emissions from production vary widely, with production of red meat releasing 2.5 times as much greenhouse gas per calorie as chicken or fish production. As stated by the authors, "Shifting less than one day per weeks' worth of calories from red meat and dairy products to chicken, fish, eggs, or a vegetable-based diet achieves more greenhouse gas reduction than buying all locally sourced food." In fact, shifting from a red meat diet to a plant-based diet cut greenhouse gas emissions almost as much as shifting from driving a Suburban to a Prius would (Brown 2009). Based on the evidence, it seems that red meat has to join coal as one of the most environmentally destructive products, and we add it to our "ABCs of Unsustainability" list, which includes autos and airplanes, beef, and coal.

13.3.3 Increase Food Supply: Sustainable Food Production

According to the Committee on Twenty-First Century Systems Agriculture and the U.S. National Research Council (2010), a sustainable farming system needs to "be sufficiently productive, robust (that is, be able to continue to meet the goals in the face of stresses and fluctuating conditions), use resources efficiently, and balance the four goals:

- Satisfy human food, feed, and fiber needs, and contribute to biofuel needs
- Enhance environmental quality and the resource base
- Sustain the economic viability of agriculture
- Enhance the quality of life for farmers, farm workers, and society as a whole"

Sustainable agriculture is an important component of sustainable development. Michael Pollan (2008) wrote that improved food policies should: (1) Strive to produce a healthful diet for all people by increasing the quality and diversity of calories rather than the per capita quantity. (2) Aim to improve the resilience, safety, and security of our food supply. (3) Reconceive agriculture as part of the solution to environmental problems like climate change. To make food production more sustainable, he recommends that we resolarize farms, reregionalize the food system, and rebuild America's food culture. By resolarize, he means that we go back to relying more on the sun than fossil fuels to grow food. Locavores in the local food movement are reregionalizing the food system which decreases food miles and associated carbon dioxide emissions. Near term, voluntary changes in food culture such as reduced meat consumption seem unlikely, but eventually these changes may become necessary.

13.3.3.1 Genetic Engineering

One approach to increasing food supply is to use **genetic engineering** to create genetically modified organisms (**GMO**s) with increased crop yields. Genetic engineering involves inserting foreign DNA into a gene to obtain some desired quality such as faster growth or pesticide resistance in a plant. Some fear that modifying the genome of an organism could have disastrous unexpected results. However, the genetic makeup of crop plants naturally changes over time; crop breeding gives direction to this genetic drift by selecting plants with desirable traits. New crops such as tomato, corn, and potato were engineered by early farmers who joined species to form genetic **hybrids** (Coleman 2005). Over time agriculture has artificially increased the rate of genetic change, first by using chemicals, and then radiation to introduce genetic mutations (Brand 2009). These techniques were imprecise because they caused much unwanted genetic modification. As a precise tool, genetic engineering replaces these brute-force techniques. With genetic engineering, scientists can produce new crops with less genetic modification than the breeding and seed mutation techniques that have produced most of our modern food crops.*

As of 2015 there are 14 GMO crops that are commercially produced, but only two genetically engineered traits are widely used in these crops—insect resistance and herbicide resistance (National Academy of Sciences 2016). Insect resistant GMO cotton and corn have a gene that causes the plant to produce the toxin of the bacteria *Bacillus thuringiensis* that kills the larvae of moths and butterflies that feed on conventional cotton and corn. Using Bt cotton and Bt corn means no pesticides need to be used; the plants kill the moths and butterflies themselves. Adoption of Bt cotton in India increased average cotton yield per acre by 24% and increased profits by 50% among small landholders, which has had a positive impact on economic and social development (Kathage and Qaim 2012). Use of Bt crops has generally increased yields and decreased the need for insecticides (National Academy of Sciences 2016).

Herbicide resistant GMO crops that can tolerate the herbicide **glyphosate** (tradename Roundup) are now widely used. Normally farmers have to eliminate weeds between plantings by plowing (tilling) the soil, a process that adds large amounts of carbon dioxide to the atmosphere in two ways: by exposing soil carbon to the atmosphere so that it can be converted to carbon dioxide through bacteria-mediated oxidation, and through burning fossil fuels to provide the energy for tilling. Farmers that use "Roundup-ready" crops can use the more environmentally friendly practice of **no-till farming**, a sustainable alternative to plowing that uses less fossil fuel energy than conventional tilling and decreases soil erosion. Rather than removing crop residues at the end of the fall harvest, farmers leave it in place over the winter to protect the soil from erosion and to gradually decompose and release nutrients to the soil. In the spring farmers use a special machine to cut slits in the residue and insert GMO seeds in the slits. When the seedlings sprout they spray the entire field with glyphosate to kill all of the weeds. Glyphosate rapidly degrades in the environment, so it is less harmful to ecosystems than most herbicides. The crop residue acts like a mulch, holding in moisture and therefore decreasing water needs (Collin and Collin 2010). No-till farming also promotes retention of soil carbon. Studies show that using glyphosate-resistant crops results in small yield increases without reducing plant diversity, although integrated pest management strategies are needed to prevent the development of glyphosate-resistance in weeds (National Academy of Sciences 2016). To date the evidence indicates that use of GMO crops with insect and herbicide resistance can increase food security and decrease the environmental impacts of agriculture.

In the past, some studies raised concerns about the use of GMO crops. For example, a 1999 study concluded that Bt corn pollen is lethal to Monarch butterflies (Losey, Rayor, and Carter 1999). However, subsequent ecological studies showed that Bt corn does not pose a significant risk to Monarch butterflies (Gatehouse, Ferry, and Raemaekers 2002). Another study that claimed rats fed GMO corn developed cancer was retracted.† Now after several decades of testing GMO foods, the results are conclusive: there is no evidence that use of GMOs in animal feeds harms livestock, nor

* Note that this section only discusses combining DNA of different species, and not the new CRISPR technology that allows scientists to edit specific genes without introducing foreign DNA. See "The DNA Revolution," National Geographic, August 2016.

† http://www.forbes.com/sites/jonentine/2014/09/17/the-debate-about-gmo-safety-is-over-thanks-to-a-new-trillion-meal-study/#1e79a47eca93, retrieved July 29, 2016.

is there evidence that foods containing GMOs cause human health problems (National Academy of Sciences 2016). For example, GMO-containing food products have been sold in the United States since 1994, and today roughly 70% of the ingredients in processed foods in the United States are from GMOs, yet no health differences have been identified between the U.S. experimental group and the EU control group, where GMOs were banned (Brand 2009).

GMO crops are being developed for resistance to insects, viruses, bacteria, and fungi; some are being developed for drought tolerance, some with nutritional supplements (e.g., beta-carotene, the precursor of vitamin A, in golden rice to prevent blindness caused by vitamin A deficiency), and some to have better photosynthetic efficiency and therefore higher yields (National Academy of Sciences 2016). Another interesting approach is to change annual crops such as wheat, rice, and corn into perennial crops, which would reduce soil erosion and water pollution while increasing yields. Non-leguminous crops could be bred to fix nitrogen as legumes do, reducing or eliminating the need for artificial nitrogen fertilizers. Farmers, agronomists, and breeders have been working toward many of these goals for centuries; genetic engineering could potentially accomplish many of them within decades.

One problem that has resulted from the proliferation of GMOs is a shift from seeds being public goods to private goods (National Academy of Sciences 2016). In the past, farmers could collect their seeds and use them in the next planting season for free. Natural seeds were nonexcludable and nonsubtractable. The adoption of hybrid varieties led to farmers depending on seed companies because seeds collected from hybrid crops have lower yields than the parent hybrid. U.S. law now allows companies to patent GMOs, and GMO seeds are therefore excludable, private goods. Many are concerned that our food security is increasingly dependent on agricultural biotechnology companies like Monsanto, which hold exclusive patents on seeds for high-yielding crops.

In summary, GMO crops can reduce hunger through their higher yields, and can reduce the environmental impacts of agriculture by allowing for no-till agriculture and decreased use of pesticides. Of course, we should always be alert to the possibility of unintended negative consequences, and to maintain the resilience of our food supply we must help preserve crop biodiversity, but banning GMO crops would make it harder to feed the growing human population.

13.3.3.2 Sustainable Farming Methods

To reduce fossil fuel consumption and mitigate climate change, we must make agriculture less centralized and energy-intensive. **Sustainable agriculture** uses the principles of ecology to reduce the environmental impacts of agriculture while maintaining high crop yields. For example, using the ecological principle of closed material cycles, farmers can reduce the need for synthetic fertilizers and thereby decrease environmental impacts by using crop and livestock waste as fertilizers (Figure 13.3). U.S. government policies should promote the rapid adoption of sustainable agriculture practices.

The U.S. industrial agriculture system depends on cheap, abundant energy from oil to grow and transport food. As a result, food production emits an average of 1.25 tons of carbon dioxide per year for every American (Hill and O'Neill 2008). Industrial agriculture uses chemicals that can wash into water supplies or accumulate in agricultural waste, leading to chemical pollution and ecosystem degradation (Halweil 2006). Industrial agriculture also causes soil erosion, due to excessive tillage, and salinization of soils, caused by irrigation without proper drainage. Finally, use of expensive machinery has given a large advantage to rich farmers and corporations, who have put many small farmers out of business. However, industrial agriculture has benefits, including higher yields and reduced risk to farmers from blights and pest outbreaks. Industrial agriculture therefore offers increased food security, which makes it attractive from a sustainability perspective, but it often relies on nonrenewable resources such as petroleum, making it unsustainable. Therefore, it makes sense to choose a "middle path" where we use organic farming methods as much as possible without sacrificing food security (Halweil 2006). Rather than adopting a sometimes risky approach of strict organic farming, farmers can adopt the organic methods that work well for their unique situation, but still have the flexibility to use synthetic nitrogen or other yield-enhancing technologies.

To sustainably achieve high crop yields, farmers must use methods that promote thick, nutrient-rich soils and supply abundant but not excessive water to plants, all while minimizing the use of fossil fuels and any chemicals that might have negative environmental impacts. Sustainable agriculture methods include **conservation tillage**, use of **cover crops**, maintenance of high crop and genetic diversity, **crop rotation**, **intercropping**, use of **best practices** for **water conservation** and for maintaining water quality, use of natural fertilizers (compost, animal manure, green manure), precision application technologies, **integrated pest management**, long-term use of **biochar**, and genetic improvement of livestock and crops through traditional breeding and genetic engineering (Committee on Twenty-First Century Systems Agriculture and National Research Council 2010). **Holistic management** is a sustainable approach to resource management in general and agricultural systems in particular. Here we look at some of these practices in more detail.

Plowing is an ancient practice that prepares soil for planting by burying weeds and aerating the soil. It also has the greatest environmental impact of all agricultural practices because it accelerates soil erosion and releases carbon dioxide to the atmosphere. Plowing of the U.S. prairies contributed to soil erosion, leading to the infamous Dust Bowl of 1931–1939 that caused massive crop failures. This led to the adoption of soil conservation measures, now overseen by the U.S. Natural Resources Conservation Service, which developed the Ecosystem Sustainability Framework that can be used by farmers to see if their farms are sustainable (Collin and Collin 2010).* Conservation tillage practices such as no-till farming greatly reduce the food production footprint by decreasing carbon emissions and soil erosion.

Organic farming is one approach to achieving sustainability in agriculture. Organic farming "includes care for the soil and the ecosystems surrounding organic farms, and relies on biological and mechanical controls to deal with pests and on organic materials for fertilizers" (Brand 2009). In the United States, organic farmers practice crop rotation, use organic seeds, minimize the use of synthetic chemicals or fertilizers, improve soil fertility by natural compost and manure, and do not use GMO crops.†

Organic agriculture offers many advantages over industrial agriculture. Particularly when no-till methods are used, organic farming conserves soil. The planting of polycultures instead of monocultures helps preserve biodiversity, and the synergetic coexistence of plants and animals closes the resource loop (Figure 13.3). Finally, adopting organic farming practices could reduce overall U.S. greenhouse gas emissions by 20%‡ because organic farming uses 50% less energy than traditional farming and 60% less than industrial farms (Hill and O'Neill 2008). Overall, the external costs of organic farming are about one-third that of industrial agriculture. In part this is because industrial farms use "one size fits all" approaches that are not custom-tailored to local environmental conditions; by ignoring local ecological balances, industrial farmers risk exceeding the local carrying capacity, thereby causing negative environmental impacts including biodiversity loss (Collin and Collin 2010).

While agriculture before the Green Revolution was completely organic, it was not as sophisticated and did not have yields as high as modern organic farming methods. Generally speaking, organic farming is more labor intensive and produces yields ~10% to 18% lower than conventional farms (Crowder and Reganold 2015). However, the environmental benefits have convinced many people to choose organic over conventional produce, even when organic costs are ~30% higher. The willingness to pay this premium means that organic farming can be more profitable, as organic farming profits match conventional farming profits when the premium is only 5% to 7%. This has led to a rapid expansion of farmland dedicated to organic farming, from one to four million acres in the United States between 1992 and 2005 (Brand 2009). In Europe and Australia, nearly one-third

* Ecosystem Sustainability Framework, http://www.nrcs.usda.gov/wps/portal/nrcs/detail/national/technical/econ/data/?cid=nrcs143_009664, retrieved September 2, 2016.

† https://www.ams.usda.gov/sites/default/files/media/Organic%20Practices%20Factsheet.pdf, retrieved August 4, 2016.

‡ http://www.pewclimate.org/.

of farmland is organic, but globally only 1% of agricultural land is organically farmed (Crowder and Reganold 2015).

Exporting modern organic farming methods to areas in the developing world where farmers still use pre-modern agricultural methods could triple crop yields and at lower cost than adopting the methods and tools of industrial agriculture, all while reducing the environmental impact of agriculture (Halweil 2006). For farmers in developing countries, adopting modern organic farming methods is a win-win situation: it increases food production and profits, without causing environmental harm or human health problems. Furthermore, because modern organic farming can triple yields, farmers require only one-third as much land to feed the population. Thus, adopting organic farming methods can help preserve some of the world's most ecologically sensitive and biologically diverse forests, which are often at risk because they are near subsistence farms in developing countries.

However, organic farming falls short in one category. By banning the use of GMO crops, organic farmers put themselves at a disadvantage. NGOs are avoiding this problem by putting GMO crops in the hands of poor farmers and teaching them sustainable farming methods. Besides the previously mentioned golden rice, GMO crops that are being developed and distributed by humanitarian agencies such as the Bill and Melinda Gates Foundation include BioCassava Plus, which has more protein and vitamins and less cyanide than conventional cassava; African Biofortified Sorghum, which like conventional sorghum is drought-tolerant but is more easily digested and adds vitamins, minerals, and amino acids; and GMO bananas that have more vitamins than conventional bananas (Brand 2009). In most cases, GMO crops are developed in the United States and Europe but become wholly owned and implemented by African countries, thereby avoiding the problem of multinational corporations charging high prices for food needed to keep people in Africa from starving.

Evidence suggests that U.S. agriculture is experiencing a Renaissance, with increasing numbers of farmers adopting sustainable farming methods. Sustainable management of agricultural lands can reduce costs and environmental impacts while maintaining high crop yields indefinitely.

13.3.3.3 Aquaculture

Agricultural protein production using livestock and soybeans is supplemented in many areas by wild-caught and farmed fish and seafood. Consumption of aquaculture products exceeded that of wild fisheries for the first time in 2014 (Figure 4.6), and aquaculture production is increasing 5% annually.* Demand for seafood has risen due to population growth but also because of urbanization, which increases access to seafood, and increased prosperity, which increases the proportion of meat and seafood in people's diets.

To supplement the falling per capita wild catch harvest, aquaculture has expanded. The shift from fish hunting to fish farming mirrors the transition on land from hunter-gatherer to agricultural societies that occurred thousands of years ago in Asia and the Middle East. But is it a solution? Meadows et al. (2004) argue that it is not for three reasons: fish increasingly feed the rich rather than the poor, as increasing scarcity is causing rising prices; fish farms can cause environmental degradation; and fish have become a food sink rather than a food source. Farmed fish are a food sink because roughly 10 kg of grain is required to produce 1 kg of fish. However, approaches to sustainable aquaculture are being developed, as discussed below.

Fish farming produced only 6.3% of seafood in 1970 but ~50% in 2009 (Naylor et al. 2009), and it is now the world's fastest-growing food production sector (World Bank 2007). More than 90% of global aquaculture production occurs in developing countries, with China alone accounting for 67% (World Bank 2007). Aquaculture employs more than 12 million people in Asia, and it is an important trade commodity. The Chinese integrate agriculture with aquaculture, and use a very efficient fish polyculture system in which four species of carp fill four niches in an ecosystem (Brown 2009). This system exploits synergies between fish and rice that produce high, stable rice

* http://academy.pittmanseafoods.com/en/2015/12/fao-farmed-fish-overtakes-wild-fish-in-global-supply/, retrieved July 28, 2016.

yields with reduced requirements for chemical inputs (fertilizer, pesticides, herbicides) (Xie et al. 2011). For example, rice paddies with fish require 68% less pesticide and 24% less fertilizer than paddies without fish.

Aquaculture has grown in popularity primarily because herbivorous fish efficiently convert feed into protein, and therefore can be produced at relatively low cost. The efficiency of animal protein production is measured by the **feed conversion ratio**, which equals the feed mass divided by the increase in body mass of the animal, with low values indicating higher efficiency. Feed conversion ratio values range from 2–3 for chicken, 2.2–5.9 for pork, and 5.5–7 for beef, but only 1.5–2.0 for herbivorous fish (Brown 2009). Combining one-part soybean meal with four parts grain can nearly double the efficiency with which grain is converted into animal protein (Brown 2009).

Aquaculture holds great promise for feeding growing populations because protein yields of herbivorous fish farms are extremely high. In part this is because the **net primary productivity** of shallow waters (the amount of plant growth per acre per year) is higher than on land. Rapidly growing plants provide the food for rapidly growing fish. Furthermore, indications are that sustainable fish stocking densities (amount of protein produced per acre per year) are much higher than sustainable livestock densities. This is because a fishpond makes use of the vertical dimension, with fish growing over a range of depths. In contrast, livestock are restricted to the land surface. Substituting farmed fish for wild-caught fish in our diets can also reduce the harvesting rate of wild fish, possibly allowing wild fish populations to recover.

Scientists are developing new sustainable aquaculture strategies to reduce pressure on wild fisheries. One approach is to feed human waste to oysters, and the oyster waste to worms that are harvested and sold as fish bait (Sohns and Crowder 2013). Holistic approaches to creating artificial ecosystems with multiple trophic levels such as this hold the promise of growing more food in smaller areas. One problem is that most growth in aquaculture is in production of shrimp and salmon, predators that are near the top of the food web, have high feed conversion ratios, and consume large amounts of fishmeal, which is often supplied by marine catches. Wild fisheries inputs per unit of farmed fish output has declined to 0.6, but for Atlantic salmon it can be as high as 5 (Naylor et al. 2009). The farming of salmon and shrimp is known to be environmentally destructive, but they account for only ~10% of farmed output; global aquaculture is dominated by herbivorous fish (carp, tilapia, and catfish) that are less harmful (Brown 2009).

To be sustainable, aquaculture should produce species such as filter-feeding oysters and clams and herbivorous fish that are lower in the food chain and therefore don't require marine fish for food (Sohns and Crowder 2013). On the supply side of the equation, sustainable aquaculture could help protect ocean ecosystems and provide food security to consumers who depend on seafood as a protein source. On the demand side, the environmental impact of seafood consumption could be decreased if consumers eat less seafood and eat lower on the food chain, substituting anchovies for tuna, for example.

However, the rise of fish farming raises new potential problems, which include pollution of marine coastal waters and of surface and groundwaters in coastal zones; overfishing of wild catch to feed farmed fish (Naylor et al. 2000); increased use of antibiotics; accidental release of genetically modified species to the environment; and the spread of disease to wild animal populations and to humans (Sohns and Crowder 2013). In many coastal developing countries without environmental regulations, export-driven fish farming is spreading into sensitive ecosystems and degrading them. Mangrove swamps, which maintain high levels of biodiversity and are important buffers that protect the mainland from storms, have been cut down in many areas and replaced by shrimp farms. The use of brackish water in unlined brine shrimp ponds on land is likely causing salinization of unconfined shallow aquifers in many regions, which would displace terrestrial farmers who rely on well water for irrigation (Paez-Osuna 2001).

Are fish farm practices unhealthy for fish? A sure sign that farmers are not treating animals properly is when the animals have abnormally high rates of illness, which farmers usually counter by administering antibiotics. High stocking densities in aquaculture ponds are causing the spread of disease. For example, infectious salmon anemia wiped out salmon fish farms in Chile and other countries, and now has spread as far north as British Columbia in Canada.* Krkošek et al. (2011) found that sea lice from salmon farms cause deaths of wild salmon; this has led to a moratorium on development of salmon farms on the central British Columbia coast. Paez-Osuna (2001) found that rapid development of shrimp farms often leads to catastrophic collapse of shrimp populations due to disease outbreaks; abandonment of the collapsed shrimp farms causes environmental damage. However, Paez-Osuna (2001) concludes that shrimp farming can be sustainable, and in most cases it causes less environmental damage than modern agriculture. Problems mostly arise when fish farmers become greedy and raise the intensity of fish farming by increasing stocking densities to unsustainable levels, causing ecological overshoot in aquaculture ponds and eventually resulting in collapse in the form of die-offs of fish populations.

The U.S. NOAA published a set of aquaculture policies in 2011 to promote sustainable aquaculture in the United States.† Because improving the fish environment decreases stress on fish and reduces the incidence of disease, fish farmers should lower stocking densities until fish no longer need antibiotics to survive. Fish farmers can also increase the sustainability of aquaculture by treating wastewater before discharging it. To avoid introducing nonnative species to the local environment, fish farmers should use native species or infertile fish (Food and Agriculture Organization of the United Nations 2005).

Environmentally conscious food distributors such as Whole Foods are promoting standards to make aquaculture more sustainable. Whole Foods does not buy farmed fish that were fed antibiotics, growth hormones, or animal by-products, were treated with preservatives, or were cloned or genetically engineered.‡ Whole Foods also requires their fish-farm suppliers to protect sensitive habitats such as wetlands and mangrove swamps, to abstain from catching wild fish to feed farmed fish, and to monitor water quality in fishponds to prevent pollution.

In summary, aquaculture can increase the resilience of the global food system if government policies provide incentives for aquaculture operations to efficiently use resources and protect the environment (Troell et al. 2014). To protect marine ecosystems and reduce competition for food resources, farmed fish should be fed with aquatic crops and wastes, not with wild fish or terrestrial crops.

13.3.3.4 Grow Your Own Food

What gardens do best is help gardeners grow.

David Wann
Simple Prosperity: Finding Real Wealth in a Sustainable Lifestyle 2007

Many people are now rediscovering how rewarding it is to grow their own food. Doing so has so many advantages. Gardening is good for your physical and mental health, and producing your own food makes you more self-sufficient. Eating food that you produce eliminates the uncertainty and risks associated with foods produced in unknown locations using unknown methods that may be unsafe, leading to the many cases of food poisoning in recent years (national recalls and deaths resulting from contaminated peppers, peanut butter, and spinach in 2007–2008, and cantaloupes in 2011). Freshly harvested produce that you grow in your garden tastes better and may be healthier than what you buy in the store. Finally, gardening helps you and your children to understand how people used to live, and appreciate how much easier we have it today.

You can decrease your ecological and carbon footprints and increase your food security by growing your own organic produce. Most people start with an organic vegetable garden. A simple approach

* "Most Feared Salmon Virus Has Arrived in BC Waters," *Food Safety News*, January 8, 2016, http://www.foodsafetynews.com/2016/01/most-feared-salmon-virus-has-arrived-in-bc-waters/, retrieved August 4, 2016.
† http://aquaculture.noaa.gov/us/aq_policies.html.
‡ http://www.wholefoodsmarket.com/farm-raised-seafood, retrieved July 29, 2016.

is to use the method of **sheet mulching**, also referred to as "lasagna gardening," that requires no construction or digging. Simply lay cardboard on your lawn, then newspapers. Thoroughly soak them, and then lay a thick pile of compost on top. Voila! Your garden is ready to plant.

Use organic methods in your vegetable garden whenever possible. Use synthetic pesticides as a last resort when trying to save a crop. Test plant crops and find which ones grow well in your climate and soil type; stop planting crops that do not thrive and therefore require pesticides and herbicides in order to survive.

When purchasing and planting edible terrestrial plants, choose perennials over annuals. Perennials do not need to be repurchased and replanted every year, and you will not have to retill the soil every year. Perennials, including fruit and nut trees, generally require a greater investment of time and energy to get started (trees may not produce fruit for several years after planting). However, once your perennials start to produce, you can harvest them for many years with little effort. Grow plants that can grow in the vertical dimension on trellises; perennial varieties include grapes, kiwi, and passion fruit, and annual varieties include pole beans, squash, cucumbers, and peas (Kellogg and Pettigrew 2008). Another approach is to garden unused sunlit areas such as roofs and unused parking lots and roads exposed to the sun. Flat roofs are easy to convert into oases, but make sure the roof you plan to use has adequate structural support, as wet soil can be very heavy. To garden areas covered by asphalt, you must first remove the asphalt, which is impermeable (Kellogg and Pettigrew 2008).

If you live in an urban setting you can practice **urban agriculture**, which is the primary food source for more than 800 million people worldwide (Halweil and Nierenberg 2011). Most urban agriculture is not commercial; rather, it is practiced by urban dwellers who wish to grow their own food. It can be challenging due to the lack of land, but it is growing in popularity.* Several different approaches can be used to solve the problem of inadequate space in urban areas. Many communities now preserve land for **community gardens**, sometimes located in floodplains where building homes would be unwise. **Vertical farming** methods increase usable space and yields by growing crops year-round at multiple levels in greenhouses or glass skyscrapers near the source of consumption. One indoor acre of land used for vertical farming can produce as much as 4 to 30 outdoor acres (Collin and Collin 2010). One concern is that vertical farming may not be energy efficient; for example, greenhouse products such as hothouse tomatoes can have much larger carbon footprints than their outdoor equivalents (Baldwin and Wilberforce 2009).

To increase your food security, breed microlivestock like chickens that are well suited to urban living because they require little space. Larger mammals such as cows are impractical because they trample the soil, produce unmanageable amounts of waste, have high feed conversion ratios, and require too much space to graze. Chickens provide meat and eggs, and they control insects and build, aerate, and fertilize soil. Because animals store nutrients into the cold months, you can eat them in the winter after plants have ceased to grow (Kellogg and Pettigrew 2008).

Perhaps the home agriculture system most compatible with the concept of sustainability is **permaculture**, short for permanent agriculture (Mars and Mars 2007). The objective of permaculture is to use a diversity of plants and animals to produce the greatest amount of food possible per acre with the least energy and effort. Permaculture goes beyond organic farming, which primarily describes what not to do (don't use pesticides, herbicides, or GMO crops). Permaculture uses integrated designs based on the principles of natural ecosystems to ensure that all plants and animals perform several functions and are complementary. The three main ethics of permaculture are Earth care, people care, and surplus share, in which practitioners donate time and money to further the care of Earth and people (Mars and Mars 2007).

---

* The online magazine *Urban Sustainable Living*, written by "The Garden Girl" Patti Moreno, promotes urban permaculture; see http://urbansustainableliving.com/.

13.3.4 Conserve and Create Soil

Soil is an extremely valuable resource that we often take for granted. Without soil, we cannot grow food. To maintain food security, we must ensure that we have adequate supplies of soil in the future. Unfortunately, soil is often lost to erosion much faster than it can be replaced by natural processes. One inch of soil can take 600 to 1500 years to form (Collin and Collin 2010), depending on the climate, so soil is effectively a nonrenewable resource. This means we must take great pains to protect our soil from erosion or contamination.

Soil loss and degradation are serious global problems. Soil loss through erosion usually results from deforestation and overgrazing. The United States is the only country that regularly monitors soil erosion, so it is difficult to estimate the global soil erosion rate (Smil 2002). However, satellite photos of dust clouds indicate that soil erosion is most severe in Africa.

Besides the decrease in soil quantity due to erosion, soil quality may also deteriorate. Soil salinization can result from improper irrigation practices in arid environments (Postel 1999). In developed countries, over-application of agrochemicals and soil compaction by heavy machinery are concerns (Smil 2002). Loss of soil organic carbon, which is essential to soil fertility, results from over-tillage and inadequate application of manure. Soil organic carbon is known to filter pollutants, reduce erosion, and decrease eutrophication in surface water bodies (Lal 2004). Enhancing carbon sequestration in soils can increase soil fertility by increasing retention of water and nutrients, resulting in higher crop yields, in addition to removing greenhouse gases from the atmosphere. Soil organic carbon can be preserved by reducing runoff from agricultural fields, and can be restored by using conservation tillage (Papanicolaou et al. 2015).

On a small scale, you can produce soil and reduce waste by **composting**, even in urban environments. Household garbage often contains large amounts of organic debris that contain stored energy. You can use that energy to create valuable soil rather than wasting it by sending it to a landfill, where it will decompose and release methane, a potent greenhouse gas (Kellogg and Pettigrew 2008). One of the easiest and most satisfying ecological practices is to compost organic material and use it to produce valuable **humus**, the organic-rich component of soil that is rich in nutrients and microbes and is essential for fertile soil. You can use compost to amend sandy soils, which helps them hold nutrients and water. Composting closes the resource loop, returning materials to the Earth where they are still valuable resources that the ecosystem can use. Creating your own soil by composting is another way to move yourself toward sustainability and independence (Kellogg and Pettigrew 2008).

13.4 CONCLUSIONS

Human population increases and shifts in diet have led to increased global food demand. Further expansion of agricultural lands was no longer feasible in developed countries, so they had to resort to agricultural intensification to increase crop yields. Countries whose supply could not meet demand started to import food or engaged in land grabbing to expand agricultural production in foreign countries.

Factory farms in the developed world have greatly increased crop yields by using chemical fertilizers, irrigation, new varieties of crops, pesticides, and herbicides, and industrialized systems. An 11-fold increase in fertilizer use combined with a threefold increase in irrigated area and the adoption of high-yielding hybrids of corn, wheat, and rice led to a tripling of world grain harvest (Brown 2009). However, this industrial system of agriculture is unsustainable because it relies on energy from a nonrenewable resource—oil—and it has a very large carbon footprint (Greer 2009; Bomford 2010).

Currently hunger is mostly caused by political and economic issues rather than food shortages. That will likely change in the near future as population increases and shifts to more resource-intensive diets increase demand, and shortages of land, soil, water, and energy and changing climate threaten to reduce food supply. Sustainably intensifying food systems, which involves increasing agrosystem productivity and efficiency through use of regulations and policy incentives, can provide many benefits, including decreased greenhouse gas emissions, land-sparing, and decreased land degradation. Food security can be increased by stopping deforestation, increasing agricultural yields in developing countries, using resources more efficiently, decreasing food waste (~1/3 currently wasted), decreasing meat consumption, decreasing biofuel production, and practicing sustainable agriculture (National Academy of Sciences 2012).

The environmental impacts of agriculture can be reduced by using organic farming and holistic management methods such as permaculture. Sustainable solutions to food supply include conservation tillage, use of cover crops, crop rotation, use of best practices for water conservation and for maintaining water quality, nutrient and livestock manure management, integrated pest management, use of natural fertilizers (compost, animal manure, green manure), and genetic improvement of livestock and crops. More money should be spent on agricultural research and development to ensure that increases in crop yields keep pace with demand, and on technology transfer to increase crop yields in developing countries. Global adoption of sustainable, high-yield agricultural practices can reduce undernourishment, unemployment, desertification, water pollution, and climate change (Nierenberg 2013).

Individuals can increase their food security by growing and storing their own organic food using urban agriculture and permaculture methods and by purchasing locally produced organic food. Food demand can be reduced by decreasing food waste and meat consumption. Universal adoption of sustainable food production and consumption could ensure high levels of food security for everyone.

WEB RESOURCES

- 49 Good Reasons For Being A Vegetarian: http://www.britishmeat.com/49.htm
- A Sustainability Challenge: Food Security for All (book): http://www.nap.edu/catalog.php?record_id=13378
- United Nations Food and Agriculture Organization: http://www.fao.org/
- Full Planet, Empty Plates: The New Geopolitics of Food Scarcity: http://www.earth-policy.org/books/fpep/fpep_presentation
- http://www.ted.com/talks/allan_savory_how_to_green_the_world_s_deserts_and_reverse_climate_change.html
- Measuring Food Insecurity and Assessing the Sustainability of Global Food Systems (book): http://www.nap.edu/catalog.php?record_id=13290

HOMEWORK PROBLEMS

1. Summarize the pros and cons of aquaculture, and take a side on whether it should be expanded. Elaborate.
2. Describe a scheme to diversify your food supply portfolio to reduce risk.
3. Research topics to write short papers or give short presentations on:
   a. Sustainable livestock production
   b. Community gardens
   c. Community supported agriculture
   d. Permaculture
   e. Biochar
4. Use the calculator at www.myfootprint.org to compare the food footprint of vegan, vegetarian, omnivore, carnivore, and "top of food chain" diets.

5. Use the calculator at www.myfootprint.org to compare the food footprint associated with different food sources:
   a. Farmers markets, gardens, cooperatives, and other local and fresh sources
   b. Natural foods markets
   c. Supermarkets for some items, natural food stores for others (average American)
   d. Supermarkets, convenience stores, and prepared foods from restaurants
   e. Restaurants, fast food, and take out

14 Waste and Pollution

> Pollution is nothing but the resources we are not harvesting. We allow them to disperse because we've been ignorant of their value.
>
> **R. Buckminster Fuller***

Up to now we have examined limits associated with material sources, that is, we started at the beginning of linear material cycles (Figure 3.10). Now we will examine material sinks and approaches to closing resource loops. Planetary sinks include the atmosphere, hydrosphere, and lithosphere; ecosystem degradation and negative health effects occur when the biosphere (which includes our bodies) becomes a sink for wastes and pollutants. A material is called a **resource** if it has value to humans, a **waste** if it has no value, and a **pollutant** if it has undesired effects on people, ecosystems, or other resources. There are limits to the rates at which sinks can absorb waste flows without harm to people, the economy, and the environment. When we are in ecological overshoot, material flow rates exceed the ability of the Earth and its ecosystems to self-regulate and regenerate. Local pollution problems caused by reaching local physical limits have now become global pollution problems such as ozone depletion and climate change.

In most cases, pollutants are resources that are out of place, that is, they are not fully utilized (Keller 2011). Phosphorous is an excellent example: In Section 6.3.1 we showed that it is an essential nutrient and a limiting factor on crop yields, that use of phosphorous in fertilizer is making it scarce, and that much of the phosphorous applied to crops is lost as runoff. The same is true of nitrogen, which is a limiting nutrient in marine systems. The excess phosphorous and nitrates in runoff end up in lakes or the oceans where they cause **nutrient pollution** and eutrophication (Section 12.3.2). We need to find ways to use nutrients more efficiently, which would reduce the amount of fertilizer lost to runoff and mitigate nutrient pollution. One approach is to close the resource loop by recovering nutrient waste. For example, a method was recently developed to recover phosphorous from the phosphate mineral struvite, which clogs pipes in wastewater treatment plants (Gilbert 2009). Several treatment plants now use this technology to recover the phosphorous and sell it as "green fertilizer."

Sustainable living requires that we not waste resources. We must find ways to use waste, which increases efficiency and closes resource loops. Unusable waste must be safely disposed of so it does not become a pollutant, as some pollutants are very toxic and can cause severe health problems.

14.1 BACKGROUND

To decrease waste and prevent pollution we must identify inefficiencies in material resource cycles. Using the analogy of an organism to represent an ecosystem or socioecological system, a consumed resource can be represented as food that enters the organism on one end. Food that the digestive system of the animal could not use comes out the other end of the organism as waste. This waste is an unutilized resource. The larger the organism, the more food it must consume to provide energy for its metabolic processes, and the more waste it produces. This analogy describes an inefficient open resource loop: Virgin resources enter one end, pass linearly through the system, and then exit the system as waste (Figure 3.10). Viewing the organism through the lens of system dynamics, each type of material (elements, compounds) has inflow and outflow rates, and the stock is the amount of that material in the organism.

* http://www.goodreads.com/quotes/2387810-pollution-is-nothing-but-resources-we-re-not-harvesting-we-allow, retrieved August 5, 2016.

Similarly, cities and households behave as organisms or systems that consume resources and produce waste. In the past when energy was scarce and the population was low, humans consumed fewer resources and produced less waste. Although human communities in the past produced less waste, both as a whole and per person, they did not use resources efficiently because resource loops were open.

The environmental ImPACT from use of a specific good can be calculated at each of the four stages (RPUD) of its life cycle. The environmental impact can be reduced through conservation and increased efficiency. Conservation reduces $A \times C$, where A is the amount of money spent on the good and C is the material intensity of the good, which is the mass of material used per unit of value, usually measured as dollars of GDP. Technology can reduce T, the resource use inefficiency, which measures the utility or benefit of the use of the good per mass of material.

The environmental ImPACT can be expressed as a mass of an essential resource such as water or energy that is consumed, or of a waste product such as CO_2 that is produced, over the entire life cycle of the good, referred to as the resource footprint. The environmental impact of CO_2 emissions is measured by the carbon footprint, which is calculated using the Kaya identity, a variant of the more general ImPACT formula (Section 3.2). Provision of goods and services to fuel consumption is the largest direct source of greenhouse gas emissions in the United States (Sheehan and Spiegelman 2010). Using the ImPACT formula to calculate the ecological footprint shows that North Americans consume four times more resources than the continent can provide, that is, $EF/B = 4$ (Sheehan and Spiegelman 2010). This is possible because we appropriate resources from other parts of the world by importing them.

In 2013, Americans generated 254 million tons of trash, or 2 kg (4.4 lb) per person per day (EPA 2015). This is only the mass of material in the last phase of the life cycle—disposal. Integrated over the full life cycle Americans annually consume roughly 100 times that amount or 70 metric tons of solid materials from the environment, while Japanese consume only 40, suggesting that Americans can reduce the amount of waste they produce by nearly half without reducing quality of life (Schmidt-Bleek 2007). **Dematerialization** aims to minimize the amount of material used to provide the same benefit or functionality of a good. On average, industrial countries must dematerialize by a factor of 10 in order to become sustainable.* To do so requires achieving significant increases in resource productivity.

14.2 PROBLEMS

14.2.1 Health Impacts of Pollution

When the waste absorption capacity of the environment is exceeded, pollutants accumulate in the environment and can become health risks. Risk was defined in Chapter 4 as a probability from harm, and we briefly discussed physical risks in the environment such as earthquakes. Chemical risks from the environment are harder to identify. A **poison** is a material that is toxic to cells and organisms. In epidemiologic studies, risk = hazard × exposure, where exposure is the amount of time a person is exposed to a health risk. Exposure to high toxin levels for short periods of time is acute exposure, while exposure to low toxin levels over long periods of time is chronic exposure (Manahan 2013). The latter is often an occupational exposure, that is, one that occurs in the workplace. Exposure pathways include breathing, drinking, eating, and direct exposure (absorption through skin).

According to the World Health Organization, estimated annual deaths from water pollution are 5.1 million, outdoor air pollution 2.5 million,† and indoor air pollution ~2 million,‡ mostly from

* www.factor10-institute.org.

† Burden of disease: Deaths by country: http://apps.who.int/gho/data/view.main.34300.

‡ http://apps.who.int/gho/data/node.main.140?lang=en, retrieved May 2, 2013.

unsafe cooking methods. More recent data indicate that ~3.3 million die from outdoor air pollution each year, mostly in Asia from airborne particulate matter, and projections indicate this number could double by 2050 (Lelieveld et al. 2015). In the United Kingdom, air pollutants reduce human longevity by an average of 8 to 14 months, and by up to nine years for the most vulnerable groups (Ludden, Peach, and Flight 2015).

Health risks vary substantially between developing and developed countries. The main causes of death in developing countries are undernutrition, exposure to environmental risks, and unsafe sexual practices, while in developed countries diet-related risks are the leading cause of death (Frumkin 2005; World Health Organization 2002). In developing countries, the greatest environmental health risk is water pollution, followed by indoor air pollution and urban air pollution. In developed countries, urban air pollution presents the greatest environmental health risk.

Much of the public focus in the United States has been on trying to find cures for diseases like cancer. However, no amount of "race for the cures" will eliminate cancer if we continue to release carcinogenic (cancer-causing) pollutants to the environment. In the United States, President Nixon started the "War on Cancer" in 1974, but since that time the cancer incidence rate has not changed and remains the highest in the world (Aggarwal et al. 2009). Research has shown that cancer is a preventable disease. Roughly one-third of cancer is attributed to environmental factors, one-third to diet, and one-third to tobacco. Lifestyle changes and pollution reduction can significantly reduce cancer risk at relatively low cost. In contrast, roughly $20 billion is spent annually on cancer research with no cures discovered to date (Aggarwal et al. 2009). Just as for other areas of health and medicine, an ounce of prevention is worth a pound of cure, that is, it is much more cost-effective to prevent cancer by stopping the release of carcinogens to the environment than to treat large numbers of cancer patients. For example, reducing tobacco smoking costs far less than treating all of the people who develop lung cancer because of smoking. So while research focused on finding a cure for cancer should continue, more emphasis should be placed on preventing diseases like cancer by reducing environmental pollution.

14.2.2 Hazardous Chemicals

> We are rightly appalled by the genetic effects of radiation; how then, can we be indifferent to the same effect in chemicals we disseminate widely in our environment?
>
> **Rachel Carson**
> *Silent Spring*

> There are thousands of different chemicals in the environment that may cause adverse human health effects. Little is known about the toxicological properties of most of these chemicals...
>
> **U.S. EPA**
> *Unfinished Business: A Comparative Assessment of Environmental Problems, 1987*

The U.S. EPA has approved more than 80,000 chemicals for use in the United States (Hill and O'Neill 2008). Some chemicals that are widely used such as the **aromatic hydrocarbon benzene** are highly toxic, and for this reason the less toxic compound **toluene** is often used in place of benzene (Manahan 2013). However, the effects on human health of most other chemicals have not been investigated. Thus, it makes sense to use the precautionary principle, and take every step to reduce human exposure, even if a health risk has not yet been demonstrated.

Safe disposal of hazardous chemicals is challenging and expensive. As a result, many hazardous chemicals have been deliberately or inadvertently released into the environment, where they can damage ecosystems and cause the loss of ecosystem services and premature deaths. Much **hazardous waste** that we must deal with today is legacy waste, material that was improperly disposed of before the advent of environmental regulations.

14.2.2.1 Toxicity

Chemical **toxicity** is usually determined through animal studies combined with data from occupational exposures. The data define a **dose–response relationship** that when plotted on a chart defines a **dose–response curve** (Figure 14.1), where dose is equivalent to exposure. The response can be any adverse health effect such as breast cancer. Sometimes the dose–response curve is assumed to be linear and to pass through the origin; this is equivalent to assuming that there is no threshold dose (a maximum dose below which there is no adverse health effect, see Figure 14.2) or background response (the health effect occurs at a lower rate in individuals not exposed to the chemical, perhaps because other environmental contaminants cause the same effect). Note that a compound with high toxicity can cause a high response even when present at low concentration (Figure 14.1).

For a given compound the EPA sets the **maximum contaminant level** at or below the threshold dose, although in reality dose–response curves have not been derived for most chemicals. To complicate matters further, dose–response curves are usually not linear (Figure 14.3). However, as a rule of thumb, "dose makes the poison," that is, almost any chemical can become toxic if a person is exposed to very high levels. The higher the dose, the more probable there will be a negative health

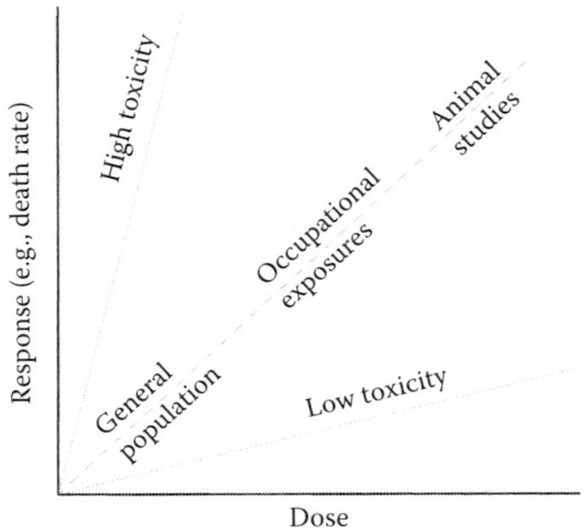

FIGURE 14.1 Schematic diagram illustrating how responses such as cancer rate can depend on the dose of a toxic chemical.

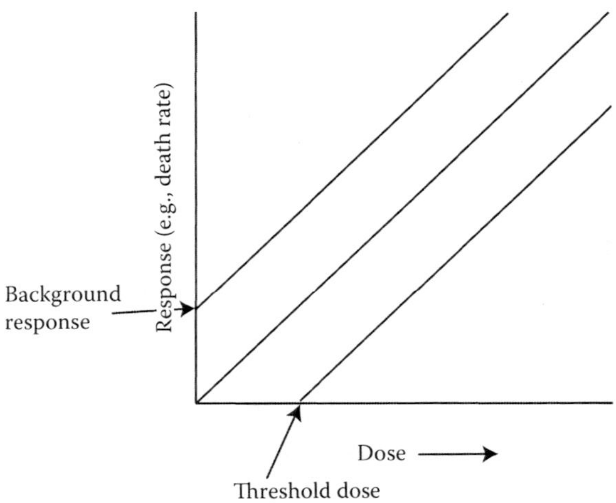

FIGURE 14.2 Different responses at low dose of a toxic chemical.

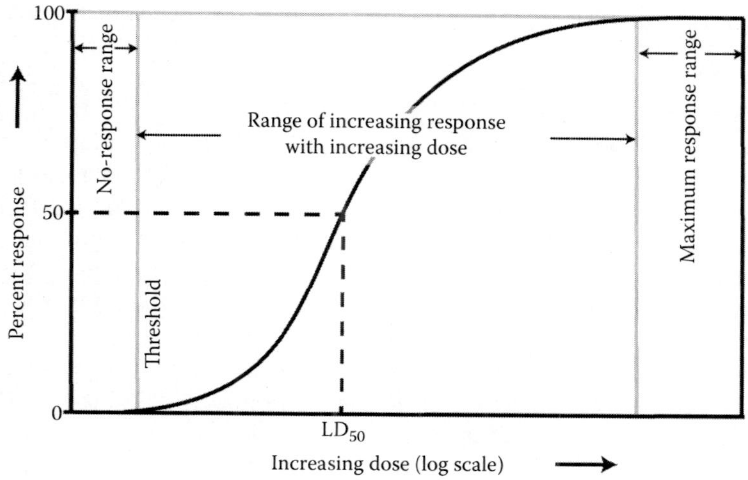

FIGURE 14.3 A nonlinear dose–response curve. If the response is death, then the term LD_{50} is used to refer to the dose at which 50% of the population die.

effect. Conversely, the probability that an environmental carcinogen caused an individual's cancer increases with the dose. Sometimes high doses can result from natural causes. For example, people may be poisoned by drinking well water with unsafe levels of heavy metals that were leached from the aquifer, or mineral dusts can contain asbestos minerals that cause lung cancer. More commonly high doses result from anthropogenic contamination.

Compounds with high toxicity cause high response rates at low doses (Figure 14.1) and have low threshold doses (Figure 14.2). They also have low lethal dose values (Figure 14.3), which are usually reported as LD_{50}, the concentration that causes 50% mortality (Manahan 2013). As the concentration of a toxin increases, the hazard that it poses to an organism increases. The toxicity of metals correlates well with their average concentration in streams (Langmuir 1997), suggesting that over long times humans have adapted to their natural environment. The process of evolution takes millions of years, but we are changing concentrations of pollutants in streams at a rapid rate, and we are introducing new chemicals to which humans have not adapted.

The dose or total exposure to an environmental contaminant depends on the concentration of the chemical in the environment and the duration of exposure. Both factors can depend on the stability of the compound as measured by the compound's half-life, the time it takes for half the molecules to break down in the environment. Unstable compounds will have short half-lives, but stable compounds with long half-lives are said to be environmentally persistent. Compounds like DDT that are persistent and highly toxic are the most dangerous (Eby 2004).

Keep in mind a few important points. First, toxicity is often determined from animal studies, and chemicals that harm animals do not necessarily harm humans. Second, association is not (necessarily) causation. Many potential causes of cancer exist, and many carcinogens occur naturally. As a result, some individuals will contract a specific type of cancer even if not exposed to the carcinogen in question (the "background response"). **Epidemiology** studies try to establish cause and effect by comparing response in an exposed group to the background response in a large, unexposed (control) group. For example, certain forms of asbestos are known to cause lung cancer. If a group that works in an asbestos mine has a lung cancer rate that is significantly higher than in the general population, then concluding that exposure of the workers to asbestos is responsible for some of their lung cancer cases is reasonable. However, lung cancer in some of those workers may not have resulted from their occupational exposure to asbestos. With these caveats in mind, managers should make every effort to protect their workers from occupational exposure risks, and individuals should protect their health by avoiding toxins. Below we examine some toxins and give recommendations for reducing exposure.

14.2.2.2 Trace Elements and Heavy Metals

Heavy metals are **transition elements** that are usually toxic. They are referred to as **trace elements** because they occur at low concentrations in the environment. Examples include **lead**, **mercury**, **arsenic**, **cadmium**, and **selenium**. Heavy metals are used in many products, and industrial processes and improper disposal have led to heavy metal pollution of the environment. The most common heavy metal exposure pathway is ingestion of drinking water, although other exposure pathways such as inhalation of atmospheric particles can present significant risk. Sustainability requires that we reduce risk by safely managing toxic metals and preventing them from contaminating our supporting ecosystems. Here we briefly examine health effects, identify products that often contain these metals, and discuss how to safely dispose of or recycle them.

Many trace elements are essential inorganic nutrients and are referred to as **micronutrients** (Manahan 2009). Examples of elements (sometimes incorrectly called minerals) that are essential for humans and the form in which they are usually supplied include **selenium** from yeast, **copper** as oxide or gluconate, **cobalt** as cyanocobalamin (vitamin B-12), and **iodine** as potassium iodide. Other elements are beneficial, such as fluorine, which toughens mammalian bones and tooth enamel by replacing **hydroxyapatite** with the less soluble **fluorapatite**. Some elements such as strontium and barium are nonessential and not highly toxic. The elements we should reduce exposure to are those that are nonessential and toxic. For example, overexposure to the heavy metal **cadmium** can cause **Itai-Itai disease**, which causes painful softening of the bones and kidney failure (Manahan 2013).

Micronutrients must be ingested at the appropriate concentrations: too little is debilitating or fatal, the right amount is essential for good health and successful reproduction, and too much is detrimental, toxic, and ultimately fatal (Manahan 2009). The concentrations of trace elements are much more highly variable in the natural environment than those of major elements, resulting in the "too little" or "too much" syndrome. "Too little" often results from natural processes, whereas "too much" more commonly results from anthropogenic inputs (pollution).

In the past when people obtained all of their food locally, local geology primarily determined the exposure to trace elements. Concentrations of trace elements (minerals and vitamins) in foods depend on concentrations in soils, which in turn depend on concentrations in bedrock (Faure 1998). Well-balanced trace element abundances are found in rocks with compositions close to average continental crust. Muds and mud-derived rocks are clay-rich, and because trace elements adsorb onto (bind to the surface of) clay minerals, they are also trace element-rich. For example, crops grown in soils derived from shales usually provide abundant essential nutrients.

The most important factor controlling plant and animal health effects associated with environmental deficiencies and excesses of trace elements is **bioavailability**. When an element has a high solubility in soil water, it has high chemical mobility and therefore potentially high bioavailability. Algae and plants concentrate bioavailable elements, ensuring that they move up the food chain. If trace elements are otherwise available (not bound in insoluble minerals), soil moisture pH usually determines bioavailability. Elements usually dissolve in water as ions; most metals are positively charged ions called **cations**, while negatively charged species are **anions**. Cations of metals such as copper, cobalt, zinc, and nickel have enhanced bioavailability under acidic (low **pH**) soil conditions and low bioavailability in alkaline (high pH) soils. For example, corn may show damaging effects from excessive metal uptake when grown in acidic soil, but is healthy when grown in naturally alkaline or limed soils. **Liming** the soil increases the pH and makes most heavy metals less bioavailable (Manahan 2009).

The "too little, too much" problem is well illustrated by the element selenium. Too much selenium sickens and kills grazing animals, and may cause locally high rates of birth deformities. Mobilization and hazardous exposures of the heavy metal selenium can result from irrigation in arid regions. For example, when selenium-rich drainage water from irrigated fields in the San Joaquin Valley, California, drained into the **Kesterson National Wildlife Refuge**, it caused high rates of birth deformities in the waterfowl (Keller 2011). In contrast, too little selenium causes **Keshan disease**, which can cause sudden death.* It is named after a city in northern China having low selenium concentrations in local soils and foodstuffs. Keshan disease can be cured by administering selenium supplements. A healthy diet is usually sufficient to eliminate vitamin and mineral deficiencies, but multivitamin supplements can eliminate the small risk.

Environmental lead (Pb) causes high rates of miscarriage, infant mortality, and diminished mental capacity (Manahan 2013). Lead was added to gasoline as an anti-knocking agent until the early 1970s, which caused widespread lead contamination of air and soil (Faure 1998). Geochemist Claire Patterson recognized this problem in the 1970s, and had to fight against the chemical industry for several decades before convincing the government to phase out leaded gas.† Plumbers widely used lead pipes and lead-based solder to connect copper pipes, which led to unsafe levels of lead in drinking water (Section 12.3.2). Lead was also used as a pigment in house paint until 1978, which caused lead poisoning in many young children. Information from the EPA about drinking water quality and concentrations of toxins like lead in municipal water supplies is available at www.epa.gov/ccr. Information for private wells is at www.epa.gov/privatewells.‡

Mercury is another very dangerous heavy metal (Manahan 2013). It is highly toxic and causes severe nervous system and brain impairment. Mercury is the only metal that occurs in liquid form at ambient conditions. It has a high vapor pressure, causing it to evaporate, which introduces another pathway to exposure (from burning of fossil fuels, especially coal). Mercury also efficiently adsorbs to sediments; bacterial action converts it into soluble methylated mercury compounds, which become concentrated in bottom feeders and then bioaccumulates in the food chain, reaching unsafe levels in tuna, trout, and waterfowl. Both lead and mercury can inflict serious damage to fetuses and young children even at very low concentration. Mercury levels in the ocean's surface waters have risen threefold since the Industrial Revolution due to anthropogenic activities such as burning fossil fuels and mining (Lamborg et al. 2014), which has increased human exposure through seafood consumption. To reduce your exposure to mercury, avoid eating shark, swordfish, king mackerel, or tilefish.§

Arsenic (As) is a metalloid that is toxic, being a known carcinogen, and can cause **arsenic poisoning** when present in drinking water (Mitchell 2014b). The use of arsenic in pesticides and in pressure-treated wood as a preservative sometimes exposed people to unsafe levels. Fortunately, the government recently banned the use of arsenic as a wood preservative. In developing countries, the main cause of arsenic poisoning is ingestion of contaminated drinking water from groundwater wells (Charlet and Polya 2006). For example, since around 1980 most people in populated Bangladesh have obtained their drinking water from wells because nearly all surface water bodies are contaminated with human or animal waste (Zhu and Schwartz 2011). By the 1990s **arsenicosis**, which causes skin lesions and can be fatal, became widespread. By 2002, 30 million people were exposed to arsenic concentrations in their drinking water higher than the maximum contaminant level (Frumkin 2005). Now roughly 21 million people in Bangladesh and eastern India are affected by arsenic contamination of their water supply (Zhu and Schwartz 2011). Anthropogenic processes such as leakage of sewage into the shallow aquifers may mobilize arsenic in aquifer sediments, contaminating the groundwater (Ayers et al. 2016). Now communities are drilling deeper wells in hopes of finding uncontaminated drinking water. Rice paddies are often irrigated with water contaminated with arsenic that can end up in the rice. The animal drug **Roxarsone** that is used in

* http://pubs.acs.org/cen/80th/print/selenium.html, retrieved August 5, 2016.

† See the excellent video "The Clean Room," episode 7 of the 2014 TV series *Cosmos*.

‡ For more information, see http://www.cnn.com/2016/01/21/health/lead-testing-home-drinking-water/, retrieved August 11, 2016.

§ http://www.webmd.com/food-recipes/tc/avoiding-mercury-in-fish-topic-overview#1, retrieved August 11, 2016.

chicken feed also contains arsenic, and studies have shown that chickens that ingest Roxarsone have higher arsenic contents in their livers,* although at levels too low to be of concern.†

Being aware of what products contain toxic heavy metals can help you avoid exposure to them and reduce your health risk. It can also alert you to the need for safe disposal. In the United States, heavy metal-bearing electronics and hazardous waste is usually collected at municipal drop-off sites.

14.2.3 Air Pollution

Human activity cannot significantly affect the concentrations of the atmosphere's main components, nitrogen (N_2, 78%) and oxygen (O_2, 21%). For example, complete combustion of all fossil fuel reserves would decrease the oxygen concentration in the troposphere by less than 2% (Smil 2002). However, humans have greatly increased the concentrations of some trace compounds in the atmosphere, sometimes with devastating effects. For example, atmospheric pollution of CFCs led to stratospheric ozone depletion (Section 8.3.4), and greenhouse gas pollution is causing global climate change (Chapter 7). The Tragedy of the Commons has led to unsafe levels of air pollution, especially in cities. The atmosphere is a public good, but most governments do not adequately protect it from polluters, who externalize their costs by using the atmosphere as a pollutant sink.

Volatile organic compounds, from petroleum products including oil-based paints and incomplete combustion of fossil fuels, are often carcinogenic and contribute to indoor air pollution, which can cause **sick building syndrome** (Manahan 2013). Sulfur dioxide (SO_2) and nitrous oxides (NOx, including NO and NO_2) from fossil fuel combustion contribute to acid rain. Coal ash released from smokestacks introduces cadmium and mercury to the atmosphere (Keller 2011). Other harmful outdoor air pollutants include asbestos and lead (Keller 2011). Ground level ozone (O_3) forms through reactions involving nitrous oxides, volatile organic compounds, and carbon monoxide in the presence of sunlight and is the major component of **photochemical smog** (Smil 2002), along with PM and volatile organic compounds (Grobéty et al. 2010). Atmospheric particulate carbon or black carbon contributes to global warming, and by sorbing other pollutants and catalyzing pollutant-forming reactions it exacerbates other pollution problems (Manahan 2013). Fly ash from minerals in fossil fuels, especially coal, often escapes from power plant smokestacks and causes damage to human health and ecosystems and reduced visibility (Manahan 2013). Burning of coal is the greatest contributor to outdoor air pollution, as it releases PM, toxic volatile elements such as mercury, and greenhouse gases.

Radon, a radioactive gas formed by natural decay of uranium, is a common indoor air pollutant because it often seeps into building foundations and accumulates in basements (Eby 2004). Inhaled radon can increase the risk of lung cancer, especially for smokers (Keller 2011). If a radon home test kit shows an unsafe level of radon in a basement, improving home ventilation can usually fix the problem.‡

Chemically reactive pollutants have relatively short atmospheric residence times, so they only contribute to local pollution. Global air pollution is caused by air pollutants with longer residence times that become thoroughly mixed in the atmosphere. Most greenhouse gases fall in this category. As mentioned previously, the greenhouse gas carbon dioxide is released during fossil fuel combustion and deforestation, which have increased atmospheric carbon dioxide concentrations more than 30% over preindustrial levels. Other greenhouse gases are present in smaller amounts in the atmosphere, but they absorb more infrared radiation per molecule than carbon dioxide, and their atmospheric concentrations are increasing more rapidly (Smil 2002). Anthropogenic methane

---

* http://www.fda.gov/AnimalVeterinary/SafetyHealth/ProductSafetyInformation/ucm257540.htm, retrieved August 11, 2016.

† https://health.clevelandclinic.org/2015/02/how-you-can-avoid-low-level-arsenic-in-rice-and-chicken/, retrieved August 11, 2016.

‡ https://www.epa.gov/radon, retrieved August 8, 2016.

(CH_4) is released from landfills, flooded agricultural fields like rice paddies, livestock flatulence, and directly from coal mines and natural gas wells and pipelines, causing methane concentrations to double during the industrial era (Wuebbles and Hayhoe 2000). The greenhouse gas **nitrous oxide** (N_2O, known as "laughing gas") is released from nitrogen fertilizers, biomass and fossil fuel combustion, and industrial processes. In the atmosphere it can break down to form nitric acid (HNO_3), a component of acid rain, and nitric oxide (NO), which can catalyze the destruction of ozone in the stratosphere (Manahan 2013).

Local air pollution is a serious problem in densely populated areas, especially large cities like Los Angeles, California with high population density, traffic congestion, heavy industrialization, and poor air circulation. Indirect effects of air pollution, such as loss of ecosystem services, cause increased mortality, decreased worker productivity, and large economic losses. Air pollution is estimated to cause over 130,000 deaths in the United States annually; considering that the value of one statistical life in the United States is usually estimated at $5 million, deaths due to air pollution cost the United States about $650 billion per year (Diamond 2005).

14.2.3.1 Acid Rain

Acid rain affects humans indirectly through damage to supporting ecosystems. It is created when sulfur and nitrous oxides emitted during burning of fossil fuels react with water to create strong acids in the atmosphere (Langmuir 1997). Sulfuric and nitric acids typically precipitate within tens to hundreds of kilometers downwind of the source. Acidic salts that contribute to **acid deposition** can be transported farther and can harm the lungs when inhaled (Manahan 2013). Acids concentrate in the rain that falls first, causing the pH to be as low as 2.4,* which is harmful to vegetation, aquatic life, and human structures (Keller 2011). Coal-fired power plants have high smokestacks so they release sulfur oxides at high elevations, dispersing them so that acid rain precipitates far downwind.

Acid rain is most serious in areas that are downwind of coal-fired power plants and that don't have alkaline bedrock such as **limestone** to neutralize the acidity; this includes large portions of the northeastern United States and eastern Canada.† Acid rain seriously damages structures made of limestone and **marble**, which dissolve in the acid or react to form **gypsum**, which easily flakes off. Valuable statues can be protected from acid deposition by coating them with waxes that are impermeable and acid-resistant. However, the damage to ecosystems is harder to avoid. Acid rain can leach nutrients from soils and harm crop plants (Manahan 2013). Fortunately, acid rain has abated due to decreased coal use, the use of low-sulfur coal, and the use of **scrubber** systems in coal-fired power plants: between 1970 and 2008 rainfall acidity decreased by 77% at one representative site in the White Mountains of New Hampshire, but it is still too acidic in most of the northeastern United States. ‡ Acid rain will not completely disappear until we stop burning fossil fuels, especially coal.

14.2.3.2 Case Study: Ducktown, Tennessee

Acid rain caused by copper ore smelting caused much damage to ecosystems surrounding Ducktown, Tennessee. Miners discovered native copper in Ducktown in 1843, and mined metal sulfides there from 1854 to the 1960s (Keller 2005; Kaufman and Noller 1999). They smelted the metal sulfides in ovens to separate the copper. Sulfur oxides emitted from the ovens combined with water in the air to form sulfuric acid, which fell as acid rain, killing local vegetation and leaching nutrients and trace metals from the soil. In local streams, concentrations of metals such as aluminum, copper, zinc, and arsenic were high enough to exceed toxicity guidelines and harm stream biota (Hammarstrom et al. 2003). These metals mostly originated from weathering of mine tailings.

* Note that the pH of unpolluted rainwater is 5.7 because atmospheric CO_2 dissolves to form the weak acid H_2CO_3, carbonic acid (Langmuir 1997).

† http://www.epa.gov/acidrain/effects/surface_water.html.

‡ http://www.newsweek.com/photo/2010/04/21/earth-day-progress.html.

Miners used trees from the local hardwood forests to fuel the smelting ovens. Deforestation led to soil erosion, ultimately leaving a thin layer of hard, red, infertile soil covering the rocks. Thus, the area surrounding Ducktown looked like the surface of Mars for many decades; U.S. astronauts in space said it was one of the most recognizable features on Earth's surface (Keller 2005). Since the 1930s the government has been trying to revegetate the Ducktown area to reduce erosion and the toxic heavy metal load in local streams (Kaufman and Noller 1999). However, the soil is so acidic and infertile that almost nothing will grow in it except a few hardy pine species.

Fortunately, in 1903 the mining companies figured out how they could reduce the environmental damage caused by smelting and make more money (Keller 2005). They simply collected the sulfur oxides released during smelting, added water to make sulfuric acid, and then sold the acid for more money than they made from selling the copper. Businesses should always look for "win-win" situations like this that minimize environmental damage and maximize profits.

14.2.3.3 Particulate Matter

Atmospheric particulate matter consists of fine particles that can remain airborne for long periods of time. Natural sources include volcanic eruptions, dust storms, forest and peat fires, salts from sea spray, and biogenic particles such as pollen and fungi (Gieré and Querol 2010). Anthropogenic sources of PM include fossil fuel and biomass combustion, surface mining, ore smelting, waste incineration, traffic, construction and demolition sites, and many industrial operations (Keller 2011). Particulate matter pollution, especially of mineral dusts, can be intensified by desertification, and can become severe during volcanic eruptions and large wildfires. Fine particles with diameters less than 2.5 microns ($PM_{2.5}$) are linked to increased mortality from cardiovascular and respiratory diseases due to inhalation. Coarser particles may also cause adverse health effects. Chronic exposure through inhalation of mineral dusts can cause **silicosis** (**desert lung syndrome**) from fine silica dust and **asbestosis** from asbestos particles (Fubini and Fenoglio 2007). Asbestos is also believed to cause lung cancer and **mesothelioma** (Manahan 2013). The EPA has banned the use of asbestos in most products.*

14.2.4 Water Pollution

Filthy water cannot be washed.

West African Proverb

Many examples illustrate the pervasive environmental and health impacts of anthropogenic water pollution. The collapse of a bauxite tailings impoundment in Hungary in 2010 allowed 200 million gallons of sludge with a pH of 13 to flow into the Danube River and its tributaries, killing most wildlife and 9 people and injuring 122 people (Zhu and Schwartz 2011). Rivers draining areas containing mountaintop removal coal mines have elevated concentrations of selenium, sulfate, magnesium, and other inorganic solutes (Lindberg et al. 2011). Nutrient pollution affects many surface water bodies in developed countries, where it causes eutrophication and the development of dead zones in the Gulf of Mexico and other coastal waters (Howarth et al. 2000). Highly toxic **cyanide** used in **heap leaching** to extract gold from ore often ends up in streams where it causes fish kills (Manahan 2013).

The limits to human population growth may lie not in resource depletion, but in the waste absorption capacity of the environment (Daly and Farley 2011). We can understand this using the following analogy. Water purification filters usually contain a resin that turns color when it becomes saturated, that is, it cannot absorb any more pollutants. The interface between the two colors of resins (the reaction front) will migrate through the column from the inlet toward the outlet. The filter will purify water until the interface reaches the end of the column, at which point the column resin is

* https://www.epa.gov/asbestos/us-federal-bans-asbestos, retrieved August 11, 2016.

saturated in pollutants and cannot absorb any more. Pollution exceeded the waste absorption capacity of the filter, and from that time on the outlet water will be just as polluted as the inlet water. Our environment acts as a filter, purifying water that passes through it, but eventually the environmental filter becomes saturated.

Let's examine this in a little more detail. What happens when the concentration of a pollutant in a sediment-water system (lake or stream) keeps increasing? Imagine that we pour uranium (U) into a beaker containing water and sediment. In this closed system some U will dissolve in the solution, but some adsorbs onto the surface of mineral grains in the sediment (Figure 14.4). At first, the proportions of U in solution and adsorbed to sediment will be relatively constant as the total U concentration increases (move away from the origin along the curve labeled "Uranium adsorption isotherm"). As the uranium concentration increases, the number of available sites for U to sorb onto mineral surfaces begins to decrease, and a greater proportion of U enters the fluid. This causes the adsorption isotherm to level off and approach a slope of zero when the adsorption sites become "saturated." Eventually even the solution becomes saturated, that is, it can't dissolve any more U. What happens then? Any additional U added to the system will precipitate as the U-rich mineral **schoepite** ($(UO_2)_8O_2(OH)_{12}{\bullet}12(H_2O)$), which is added to the sediment. This causes the sediment concentration of uranium to begin increasing again. Note that if the solution remains saturated in schoepite, any additional U we add will go into the sediment, increasing the U concentration in the sediment. Conversely, no matter how much additional U we add, the concentration of U in the solution is fixed at its highest possible concentration.

Now consider a more complicated open-system model. Polluted water enters a beaker with sediment, equilibrates with the sediment, and then exits the beaker, and the process repeats. At first, most of the pollutant will sorb onto the sediment, causing the concentration in the solution to

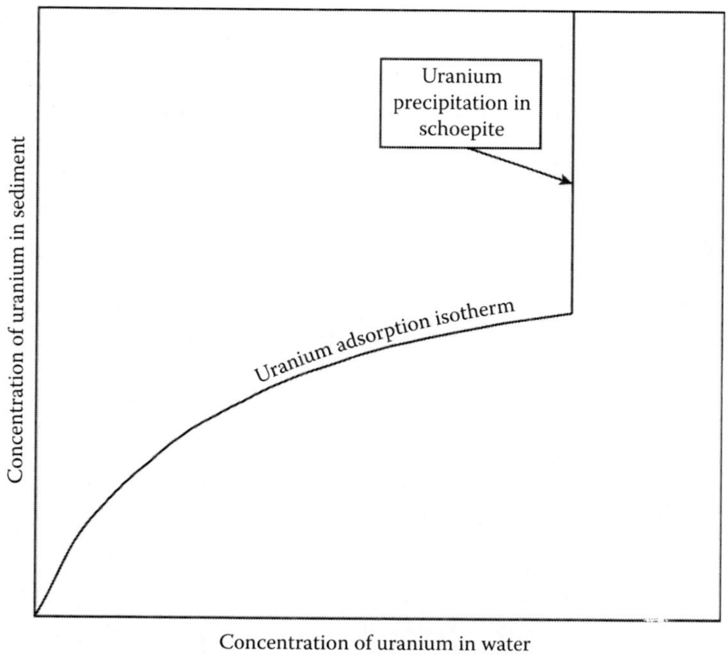

FIGURE 14.4 As the concentration of uranium dissolved in water increases (x-axis), the concentration of uranium adsorbed on the surface of solids increases (y-axis). The fluid, at pH = 7.23, eventually becomes saturated in the mineral schoepite. (Reprinted from *Geochimica et Cosmochimica Acta*, Vol. 42(6), Donald Langmuir, Uranium solution-mineral equilibria at low temperatures with applications to sedimentary ore deposits, 23 pp., Copyright 1978, with permission from Elsevier.)

decrease substantially. As more batches of polluted water equilibrate with the sediment, the concentration of pollutant in the sediment will increase, and therefore the concentration of pollutant in the water that exits the beaker will increase in direct proportion. As the sediment approaches "saturation," it can sorb less pollutant, so most of the pollutant remains in solution, and our sediment filter becomes increasingly ineffective.

Can the system recover if we stop polluting? It can with the gradual replacement of contaminated water with fresh water. Water that exits the beaker would at first have high concentrations of pollutant because the sediment filter was saturated in pollutants. Over time, the concentration of pollutant in the sediment and in the exiting fluid would decrease and eventually go to zero. Thus, we can "flush" pollutants out of a sediment-water system such as a stream or lake. However, it may take a long time and much fresh water to remove all of the pollutant, especially if the pollutant strongly sorbs to the sediment (which is why PCBs are still in Hudson River sediments after many decades [Keller 2011]).

Now imagine a stock such as a wetland with one stream entering and one stream exiting. If the stream entering the wetland is polluted, sediments near its entrance point will strip pollutants out of solution. With time, a concentration gradient will develop across the wetland, with high pollutant levels near the input stream and low levels near the output stream (Figure 14.5). As polluted water flows across the wetland, it encounters sediments with decreasing pollutant concentrations, so the concentration of the pollutant in the solution will continuously decrease. The water becomes increasingly pure as it traverses the wetland. In nature, wetlands do an excellent job of filtering pollutants from water. However, if pollutants continue to enter the wetland, the total pollutant concentration in the wetland will keep increasing. Eventually sediments near the input stream will become saturated, and that "saturation front" will slowly migrate across the wetland until it reaches the output stream. At that point, the entire wetland system has become saturated, and the output water will be just as polluted as the input water. As in our beaker example, if we stop polluting and the water in the input stream becomes pure again, then over time the process will be reversed. The pollutants will slowly be flushed out of the wetland.

Lake Erie in the northeastern United States is one of many examples of pollution and subsequent recovery of a lake. Until 1972 rivers that drained into Lake Erie were dumps for industrial, agricultural, and municipal waste. Eutrophication was a serious problem at that time (see Section 12.3.2). Pollution was so bad that the Cuyahoga River caught fire multiple times between 1868 and 1969 (Jordan et al. 2010). The fire in 1969 was a social tipping point that in part led to the establishment of the U.S. EPA in 1970 and the Federal Water Pollution Control Amendments of 1972. New policies on both the U.S. and Canadian sides of Lake Erie led to decreased pollutant runoff, declining nutrient runoff, increasing oxygen levels and water clarity, and rising fish productivity. Lake Erie is no longer a dead lake; it has shifted into a new regime and is once again

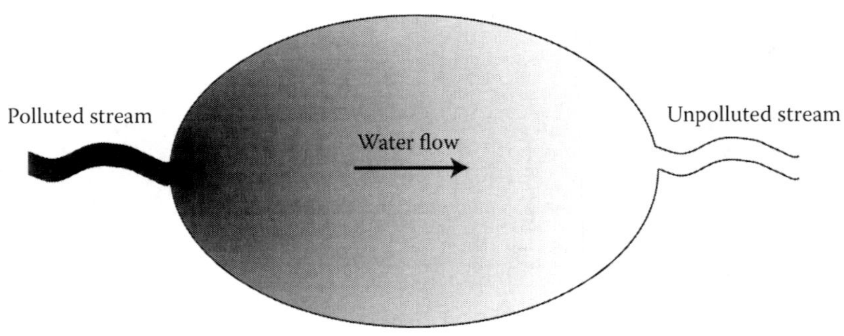

FIGURE 14.5 Filtration of water pollutants in a wetland. Polluted stream enters the wetland on the left. Grayscale is proportional to pollutant concentration, with black representing the highest concentration. Pollutants are removed through adsorption onto wetland sediments. Unpolluted water exits the wetland on the right.

popular for recreation, although there has been a resurgence in algae blooms in western Lake Erie in recent years.*

14.2.5 Case Study: Deep Well Injection of Chemical Wastes

Until the 1990s the Chemours (previously DuPont) Plant in New Johnsonville, Tennessee, manufactured titanium dioxide (TiO_2) by mining the mineral ilmenite ($FeTiO_3$) and reacting it with hydrochloric acid (HCl) as follows: $FeTiO_3 + 2HCl = FeCl_2 + TiO_2 + H_2O$. The titanium dioxide is a pigment that gives Kilz paint, Oreos, and many types of toothpaste their brilliant white color and it is used in many coatings and plastics. The waste solution is very acidic, and contains high concentrations of heavy metals. In the 1960s when people didn't know better, they allowed DuPont to dispose of hundreds of thousands of gallons of this toxic acid solution directly into the Tennessee River, which killed all fish and bottom feeders downstream. Later DuPont switched to the more environmentally friendly but more expensive process of deep-well injection. They drilled wells between 1,000 and 2,000 feet deep and then pumped the acidic waste into a confined, deep limestone layer. The thinking was that the limestone would neutralize the acid. The confining (impermeable) layer above would keep the waste isolated from shallow aquifers that supplied drinking water.

Deep well injection of acidic waste into limestone has two problems. First, the acidic solution dissolves the limestone, which results in the formation of large caves deep underground. Eventually the weight of the overlying rock layers causes them to collapse, break into pieces, fall, and fill the caves, resulting in formation of **sinkholes** and **land subsidence**. This shatters the confining layer and makes it permeable, so that the wastes can rise into the aquifers. The other problem is that limestone dissolution produces CO_2 gas, and the pressure of that gas can build until it shatters the overlying rock and escapes. Either way, it seemed likely that dissolution would eventually compromise the confining layer. So, to their credit, DuPont came up with a new solution that was more environmentally friendly but more expensive. DuPont now reacts ilmenite with sodium carbonate to produce TiO_2, and according to the DuPont engineers the only by-product is harmless $FeCO_3$ (the mineral siderite), which they use to make bricks for construction. However, we now know that this process produces **dioxin** as a by-product.† Pure dioxin, which is environmentally persistent and bioaccumulates, is highly toxic to some animals (Manahan 2013). People exposed to high levels of dioxin, which was a contaminant in the defoliant **Agent Orange** used in the Vietnam War, have increased risk of skin lesions, liver disease, and possibly cancer (National Research Council 2006). The entire town of **Times Beach, Missouri**, was abandoned in 1983 due to dioxin contamination and is now a **ghost town** (Manahan 2013). In 2002, the New Johnsonville Plant was ranked in the worst 10% of facilities in the United States for total environmental releases and cancer risk score,‡ although the annual amount of dioxin released has since decreased significantly.§

This case study illustrates many different points. First, anticipating all of the potential outcomes of a complex industrial process is difficult. That is why ecologists advocate the precautionary principle. Second, industrial chemistry sorely needs to be "greened." **Green chemistry** is a new field that has the potential to greatly reduce the environmental impact of the chemical industry by using safe and nonpolluting methods, minimizing material and energy consumption, and producing little or no waste (Manahan 2013). Third, despite repeated attempts at trying to "green" the chemical process, the production of titanium dioxide still causes serious environmental problems.

---

* http://msue.anr.msu.edu/news/lake_erie_harmful_algal_bloom_early_season_projection_report_may_17_2016, retrieved September 16, 2016.

† Dioxins are actually a group of compounds. TCDD is the most toxic of these, and is simply called dioxin (National Research Council 2006).

‡ http://scorecard.goodguide.com/env-releases/facility.tcl?tri_id=37134DPNTJ1DUPO#major_chemical_releases, retrieved August 8, 2016.

§ http://www.reliableplant.com/Read/5092/dupont's-titanium-technologies-cuts-dioxin-by-63, retrieved August 8, 2016.

This example also raises many questions: Should we allow chemical companies to manufacture goods like titanium dioxide that are nonessential (it is simply used for aesthetic reasons) but that cause harm to human health and the environment? Or should we close the plants, even if it means that thousands of people would lose their jobs? The plants in New Johnsonville and DeLisle are by far the largest local employers, so closing them would be an economic disaster for those communities. Like many issues, the benefits of manufacturing have to be weighed against the costs, and there are always tradeoffs.

14.3 SOLUTIONS

14.3.1 Waste Management

As discussed in Section 3.6.2, we can increase efficiency and decrease waste by closing resource loops (Figure 3.10). The amount of waste our society produces is a good measure of the amount of resources we consume because most resource loops are open-ended (linear). Resource depletion can be slowed by adopting conservation measures, and stopped by closing resource loops. Section 12.3.2 covered liquid wastes, and Section 13.3.2.1 covered organic (food) waste; in this section we cover inorganic solid waste.

In the past, **waste management** and **pollution control** followed the dictum, "The solution to pollution is dilution," that is, bury solid waste in the ground, pour liquid wastes into large bodies of water, and release gaseous wastes to the atmosphere. Solid wastes were often buried in underground drums at sites like **Love Canal** and the **Valley of the Drums** under the mistaken premise that nothing happens underground: "Out of sight, out of mind." We now know that the soil zone and underlying sediments and permeable rocks are a dynamic system through which water flows and reacts with buried wastes. Many buried drum sites are now EPA Superfund sites (Keller 2011). Billions of dollars have been spent to dig up and treat the wastes. Usually, using smart technologies to reduce the volume and toxicity of pollutants at the source is more cost-effective and environmentally friendly than cleaning up afterward (Manahan 2013).

Most communities dispose of solid waste in a **landfill** by dumping it on the ground and covering it each night with a fresh layer of soil (Keller 2011). Rainwater can infiltrate the waste and dissolve material to form what is called a **leachate** solution.* Modern landfills have liners and leachate collection systems that prevent the leachate from contaminating the underlying groundwater. However, unregulated and preregulation landfills have extensively contaminated aquifers. Most people don't want a landfill sited near their home, an example of the NIMBY syndrome. Also, because expansion of urban centers and overall population growth make it difficult to find suitable locations for landfills, the number of open landfills in the United States has been decreasing, although average size is increasing (EPA 2015). Some cities are finding it hard to find places that can accept their waste. The "poster child" of the landfill shortage problem was the garbage barge that in 1987 could not find a facility to accept its waste. After traveling more than 5,000 miles over 112 days, it finally unloaded its waste at an incinerator in Brooklyn, New York.†

The **waste hierarchy** is the order of effectiveness of waste and pollution reduction strategies, which from highest to lowest are prevent, reduce, reuse, recycle, energy recovery, and disposal. The most effective and economical approach, **pollution prevention** (P^2), seeks to increase the efficiency of processes such as industrial production and consumer consumption to reduce pollution (see http://www.epa.gov/p2/). It includes the use of green chemistry to prevent pollution by substituting nonhazardous for hazardous chemicals in industrial processes. **Cradle-to-cradle design** is used to create resource efficient products with minimal waste production over the entire life cycle (McDonough and Braungart 2002). Pollution prevention can also use methods from **industrial**

* https://www.epa.gov/landfills/municipal-solid-waste-landfills, retrieved August 5, 2016.

† http://www.nytimes.com/1987/07/11/nyregion/trash-barge-to-end-trip-in-brooklyn.html.

ecology, which studies material and energy flows in industrial systems and finds holistic methods for increasing material and energy use efficiency.

After pollution prevention, the mantra for further waste reduction is the three Rs: Reduce, Reuse, and Recycle, in order of decreasing importance. It's best to reduce by consuming less. Your effort to reduce waste must start at the beginning, when you are in the purchasing phase (Jeffery, Barclay, and Grosvenor 2008). Consume less, and when you do have to buy, choose the option that uses less or recyclable packaging (Hill and O'Neill 2008). For example, buy concentrated laundry detergent that uses less packaging and less carbon-based energy to transport. Download an album instead of buying the CD and case. Replace older items only when necessary, and switch from the old model of product ownership to the more efficient approach of sharing enabled by online tools and smartphone apps. Avoid using disposable products. Many countries and municipalities have started banning disposable plastic shopping bags, which present a risk to wildlife and can be replaced easily by reusable bags.*

The opportunities to reuse and repurpose materials are nearly endless. For example, Jeffery et al. (2008) suggest that we reuse plastic bags after washing, use empty glass jars as storage containers, shop at second-hand clothing and bookstores, reuse wrapping paper, and borrow books from the library to save both money and paper. Many charities accept used goods for reuse. **Goodwill** accepts many types of items including clothing and resells them in their stores. **Amvets** accepts furniture, electronics, and even used cars—you can claim a tax deduction for donating an old car instead of taking it to the dump. Reusing and repurposing items is also an opportunity to think "outside the box" and be creative. For example, plastic shopping bags can be used to line small garbage cans.

Recycling is an attempt to close the resource loop by recovering material from the waste stream and converting it back into a usable form. While it is the least effective of the 3 Rs at reducing pollution and consumption of energy and raw materials, for some materials such as aluminum it can be highly effective. Because most waste is produced in the raw materials and production stages of the material lifecycle, primary materials produced from raw materials have larger footprints than secondary (recycled) materials. For example, recycled copper requires 50 times less material than primary copper (Schmidt-Bleek 2007). However, because most waste is produced in the $R + P$ stages, and only materials that reach the use stage can be recycled, recycling cannot greatly reduce waste. Moreover, recycling rarely comes close to completely closing material loops. Even when the return rate is 100%, material is lost during reprocessing. If 75% of material is recovered after one cycle through the resource loop, less than 1% would remain after 15 loops, and the only material that comes close to this level of efficiency is aluminum, which is recycled at a rate of 70% (Schmidt-Bleek 2007).

Many products such as fossil fuels cannot be reused or recycled because they are completely consumed or dispersed during use, and other products degrade in quality when we recycle them (Hill and O'Neill 2008). For example, recycled paper is generally inferior to first-use paper, so different uses must be found for it. The degradation in quality of a material during repeated recycling is called **downcycling**. A preferred approach is **upcycling** or **creative reuse** that takes used material and reuses it to make new products of better quality or lower environmental impact (McDonough and Braungart 2002). Fortunately, clever people are thinking of many new uses for reused and recycled products. In addition, it is becoming easier and less expensive to recycle many items, as U.S. companies adopt the cradle-to-cradle policies started by EU countries. For example, new cell phones and printer cartridges come with postage paid plastic bags for recycling. Compared to recycling, the more sustainable solution to reducing waste is to buy and use fewer goods, and to produce goods more efficiently by using less material in all stages of the lifecycle. The other thing to keep in mind is that it only makes sense to recycle materials when a market for the recycled product exists.

* See "Waste/Recycling: The Downfall of the Plastic Bag: A Global Picture," Janet Larsen and Savina Venkova, www.earth-policy.org/plan_b_updates/2014/update123 and "Plastic Bag Bans Spreading in the United States," www.earth-policy.org/plan_b_updates/2014/update122.

However, for most materials recycling is energy efficient, that is, more energy is consumed and more greenhouse gases are emitted when making a product from new rather than recycled materials. In 2008, Americans recycled 83 million tons of material that saved 293 billion kWh, the energetic equivalent of 10.2 billion gallons of gasoline.* Recycling also reduces the amount of waste and the risk of pollution from hazardous materials. Finally, recycling is important because it is usually easy to do, and actions that do not require lifestyle changes can have relatively high adoption rates. Thus, the EPA concludes that "recycling boosts the economy, conserves natural resources, and reduces solid waste" (EPA 1998). The good news is that in the last 20 years Americans have greatly increased how much municipal solid waste they recycle, from 10% in 1985 to 34% in 2013 (EPA 2015), but we could increase it to as much as 75%.

Not everything can be recycled, especially when cost is considered. Moreover, we don't need to recycle everything to be sustainable. For example, glass is harmless to humans and other animals, and it is costly to recycle. It is made by melting beach sand containing silicate minerals like quartz (Ayers 2012). The sand melts at high temperatures, so it takes large amounts of energy to melt the glass during recycling, and because glass is inert and won't cause environmental damage when disposed of in a landfill, forgoing recycling is not a terrible choice. Glass recycling uses less energy than manufacturing glass from sand, but the best option is to reuse the glass so it doesn't have to be melted all over again.

The material that saves the most energy by recycling is aluminum. Aluminum ore contains aluminum oxides that we must convert to metal. This means the Al^{3+} in the oxide must be reduced to metallic Al^0 by adding three electrons, which requires a large amount of energy because aluminum prefers to be in the +3 state. This also means that aluminum metal will oxidize when in contact with oxygen in the atmosphere, but fortunately the process is very slow. Aluminum metal has a very low melting temperature, so it takes much less energy to recycle the aluminum by heating and melting it than it would to mine more aluminum ore and convert the oxide to the metal. Recycling aluminum is ~95% more energy efficient than making it from raw materials, meaning that you can make 20 cans from recycled aluminum using the same amount of energy it takes to make one from aluminum ore (Hill and O'Neill 2008).

The least effective waste reduction strategy is energy recovery, which for solid waste is called **waste incineration**. This approach takes waste that cannot be reused or recycled and burns it to reduce volume and recover energy as heat. First, noncombustible (inorganic) waste components are removed and landfilled. The remaining organic materials are biofuels, which have much lower carbon footprints than fossil fuels (Section 10.3). Arguments against waste incineration include that it releases small amounts of heavy metals to the atmosphere, that older incinerators can release trace amounts of dioxin, and that toxic wastes can be concentrated in the residue, which must be landfilled. Insufficient research has been done on these potential health risks. However, the overall risks to the environment and human health from these emissions seem likely to be lower than for coal, with the added benefit of reducing pollution from solid waste, which is a pollutant even if it is in a landfill because it is a resource out of place. All materials, even waste materials, should be viewed as potential resources from which we can derive benefits, but we obtain no benefits from landfilling solid waste. The EU has less available land to use for landfills than the United States due to its higher population density, so it incinerates a large portion of its waste to produce energy, with Denmark leading the way at 1.1 kg/d/p compared with < 0.3 kg/d/p in the United States (MacKay 2009).

Several alternatives to waste incineration exist for **biodegradable waste**. Composting can be used to convert organic waste into fertile soil, and in large-scale operations methane can be recovered for use as fuel. **Waste-to-energy** technologies for waste biomass include **pyrolysis**, which is thermal decomposition in the absence of oxygen. It is used to recover energy from woody materials, with the added bonus that creates **biochar**, which is used as a soil amendment for carbon

* http://www.newsweek.com/photo/2010/04/21/earth-day-progress.html.

sequestration and for increasing soil fertility. **Anaerobic digestion** uses fermentation to produce methane from waste biomass. The National Academy of Sciences recently suggested that phosphorous could be recovered from treated animal manure produced in CAFOs using anaerobic digestion, which would also produce methane for use as a fuel (Committee on Science for EPA's Future et al. 2012). **Gasification** is used to produce the renewable energy source **syngas** by reacting biomass at high temperature without combustion. All of these technologies are preferable to disposal of waste biomass in landfills. Comparison of these waste-to-energy technologies using LCA is made difficult because of inconsistent methodologies and criteria/objectives, with some studies concluding that waste incineration is preferable, while others favor anaerobic digestion (Astrup et al. 2015).

14.3.2 Safe Hazardous Waste Disposal

One of the biggest environmental problems is the improper disposal of waste. Because government now highly regulates the chemical industry, it is less common nowadays for corporations to dump waste illegally; mostly gone are the days of "midnight runs" where they secretly dumped toxic wastes into water bodies. In the 1960s the public might turn a blind eye to illegal dumping by corporations, but today the risks of fines, damaged reputation, and loss of revenue are too great. Many industrial processes are used to treat hazardous waste (Manahan 2013), but these are beyond the scope of this book. Here we focus on educating consumers on how to dispose of household hazardous waste safely.

Examples of unsafe hazardous waste disposal practices include pouring paint thinner into drainage ditches, and stale gas into storm drains. These liquids contain toxic, carcinogenic organic compounds, and all of them can end in surface water or groundwater where they can contaminate drinking water supplies or disrupt ecosystems. Because they are organic, these liquids have stored energy that we should not waste but rather put to use. Similarly, waste oil should not be dumped down the drain, but instead taken to public recycling centers, auto mechanic shops, or oil change centers, which often recycle the oil or burn it as fuel (Manahan 2013).

Some products like batteries contain toxic heavy metals. You can recycle car batteries at any automotive repair or parts store. Disposable batteries contained toxic mercury until 1996 when the United States banned it, so it's OK to place them in the trash, though you might feel better if you drop them off at your local electronics or battery store for recycling. Newer rechargeable batteries usually come with instructions for returning to the manufacturer for recycling, usually by dropping it off where you purchased it.

Most communities collect hazardous waste at central drop-off sites and then safely dispose of it. Although the homeowner may consider it a hassle to annually collect hazardous materials and drive them to the drop-off site, the benefits greatly outweigh the effort. First, having these unsafe chemicals in your home increases the health risk of your family, particularly of children, who are often more sensitive and therefore more susceptible to the negative health effects of toxins because their brains are still developing. Second, improper disposal can expose you, your family, and your community to unsafe pollutants. Improper disposal can cause tremendous economic losses from damage to ecosystem support services and to people's health. It's important to remember that waste that is improperly disposed of down drains, storm sewers, on lawns, and in garbage that ends up in landfills eventually ends up in our water and air.

Compared with safe disposal, a better approach to reducing the harmful effects of hazardous waste is not to use products containing hazardous chemicals in the first place. For example, latex paints now perform just as well as oil-based paints, but unlike oil-based paints they don't contain volatile organic compounds that are notoriously unsafe. When you use oil-based paints, you expose yourself to noxious fumes during painting and drying. However, if a market exists for unsafe products like oil-based paints, companies will continue to manufacture them, causing harm to people's health and the environment.

14.3.3 Policies to Reduce Waste and Pollution

Waste is a tax on the whole people.

Albert W. Atwood

The economic costs of pollution are staggering. For example, the U.S. EPA oversaw the cleanup of hundreds of **Superfund** sites that contain legacy toxic waste; now the **polluter pays principle** is used to cover much of the cost of around $1 billion annually (Keller 2011; Manahan 2013). Taxpayers pay for the disposal phase of the life cycle through **municipal waste** collection, instituted to protect public health. Waste management costs have continued to increase because the primary waste producers (think $R + P$ stages of the life cycle) have externalized their costs, so there is little incentive to reduce waste, and because population and per capita consumption levels continue to rise. The opportunity cost represented by tax dollars lost to waste management is money that would be better spent on public safety and education, for example (Sheehan and Spiegelman 2010).

Because pollution costs are high, there is great potential to save money through pollution prevention. For example, the U.S. **Clean Air Act** of 1970 saves the United States roughly $1 trillion per year by saving lives and reducing health care costs (Diamond 2005). We already have the technology needed for further substantial reduction of waste and pollution in the United States, which could save billions of U.S. dollars each year.* Further reductions in pollution would also improve the health and quality of life for our children and future generations.

Perhaps the best strategy for reducing many forms of air pollution is to eliminate the burning of fossil fuels, particularly coal. Policies that encourage the transition to clean sources of energy such as transferring federal subsidies from fossil fuels to renewable energy sources would save money, improve health, and mitigate climate change. All forms of waste can be reduced by using incentives or penalties, but usually the carrot works better than the stick. Policies that encourage dematerialization, especially by reducing material flows in the raw material and production stages of the life cycle, will reduce pollution risks and disposal costs. One radical approach that might be very effective is for government to shift taxes from labor to companies that consume natural resources. This type of **tax shifting** would make resource-intensive products more expensive, causing a negative feedback that would reduce consumption and waste generation, and the reduced labor costs would reduce unemployment (Schmidt-Bleek 2007).

It is beyond the scope of this book to examine all of the potential policies that could reduce waste in each sector of the economy, but we will examine a few examples. To reduce residential solid waste, municipal governments can adopt a **pay as you throw** program in which residents pay a garbage pick-up fee that is based on the amount of garbage; most communities charge a flat rate, meaning there is no financial incentive to reduce how much waste you dispose of. Note, however, that pay as you throw programs only make sense in well-to-do communities; in poor neighborhoods they only encourage residents to improperly dispose of their waste.

One way to increase recycling rates is to require a deposit on containers that can be recycled. When the empty bottles are returned the deposit is refunded. The first bottle deposit law was passed in Oregon in 1971, and it increased the recycling rates for glass and aluminum from 25% to 90% in one year. Today recycling rates for the 11 states with bottle bills range from 65% to 95%, much higher than for the states without bottle bills (Gleick 2010). The beverage and grocery store lobbies have defeated attempts to expand bottling bills to other states or to other beverage types. In most cases bottle bills were passed decades ago when bottled water was not yet widely sold, so water bottles are excluded in 5 of the 11 states. Furthermore, the deposit rates have not kept pace with inflation, so recycling rates are decreasing in states with bottle bills. Some countries have much higher recycling rates than in the United States. One approach is to increase the convenience of recycling; this has increased the recycled rate for water bottles to 80% in Switzerland. A second approach is

* http://www.ucsusa.org/publications/catalyst/member-issue/fall11-climate-change-health.html.

to institute **extended producer responsibility** programs in which producers are required to accept or collect the packaging waste from their products (Gardner 2013). A third approach is to require that products or their packaging contain a minimum amount of recycled material (Gleick 2010). Finally, reducing the volume or weight of packaging means that less waste must be recycled. Plastic water bottles have become thinner and bottle caps smaller in the United States, which has the added benefit of reducing embodied energy and transportation energy. Recycling rates will likely only increase significantly in the United States if the responsibility for reducing and recycling packaging waste is placed on producers rather than consumers.

If hazardous waste contamination of the environment occurs, it is important to balance costs and benefits when formulating policies for **environmental remediation**. The public often demands that the concentrations of harmful pollutants be reduced to zero. Unfortunately, that is not possible, and even if it was, the cost would be prohibitive due to diminishing returns (Meadows, Randers, and Meadows 2004). Because the costs of environmental remediation are so high, it is much more cost-effective to prevent pollution than to cleanup afterward.

14.4 CONCLUSIONS

To be sustainable we must maximize the utility and minimize waste of all resources. Dematerialization can reduce material intensity at all stages of the life cycle, which would also reduce energy consumption, and decrease the vulnerability to natural resource depletion and supply disruptions. When resource loops are not closed and waste is not properly disposed of, pollutants are added to the environment. Reducing the material throughputs of our economy would reduce pollution and greenhouse gas emissions, thereby saving lives and mitigating climate change. Waste management seeks first to prevent, then reduce, reuse, and recycle waste. Energy can be recovered from the remaining waste before disposal in a landfill. Policies that provide incentives for consumers and corporations to dematerialize would protect public goods like the atmosphere and common pool resources like surface water bodies and groundwater from pollution.

RESOURCES

- http://www.ted.com/talks/dianna_cohen_tough_truths_about_plastic_pollution
- ScoreCard: Environmental Pollution: http://www.scorecard.org/

HOMEWORK PROBLEMS

1. Make lists of examples of the 3 Rs: reduce, reuse, recycle.
2. Give examples of commons-based peer production that reduce waste and pollution.
3. Find a dose–response curve for an anthropogenic pollutant. Is there a linear relationship? Is there a threshold dose?
4. Use Scorecard (http://www.scorecard.org/) to find the major pollutants and polluters in your neighborhood.

15 The Biosphere

A thing is right when it tends to preserve the integrity, stability and beauty of the biotic community. It is wrong when it tends otherwise.

Aldo Leopold

15.1 INTRODUCTION

Humans have always been integral parts of ecosystems, but over time human–environment interactions have intensified, often leading to environmental degradation. The intensification of human impacts is primarily due to increases in population and consumption, which combined with a focus on short-term economic benefits rather than long-term environmental impacts has caused human overexploitation of natural resources. When competing with other species for scarce resources humans almost always win, but at a cost: the ability of our environment to provide ecosystem services and natural resources is often undermined, reducing the security and resilience and therefore the sustainability of human societies.

In many regions humans have dramatically changed the landscape, altering the structure and components of ecosystems, and changing the magnitudes of material and energy flows. The **Anthropocene** is a proposed geologic epoch that refers to the last few decades to thousands of years when humans have changed the Earth on a global scale (Zalasiewicz et al. 2010; Lewis and Maslin 2015). The advent of the Anthropocene is marked in the stratigraphic record by the appearance of manufactured and synthetic materials such as plastics, radioactive material from **nuclear fallout**, particulates from fossil fuel burning, and increased erosion and sedimentation rates (Waters et al. 2016).

Human-altered ecosystems often bear little resemblance to the natural ecosystems they replaced. New types of landscape include urban, suburban, and rural/agricultural. These unnatural landscapes are part of **novel ecosystems** or more generally socio-ecological systems (SESs). Like ecosystems, SESs are dynamic systems with components that are living (species) and nonliving (soil, water, rock, air). Species in an SES are linked to each other by a complex web of interactions, forming a network of linkages in which energy and nutrients are exchanged.

An SES contains nested elements with the economy contained within human society, which in turn is embedded in the environment (Figure 2.1). Because they are part of the environment, society and the economy benefit when ecosystems are healthy. Thus, ecosystem impacts must be considered when planning and managing natural resource use (McLellan et al. 2014). Impacts can be quantified by monitoring changes in ecosystem health measures such as the population size of species or groups of species.

The health of an ecosystem depends on the level of **biodiversity**, defined as "the variability among living organisms from all sources including terrestrial, marine and other aquatic systems and the ecological complexes of which they are part; this includes diversity within species (genetic diversity), between species and of ecosystems" (U.N. Convention on Biological Diversity 1992). Biodiversity includes species diversity, ecosystem diversity, and genetic diversity (Primack and Ellwood 2012a). Our discussion will be limited to species and ecosystem diversity.

Biodiversity varies greatly over the surface of the Earth. There are **latitudinal gradients in species biodiversity**, with biodiversity being highest near the equator and lowest near the poles (Chiabai 2012). Biodiversity is highest on land in tropical jungles and in the ocean in coral reefs near shorelines where there are steep gradients in depth and temperature (Wilson 2002). Global biodiversity is not accurately quantified because we have discovered only 1.5 million to 1.8 million species, less than half of the total estimated amount of between 3.6 million and 100 million.

A given species will show preferences for certain types of climate, geography, and plant types. Other species with similar preferences will tend to be associated with that species. Ecosystems are associations of species and their habitats. When spread over a contiguous area of similar climate and geography, these associations are referred to as a **biome**, which is a major habitat type. Whereas a biome is mainly defined by geography and climate, an ecosystem is defined by the stocks of species and nutrients and by the flows of nutrients and energy in the system.

Energy and nutrients are generally cycled through an ecosystem, creating a **food web** (Figure 15.1a). In a food web, a node can represent a single species, or a collection of species that fill similar **ecological niches**, which are defined by where an animal lives and how it behaves. Each node or niche is defined by the ecosystem setting (soil, underbrush, tree canopy), the interactions with other species (node connections), and the associated flows of nutrients and energy. Each node also corresponds to a specific trophic level, with secondary or higher-level consumers at the top of the **ecological pyramid** and producers (plants) and decomposers (detrivores) at the bottom (Figure 15.1b). Each trophic level has only ~10% of the biomass of the trophic level beneath it (Wright 2005). In the northern Boreal terrestrial food web shown in Figure 15.1, the wolf occupies a specific ecological niche: it is a secondary consumer that feeds on squirrels and prairie dogs.

In the food web each node may have multiple connections to other nodes corresponding to specific species interactions. Each connection represents an exchange of nutrients; if drawn as an arrow, the arrowhead points to the consumer, and the species on the other end is the food. The sustainability of a food web depends on the maintenance of its nodes and connections (Miller and Schmitz 2012). The number and strength of connections and nodes is a measure of the resilience of an ecosystem, with more diverse and interconnected food webs more resilient to the decline or absence of any one species. Note that food webs have a distributed network structure (compare Figure 15.1a to Figure 4.7c), which gives them greater resilience.

Through evolution, species adapt to fill available ecological niches in a mature ecosystem. The interactions of species are generally mutually beneficial and help to keep the ecosystem stable. For example, trees draw nutrients and water from the soil. Beetles help living trees by eating their dead wood and releasing the stored nutrients. Woodpeckers expose dead wood by pecking and keep the

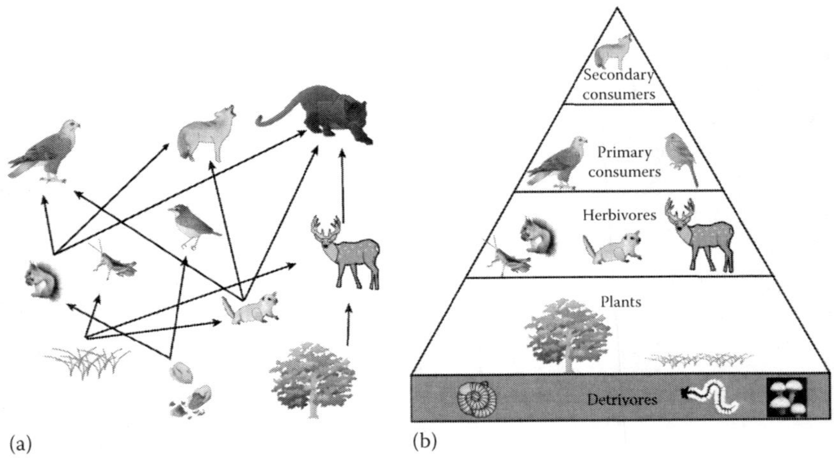

FIGURE 15.1 (a) A simplified community food web and (b) an ecological pyramid illustrating ecological relations among creatures that are typical of a northern Boreal terrestrial ecosystem. The size (area) of each trophic level is proportional to its contained biomass. (Symbols courtesy of the Integration and Application Network, University of Maryland Center for Environmental Science, ian.umces.edu/symbols/.)

beetle population in check by eating them. Each species is uniquely adapted to its environment, for example, the woodpecker's pointed beak allows it to extract beetles from dead wood (Weiner 1994). There are also many interactions that are hard to see, such as mycorrhizal associations with trees where nitrogen is often provided to trees in exchange for carbon.

Over time, ecosystems typically undergo cyclical changes. They spend relatively long periods in stable states that correspond to stages of the adaptive cycle (Walker and Salt 2006). An internal or external perturbation may push the ecosystem over a tipping point and into a new stage of the adaptive cycle (Figure 4.4). The classic example is the **ecological succession** of a forest (Figure 4.5). Imagine a bare patch of land in a temperate climate. **Pioneering plants and animals** such as grasses, shrubs, and small rodents spread in the rapid growth phase. Biomass increases and the flora grow in size until a **climax community** of large trees with little undergrowth is established. In this conservation phase, the mature forest is relatively stable and it stores much energy. There are strong food web connections between the species and the ecosystem efficiently utilizes resources such as nutrients. However, over time the resilience of the forest may decrease: Fewer species are present (old growth forests can have only a handful of tree species), so there is less redundancy in the system. The longer the ecosystem spends in this conservation phase, the less resilient and the more vulnerable it becomes to pests, wildfires, drought, and disease. Eventually a disturbance such as a fire can cause collapse of the ecosystem and release of the nutrients that were bound up in the trees. The release phase is followed by the reorganization phase in which the structure of the forest (species and their connections) changes: Buried seeds germinate and grow into shrubs, and the cycle begins again.

In the absence of environmental change ecosystems will continuously cycle through the same stages. The response of an ecosystem to a change depends on the magnitude of the change and the resilience of the ecosystem. In the Anthropocene it is usually anthropogenic changes to the environment that undermine the resilience of ecosystems and push them toward tipping points, making them more susceptible to regime changes such as the switch from sawgrass to cattail dominated wetlands caused by anthropogenic phosphorous pollution (Section 4.4). Methods are being developed to estimate how close an ecosystem is to a tipping point. For example, Dakos and Bascompte (2014) found that population declines of specialist species were the best indicators of how close an ecosystem is to collapse.

While high biodiversity alone does not make a system stable and resilient (McCann 2000), high diversity is one essential element of resilient systems (Section 4.4). For example, ecological studies have shown that increasing the number of species growing together in gardens and grassland plant communities increases biomass production and drought resistance (Primack and Ellwood 2012a). Maintaining the biodiversity of an ecosystem helps ensure that it can continue providing the same level and variety of ecosystem services, including climate regulation and provision of food, medicine, fiber, fuelwood, and freshwater.

Indicator species have a trait that indicates the health of the environment. For example, miners used to bring canaries into mines; if a canary died, the miners knew to leave the mine to avoid asphyxiation or toxic levels of carbon dioxide or methane (Majer 2012). Indicator species play the role of the "canary in the coal mine" by giving us early warning signs of environmental degradation. Fish are used to monitor river health, lichen are used to monitor air pollution, and various insects are used to evaluate fire impact and environmental quality (Majer 2012). Plant species have been genetically engineered for use as bioindicators of nuclear radiation and land mines (Majer 2012).

Another recent example of an indicator species signaling ecosystem impairment is **colony collapse disorder** in bees. Bees are indicator species that provide critically important ecosystem services, the most important being pollination but also the provision of honey. Colony collapse disorder became a recognized problem during the winter of 2006–2007 when beekeepers reported losing 30% to 90% of their hives. Some studies have linked colony collapse disorder in bees to the use of neonicotinoid pesticides (Henry et al. 2012; Whitehorn et al. 2012), and others to rising atmospheric carbon dioxide levels and subsequent reduction in protein content of plants and pollens (Palmer 2016). Fortunately, colony collapse disorder has recently declined, with loss rates averaging ~29% but dropping to 23% in the winter of 2014–2015. Preserving indicator species makes it more likely that we can detect threats to ecosystem health and remedy them before they cause irreversible damage and permanent loss of ecosystem services.

In the following section we will review some of the many ways that humans damage ecosystems, and then we will explore solutions to the problems of biodiversity loss and ecosystem degradation.

15.2 PROBLEMS

> Destroying a tropical rainforest and other species-rich ecosystems for profit is like burning all the paintings of the Louvre to cook dinner.
>
> **Edward O. Wilson**

15.2.1 Global Biodiversity Loss and Species Extinction

Species have three options to respond to environmental change: move, adapt, or go extinct (locally or at the species level). While adaption typically requires much time, species with more plastic diets or behaviors may be better able to adapt on shorter time scales. Similarly, mobile species can move farther than species whose mobility is limited be seed dispersal (such as many tree species). Immediately before the Anthropocene (but not during mass extinctions long ago), the rate of environmental change was low enough for species to successfully adapt using these strategies. However, the rapid expansion of humans into new geographic areas has caused many species to decrease in abundance or go extinct. For example, the World Wildlife Federation measures the health of the biosphere using the Living Planet Index, which sums the populations of thousands of vertebrate species worldwide (McLellan et al. 2014). They found that between 1970 and 2010 the Living Planet Index decreased 52%. Decreases were largest in South America and the Asia-Pacific region, and were much larger for freshwater species than terrestrial or marine species.

The rate of change in the biosphere is higher today than at any time in Earth's history, causing species to go extinct at an increasingly rapid rate (Vitousek et al. 1997). Wilson (2002) estimated the annual extinction rate at 27,000 species, or one every 20 minutes, while a more recent assessment estimated 8,700 species per year, or ~24 per day (Millennium Ecosystem Assessment 2005b).

Of species studied to date, roughly 40% are in danger of going extinct, including ~50% of insects and reptiles and over 70% of flowering plants (Whitty 2007). Many species go extinct before they are ever identified or studied (Primack and Ellwood 2012b). Species that have recently been declared extinct include the golden toad in 1989, the Zanzibar leopard in 1996, and the Pyrenean Ibex in 2000.* Millennium Development Goal 7, ensure environmental sustainability, sets target 2 as "reduce biodiversity loss, achieving, by 2010, a significant reduction in the rate of loss." This goal, like many other Millennium Development Goals, has remained out of reach (Sachs and McArthur 2005).

Before they become extinct, species may be affected by three types of partial extinction. Species may experience a **local extinction** but persist in other parts of their range. **Ecological extinction** occurs when a species population becomes low enough that it no longer plays a functional role in its ecosystem even though it is not yet globally extinct. This phenomenon can result in "empty forests" on land and "empty reefs" in the oceans. When a commercially exploited species is hunted to near extinction it can no longer be economically harvested, resulting in **commercial extinction**. The collapse of a marine fishery such as the cod fishery off the coast of New England results in local and commercial extinction and possibly ecological extinction (Auth 2015).

* http://extinctanimals.petermaas.nl/.

Why should humans be concerned about species extinction? Species are fundamental parts of ecosystems, and ecosystems provide us with countless services that are valued at $33 trillion (2008 dollars) annually (Costanza et al. 1997). Species that provide ecosystem services for free include earthworms regenerating soil and maintaining its texture, soil bacteria that fix the essential crop nutrient nitrogen, bees and other insects that pollinate plants, birds and mammals that disperse fruit seeds, and wild animals and plants that decompose wastes and recycle nutrients (e.g., fungi such as mushrooms). Our ability to grow food would be greatly impaired if other species did not provide these ecosystem services. Sustainability therefore requires the conservation of biodiversity and protection of ecosystems. The importance of this issue is highlighted by the U.N. naming the years 2011–2020 the "United Nations Decade on Biodiversity."

Once a species goes extinct, it is lost forever. With each lost species, we lose valuable scientific information and potential products including life-saving pharmaceuticals. Hundreds of traditional medicines and modern pharmaceuticals are derived from plants or animals, including **aspirin**, **morphine**, **ipecac**, and **pseudoephedrine**. The rosy periwinkle of Madagascar was the source of the drug **vincristine** used to treat childhood leukemia and the drug **vinblastine** used to treat Hodgkin's lymphoma.* ACE inhibitors used to treat high blood pressure come from the venom of the Brazilian pit viper. Research on a chemical found in the stomach of gastric brooding frogs, which showed potential as a treatment for human peptic ulcers, ended when the species went extinct.†

When one species goes extinct, it often causes other species that are codependent on it to go extinct. The fewer species present in an ecosystem, the lower the probability that a species that can adapt to a changing environment will be present (Friedman 2008). Species diversity means adaptation diversity. Thus, the more species that go extinct, the less resilient an ecosystem becomes, until eventually it collapses. In many cases, ecosystems become less resilient and less stable as diversity decreases (Ives and Carpenter 2007), and less effective at providing ecosystem services such as cleaning our air and water and enriching our soil. While human use of almost all kinds of ecosystem services is expanding, the health of provisioning and regulating services is declining, and the latter will likely cause future declines in other ecosystem services (Carpenter et al. 2009).

The factors that cause extinction of species and biodiversity loss are summarized by the acronym **HIPPO: habitat destruction**, **invasive species**, pollution, population, and **overharvesting** (Wilson 2002). Habitat destruction is the leading cause of species extinction (Sala et al. 2000). Significant habitat destruction began when humanity's environmental impact started to grow following the **Neolithic Revolution** (formerly known as the "Agricultural Revolution") about 10,000 YBP and then accelerated during the **Industrial Revolution** of the eighteenth century. Human activities have transformed about 15% of all ice-free land surface from natural ecosystems to fields, pastures, and settlements (Smil 2002). Humans have also impacted another 40% of available land surface through activities like burning **scrublands** to prevent encroachment of shrubs and trees on grazing lands and planting of tree plantations for timber and other goods. Thus, a total of at least 70 million km^2 or 55% of ice-free land has been transformed or impacted by human activity (Smil 2002). This has inevitably led to degradation of ecosystems and loss of biodiversity. For example, the near extinction of the Vancouver Island marmot is a result of clear-cutting of forest to harvest timber. Humans either wipe out species' habitats during the process of extraction of natural resources (in this case, timber) or occupy the land, killing or displacing species from their ecosystem (Wilson 2002).

Invasive or introduced species that are not native to a specific region often spread rapidly, causing environmental destruction and the displacement and extinction of other species. Invasive species are the primary cause for roughly 42% of species being listed as threatened or endangered in the United States (Pimentel, Zuniga, and Morrison 2005). They can disrupt ecosystem structure and function, reducing the resilience of ecosystems and their ability to provide ecosystem services and potentially causing them to pass thresholds into new regimes. In some cases, humans

* "Medicines From Nature," http://www.chgeharvard.org/topic/medicines-nature, retrieved August 15, 2016.
† http://www.nytimes.com/slideshow/2011/02/06/opinion/specimens_extinct-2.html.

have deliberately introduced species to new environments. **Rabbits in Australia** is a well-known example of a deliberate species introduction that went awry. First brought to the continent in 1788 as a food source, rabbits have proliferated and become a serious blight to farmers, causing millions of dollars of crop loss each year. Examples of environmentally destructive invasive species include **feral pigs**, **avian malaria**, and the **water hyacinth** (Lowe et al. 2000). The number of invasive species in Europe now exceeds 11,000 (PyŠek and Richardson 2012), and in the United States 50,000 (Pimentel, Zuniga, and Morrison 2005), and rising levels of global travel and trade will lead to increasing rates of introduction of invasive species. In general, once an invasive species takes hold in an ecosystem it becomes almost impossible to eradicate (Smil 2002). Annual economic losses from invasive species in the United States are ~$120 billion (Pimentel, Zuniga, and Morrison 2005).

We have previously discussed the problems of human population growth (Section 3.4), which is the driving force for most modern extinctions because it directly causes habitat destruction, the leading cause of species extinction. Human population growth also exacerbates the problems of invasive species (more people moving species around the world), pollution (Chapter 14), and resulting global climate change (Chapter 7). Finally, humans are responsible for the overharvesting of species (Section 13.2.3) and renewable resources such as water that other species depend on (Section 12.3.1). For example, in the nineteenth century North America had over six billion passenger pigeons, but by 1914 the species was hunted to extinction (Prosek 2010). Thus, humans are responsible for all five causes of species extinction represented by HIPPO.

Data from North America show that the rate of species extinctions increased dramatically shortly after humans arrived, decreasing mammal diversity 15% to 42% (Carrasco, Barnosky, and Graham 2009). Rates of extinction have increased further during the Industrial era. For example, surveys of amphibian habitats in the United States between 2002 and 2011 found that occupancy declined almost 4% annually, a shockingly high rate of change (Adams et al. 2013).

Biomes most at risk for biodiversity loss include Mediterranean climate and grasslands (Sala et al. 2000). Many species extinctions occur in biomes with the highest biodiversity: tropical forests, coral reefs, and wetlands (Meadows et al. 2004). Extinction rates are especially high on small, isolated islands, in rivers, and in arid ecosystems. In the 514 years preceding 2015, the International Union for Conservation of Nature (IUCN) Red List of Threatened Species (http://www.iucnredlist.org/) recorded 514 extinctions of terrestrial animals but only 15 extinctions of marine animal species. Marine biodiversity loss lags behind terrestrial loss because the oceans are less accessible to humans (McCauley et al. 2015), but also because marine species have larger geographic ranges, and because in a warming ocean they can maintain optimal temperatures by migrating to greater depths or higher latitudes where the water is cooler.

How do current extinction rates compare to those in Earth's past? Extinction events are characterized by their magnitude (percentage of species that go extinct) and rate (usually extinctions per million species-years). Mass extinction events are defined as times when the rate of extinction exceeded the rate of creation of new species long enough to decrease the global number of species by 75% or more (Barnosky et al. 2011). Five mass extinction events in the last 540 million years are recorded in Earth's fossil record, and these events typically lasted less than two million years. Current extinction rates are higher than during those five mass extinction events (Barnosky et al. 2011). These elevated rates have only persisted over a few hundred years, causing the extinction of several percent of identified species. However, if these elevated rates persist or increase, more than 75% of species could go extinct in as little as three centuries.

Given that the HIPPO stressors that cause extinction are currently becoming more intense, and synergies or positive feedbacks between stressors can amplify the negative impacts on biodiversity, it appears that a "perfect storm" might lead to even higher extinction rates and magnitudes. So while humans still coexist with more than 90% of the species that were present a few hundred years ago, we must decrease the intensity of extinction stressors, or the extinction event that humans started in the Anthropocene will officially become the sixth mass extinction. Previous mass extinctions show that evolution requires ~10 million years to restore diversity to predisaster levels (Sepkoski 1998), so a sixth mass extinction would effectively push most or all of Earth's ecosystems past tipping points into new regimes, and these changes would be irreversible on a human timescale. Thus, future generations will suffer from a loss of biodiversity.

15.2.2 Global Climate Change: A New Threat to Ecosystems

Rapid human-induced climate change presents a new challenge to species that could lead to a drastic decrease in biodiversity. It is already causing species to migrate or go extinct (Thomas et al. 2004). The Intergovernmental Panel on Climate Change (IPCC) Fourth Assessment Report published in 2007 suggests that a 2°C increase in average global temperature could put 20% to 30% of plants and animals at risk of extinction, and a 4°C increase will put 40% to 70% at risk (IPCC 2007). More recent estimates are that 6% to 9% of birds, 11% to 15% of amphibians, and 6% to 9% of coral species are highly vulnerable to climate change and threatened with extinction (Foden et al. 2013).

Given enough time, species can adapt to even extreme environmental conditions. **Extremophiles** are known to survive in water with extremely high pressure, temperature, salinity, or acidity. However, when change occurs over short time spans, creatures do not have time to adapt through evolution, and therefore go extinct. Humans were able to survive the Pleistocene glaciation because they were smart enough to wear clothing to keep warm; other surviving species used different adaptation strategies. Our brains give us an adaptive capacity that allows us to survive rapid change that may cause other species to go extinct. During rapid or extreme environmental change, humans are therefore more likely than other large animals to survive, but probably less likely than species with high reproductive rates (typically small species) and extremophiles.

During climate change, animals and plants most at risk of extinction will be those with limited mobility and high temperature sensitivity. Mobile species will likely adapt to rapidly increasing temperatures by migrating to higher latitudes or elevations to maintain their optimal average environmental temperature; however, this is contingent on the presence of those environments at higher latitudes and/or elevations (e.g., if you can't move up a mountain or to higher elevations, then movement is no longer a viable solution). During global warming, the optimal temperature of a species will migrate poleward, and the death rate at lower latitudes will increase (Walther et al. 2002). This process, which leads to a passive poleward migration of species through differential survival rates, has been documented for plants, which were observed to increase their numbers "in newly favorable areas and decline in increasingly hostile locations" (Kelly and Goulden 2008). Migration of various species of birds, butterflies, trees, shrubs, and foxes to higher latitudes or elevations has been documented (Walther et al. 2002; Freeman and Class Freeman 2014). In some cases, the migrating species become invasive species, displacing other species in their new geographic range. For example, as the number of frost days in southern Switzerland decreased from 1950 to 2000, the number of exotic plants in forests increased exponentially (Walther et al. 2002). Species that are less mobile may not migrate fast enough to survive: If the optimal temperature migrates poleward faster than the species, the total number of individuals will decrease as the death rate exceeds the birth rate until eventually the species becomes extinct. Models show that ~9% of mammals in the western hemisphere will not be able to migrate fast enough to keep pace with climate change, and 87% of mammals will experience range size reductions (Schloss, Nuñez, and Lawler 2012). The problem is made worse by the fact that humans have constructed many barriers to migration including cities, fenced farmlands, roads, and dams. These barriers have led to habitat fragmentation and restricted or prevented migration, making species more vulnerable to the effects of climate change (Wilson 2002). Besides the adaptive response of migration, species **phenology**, the timing of seasonal activities of species, also changes. Studies have shown that plants bloom and birds migrate and breed earlier in the spring in response to regional warming (Walther et al. 2002). For example, cherry blossoms in Japan now bloom significantly earlier than was the case hundreds of years ago. Rapid climate-induced changes in species range, phenology, and ecosystem composition and structure are likely making ecosystems less resilient.

15.2.3 Threats to Marine Ecosystems

The world's oceans are the largest water reservoir on Earth (Table 12.1). They are the primary regulator of global climate and an important sink for greenhouse gases (UNEP 2007). The oceans provide us with food and oxygen and act as a giant carbon and heat sink, both of which have slowed the rate of global warming. The oceans are also home to **phytoplankton** that, due to photosynthesis, are responsible for most of the world's **primary production** and are the foundation of the marine ecological pyramid. The oceans provide a source of employment, revenue, and food for many people around the world. For example, because one in six U.S. jobs is marine related, the oceans added $223 billion to the U.S. economy in 2009 (Sohns and Crowder 2013). Mangroves and coral reefs are examples of particularly valuable coastal marine ecosystems that are endangered. Mangroves are valued at $200,000–$900,000 per km^2 and coral reefs $100,000–$600,000 per km^2 (Sohns and Crowder 2013).

Human impacts in the marine environment are greatest along shorelines and in estuaries, where populations have declined for >90% of species and >65% of wetland and sea grass habitats have been destroyed (Lotze et al. 2006). The collapse of marine fisheries was discussed in Section 13.2.3. Another major problem faced by coastal marine ecosystems is eutrophication caused by nutrient pollution. The oxygen depleted "dead zone" in the Gulf of Mexico results from fertilizer runoff in the Mississippi river drainage basin. These excess nutrients cause algae blooms and consumption of dissolved oxygen when the algae die and decompose. In 2010 this dead zone covered 7,722 square miles (Zimmer 2010). Over 400 dead zones have been identified near coastlines worldwide.

The oceans stand to lose even more oxygen due to global climate change. Seawater holds less oxygen as temperature increases, so the observed trend of increasing average temperatures of sea surface waters is causing a decrease in the oxygen content of the oceans' surface waters worldwide. Models predict that the amount of oxygen dissolved in the world's oceans will decrease 1% to 7% in the next century (Zimmer 2010). Since much of the loss will occur in shallow coastal waters where most fish are caught, and fish need dissolved oxygen to survive, the marine catch may decrease dramatically.

As mentioned in Chapter 7, increasing atmospheric carbon dioxide concentrations does more than increase global temperatures; it also leads to increasing carbon dioxide concentration in seawater. Dissolved carbon dioxide reacts with water to form carbonic acid H_2CO_3, a weak acid that is causing acidification (decreasing pH) of the world's oceans. Ocean acidity has increased ~30% since the Industrial Revolution (Auth 2015), with seawater pH decreasing from 8.11 to 8.01 between 1992 and 2007 (Sohns and Crowder 2013). **Ocean acidification** is a major stressor for organisms like coral reefs that extract calcium carbonate ($CaCO_3$) from seawater to build shells, because calcium carbonate dissolves readily in acidic water. Furthermore, if atmospheric carbon dioxide concentration becomes too high, then carbonate shell-forming phytoplankton will stop growing and start dying and decomposing, changing the oceans from a carbon sink to a carbon source, a potential positive feedback. Phytoplankton produce almost half of the Earth's biomass (Sohns and Crowder 2013); if they were to stop removing carbon from the atmosphere through photosynthesis, the global carbon cycle would become strongly unbalanced.

The combined effects of acidification and oxygen depletion mean that a 2°C increase in global temperature by 2050 is estimated to cause annual economic losses of $17billion to $41 billion from commercial fisheries (IPCC 2007). In part this is because coral reef ecosystems are host to most major marine fisheries. Corals are considered **keystone species** because they have a disproportionately large influence on coral reef ecosystem structure, composition, and function (Nuáez and Dimarco 2012). That is to say, corals are essential components of coral reef ecosystems, which are **biodiversity hotspots** on which many local coastal and island communities and commercial fisheries depend. Hotspots are geographical areas with high conservation value due to disproportionately high biological diversity and a high risk of species extinctions (Bode et al. 2012). Unfortunately, corals are also indicator species that are particularly sensitive to environmental change, and widespread **coral bleaching** events that destroy coral reef ecosystems appear to be early warning signs that we are now in global ecological overshoot. Coral bleaching is believed to be caused primarily by climate change-induced ocean warming (Baker, Glynn, and Riegl 2008). The irony is that many coral species have survived environmental challenges in the oceans for hundreds of millions of years, only to face extinction because of changes in the concentration of a trace gas in the atmosphere occurring over a few decades. If the world's coral reef ecosystems collapse, so will most of the world's coastal fisheries, leading to the loss of the primary protein source for most low-income coastal communities.

Acidification occurred in the world's oceans during the **Paleocene-Eocene Thermal Maximum** (PETM) at 55 Ma, when seawater pH was 0.8 units lower because the atmospheric carbon dioxide concentration was five times higher than today. During the PETM about 6.8 trillion tons of carbon entered the Earth's atmosphere in just 10,000 years, causing temperature to increase 5°C to 9°C (9°F to 16°F) (Panchuk, Ridgwell, and Kump 2008). The decrease in pH caused calcium carbonate in shells to dissolve, resulting in the extinction of deep-water species that built carbonate shells. Today acidification is happening roughly 10 times faster than during the PETM, faster than at any other time in the last 65 million years. Computer modeling suggests that current trends will cause calcium carbonate to become even more soluble than during the PETM (Ridgwell and Schmidt 2010), potentially causing widespread extinction of marine species. For example, studies of vents discharging carbon dioxide into the Mediterranean Sea found that carbon dioxide–rich acidic waters supported benthic marine communities with decreased diversity, biomass density, and trophic complexity relative to areas far from the vents that did not have elevated carbon dioxide contents (Kroeker et al. 2011). This suggests that extreme ocean acidification will damage marine ecosystems and cause biodiversity loss.

There is also the possibility that increased ocean surface temperatures could eventually shut off oceanic circulation (Broecker 1997). Currently warm surface waters move toward the poles and lose heat, causing them to become denser and eventually sink to the bottom of the ocean, displacing cold, deep, oxygen-poor waters that rise to the surface and become reoxygenated. Now, however, rising surface temperatures are causing the ocean's surface waters to become less dense, potentially disrupting **thermohaline circulation**. Furthermore, melting glaciers are adding freshwater to the oceans, which mixes with surface waters and make them even less dense. If ocean surface waters become so light that they no longer sink at the poles, the oceans will no longer be well-mixed, and an oxygen-depleted dead zone will develop in deep waters around the globe. This is what happens in freshwater lakes in temperate zones, which become stratified and develop a deep oxygen-depleted layer in the summer, but in the case of the oceans the oxygen-depleted dead zone would be permanent rather than seasonal. Further, this would dramatically change regional climates—some regions would get warmer while some would get colder (e.g., the United Kingdom, where currently thermohaline circulation is responsible for warm waters from the Gulf of Mexico moving northward and yielding more mild climates).

The combined effects of ocean warming and acidification, pollution, overfishing, and habitat destruction have caused marine populations to decrease roughly 50% between 1970 and 2012 (WWF International 2015). Depopulation rates vary geographically, being highest in tropical and subtropical regions. They also vary over time, with populations declining most rapidly from 1970 to the mid-1980s before stabilizing until 2007. Numbers began declining again in 2007, raising concerns about the health of marine ecosystems (WWF International 2015). These numbers should serve as wake-up calls to the international community, as loss of the services provided by marine ecosystems could irrevocably decrease global biocapacity and cause widespread human starvation.

15.2.4 Effects of Ecosystem Degradation on Human Health

Humans are an integral part of the biosphere, and as a result, ecosystem degradation can harm human health. Pollution presents many potential health risks, including exposure to toxic materials, pathogens, and carcinogens (Chapter 14). Decreased biodiversity has been correlated with increased rates of chronic inflammatory diseases in human populations (Hanski et al. 2012). Climate change raises the specter of a host of many new potential health risks, many of which are interrelated. For example, increasingly intense hurricanes coupled with sea level rise can lead to flooding and outbreaks of cholera and the spread of disease-carrying mosquitoes and rodents. When Hurricane Mitch hit Central America in October 1998, it caused epidemics of **cholera**, **malaria**, **dengue fever**, and **leptospirosis** (Starke 2009).

The draining and destruction of wetland ecosystems followed by human settlement has caused increased mortality and economic losses from hurricanes and earthquakes. Loss of wetlands in the Mississippi delta caused an increased storm surge when Hurricane Katrina struck New Orleans in 2005, costing 1836 lives and over $100 billion in damages (Keller 2011). Building on coastal wetlands surrounding San Francisco Bay magnified the effects of the 1989 **Loma Prieta earthquake**, causing **soil liquefaction** that cost 62 lives and $5 billion in property damage (Keller 2011). Most natural disasters affect the availability of food, uncontaminated water, and medicines, so epidemics frequently occur after disasters (Becker 2014).

Infectious diseases may be the greatest health risk with a strong environmental influence. Infectious (communicable) diseases result from **infection** and growth of pathogens, including bacteria, viruses, fungi, and parasites, in the body. Lower respiratory infections (**pneumonia**) are the leading cause of death from communicable diseases, and are often caused by **influenza** viruses. For example, the **1918 flu pandemic** killed roughly 50 million people, or 3% of the global human population (Becker 2014). Bacteria cause communicable diarrheal diseases such as cholera and **salmonella**, which are the second leading cause of death by infectious disease, while the viral disease **HIV/AIDs** is the third deadliest (Becker 2014). **Tuberculosis** and malaria are also deadly infectious diseases.

Infectious diseases progress from an introduction phase, where they are first introduced to a new population, to an adoption phase, where the disease spreads in the new population (Morse 2001). Infectious diseases can be spread by physical contact, inhalation or ingestion, or through a vector such as a fly or mosquito (Becker 2014). In 2015 there was an outbreak of the mosquito-transmitted Zika virus in Brazil, which spread to other countries, causing the **2015–2016 Zika virus epidemic**. The Zika virus can cause **Guillain–Barré syndrome**.* Zika can also be passed from a pregnant woman to her fetus, which can cause birth defects including **microcephaly**.

Infectious diseases are sometimes transmitted from animals to humans, a process called **zoonosis**† or a spillover event. Influenza is often transmitted as a virus from birds to humans, especially in areas in Asia where humans and **fowl** cohabitat. The **Ebola virus** likely spread from fruit bats or primates to humans, and the **2013–2015 Ebola virus epidemic in West Africa** caused over 11,000 deaths.‡ The spread and persistence of viruses is strongly influenced by environmental conditions and by the proximity of hosts and potential hosts. Infected individuals are quarantined to reduce the risk of the disease spreading. Epidemics are usually contained and dissipate within a few months or years of the initial outbreak.

Many scientists believe that spread of infectious disease presents an ever-growing risk to global human health and sustainability, and that all diseases are affected by the health of ecosystems, particularly infectious diseases (National Academy of Sciences 2012). The factors that cause infection introduction are on the rise, including increased population density and human-animal interactions (Morse 2001). The factors that cause infection adoption (transmission) such as increased population density and increased mobility (plane travel, etc.) are also on the rise. Plane travel may now be the primary factor influencing the global spread of infectious diseases (Colizza et al. 2006).

---

* http://www.cdc.gov/zika/about/index.html, retrieved August 16, 2016.

† http://www.who.int/zoonoses/diseases/en/, retrieved August 16, 2016.

‡ http://www.cdc.gov/vhf/ebola/outbreaks/2014-west-africa/case-counts.html, retrieved August 16, 2016.

Climate change is also increasing the rate of infection introduction by, for example, causing the geographic ranges of infectious disease vectors such as mosquitoes to expand. In New Guinea both birds and malaria-bearing mosquitoes have moved their elevation ranges up mountains so as to stay within their maximum and minimum temperature limits. This will affect the health of millions of New Guinea Highlanders who moved to elevations above 1,500 m in part to avoid malaria (Diamond 2014).

In addition, we may be making ourselves more susceptible to infectious bacterial and fungal diseases by accelerating the rate of microbial evolution through the use of antibiotics. The appearance and spread of **drug resistance** in bacteria is well documented (Smil 2002). Prophylactic use of antibiotics for livestock is now causing resistant bacteria to show up in soils and water (Witte 1998), and use in aquaculture is leading to the appearance of antibiotic-resistant bacteria in water and sediments (Cabello 2006).

Most factors that can aid the spread of infectious diseases are on the upswing: reductions in biodiversity, including loss of predators of disease-carrying insects; increased interactions between disease-carrying livestock and humans in urban settings that cause zoonosis; increased long-distance transport of wild animals; overuse of pesticides and antibiotics that cause the development of resistant strains; the spread of humans into new natural environments harboring new diseases; climate change that expands the geographic range of disease-carrying insects such as mosquitoes spreading malaria; increased human mobility; and the spread of human waste carrying infectious diseases (Corvalan, Hales, and McMichael 2005). We can expect that the risks associated with infectious diseases will continue to increase, and that pandemics in particular are likely to limit, or at least periodically greatly reduce, human population.

The effects of climate change on human health are not well understood, but are likely to be large. Increases in temperature, changes in precipitation, rising sea levels, and increases in the frequency and intensity of extreme weather events and resulting natural disasters will all present health risks by affecting the air we breathe, the weather, our food and water sources, and our interactions with the environment (Crimmins et al. 2016). Specific health risks that are expected to increase with continued climate change include extreme heat, worsened air quality, flooding, vector-borne infections, water-related infections, and food-related infections (Crimmins et al. 2016). For more information, see health2016.globalchange.gov.

15.3 SOLUTIONS

> There can be no purpose more enspiriting than to begin the age of restoration, reweaving the wondrous diversity of life that still surrounds us.
>
> **Edward O. Wilson**
> *The Diversity of Life*

15.3.1 Wildlife Conservation

Wildlife conservation to preserve biodiversity is essential to sustainability. To be successful, it must focus on maintaining the optimal abundance of high-level predators, because they play a particularly important role in food webs and in determining ecosystem stability. When a tertiary consumer decreases the population of secondary consumers through predation, it relieves predation pressure on primary consumers (herbivores) and plants, and their abundance increases. This type of top-down control by apex predators on an ecological pyramid is called a **trophic cascade** (Miller and Schmitz 2012). Unfortunately, apex predators are often hunted down to protect livestock.

Some ecologists and biologists have proposed comprehensive plans for conserving biodiversity. Biologist E.O. Wilson advocates the purchase by non-governmental organizations (NGOs) of large undeveloped contiguous tracts of land in areas that have high biodiversity (Wilson 2002). These lands would be set aside as reserves or **protected areas** in an effort to preserve as many species as possible. Global conservation hotspots are a high priority to purchase because they are at risk and have high concentrations of species (Chiabai 2012). We can focus our efforts on preserving global biodiversity on 25 biodiversity hot spots that add up to 1.4% of land surface but contain 35% of all species in four vertebrate groups and 44% of all species of vascular plants (Smil 2002). Alternatively, efforts can be focused on keystone species or on conserving corridors allowing for species migrations and connecting areas of critical conservation importance. Recent studies have identified the areas where biodiversity conservation is most severely underfunded (Waldron et al. 2013). The geographic region including Malaysia, Indonesia, and Australia has a large amount of threatened biodiversity but inadequate levels of biodiversity conservation funding. Modest levels of investment in areas like this could significantly reduce global rates of biodiversity loss.

Wilson (2002) argues that wildlife reserves should be large in area because the number of species a reserve can support is roughly proportional to the fourth root of its area; also, large size makes them less vulnerable to human activities and invasion of alien species. The reserves should be implemented in three steps to maximize their effectiveness: (1) creation of reserves, (2) restoration by reclaiming developed land to enlarge reserves, and (3) connect reserves using large natural corridors. Other elements of the plan advocated by Wilson (2002) include preserving existing frontier forests, ceasing all logging of old-growth forests, protecting freshwater and marine ecosystems, continuing scientific and mapping studies of species and ecosystems, using biodiversity to improve health and make money, and supporting population planning to reduce the rate of increase of human population. Together these changes will help reduce the drivers of species extinction represented by the letters in "HIPPO." The plan is economically feasible because the total cost was estimated in 2002 to be only 1/1000 of the annual world domestic product. It is politically feasible because it relies on NGOs and private donations.

One of the arguments against this conservation plan is that it is just another example of wealthy developed countries using their money to steal land from poor countries. How can this plan be made attractive to the governments and citizens of developing countries, and how could it actually benefit them? In general, people in developing countries want to raise their standard of living, and land is usually essential to accomplish that goal. They will resent the purchase of land in their countries by foreign concerns unless they actually profit, not just in the short term by a lump sum payment, but in the long term. Conservation must be made profitable for native peoples, perhaps by promoting ecotourism or by identifying or growing plants for pharmaceuticals. Once the native people recognize that the preserved land is a long-term source of income, they will be motivated to protect the land. Involving natives in the process of making decisions that affect the reserve, and guaranteeing that the reserve will be a source of jobs and income, gives natives a stake in conservation (Chiabai 2012). Because the ecosystem is a CPR, a commons-based approach is needed to effectively manage it and to ensure that all stakeholders benefit (Dietz, Ostrom, and Stern 2003).

Unfortunately, areas with the highest biodiversity are often the least protected (Chiabai 2012). This is primarily because they are located in developing countries that lack the resources for wildlife conservation. In contrast, developed countries have the resources to protect biodiversity, but have less to protect because historical development increased the HIPPO drivers of biodiversity loss, and because many of these countries are located at higher latitudes with fewer species. Evaluations of the effectiveness of wildlife conservation show that failures, evidenced by declining populations and species extinctions, usually result from political instability, socio-economic issues, organized crime (poaching), and inadequate governance. For example, lion populations are declining in Africa except in areas where adequate resources and political will allow for the protection (fencing off) of intensively managed wildlife areas (Bauer et al. 2015). Recent research shows that **land sparing**, in which land is set aside for wildlife, is a more effective approach to wildlife conservation than "land sharing," where wildlife exists on land used for agriculture (Phalan et al. 2011). Thus, agricultural intensification combined with land sparing is a preferred approach to wildlife conservation because it minimizes the negative environmental impacts of food production.

For marine species, **marine protected areas** (MPAs) provide varying levels of local protection, ranging up to **marine reserves** that prohibit all fishing. In 2010, 5,800 MPAs covered 1.2% of the ocean (Grorud-Colvert and Lester 2012). MPAs are most effective at achieving their biodiversity conservation goals when they are designed by stakeholders using the best available scientific knowledge, when they receive official designation and government financial support and enforcement of compliance, and when local stakeholders are involved in ongoing monitoring and enforcement (Grorud-Colvert and Lester 2012). For example, a study of marine reserves in the Caribbean found that compliance with reserve rules was correlated with fish biomass (Pollnac et al. 2010). Like terrestrial protected areas, MPAs are also more effective when they are connected so as to allow for species migration, and when adaptive management is used. Unfortunately, marine conservation efforts have failed to restore ecosystem structure and function in most cases, although populations of species in upper trophic levels have partially recovered (Lotze et al. 2006). Successful marine ecosystem restoration has occurred in the Enewetak and Bikini coral atolls, where the United States tested nuclear weapons in the 1940s and 1950s (McCauley et al. 2015).

Marine reserves in the Gulf of California, Mexico, provide an excellent example of the need for government regulation and enforcement of protected areas for biodiversity conservation (Cudney-Bueno et al. 2009). When they observed declining populations of snails and scallops that they harvested, fishermen in Puerto Penasco banded together to establish a marine reserve as a breeding ground. However, because marine ecosystems are open access, meaning they have low excludability, the fishermen could not prevent takings by fishermen from other communities. Without government support for their marine reserve, the protections provided by the reserve evaporated, leading to overharvesting, a typical example of the Tragedy of the Commons (Cudney-Bueno et al. 2009).

So how can we as individuals help conserve biodiversity? Just as high biodiversity helps us live sustainably, living sustainably helps conserve biodiversity. The fewer resources you use, the less pollution you produce. The less carbon dioxide you emit from the use of fossil fuels, the fewer species that may go extinct from climate change. The more plants you grow, the more carbon dioxide they will remove from the atmosphere through photosynthesis, aiding us in our battle against climate change. Encouraging plant and animal diversity in your backyard helps preserve small ecosystems. You can also financially support wildlife conservation societies such as the World Wildlife Federation* or nonprofits aimed at land conservation like the Nature Conservatory, and support political candidates who favor environmental protections for species and ecosystems.

15.3.2 Policies That Promote Sustainability

The **Endangered Species Act** of 1973 was designed to protect species at risk of extinction (Nagle 2011). The Act is administered by the **U.S. Fish and Wildlife Service** (FWS) for terrestrial and freshwater species and the National Marine Fisheries Service for species that are marine or that migrate between marine and freshwater. Species on the endangered species list (https://www.fws.gov/endangered/) receive special legal protections; for example, the Act bars construction projects that could harm endangered species. In the United States there are currently 1,593 species on the endangered species list, including 1,226 endangered and 376 threatened species.†

Unfortunately, the Endangered Species Act has been a lightning rod for controversy since its beginning. One reason is that the Fish and Wildlife Service is not allowed to consider the economic impact when choosing whether to list a species (Nagle 2011). Another reason was that environmentalists used it as a tool to fight larger battles. For example, immediately after the Act was passed in 1973, it played a role in the **snail darter controversy**, in which it was used in a lawsuit to temporarily prevent completion of the Tellico Dam by the Tennessee Valley Authority in order to protect the snail darter (Nagle 2011). In another example, environmentalists wanted to stop logging

* http://www.worldwildlife.org/.

† http://ecos.fws.gov/ecp0/reports/box-score-report, retrieved August 17, 2016.

of **old-growth forests** in the Pacific Northwest, but had no legal recourse until they realized that the Act could be used to protect the habitat of the threatened **northern spotted owl**. Legal battles between the Fish and Wildlife Service and ranchers and loggers have raged ever since. Political battles have prevented its reauthorization since 1988.

The Endangered Species Act has had some successes. Between 1973 and 2009 a total of 22 species have been delisted due to population recovery, including the nation's symbol, the bald eagle, and other famous animals such as the whooping crane, peregrine falcon, gray wolf, and brown pelican.* However, this is a small percentage of the nearly 1,600 species currently on the list.

A relatively new approach to managing ecosystems involves **payment for ecosystem services**. In contrast with the polluter pays approach, in a payment for ecosystem services program the beneficiaries pay (Engel, Pagiola, and Wunder 2008). Beneficiaries can pay individuals or communities who take actions to provide or protect ecosystem services such as water purification or carbon sequestration (Jack, Kousky, and Sims 2008). For example, farmers upstream of the main water reservoir for Beijing, China are paid by downstream water consumers to convert from rice paddy to dry land production of corn (Zheng et al. 2013). This switch reduces upstream water use and runoff of nutrients, improving both the quantity and quality of water in the reservoir. Downstream water consumers pay upstream consumers 1.2 times the opportunity cost, which is the amount upstream farmers lose from switching from rice to corn production. The benefit–cost ratio of this system is 1.5, and both providers and beneficiaries of the ecosystem services are better off with the system in place. The switch from labor-intensive rice production to corn production also decreased the amount of labor to farm the plots. Some farmers invested the saved time and income in education, compounding the benefits of the system. Thus, short-term improvements in human well-being do not have to come at the expense of the environment, nor does protecting the environment require reducing human well-being (Zheng et al. 2013). Effective policies can protect ecosystems and the services they provide and improve human well-being in both the short and long term.

15.4 PUTTING IT ALL TOGETHER

Humans affect ecosystems in many different ways. We compete with other species for food and water, and usually win because we have harnessed the energy of fossil fuels to do work for us. However, these short-term victories come with long-term consequences that we are just starting to come to grips with. Few studies have used a holistic approach to studying the interactions between human activities and ecosystems. Vorosmarty et al. (2010) describe a high spatial resolution analysis of human water security and biodiversity threat along rivers throughout the world that illustrates how humans compete with other species for vital resources such as water. They examined four types of stressors: watershed disturbance, pollution, water resource development (dams, canals, etc.), and biotic factors. In developing countries human water security is often low, but it is positively correlated with biodiversity. Both are negatively impacted by pollution and water resource development, which tend to occur in highly populated areas and are the dominant stressors. Reducing pollution and the negative impacts of water resource development can benefit both humans and biota.

In summary, human well-being and ecosystem well-being are tightly coupled. Sustainable development requires that ecosystems be protected. Biodiversity loss is both an indicator and a cause of ecosystem degradation and collapse. Effective conservation policies are needed to ensure that our rich biological heritage and the ecosystem services it provides are preserved for posterity.

---

* http://ecos.fws.gov/tess_public/DelistingReport.do, accessed December 8, 2009.

WEB RESOURCES

- IUCN Red List of Endangered Species: http://www.iucnredlist.org/
- Millennium Ecosystem Assessment Reports: http://www.millenniumassessment.org/en/index.aspx
- Nature Conservation: International Union for Conservation of Nature: http://www.iucn.org/
- TED talk by Jason Clay: How Big Brands Can Help Save Biodiversity: http://www.ted.com/talks/lang/en/jason_clay_how_big_brands_can_save_biodiversity.html
- TED talk by Pavan Sukhdev: Put a Value on Nature: http://www.ted.com/talks/pavan_sukhdev_what_s_the_price_of_nature.html
- Twenty-First Century Ecosystems: Managing the Living World Two Centuries After Darwin: http://www.nap.edu/openbook.php?record_id=13109
- World Wildlife Federation http://www.worldwildlife.org/

HOMEWORK PROBLEMS

1. Describe an example of a problem caused by biodiversity loss in the United States.
2. Describe a problem caused by an invasive species in the United States.
3. Give an example of an ecosystem, and sketch a realistic food web for that ecosystem.
4. Describe an adaptive cycle of an ecosystem other than the ecological succession of a forest described in Section 15.1.
5. Go to the IUCN Redlist (http://www.iucnredlist.org/) and choose a species that is now extinct. Do some research on that species, and write a short description of how the species became extinct, relating the factors to the HIPPO model.
6. Choose a species that was on the endangered species list (https://www.fws.gov/endangered/) and is now delisted. Do some research on that species, and write a short description of the methods that were used to restore the species population.
7. In the United States wildlife is protected in terrestrial wildlife refuges (https://www.fws.gov/refuges/) and marine protected areas (http://marineprotectedareas.noaa.gov/nationalsystem/nationalsystemlist/). Give a short class presentation on a refuge or MPA of your choosing. What methods were used to preserve wildlife? Were the conservation goals achieved?

16 The Future

If we cannot envision the world we would like to live in, we cannot work towards its creation. If we cannot place ourselves in it in our imagination, we will not believe it is possible.

Chellis Glendinning
My Name Is Chellis and I'm in Recovery from Western Civilization

We shall require a substantially new manner of thinking if mankind is to survive.

Albert Einstein

In the past human populations grew until they became limited by the scarcest essential resource during times of greatest scarcity, according to Leibig's Law of the Minimum. In many cases the limiting resource was food during a famine or water during a drought. Shortages of these essential resources caused death rates to increase, slowing or temporarily reversing population growth. The Green Revolution, particularly use of the limiting nutrient nitrogen in fertilizer, temporarily relieved the limitation of food availability, resulting in unconstrained logistic growth in human population (Smil 2004). Logistic growth can continue until the next scarcest essential resource slows down growth (DeVries 2013). Fortunately, in countries that have passed through the demographic transition, population is now limited by personal choice rather than by increased death rates caused by resource shortages. Thus, there is hope that we can transition smoothly to a sustainable world.

However, humanity is now pushing up against global biophysical limits (Steffen et al. 2015). Of the nine global systems operating on Earth, safe operating limits (read sustainable limits) have been exceeded for biochemical flows, which includes phosphorous and nitrogen, and the genetic diversity (biodiversity) component of biosphere integrity (Figure 16.1). The systems atmospheric aerosol loading (Section 14.2.3.3), novel entities (new substances or life forms), and functional diversity of biosphere integrity have not yet been quantified. We are still within the planetary boundaries for ocean acidification (Section 15.2.3), freshwater use (Chapter 12), stratospheric ozone depletion (Section 8.3.4), and land system change, although some of these are approaching planetary limits.

The good news is that the global average per capita ecological footprint EF is roughly the same now as it was in 1973 (Figure 16.2), despite increases in affluence and consumption. The bad news is that the global average per capita biocapacity B has decreased steadily since the early 1960s, primarily due to increased population. As a result, the biocapacity surplus of +0.96 gha in 1961 became a deficit of −1.11 gha in 2012. We must reverse this trend of growing biocapacity deficits in order to maintain high levels of well-being in the future.

16.1 FUTURE SCENARIOS

You never change things by fighting the existing reality. To change something, build a new model that makes the existing model obsolete.

R. Buckminster Fuller

The future cannot be predicted, but futures can be invented.

Dennis Gabor
Hungarian-British electrical engineer and physicist

Many reports and studies have used the scenarios approach to developing educated guesses about what the future might be like. Scenarios combine quantitative models (usually system dynamics

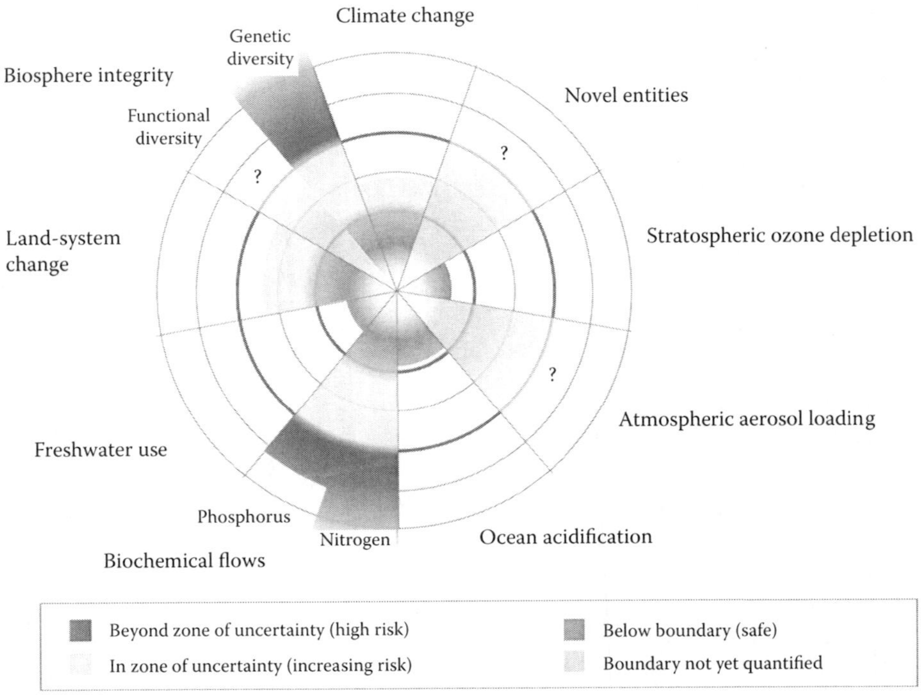

FIGURE 16.1 The current status of planetary boundaries. The zone within the innermost bold circle is the safe operating space, the space between the inner and outer bold circles represents the zone of uncertainty (increasing risk), and the space outside the outermost bold circle is a high-risk zone. (From Steffen, Will, Katherine Richardson, Johan Rockström, Sarah E Cornell, Ingo Fetzer, Elena M Bennett, Reinette Biggs et al. 2015. "Planetary Boundaries: Guiding Human Development on a Changing Planet." *Science* 347 (6223). doi:10.1126/science.1259855. Reprinted with permission from AAAS.)

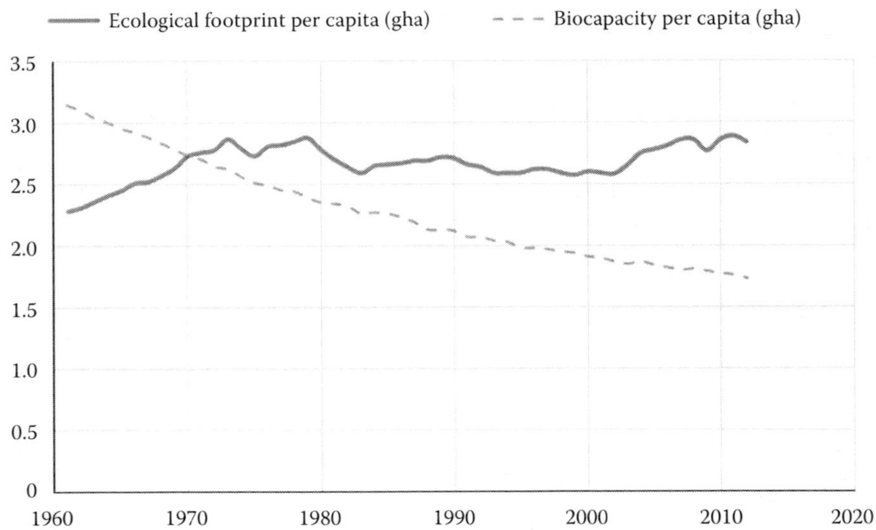

FIGURE 16.2 Global average per capita ecological footprint and biocapacity from 1961 to 2012. (Data from © Global Footprint Network. National Footprint Accounts, 2016 Edition. Licensed and provided solely for noncommercial informational purposes. Contact Global Footprint Network at www.footprintnetwork.org to obtain more information or obtain rights to use this and/or other data.)

models that incorporate all relevant scientific knowledge of the system being modeled) with qualitative stories. They can be used to construct alternative models of the future, which then can be used to develop strategies and policies that will help prepare for desired futures or prevent undesired futures (de Vries 2013). The scenario-based approach to policy development takes into account uncertainties in temporal trends of system variables like population and GDP, and the multiple values/perspectives of diverse stakeholders. The goal of policy makers should be to use such models to devise robust policies that lead society to a desired future despite changes in the economic, social (political), and environmental landscapes. As Benjamin Franklin said, "Failing to prepare is like preparing to fail."

One example of the scenario approach is the IPCC greenhouse gas emissions scenarios (Nakicenovic et al. 2000). Another is **integrated assessment models**, which include the Limits to Growth system dynamics models discussed in Section 4.6. The most recent Limits to Growth study concluded that global sustainability is still achievable, but not by markets and technology alone, that is, lifestyle changes will be required (Randers 2012). Another example of scenario-building is the 2009 Growing Within Limits report* (de Vries 2013).

Modeling efforts often focus on two future scenarios based on our current scientific knowledge. One path is "business as usual," and the other is a truly sustainable path. Current policy decisions will determine what path society will take, and which model of the future will become reality. In the future we may see abrupt collapse or a smooth transition to sustainability. However, we will not see indefinite growth in physical throughput—that is not an option on a finite planet (Meadows et al. 2004).

Based on what we've learned in previous chapters, we can make educated guesses about future trends in system variables:

1. Population will grow to between 10 and 12 billion before stabilizing around the year 2100 (Chapter 3).
2. Increasing urbanization, globalization, and technology development will continue, which will decrease or keep stable the average per capita ecological footprint (Chapters 3 and 5).
3. Climate change will accelerate and result in rising economic costs and environmental migration (Chapters 7 and 8).
4. Energy consumption will shift from fossil fuels to renewable energy, first wind and then solar (Chapters 9 through 11).
5. Water security (Chapter 12) and food security (Chapter 13) will decrease in many parts of the world but will remain steady in most developed countries.
6. Global biodiversity will continue to decline, but may stabilize as wildlife conservation measures take hold, and as human populations migrate to cities and allow rural areas to revert to their natural state (Chapter 15).
7. The pace of economic, environmental, social, and technological change will continue to accelerate, making it increasingly difficult for individuals and societies to make timely decisions.
8. The physical limits on economic throughput will cause long-term economic growth to slow until growth rates decline to near zero.
9. Average human health and longevity will continue to improve as the environmental Kuznets effect results in declining pollution and increasing proportions of GDP are invested in healthcare and medical research.
10. However, human well-being will not increase in failed states, and may even decline in states where serious environmental degradation makes environmental and economic recovery impossible.

* http://www.pbl.nl/en/publications/2009/Growing-within-limits.-A-report-to-the-Global-Assembly-2009-of-the-Club-of-Rome, retrieved August 25, 2016.

Given these trends, is it possible for global society to make the sustainability transition and avoid collapse? Many economic models show global average income doubling between 2010 and 2050, while population will increase from seven to nine billion. This will present many challenges and two major risks (de Vries 2013). The external risk involves the environment: If business as usual continues and environmental impacts are not decoupled from income, we can expect severe environmental degradation to slow economic growth through negative feedbacks. The internal risk involves the other two components of sustainability, economic and social: public goods and common pool resources may be appropriated by the powerful, leading to rising income inequality and social unrest. These risks can only be reduced by using scenarios to devise and implement strategies that protect the environment and guarantee the equitable distribution of resources necessary to achieve society's objective of maintaining a high level of human well-being for all. Choosing the sustainable path will require hard work and planning, and effective policies and regulations at local, regional, national, and global scales. We can choose the easy route and assume that a free market, and the technological advances it spawns, is all that we need. However, that choice would likely lead to unnecessary suffering and declining human well-being and longevity.

As noted by Hall and Day (2009), "For large environmental and health issues, from smoking to flooding in New Orleans, evidence of negative impacts has historically preceded general public acceptance and policy actions by several decades." The lack of progress on addressing environmental problems on a national and global scale is a result of **sustainability and systemic change resistance**. Evidence of global climate change, biodiversity loss, and other forms of environmental degradation have not induced drastic lifestyle or policy changes, and seems unlikely to anytime soon, which means that future generations will have to deal with serious environmental problems. However, future generations won't be able to restore the Earth and its ecosystems to their pristine conditions. Many systems will have passed tipping points, and the resulting regime changes may be irreversible. This invalidates the argument that we should grow the economy at all costs, and then in the future use the economic capital to restore environmental capital. In Chapter 2 we argued that the three forms of capital (economic, social, and environmental) cannot always be substituted for each other. While the environmental Kuznets curve suggests that economic capital can in some cases, such as for air pollutants, reduce environmental impacts, we know that we do not have sufficient knowledge or skill to rebuild ecosystems that have collapsed (Adams 2006). We also lack the knowledge needed to predict the timing of environmental changes, so we cannot confidently delay remedial actions. Furthermore, the costs of inaction rise dramatically over time. Thus, we must favor strong sustainability over weak sustainability, observe the precautionary principle, and act now to preserve ecosystems and the global environment.

Currently the business as usual trend in developed countries like the United States is toward rising economic inequality, that is, the rich get richer and the poor get poorer. If left unchecked this trend could result in extreme inequality where wealth is concentrated in a very small minority, who would exploit the rest of humanity, much as we see in Haiti. This post-modern world would pit the haves against the have-nots in a desperate battle over resources. In these situations, most human capital is wasted because, without proper education or healthcare, most people do not have the opportunity to realize their full potential. Such extreme inequality is highly unsustainable and leads to inefficient use of resources, especially human resources. Recent research shows that rich people are less willing to give to charity when economic inequality is high (Côté, House, and Willer 2015). Thus, when economic inequality rises, as it is today, and the negative feedback of charity weakens, a social tipping point can be reached where the change to high levels of economic inequality become irreversible. Society must reverse the trend of rising economic inequality before this tipping point is reached.

On the bright side, it is unlikely that there will be an apocalypse. As noted by Diamond (2005), "Much more likely than a doomsday scenario involving human extinction or an apocalyptic collapse of industrial civilizations would be 'just' a future of significantly lower living standards, chronically higher risks, and the undermining of what we now consider some of our key values." In some areas trends are very positive. For example, our modern world has the lowest rate of violence in human history (Pinker 2011). Average income has increased 13 times since 1820 (de Vries 2013). People are healthier and live longer (World Health Organization 2015). In 2015 the estimated percentage of people living in extreme poverty fell below 10% for the first time in history.* Suffering from malnutrition is declining, as is discrimination based on sex and race.† So in most places, in most cases, human well-being continues to rise. However, we must preserve the environment for these trends to continue. Science and the market will likely find solutions to most, if not all, environmental problems, and substitutes will be found for most resources that become depleted. However, for really big problems that lack simple solutions like global climate change, we must exercise caution.

16.2 CASE STUDY: CHINA

Our final case study focuses on the country of China, which faces greater environmental challenges than any other country due to its large size and population. China is expanding its economy and changing land use at a scale and pace greater than at any time in human history. China's environmental ImPACT is enormous due to a population P = 1.37 billion in 2016,‡ more than four times as many people as in the United States in a country that is slightly smaller (Vermeer 2012). China's environmental impact is also rapidly rising because, until 2015, its economy was growing at a rate of 10% annually, causing sharp increases in affluence A and consumption C, with a per capita GDP in 2012 of $5,561.§ However, even the Chinese know that this growth cannot be maintained indefinitely, as stated by Pan Yue, China's deputy minister of the environment: "This miracle will end soon because the environment can no longer keep pace."¶

The unsustainability of China's growth is evident in the growing size of its ecological footprint EF. By 2012 its average per capita values had reached EF = 3.4 gha and B = 0.9 gha, for a biocapacity deficit B – EF = –2.4 gha.** This means that China's population would need EF/B = 3.4/0.9 = 3.6 Chinas to maintain its current consumption levels. This degree of ecological overshoot is apparent from rising pollution levels and overexploitation of environmental resources and ecosystem services that are undermining China's sustainability.

If you want to see what happens when an economy grows without adequate enforcement of regulations to protect the environment and human health, look at China today. By its own estimates, environmental impacts reduced GDP by nearly one-quarter in 2008: while GDP was rising ~10% per year, premature mortality and reduced biocapacity and ecosystem services reduced growth to ~7.5%.†† In 2011, China announced that cancer had become the leading cause of death, with lung cancer being the most common cause.‡‡ Rising air pollution levels, especially of particulate matter from coal, is mostly responsible. Environmental impacts are expected to continue to rise, while in 2015 GDP growth was declining. Also, industrial accidents have become commonplace in China due to lax regulations. For these reasons, business executive recruiting companies are finding it harder to recruit foreign executives to work in China, which hurts the Chinese economy.§§ Recognizing these problems, Chinese government officials have instituted many policy changes to promote sustainable development, but the pace and scale of environmental change may still lead to social instability.

---

* http://www.worldbank.org/en/topic/poverty/overview, retrieved August 26, 2016.

† https://thinkprogress.org/5-reasons-why-2013-was-the-best-year-in-human-history-392c4888e603#.8v9e5ekt7, retrieved August 25, 2016.

‡ http://www.livepopulation.com/country/china.html, retrieved August 25, 2016.

§ International Monetary Fund World Economic Outlook Database, published 2015.

¶ Der Spiegel, 2005.

** Global Footprint Network Public Data Package 2016.

†† "Can China Go Green," Bill McKibben, *National Geographic*, June 2011.

‡‡ http://www.earth-policy.org/plan_b_updates/2011/update96.

§§ http://www.tennessean.com/viewart/20130501/BUSINESS01/305010186/China-s-air-pollution-too-much-some-foreign-executives.

People in China are currently crammed into a small percentage of the country's total area, mostly along the coastline and the basins of the Yangtze and Yellow rivers (Diamond 2005). This enormous population creates pressures on the environment so extreme that around 1980 the Chinese government instituted mandatory fertility control, limiting most families to just one child and decreasing the population growth rate to 1.3% in 2001. The **one-child policy** led to the development of a **demographic window**, during which the proportion of people of working age was relatively high. This contributed to rapid economic growth in China from 1990 to 2015. The window is now ending for China, however, because the age distribution of their population is very top-heavy (Figure 3.8). This is referred to as the **4-2-1 problem** because one young worker in China now has to support 2 parents and 4 grandparents (Diamond 2005). As a result, China's government decided in 2015 to roll back its one-child policy.* The population pyramid for China (Figure 3.8) shows that currently there is a low percentage of people in the pre-reproductive age group, indicating that future population growth will be slow.[†] In fact, the World Bank estimates that China's population will actually decline slightly to 1.34 billion by 2050.[‡]

Even though the one-child policy reduced the population growth rate, rapid increases in affluence and per capita consumption have caused the total environmental impact to increase rapidly. Increasing wealth has led to a change in diet, with per capita consumption of meat, eggs, and milk increasing several-fold between 1978 and 2001 (Brown 2009). In 2011 China consumed twice as much meat as the United States, and per capita consumption of meat continues to rise.[§] Effectively the Chinese diet is moving up the food chain, increasing the amount of energy required to produce a person's food and thereby increasing the environmental impact of food production.

China has only 10% of the world's arable land, but almost 20% of the world's people.[¶] Lester Brown's warnings in 1995 about widespread food shortages in China before 2005 caused the Chinese government to change its agricultural policies and avert the impending disaster.** As stated by Meadows et al. (2004), "A prediction of disaster delivered to an intelligent audience with the capacity to act would, ideally, defeat or falsify itself by inducing action to avoid the calamity." However, while China's relentless pursuit of increased agricultural yields has increased food security, it has caused severe environmental degradation. Nutrient pollution from fertilizer use and runoff from expanding livestock production facilities have led to widespread eutrophication, and crops and soils commonly have unsafe levels of pesticides and herbicides (Vermeer 2012).

Another problem affecting agriculture is soil pollution.[††] Rapid, unregulated economic growth has led to extreme pollution from agriculture, mining, and factories, as evidenced by high levels of toxic heavy metals such as cadmium in soils and in crops. Industrialization led to the development of possibly hundreds of "cancer villages" with anomalously high cancer rates. Wastewater from mining and runoff from mine tailings was used to irrigate rice fields in Hunan province in central China, leading to widespread cadmium contamination of produced rice. A high percentage of the

---

* *The Washington Post*, October 29, 2015, https://www.washingtonpost.com/world/asia_pacific/china-lifts-one-child-policy-amid-worries-of-graying-population/2015/10/29/207fc0e6-7e2b-11e5-beba-927fd8634498_story.html.

† See "Population Pyramids: Powerful Predictors of the Future," Kim Preshoff, https://www.youtube.com/watch?v=RLmKfXwWQtE&feature=em-subs_digest-vrecs.

‡ http://databank.worldbank.org/data/home.aspx, retrieved August 25, 2016.

§ "Learning from China: Why the Existing Economic Model Will Fail," http://www.earth-policy.org/data_highlights/2011/highlights18.

¶ http://e360.yale.edu/feature/chinas_toxic_trail_leads_from_factories_to_food/2784/.

** See the documentary "Plan B: Mobilizing to Save Civilization," part of the Journey to Planet Earth series.

†† See the e360 three-part series on soil pollution in China by He Guangwei:

Part I: China's Dirty Pollution Secret: The Boom Poisoned Its Soil and Crops, http://e360.yale.edu/feature/chinas_dirty_pollution_secret_the_boom_poisoned_its_soil_and_crops/2782/.

Part II: In China's Heartland, A Toxic Trail Leads from Factories to Fields to Food, http://e360.yale.edu/feature/chinas_toxic_trail_leads_from_factories_to_food/2784/.

population of Shuangqiao in Hunan province has unsafe blood levels of cadmium or lead. Hunan province alone accounts for >20% of national emissions of cadmium, arsenic, mercury, and lead, with much of it ending up in the Xiang River.

China's local and national governments have suppressed information about soil pollution and resulting food contamination.* Part of the reason is that a preoccupation with food security has led to pressure on local government officials and farmers to increase food production, even if it means planting in contaminated fields. Only recently has the government published results of soil pollution surveys that indicate that ~16% of soil and ~19% of farmland is contaminated, with 2.5% of land now unfit for agriculture. Most government soil pollution data are deemed a "state secret," but public pressure led the government to release limited information in late 2013. Also in 2013 the national government changed official policy, which previously considered only economic growth when promoting government officials, to now include evaluation of environmental protection. In 2014, China passed a law that removed caps on fines paid by polluters for cleanup. However, soil remediation is complicated and expensive, so the legacy of soil pollution in China will likely persist for decades.†

Air pollution is an acute problem in China. Much of the air pollution comes from burning coal, which produces 80% of China's electricity.‡ China is the world's largest producer and consumer of coal, and in 2007 it passed the United States to become the world's largest emitter of carbon dioxide.§ In some cases, Chinese government policies have exacerbated the pollution problem and its associated health risks. For example, in north China the government provided free coal for winter heating. This increased particulate matter pollution, which increased the risk of cardiorespiratory disease, reducing average life expectancy by more than five years (Chen et al. 2013).

Water pollution is also widespread, with 90% of China's urban bodies of water considered polluted (Steffen 2006, citing the World Health Organization). In north China, 60% of river water is classified as unfit for human or animal use (Vermeer 2012). Only 20% of domestic wastewater is treated, compared to 80% in developed countries (Diamond 2005). However, recent passage and enforcement of environmental regulations have led to some improvements: treatment of industrial wastewater discharged by urban industries increased from 50% in 1990 to 82% in 2000, and government spending on environmental protection increased to ~1.4% of GDP (Vermeer 2012).

The unsustainable use of water as a natural resource has caused many of China's cities to have water shortages. China has only 7% of the world's freshwater, but 20% of its population.¶ Available surface water and groundwater decreased more than 10% from 2000 to 2009 (Vermeer 2012). As China enters the global marketplace and its average standard of living increases, its per capita usage of water and other resources will increase, while increased water pollution will decrease the total amount of useable water, together compounding the shortages. The problem may become worse because in north China climate change is causing increasing aridity, and groundwater withdrawals for irrigation have caused the water table to drop an average of 5 feet per year. This is one example of why China is extremely vulnerable to the effects of global climate change in business as usual scenarios.** Since the early 1980s, average temperatures have increased ~2°C in western China, and warming has led to more intense and frequent droughts in north China and flooding in south China (Vermeer 2012).

The Chinese government has taken steps to relieve its country's environmental crisis. Between 1982 and 1996 it passed new environmental laws affecting the marine environment, water and air

---

* Part III: The Soil Pollution Crisis in China: A Cleanup Presents Daunting Challenge, http://e360.yale.edu/feature/the_soil_pollution_crisis_in_china_a_cleanup_presents_daunting_challenge/2786/.

† See "Sustainable Soil Remediation," *Elements*, 6(6), 2010, http://elements.geoscienceworld.org/content/6/6.

‡ "Peak Coal: Why the Industry's Dominance May Soon Be Over," Fred Pearce, June 19, 2014, Yale Environment 360, http://e360.yale.edu/feature/peak_coal_why_the_industrys_dominance_may_soon_be_over/2777/.

§ "China Overtakes U.S. in Greenhouse Gas Emissions," *The New York Times*, June 20, 2007, http://www.nytimes.com/2007/06/20/business/worldbusiness/20iht-emit.1.6227564.html?_r=0.

¶ Christina Larson, 2010, "Growing Shortages of Water Threaten China's Development," http://e360.yale.edu/feature/growing_shortages_of_water_threaten_chinas_development/2298/.

** https://www.ipcc.ch/publications_and_data/ar4/wg2/en/ch20s20-7-3.html, retrieved August 30, 2016.

pollution, solid waste, and noise (Vermeer 2012). In the mid-1980s China passed laws for conservation of forests, grassland, fisheries, and wild animals, and by the year 2000 nature reserves covered 10% of China. The government has also increased transparency in pollution reporting.* However, government regulation has been mostly ineffective at preventing environmental degradation. Instead, resource shortages forced industries to use water and energy more efficiently. Also, improved education and increased public awareness of environmental problems raised pressure on local government leaders to improve the environment. Now emissions from residential use are becoming higher than for industry, with private cars, which increased in number from 1 million in 1990 to 46 million in 2009, becoming the worst air polluters in cities (Vermeer 2012).

China's government is investing in expanding clean, renewable energy sources to reduce air pollution. It is particularly focused on reducing greenhouse gas emissions due to China's vulnerabilities to the effects of climate change. China has become the world's leading producer of solar heating, solar PV, and wind turbines. Coal production and consumption both peaked in China in 2014 and appear to be falling rapidly.† China has also reduced its carbon intensity (the amount of carbon emitted per unit energy) by 34% and is planning to increase that number to 60% to 65%. This allows China to increase energy production without increasing carbon emissions while still allowing the economy to grow, an excellent example of how technology advances can decouple environmental impacts from economic growth.

China is an excellent example of the challenges humanity faces. Development and living practices in China are unsustainable, and the environmental effects are magnified by the enormous size of its population. Unlike western democracies in which large-scale social change requires changes in attitudes and personal choices, change in China is largely dictated by the government (Friedman 2008). This may allow China to quickly change course; only their leaders, not the entire population, need to be convinced that change is required. Until that happens, China will continue on a very risky course of development.

Will China become a long-term economic and socio-political powerhouse? Only if it can find sustainable solutions to all of its pressing environmental problems. China needs to be smart about development. It needs to learn from the mistakes made by developing countries like the United States, and try not to repeat them. China should avoid developing its cities around the automobile and avoid urban sprawl. Perhaps it can leapfrog our technology (Steffen 2006) and build a hydrogen infrastructure rather than a gasoline infrastructure that will be useless in 20 years. However, the Chinese economy keeps hurtling forward, too eager to duplicate the American dream.

16.3 ETHICAL ISSUES

> To waste, to destroy our natural resources, to skin and exhaust the land instead of using it so as to increase its usefulness, will result in undermining in the days of our children the very prosperity which we ought by right to hand down to them amplified and developed.
>
> **Theodore Roosevelt**
> *Seventh annual message, December 3, 1907*

> The ultimate test of man's conscience may be his willingness to sacrifice something today for future generations whose words of thanks will not be heard.
>
> **Gaylord Nelson**
> *Former governor of Wisconsin, co-founder of Earth Day*

---

* Christina Larson, 2010, "In China, a New Transparency On Government Pollution Data," http://e360.yale.edu/feature/in_china_a_new_transparency_on_government_pollution_data_/2352/.

† "The Year Humans Got Serious About Climate Change," September 7, 2015, http://nymag.com/daily/intelligencer/2015/09/sunniest-climate-change-story-ever-read.html.

Recall that sustainability refers in part to the long-term ability to maintain human society. Many would agree that we have an ethical obligation to ensure that our children and grandchildren can have the same quality of life that we have been privileged to have. This is a transgenerational ethical issue.

Imagine that your family is so isolated that you must be truly self-sufficient, and that you own a plot of land fixed in size. You, the parent, may choose to ignore your responsibility to provide for your children's future by consuming well water and livestock faster than nature can replenish them. You might also choose to have more children than the plot of land can support. By the time you die your children and grandchildren would resent you because you were greedy and selfish and left them without enough resources to sustain their lifestyle. This could happen on a global scale, too. The Earth and its resource base are fixed in size, and if we consume its renewable resources faster than it can replenish them, our children will suffer.

Our obligation to provide for our own offspring is stronger than to provide for our next-door neighbors, which in turn is stronger than for people in foreign countries whom we have never met. The transnational ethical questions are have we exploited our neighboring countries? Have we taken their resources without adequate compensation? Or have we tried to share our wealth and improve the quality of life in other countries? The **environmental justice** movement considers these questions on a local and global scale.

Academics and businesspeople in the wealthy United States can argue about the size and duration of shortages and whether we are approaching the physical limits of our global ecosystem. However, many people in developing countries are already experiencing the impacts of ecological overshoot. The only reason many U.S. citizens don't recognize that we are in overshoot is that our wealth buffers us from the consequences; for us, resource shortages are only temporary, and can usually be eliminated by spending more money. When the demand for a resource exceeds domestic supply, we begin to import it. In 2005 the imports of high-income countries averaged 61% of their total consumption footprint (Leape and Humphrey 2010). When a resource is fully utilized globally, then we can only satisfy our want using others' need. Living sustainably can liberate us from the guilt of using others' critical resources.

Globalization has enabled us to help the poor in developing countries. Most Americans have the resources to actively help those in the developing world raise their standard of living by providing the resources needed to live more sustainably, and by helping them build sustainable homes and communities. Americans can volunteer to work for, or donate money to, many different humanitarian organizations. The Red Cross provides emergency assistance and helps communities prepare for natural disasters so they are less vulnerable.* Engineers Without Borders helps poor communities improve human well-being through education and sustainable engineering.† Doctors Without Borders, which won the Nobel Peace Prize in 1999, provides medical care to communities affected by war and natural disasters and to families that cannot afford it.‡ Scientists Without Borders aims to use science to help build capacity in the developing world by, for example, designing sustainable agricultural systems and educating people about the importance of preserving ecosystem services and living sustainably. Americans can use the gifts of their time, money, knowledge, and skills to the greatest advantage by investing them to build economic, social, and environmental capital in poor communities in the developing world.

The U.S. federal government also provides international aid and invests in capacity-building in developing countries through its foreign assistance program, which budgeted $34 billion in fiscal year 2017.§ Some Americans argue that our government spends too much money on foreign aid, but only 0.81% of the U.S. government's budget is spent on international development assistance

* http://www.redcross.org/about-us/our-work/international-services, retrieved August 25, 2016.

† www.ewb-usa.org/, www.ewb-international.org/, retrieved August 25, 2016.

‡ http://www.doctorswithoutborders.org/, retrieved August 25, 2016.

§ http://beta.foreignassistance.gov/, retrieved August 25, 2016.

(Kwong 2005). Jared Diamond argues that "foreign aid is an act of self-interest to preserve our own economy and protect American lives." It's widely acknowledged that two of the fastest-growing problems now affecting America—terrorism and illegal immigration—result from feelings of despair in developing countries. When people have no hope for a better life, young men grow up hungry, frustrated, and resentful of those in the world who have more, and so are at risk of becoming terrorists. Parents in Central America who want to protect their children from terrorists and drug gangs rightly try to enter the United States. The ethical imperative to protect their children overrides the intangible laws of the destination country. The sustainable approach to improving our national security is to reduce the driving forces of terrorism and illegal immigration by helping to improve the quality of life in neighboring countries.

16.4 IMPROVING HUMAN WELL-BEING

> Economic systems should be managed so that we live off the dividend of our resources, maintaining and improving the asset base so that generations that follow will be able to live equally or better.
>
> **Anand and Sen**
> *2000*

Philanthropy has great potential to lift millions out of poverty so that they can live productive and fulfilling lives. Foreign aid from developed countries and private charities helped reach the first Millennium Development Goal of cutting the 1990 poverty rate in half: The number of people living in poverty (at or below $1.90 a day) decreased from 1.95 billion in 1990 to 896 million in 2012.* However, while globalization has helped the global economy to grow and has lifted many people out of poverty, the economic benefits have been unequal, with the richest countries gaining wealth faster than the poorest (de Vries 2013).

The utilitarian approach to reducing poverty is to invest money in a way that materially improves the lives of as many people as possible. This is the approach taken by the Bill and Melinda Gates Foundation,† which is the largest transparently operated private foundation in the world. It practices **effective altruism**, which uses scientific evidence to identify the investments that will have the greatest positive impact. Their web page states, "All lives have equal value: we are impatient optimists working to reduce inequity." Foundation priorities include eradicating polio and malaria and finding a cure for HIV/AIDS. One reason for focusing on these diseases is that they contribute to poverty: malaria alone is estimated to cost billions of dollars in lost productivity every year.‡ Also, these diseases primarily affect poor countries that do not have money for research, unlike cancer research which is very well-funded because it is a major health problem in the United States. Drug companies could not make much money selling a vaccine for malaria, for example, so they do not invest in developing a vaccine. By focusing on diseases that can be prevented at relatively low cost, that primarily affect poor people, and that private corporations cannot make money from, the Gates Foundation is ensuring that their money has the greatest possible impact, saving the maximum number of lives possible per dollar spent.

In the United States the Gates Foundation funds projects that expand educational opportunities and increase access to information technology, and globally it funds projects that aim to reduce extreme poverty and improve healthcare. Not only does it invest in the most cost-effective projects that help meet human needs such as stopping malaria, but also it requires that projects clearly state objectives and make measurements of progress toward meeting those objectives. For example, between 2004 and 2016 global efforts to eliminate malaria have reduced new cases by 25% and malaria deaths by 42%.‡ Projects that do not show progress after an agreed amount of time lose their

* http://www.worldbank.org/en/topic/poverty/overview, retrieved August 25, 2016.
† http://www.gatesfoundation.org/.
‡ http://www.gatesfoundation.org/What-We-Do/Global-Health/Malaria, retrieved August 30, 2016.

funding. Because the Gates Foundation is run like a business, with clearly stated criteria for success, it has a greater impact per dollar invested than other philanthropic foundations.

Another relatively new model for charities is **GiveDirectly**, which collects money from donors and gives it directly to the poor.* The charity focuses on helping people who live in extreme poverty in Kenya and Uganda, and uses aid workers to identify households that would spend donated money wisely to improve their well-being. They give each chosen household an inexpensive cell phone, and use online payment technologies to send money to recipients.† Roughly 91% of donated money is sent to recipients, who are given the freedom to choose the most effective way to spend the money. GiveDirectly transparently documents the performance of their investments,‡ and is now experimenting with a **basic income** program in Kenya in which recipients receive regular payments with no strings attached.

16.5 THE IMPORTANCE OF SCIENCE

Scientific thinking has solved many of the problems that have threatened sustainability; witness the Industrial and Green Revolutions, and the current Communications Revolution. To date the spread of scientific knowledge and its application have lifted billions out of poverty, and greatly increased human longevity. This process will last as long as we continue to generate new knowledge and make that knowledge available to everyone.

However, people are often skeptical about warnings issued by scientists because they don't see the evidence and are not told how the scientists arrived at their conclusions. As a result, warnings by scientists about global environmental problems such as climate change are often ignored. We face a predicament much like the occupants of a sinking boat: a few scientists have noticed we are sinking, and in response a few engineers have attempted to plug the leak. Nearly everyone else on the boat has remained oblivious to their potentially imminent demise. Some may have heard the scientists say that the boat was sinking, but most of them chose to ignore the scientist's warnings; even fewer thought to ask, "Why is it sinking?" or "How can we stop it from sinking?"

People are also skeptical about health warnings. Recently scientific and medical organizations admitted that, despite their recommendations over many decades, there is no evidence that flossing reduces tooth decay,§ or that eating low-fat dairy products reduces the risk of heart disease or type 2 diabetes.¶ Organizations making recommendations without adequate evidence undermines public confidence in science. For example, in 1998 a medical researcher named Andrew Wakefield published a paper purporting to show evidence that the **measles, mumps, and rubella (MMR) vaccine** causes **autism**. This paper was later found to be fraudulent and was retracted.** Evidence indicates Wakefield's motivation was expected profit from a related medical test.†† However, the **MMR vaccine controversy** has persisted, despite overwhelming scientific evidence that the MMR vaccine is safe and does not cause autism, and many parents in the United States refuse to have their children vaccinated for fear they would get autism. Vaccination rates declined in the United States and the United Kingdom,‡‡ leading to a rise in reported cases of measles and mumps. In 2014, 644 people in the United States contracted measles, and an epidemic that started in Disneyland, California in February 2015 led to 84 people contracting measles, most of them unvaccinated children.§§ Parents

* https://givedirectly.org/.
† http://www.ted.com/talks/joy_sun_should_you_donate_differently.
‡ https://givedirectly.org/quality-of-service.
§ http://www.nytimes.com/2016/08/03/health/flossing-teeth-cavities.html?_r=0, retrieved August 25, 2016.
¶ http://time.com/3734033/whole-milk-dairy-fat/?xid=emailshare, retrieved August 25, 2016.
** http://healthland.time.com/2011/01/06/study-linking-vaccines-to-autism-is-fraudulent/, retrieved August 26, 2016.
†† http://www.cnn.com/2011/HEALTH/01/11/autism.vaccines/?hpt=Sbin, retrieved August 26, 2016.
‡‡ http://healthland.time.com/2010/11/04/vaccination-rates-drop-in-wealthier-kids-the-autism-rumors-take-a-toll/, retrieved August 26, 2016.
§§ http://www.forbes.com/sites/stevensalzberg/2015/02/01/anti-vaccine-movement-causes-worst-measles-epidemic-in-20-years/#42a594657ef9, retrieved August 26, 2016.

who choose not to vaccinate their children put other people's lives at risk, such as children in chemotherapy who cannot be vaccinated. Effective science education can help make public health interventions more effective.

Some people, even scientists, allow their values to bias their scientific interpretations. Others deliberately lie about science when it comes into conflict with their values, or when they stand to make money by lying (Oreskes and Conway 2010). However, good scientists always aim to find and speak the truth; to do otherwise would compromise their personal integrity and the integrity of the scientific enterprise. Moreover, the vast majority of scientists speak the truth to the best of their knowledge, even if it comes into conflict with their values.

The scientific enterprise regularly spins off new technologies that spur economic growth. A recent example is the Internet, which was originally developed for scientists in the defense department and then in academics to exchange information.* Google was the product of an NSF grant to researchers at Stanford.† Now a large fraction of our economy is fueled by online purchases and financial transactions. Politicians should not decide what scientific research to fund, especially if the sole criterion is their assessment of economic potential, which is nearly impossible to quantify accurately. No one can anticipate the economic impact of scientific discoveries and the resulting technologies.

Besides spurring economic growth through development of new technologies, science can also be used to evaluate the effectiveness of government policies. Opponents of environmental policies often claim that they will not have the desired effect, or that they will cause economic harms that will outweigh the benefits. However, objections to environmental policies are often based on anecdotes rather than data, and usually fail to recognize that environmental protections that harm one industry or geographic region benefit others, and that overall, benefits outweigh costs.‡ History has shown that, in most or all cases of environmental legislation in the United States, the desired benefits were realized at little or no economic cost. The Clean Air Act, The Clean Water Act, and the Montreal Protocol all achieved their stated objectives without significantly hurting the economy.§ For example, between 1990 and 2020 the benefits of the Clean Air Act are projected to exceed the costs by a factor of more than 30 to 1.¶ To govern effectively, policy makers should use scientific evidence to decide whether to enact or repeal policies. From this perspective science is a public good, a resource that can be used to promote human well-being.

16.6 HOW TO PREPARE FOR THE FUTURE

> When gaps exist between knowledge and actions, anxiety (if not fear) is the result. So it's not the state of the world that creates the anxiety quite as much as it is someone's lack of action.
>
> **Chris Martenson**
>
> *Heinberg and Lerch 2010, Chapter 32*

The three main risks to sustainability are shortages of water, food, and energy. These risks can be reduced by communities and individuals through conservation, efficient resource use, and by adding redundancy to resource supply systems. Harvesting and storing rainwater protects against drought. Growing and storing food protects against food supply disruptions. Producing and storing energy using PV panels and wind turbines with a battery storage system can protect against power outages or shortages. Individuals and communities that take these steps will be more resilient. They will also be more sustainable if they reduce waste and pollution by closing resource loops, reusing

---

* http://www.encyclopedia.com/topic/the_Internet.aspx, retrieved August 30, 2016.

† https://www.nsf.gov/discoveries/disc_summ.jsp?cntn_id=100660, retrieved August 30, 2016.

‡ http://www.usnews.com/opinion/economic-intelligence/2015/06/08/epa-clean-water-rules-benefits-outweigh-its-costs, retrieved August 26, 2016.

§ http://abcnews.go.com/Technology/top-pieces-environmental-legislation/story?id=11067662, retrieved August 30, 2016.

¶ https://www.epa.gov/clean-air-act-overview/benefits-and-costs-clean-air-act-1990-2020-second-prospective-study, retrieved August 26, 2016.

materials when possible, and recycling them when not. And as Lester Brown (2009) argues, it's not enough to be personally virtuous; people must also make donations to environmental organizations and take the time to participate in environmental rallies, email their legislators, and volunteer in campaigns to elect environmentally friendly legislators.

Most of the progress in environmental protection in the past has come not from individual action but from U.S. federal legislation such as the Clean Air Act, the Clean Water Act, the 1987 Montreal Protocol that banned ozone-destroying CFCs, and the Endangered Species Act. However, the effectiveness of bottom-up approaches to change are harder to quantify, and while top-down approaches work best for large communities such as the United States, the bottom-up approach may be most effective for small communities (Diamond 2005). A combination of these approaches will likely be needed to make the United States sustainable.

Focusing on the bottom-up approach is easier because individuals have some control over how they live, but little control over the choices society makes. The ecological footprint numbers suggest that it is not hard for an individual to become sustainable. We saw in Section 3.3 that the average American EF is 8.2 gha and the U.S. average biocapacity is 3.8 gha. The number of United States required to maintain this consumption level is 8.2/3.8 = 2.16. To reduce the consumption level to one United States would require cutting consumption to 1/2.16 = 0.46 or 46%, meaning Americans must reduce their consumption by 100 – 46 = 54% to become sustainable. Decreasing consumption levels by 54% is not possible without lifestyle changes; buying green and efficient products alone won't decrease your consumption enough for you to be sustainable.

How can you decrease your consumption level by half to become sustainable? You've probably heard hundreds of tips on how to live more sustainably. Who can remember all of those tips, let alone act on them? And the prescriptions often sound complicated and time consuming. Many people feel overwhelmed when considering how to reduce their ecological footprint, so they throw their hands up in despair and give up on sustainable living. However, if you follow a simple rule of thumb you can become sustainable: Follow the rule of halves, and cut everything you own and use in half! Move to a house half the size of your current house, with a lawn half the size. Cut the number of cars in your family in half, and cut the size of your cars in half. Cut the number of miles you travel by car in half. Cut how much meat you eat in half. Buy energy star appliances that use half the electricity. Cut the number of televisions in your house in half, the number of computers and printers in half, the amount of clothing in half. For most Americans these steps should be easy because we currently have more of these things than we need. Don't throw away what you give up: donate it for others to use. Abstain from buying more stuff to replace the stuff you get rid of: you won't have room for it in your smaller house anyway. Soon you will realize that you didn't need all of that stuff, and that your life is more enjoyable because you spend half as much time shopping for and maintaining stuff. People in Europe and Japan have half the ecological footprint of Americans, yet they have the same level of well-being and are as happy. If you follow the rule of halves consistently, you will cut your energy and material use, your ecological footprint, and your costs in half and make your lifestyle more sustainable.

Here are some more ideas for living sustainably (you can explore these in the homework problems). Because population is one of the main drivers of the environmental ImPACT, don't have more than two children. If you want more, adopt a child and make their life better. Don't own any large pets, which will save you time and money.* Sell your house in a new suburb and your automobile, use the money to buy a small apartment or condominium (no more than 500 ft^2 per person) in the city near your work, and use mass transit to get there. Use the remaining money to pay off your debts or to do a green retrofit of your new home: installation of passive solar hot water heater and PV cells, rain barrels, ground source geothermal heat pump, added insulation, and so

* Note that decreasing the number of children or pets you have will not directly decrease your calculated EF, but we've noted that sustainability requires a steady state population, which is achieved by having no more than 2.1 births for every two people; the EF of pets should be included in the EF calculator but currently is not.

on. Adopt a vegetarian diet (or abstain from meat on weekdays) and buy all of your food from the local fresh market. Together, these changes would shrink your EF by ~75%, would not require any added physical effort, yet would increase the amount of free time you would have and would make you healthier. Your living expenses also would be greatly decreased: A smaller home requires less energy to heat and cool and less maintenance, and you would not have auto expenses (maintenance, insurance, gas). You would have less to worry about, and more money and time to enjoy life.

Living sustainably means growing, maintaining, and balancing your social, economic, and environmental capital to increase security and achieve happiness. Sometimes we can substitute one type of capital for another (weak sustainability). For example, you can use economic capital to increase social capital by purchasing healthcare services that improve your health or buy time by paying someone else to do your housework. You can cash in social capital when you ask friends or family to help you. These days, most of us have overspent our capital: we don't have as much money, time, health, and love as we would like. Many of us have accrued economic and social debts. Living sustainably means slowing down and taking the time to build up capital reserves. You can save money and time, and improve your health and state of mind, by living sustainably (Wann 2007).

16.7 TAKE-HOME MESSAGES

> The choice we face is not between saving our environment and saving our economy. The choice we face is between prosperity and decline.
>
> **President Obama**
> *April 2009**

We have learned that a society can be sustainable if it preserves and grows social, economic, and environmental capital for future generations by ensuring that the demand for each form of capital is not greater than the supply, and if it does not produce more waste than the environment can absorb. We also learned that the carrying capacity is the sustainable limit of human population; that fossil fuels and the Green Revolution temporarily relieved these limits, leading to a rapid increase in population; and that population has exceeded sustainable limits, leading to global climate change and biodiversity loss that if left unchecked will become the sixth great mass extinction. The five global mass extinctions that occurred in Earth's past are prime examples of global collapse that we want to avoid.

Resource availability is likely to decrease in the future, so we must practice conservation and increase use efficiency to maintain high levels of water, food, and energy security. Sustainability can be enhanced by increasing resilience, building capital reserves, using triple bottom line accounting, diversifying by adding redundancy to resource supply systems, avoiding the ABCs of unsustainability, protecting the environment through effective regulation, preserving wildlife in reserves and marine protected areas, and using science and philanthropy to improve human well-being. If society takes these actions, we can ensure that future generations will be able to meet their needs.

RESOURCES

- **Earth 2100 documentary**: https://www.youtube.com/watch?v=LUWyDWEXH8U
  The ABC News docudrama "Earth 2100" broadcast on July 2, 2009 discussed two possible future scenarios of what could happen before the year 2100. The bulk of the broadcast outlined the worst-case scenario to serve as a warning of what could happen if humanity does not take strong action to address the problems of climate change and resource shortages.

---

* http://www.nytimes.com/2009/04/23/us/politics/23obama.html, retrieved August 26, 2016.

HOMEWORK PROBLEMS

1. Use the ecological footprint calculator at www.myfootprint.org to calculate your ecological footprint. Then calculate how much it would shrink if you cut the following in half:
   a. Size of your house or apartment
   b. Number of airline miles traveled
   c. Gallons of gas consumed by your car per mile of travel
   d. The number of auto miles traveled by switching to mass transit or moving closer to work
2. Use the calculator at www.myfootprint.org to compare the average carbon footprint for each type of housing location: newer suburb, older suburb, rural, and inner city locations.
3. Use the calculator at www.myfootprint.org to compare the average housing footprint for each type of house:
   a. An estate, ranch or farm (1 acre)
   b. A free standing single family house (1 acre)
   c. A house or building with 4 or fewer units
   d. A small apartment building (5–20 units)
   e. A large apartment building (20+ units)

References

Abatzoglou, John T, and A Park Williams. 2016. "Impact of Anthropogenic Climate Change on Wildfire across Western US Forests." *Proceedings of the National Academy of Sciences* 113 (42): 11770–75. doi:10.1073/pnas.1607171113.

Abrahamse, Wokje, Linda Steg, Charles Vlek, and Talib Rothengatter. 2005. "A Review of Intervention Studies Aimed at Household Energy Conservation." *Journal of Environmental Psychology* 25 (3). Elsevier: 273–91.

Adams, Michael J, David A W Miller, Erin Muths, Paul Stephen Corn, Evan H Campbell Grant, Larissa L Bailey, Gary M Fellers et al. 2013. "Trends in Amphibian Occupancy in the United States." *PLoS ONE* 8 (5). Public Library of Science: e64347. http://dx.doi.org/10.1371%252Fjournal.pone.0064347.

Adams, W M. 2006. "The Future of Sustainability: Re-Thinking Environment and Development in the Twenty-First Century." *International Union for Conservation of Nature: Report of the IUCN Renowned Thinkers Meeting.* http://cmsdata.iucn.org/downloads/iucn_future_of_sustanability.pdf.

Aggarwal, Bharat B, Divya Danda, Shan Gupta, and Prashasnika Gehlot. 2009. "Models For Prevention and Treatment of Cancer: Problems vs Promises." *Biochemical Pharmacology* 78 (9): 1083–94. doi:10.1016/j.bcp.2009.05.027.

Ali, Abu Muhammad Shajaat. 2006. "Rice to Shrimp: Land Use/land Cover Changes and Soil Degradation in Southwestern Bangladesh." *Land Use Policy* 23 (4): 421–35. doi:10.1016/j.landusepol.2005.02.001.

Allcott, Hunt. 2011. "Social Norms and Energy Conservation." *Journal of Public Economics* 95 (9). Elsevier: 1082–95.

Allen, David T, Vincent M Torres, James Thomas, David W Sullivan, Matthew Harrison, Al Hendler, Scott C Herndon et al. 2013. "Measurements of Methane Emissions at Natural Gas Production Sites in the United States." *Proceedings of the National Academy of Sciences* 110 (44): 17768–73. doi:10.1073/pnas.1304880110.

Alston, Julian M, Jason M Beddow, and Philip G Pardey. 2009. "Agricultural Research, Productivity, and Food Prices in the Long Run." *Science* 325 (5945): 1209–10. doi:10.1126/science.1170451.

Anand, Suhir, and Amartya Sen. 2000. "Human Development and Economic Sustainability." *World Development* 28 (12): 2029–49. http://www.sciencedirect.com/science/article/pii/S0305750X00000711.

Anderegg, William R L, James W Prall, Jacob Harold, and Stephen H Schneider. 2010. "Expert Credibility in Climate Change." *Proceedings of the National Academy of Sciences.* doi:10.1073/pnas.1003187107.

Archer, D, M Eby, V Brovkin, A Ridgwell, L Cao, U Mikolajewicz, K Caldeira, K Matsumoto, G Munhoven, and A Montenegro. 2009. "The Atmospheric Lifetime of Fossil Fuel Carbon Dioxide." In *Annual Review of Earth and Planetary Science*, 37:117–34. http://geosci.uchicago.edu/~archer/reprints/archer.2009.ann_rev_tail.pdf.

Arrow, Kenneth, Maureen Cropper, Christian Gollier, B Groom, G Heal, R Newell, W Nordhaus et al. 2013. "Determining Benefits and Costs for Future Generations." *Science* 341 (6144). American Association for the Advancement of Science: 349–50.

Assadourian, Erik. 2010. *State of the World 2010: Transforming Cultures: From Consumerism to Sustainability (The Worldwatch Institute).* Edited by Linda Starke and Lisa Mastny. W. W. Norton & Company.

Astrup, Thomas Fruergaard, Davide Tonini, Roberto Turconi, and Alessio Boldrin. 2015. "Life Cycle Assessment of Thermal Waste-to-Energy Technologies: Review and Recommendations." *Waste Management* 37 (March): 104–15. doi:http://dx.doi.org/10.1016/j.wasman.2014.06.011.

Atkinson, Gail M, David W Eaton, Hadi Ghofrani, Dan Walker, Burns Cheadle, Ryan Schultz, Robert Shcherbakov et al. 2016. "Hydraulic Fracturing and Seismicity in the Western Canada Sedimentary Basin." *Seismological Research Letters* 87 (3): 631 LP-647. http://srl.geoscienceworld.org/content/87/3/631.abstract.

Attari, Shahzeen Z. 2014. "Perceptions of Water Use." *Proceedings of the National Academy of Sciences* 111 (14): 5129–34. doi:10.1073/pnas.1316402111.

Auth, Katie. 2015. "The Oceans: Resilience at Risk." In *State of the World 2015: Confronting Hidden Threats to Sustainability*, 79–90. Washington, DC: The Worldwatch Institute and Island Press.

Ayers, John C. 2012. "Sands and Silica." In *Berkshire Encyclopedia of Sustainability*, edited by Daniel E Vasey, Sarah E Fredericks, Shen Lei, and Shirley Thompson, 4:429–31. Great Barrington, MA: Berkshire.

Ayers, John C, Steven Goodbred, Gregory George, David Fry, Laura Benneyworth, George Hornberger, Kushal Roy et al. 2016. "Sources of Salinity and Arsenic in Groundwater in Southwest Bangladesh." *Geochemical Transactions* 17 (1): 1–22. doi:10.1186/s12932-016-0036-6.

Bain, Paul G, Matthew J Hornsey, Renata Bongiorno, and Carla Jeffries. 2012. "Promoting pro-Environmental Action in Climate Change Deniers." *Nature Climate Change* 2 (8). Nature Publishing Group: 600–603.

Baker, Andrew C, Peter W Glynn, and Bernhard Riegl. 2008. "Climate Change and Coral Reef Bleaching: An Ecological Assessment of Long-Term Impacts, Recovery Trends and Future Outlook." *Estuarine, Coastal and Shelf Science* 80 (4): 435–71. doi:http://dx.doi.org/10.1016/j.ecss.2008.09.003.

Bakker, Karen. 2010. *Privatizing Water: Governance Failure and the World's Urban Water Crisis.* Ithace, NY: Cornell University Press.

Bardi, U. 2009. "Peak Oil: The Four Stages of a New Idea." *Energy* 34 (3): 323–26.

Barilla Center for Food and Nutrition. 2012. *Eating Planet.* Edizioni Ambiente. www.barillacfn.com.

Barlow, Maude, and Tony Clarke. 2002. *Blue Gold: The Fight to Stop the Corporate Theft of the World's Water.* The New Press.

Barnes, P. 2006. *Capitalism 3.0: A Guide to Reclaiming the Commons.* San Francisco, CA: Barret.

Barnett, T P, J C Adam, and D P Lettenmaier. 2005. "Potential Impacts of a Warming Climate on Water Availability in Snow-Dominated Regions." *Nature* 438 (7066). Nature Publishing Group: 303–9. http://dx.doi.org/10.1038/nature04141.

Barnett, Tim P, and David W Pierce. 2008. "When Will Lake Mead Go Dry?" *Water Resour. Res.* 44 (3). AGU: W03201. doi:10.1029/2007wr006704.

Barnosky, Anthony D, Nicholas Matzke, Susumu Tomiya, Guinevere O U Wogan, Brian Swartz, Tiago B Quental, Charles Marshall et al. 2011. "Has the Earth/'s Sixth Mass Extinction Already Arrived?" *Nature* 471 (7336). Nature Publishing Group. 51–57. http://dx.doi.org/10.1038/nature09678.

Bates, Albert. 2006. *The Post-Petroleum Survival Guide and Cookbook: Recipes for Changing Times.* New Society Publishers.

Bauer, Hans, Guillaume Chapron, Kristin Nowell, Philipp Henschel, Paul Funston, Luke T B Hunter, David W Macdonald et al. 2015. "Lion (Panthera Leo) Populations Are Declining Rapidly across Africa, except in Intensively Managed Areas." *Proceedings of the National Academy of Sciences* 112 (48). National Acad Sciences: 14894–99.

Bauman, Yoram, and Grady Klein. 2014. *The Cartoon Introduction to Climate Change.* Island Press.

Becker, Per. 2014. *Sustainability Science: Managing Risk and Resilience for Sustainable Development.* Elsevier.

Beecher, Janice A. 2003. "Privatization of Water Management." Edited by E Julius Dasch. *Water: Science and Issues, Vol. 4 of Global Reference on the Environment, Energy, and Natural Resources.* New York: Macmillan Reference USA.

Benneyworth, Laura, Jonathan Gilligan, John C Ayers, Steven Goodbred, Gregory George, Amanda Carrico, Md. Rezaul Karim et al. 2016. "Drinking Water Insecurity: Water Quality and Access in Coastal South-Western Bangladesh." *International Journal of Environmental Health Research*, June. Taylor & Francis, 1–17. doi:10.1080/09603123.2016.1194383.

Bettencourt, Luis, and Geoffrey West. 2010. "A Unified Theory of Urban Living." *Nature* 467 (7318). Nature Publishing Group: 912–13.

Biello, David. 2010. "CO_2 Capture and Storage Gains a Growing Foothold." *Yale Environment* 360. http://www.e360.yale.edu/content/print.msp?id=2240.

Bode, Michael, Kerrie A Wilson, Takuya Iwamura, and Hugh P Possingham. 2012. "Biodiversity Hotspots." In *Berkshire Encyclopedia of Sustainability*, edited by Robin Kundis Craig, John Copeland Nagle, Bruce Pardy, Oswald J Schmitz, and William K I B Smith, 5:26–30. Ecosystems Management and Sustainability. Great Barrington, MA: Berkshire.

Bomford, Michael. 2010. "Getting Fossil Fuels Off the Plate." In *The Post Carbon Reader: Managing the 21st Century's Sustainability Crises*, edited by Richard Heinberg and Daniel Lerch, 119–27. Watershed Media and the Post Carbon Institute.

Borken-Kleefeld, Jens, Terje Berntsen, and Jan Fuglestvedt. 2010. "Specific Climate Impact of Passenger and Freight Transport." *Environmental Science & Technology* 44 (15). American Chemical Society: 5700–5706. doi:10.1021/es9039693.

BP. 2015. "Statistical Review of World Energy 2015." http://www.bp.com/content/dam/bp/pdf/Energy-economics/statistical-review-2015/bp-statistical-review-of-world-energy-2015-full-report.pdf.

Bradshaw, Corey J A, and Barry W Brook. 2014. "Human Population Reduction Is Not a Quick Fix for Environmental Problems." *Proceedings of the National Academy of Sciences* 111 (46): 16610–15. doi:10.1073/pnas.1410465111.

Brand, Stewart. 2009. *Whole Earth Discipline: An Ecopragmatist Manifesto*. Viking.

Bright, Chris. 2000. "Anticipating Environmental Surprise." In *State of the World 2000: A Worldwatch Institute Report on Progress Toward a Sustainable Society*, edited by Lester R Brown et al., 368. W. W. Norton.

British Geological Survey, and Bangladesh Department of Public Health and Engineering. 2001. "Arsenic Contamination of Groundwater in Bangladesh." http://nora.nerc.ac.uk/11986/.

Broecker, Wallace S. 1997. "Thermohaline Circulation, the Achilles Heel of Our Climate System: Will Man-Made CO2 Upset the Current Balance?" *Science* 278 (5343). American Association for the Advancement of Science: 1582–88.

Broecker, W S. 2008. "CO_2 Capture and Storage: Possibilities and Perspectives." *Elements* 4: 295–97.

Brown, Lester. 2009. *Plan B 4.0: Mobilizing to Save Civilization*. New York: W. W. Norton & Co., Inc.

———. 2011. *World on the Edge: How to Prevent Environmental and Economic Collapse*. Earth Policy Institute. http://www.earth-policy.org/books/wote.

Brown, Lester R. 2012. *Full Planet, Empty Plates: The New Geopolitics of Food Scarcity*. W. W. Norton.

Brown, Lester R, E Adams, J Larsen, and J M Roney. 2015. *The Great Transition, Shifting from Fossil Fuels to Solar and Wind Energy*. New York: W. W. Norton. earth-policy.org.

Budds, Jessica, and Gordon McGranahan. 2003. "Are the Debates on Water Privatization Missing the Point? Experiences from Africa, Asia and Latin America." *Environment and Urbanization* 15 (2): 87–114. doi:10.1177/095624780301500222.

Bunch, A G, C S Perry, L Abraham, D S Wikoff, J A Tachovsky, J G Hixon, J D Urban et al. 2013. "Evaluation of Impact of Shale Gas Operations in the Barnett Shale Region on Volatile Organic Compounds in Air and Potential Human Health Risks." *The Science of the Total Environment* 468–469C (September) 832–42. doi:10.1016/j.scitotenv.2013.08.080.

Burgess, Matthew G, Stephen Polasky, and David Tilman. 2013. "Predicting Overfishing and Extinction Threats in Multispecies Fisheries." *Proceedings of the National Academy of Sciences* 110 (40): 15943–48. doi:10.1073/pnas.1314472110.

Cabello, Felipe C. 2006. "Heavy Use of Prophylactic Antibiotics in Aquaculture: A Growing Problem for Human and Animal Health and for the Environment." *Environmental Microbiology* 8 (7). Blackwell Publishing Ltd: 1137–44. doi:10.1111/j.1462-2920.2006.01054.x.

Cain, Nicholas L. 2010. "A Different Path: The Global Water Crisis and Rainwater Harvesting." *Consilience: The Journal of Sustainable Development* 3 (1): 187–96. http://journals.cdrs.columbia.edu/consilience/index.php/consilience/article/view/118.

Caldecott, Ben. 2015. "Avoiding Stranded Assets." In *State of the World 2015: Confronting Hidden Threats to Sustainability*, 51–64.

Cardinale, Bradley J, J Emmett Duffy, Andrew Gonzalez, David U Hooper, Charles Perrings, Patrick Venail, Anita Narwani et al. 2012. "Biodiversity Loss and Its Impact on Humanity." *Nature* 486 (7401). Nature Publishing Group. 59–67. http://dx.doi.org/10.1038/nature11148.

Carleton, Tamma A, and Solomon M Hsiang. 2016. "Social and Economic Impacts of Climate." *Science* 353 (6304). http://science.sciencemag.org/content/353/6304/aad9837.abstract.

Carpenter, Stephen R, Harold A Mooney, John Agard, Doris Capistrano, Ruth S DeFries, Sandra Díaz, Thomas Dietz et al. 2009. "Science for Managing Ecosystem Services: Beyond the Millennium Ecosystem Assessment." *Proceedings of the National Academy of Sciences* 106 (5): 1305–12. doi:10.1073/pnas.0808772106.

Carpenter, Steve, Brian Walker, J Marty Anderies, and Nick Abel. 2001. "From Metaphor to Measurement: Resilience of What to What?" *Ecosystems* 4 (8). Springer: 765–81.

Carrasco, Marc A, Anthony D Barnosky, and Russell W Graham. 2009. "Quantifying the Extent of North American Mammal Extinction Relative to the Pre-Anthropogenic Baseline." *PloS One* 4 (12). Public Library of Science: e8331. doi:10.1371/journal.pone.0008331.

Carter, Luther J, and Thomas H Pigford. 1998. "Getting Yucca Mountain Right." *The Bulletin of the Atomic Scientists* March/April.

Casterline, John B. 2010. "Determinants and Consequences of High Fertility: A Synopsis of the Evidence—Portfolio Review." Washington, DC. http://documents.worldbank.org/curated/en/2010/06/14500903/determinants-consequences-high-fertility-synopsis-evidence-portfolio-review.

Cavallo, Alfred J. 2004. "Hubbert's Petroleum Production Model: An Evaluation and Implications for World Oil Production Forecasts." *Natural Resources Research* 13 (4): 211–21. doi:10.1007/s11053-004-0129-2.

Chait, Jonathan. 2015. "This Is the Year Humans Finally Got Serious About Saving Themselves From Themselves." *New York Magazine*. http://nymag.com/daily/intelligencer/2015/09/sunniest-climate-change-story-ever-read.html.

Charlet, Laurent, and David A Polya. 2006. "Arsenic in Shallow, Reducing Groundwaters in Southern Asia: An Environmental Health Disaster." *Elements* 2 (2): 91–96. doi:10.2113/gselements.2.2.91.

Chen, Yuyu, Avraham Ebenstein, Michael Greenstone, and Hongbin Li. 2013. "Evidence on the Impact of Sustained Exposure to Air Pollution on Life Expectancy from China's Huai River Policy." *Proceedings of the National Academy of Sciences of the United States of America* 110 (32): 12936–41. doi:10.1073/pnas.1300018110.

Cheney, Eric S, and Marianne W Hawkes. 2007. "The Future of Hydrocarbons: Hubbert's Peak or a Plateau?" *GSA Today* 17 (6): 69. doi:10.1130/GSAT01706GW.1.

Chevalier, Manuel, and Brian M Chase. 2016. "Determining the Drivers of Long-Term Aridity Variability: A Southern African Case Study." *Journal of Quaternary Science* 31 (2): 143–51. doi:10.1002/jqs.2850.

Chiabai, Aline B T. 2012. "Biodiversity Conservation." In *Berkshire Encyclopedia of Sustainability*, edited by Louis Kotzé and Stephen Morse, 9:39–42. Afro-Eurasia: Assessing Sustainability. Great Barrington, MA: Berkshire.

Cho, Adrian. 2010. "Energy's Tricky Tradeoffs." *Science* 329 (5993): 786–87. doi:10.1126/science.329.5993.786.

Chowdhury, Md, Yahya Khairun, Md Salequzzaman, and Md Rahman. 2011. "Effect of Combined Shrimp and Rice Farming on Water and Soil Quality in Bangladesh." *Aquaculture International* 19 (6). Springer Netherlands: 1193–1206. doi:10.1007/s10499-011-9433-0.

Coleman, Gerald D. 2005. "Is Genetic Engineering the Answer to Hunger?" *America* 192 (February).

Colizza, Vittoria, Alain Barrat, Marc Barthélemy, and Alessandro Vespignani. 2006. "The Role of the Airline Transportation Network in the Prediction and Predictability of Global Epidemics." *Proceedings of the National Academy of Sciences of the United States of America* 103 (7): 2015–20. doi:10.1073/pnas.0510525103.

Collin, Robin Morris, and Robert William Collin. 2010. "Environment and Ecology." *Encyclopedia of Sustainability*. Greenwood Press.

Committee on Science for EPA's Future, Board on Environmental Studies and Toxicology, Earth and Life Studies, and National Research Council. 2012. *Science for Environmental Protection: The Road Ahead*. The National Academies Press. http://www.nap.edu/openbook.php?record_id=13510.

Committee on Twenty-First Century Systems Agriculture, and National Research Council. 2010. *Toward Sustainable Agricultural Systems in the 21st Century*. National Academies Press.

Cordell, Dana, and Stuart White. 2014. "Life's Bottleneck: Sustaining the World's Phosphorus for a Food Secure Future." *Annual Review of Environment and Resources* 39. Annual Reviews: 161–88.

Corvalan, Carlos, Simon Hales, and Anthony McMichael. 2005. *Ecosystems and Human Well-Being: Health Synthesis: A Report of the Millennium Ecosystem Assessment*. World Health Organization. http://www.millenniumassessment.org/documents/document.357.aspx.pdf.

Costanza, Robert, Gar Alperovitz, Herman Daly, Joshua Farley, Carol Franco, Tim Jackson, Ida Kubiszewski et al. 2013. *Building a Sustainable and Desirable Economy-in-Society-in-Nature: Report to the United Nations for the 2012 Rio+ 20 Conference*. ANU Press.

Costanza, Robert, Ralph d'Arge, Rudolf de Groot, Stephen Farber, Monica Grasso, Bruce Hannon, Karin Limburg et al. 1997. "The Value of the World's Ecosystem Services and Natural Capital." *Nature* 387 (6630): 253-. doi:10.1038/387253a0.

Côté, Stéphane, Julian House, and Robb Willer. 2015. "High Economic Inequality Leads Higher-Income Individuals to Be Less Generous." *Proceedings of the National Academy of Sciences* 112 (52). National Acad Sciences: 15838–43.

Creyts, J C. 2007. *Reducing US Greenhouse Gas Emissions: How Much at What Cost?: US Greenhouse Gas Abatement Mapping Initiative*. McKinsey & Co.

Crimmins, A, J Balbus, J L Gamble, C B Beard, J E Bell, D Dodgen, R J Eisen et al. 2016. "The Impacts of Climate Change on Human Health in the United States: A Scientific Assessment." Washington, DC. http://www.globalchange.gov/browse/reports/impacts-climate-change-human-health-united-states-scientific-assessment.

Crowder, David W, and John P Reganold. 2015. "Financial Competitiveness of Organic Agriculture on a Global Scale." *Proceedings of the National Academy of Sciences* 112 (24): 7611–16. doi:10.1073/pnas.1423674112.

Cudney-Bueno, R, L Bourillón, A Sáenz-Arroyo, J Torre-Cosío, P Turk-Boyer, and William W Shaw. 2009. "Governance and Effects of Marine Reserves in the Gulf of California, Mexico." *Ocean & Coastal Management* 52 (3). Elsevier: 207–18.

Cuéllar, Amanda D, and Michael E Webber. 2010. "Wasted Food, Wasted Energy: The Embedded Energy in Food Waste in the United States." *Environmental Science & Technology* 44 (16). American Chemical Society: 6464–69. doi:10.1021/es100310d.

Daily, Gretchen C, and Paul R Ehrlich. 1992. "Population, Sustainability, and Earth's Carrying Capacity." *BioScience* 42 (10). JSTOR: 761–71.

Dakos, Vasilis, and Jordi Bascompte. 2014. "Critical Slowing down as Early Warning for the Onset of Collapse in Mutualistic Communities." *Proceedings of the National Academy of Sciences* 111 (49): 17546–51. doi:10.1073/pnas.1406326111.

Daly, Herman E. 1990. "Toward Some Operational Principles of Sustainable Development." *Ecological Economics* 2 (1). Elsevier: 1–6.

Daly, Herman E, and Joshua Farley. 2011. *Ecological Economics: Principles And Applications.* 2nd ed. http://www.netlibrary.com/AccessProduct.aspx?ProductId=118237.

Daniel, E Vasey. 2012. "Introduction to Natural Resources and Sustainability." In *Berkshire Encyclopedia of Sustainability*, edited by Daniel E Vasey, Sarah E Fredericks, Shen Lei, and Shirley Thompson, 4:XXI–XXV. Great Barrington, MA: Berkshire.

Dawson, Brian, and Matt Spannagle. 2009. *The Complete Guide to Climate Change.* Routledge.

de Sherbinin, A, M Castro, F Gemenne, M M Cernea, S Adamo, P M Fearnside, G Krieger et al. 2011. "Preparing for Resettlement Associated with Climate Change." *Science* 334 (6055): 456–57. doi:10.1126/science.1208821.

de Vries, B J M. 2013. *Sustainability Science.* Cambridge University Press.

Deffeyes, Kenneth S. 2001. *Hubbert's Peak: The Impending World Oil Shortage.* Princeton, NJ: Princeton University Press.

Delmas, Magali A, Miriam Fischlein, and Omar I Asensio. 2013. "Information Strategies and Energy Conservation Behavior: A Meta-Analysis of Experimental Studies from 1975 to 2012." *Energy Policy* 61. Elsevier: 729–39.

DeOreo, W B, P W Mayer, B Dziegielwski, and J C Kiefer. 2016. "Residential Uses of Water 2016." Denver, CO.

Diamond, Jared. 2005. *Collapse: How Societies Choose to Fail or Succeed.* New York: Viking Penguin.

———. 2014. "Birds and Grapes on Mountaintops." *Proceedings of the National Academy of Sciences* 111 (12): 4349–50. doi:10.1073/pnas.1403168111.

Diaz, Robert J, and Rutger Rosenberg. 2008. "Spreading Dead Zones and Consequences for Marine Ecosystems." *Science* 321 (5891). American Association for the Advancement of Science: 926–29.

Dietz, Thomas, Gerald T Gardner, Jonathan Gilligan, Paul C Stern, and Michael P Vandenbergh. 2009. "Household Actions Can Provide a Behavioral Wedge to Rapidly Reduce US Carbon Emissions." *Proceedings of the National Academy of Sciences* 106 (44): 18452–56. doi:10.1073/pnas.0908738106.

Dietz, Thomas, Elinor Ostrom, and Paul C Stern. 2003. "The Struggle to Govern the Commons." *Science* 302 (5652): 1907–12. doi:10.1126/science.1091015.

Diffenbaugh, Noah S, Daniel L Swain, and Danielle Touma. 2015. "Anthropogenic Warming Has Increased Drought Risk in California." *Proceedings of the National Academy of Sciences* 112 (13): 3931–36. doi:10.1073/pnas.1422385112.

Dinda, Soumyananda. 2004. "Environmental Kuznets Curve Hypothesis: A Survey." *Ecological Economics* 49 (4): 431–55. http://www.sciencedirect.com/science/article/B6VDY-4CVR6PG-1/2/2cc1fbead4d0d8d07f7b6f9563807a7a.

Easterlin, Richard A. 2003. "Explaining Happiness." *Proceedings of the National Academy of Sciences* 100 (19): 11176–83. doi:10.1073/pnas.1633144100.

Eby, G Nelson. 2004. *Principles of Environmental Geochemistry.* Brooks/Cole-Thomson Learning.

Ehrlich, P R, and J P Holdren. 1972. "A Bulletin Dialogue on 'The Closing Circle,' Critique." *Bulletin Atomic Science* 28: 16–27.

Ehrlich, Paul R, and John Harte. 2015. "Opinion: To Feed the World in 2050 Will Require a Global Revolution." *Proceedings of the National Academy of Sciences* 112 (48): 14743–44. doi:10.1073/pnas.1519841112.

EIA. 2015. "Annual Energy Outlook 2015." http://www.eia.gov/forecasts/aeo/pdf/0383(2015).pdf.

———. 2016a. "Annual Energy Outlook 2016." https://www.eia.gov/forecasts/aeo/tables_ref.cfm.

———. 2016b. "International Energy Outlook 2016." www.eia.gov/forecasts/ieo/pdf/0484(2016).pdf.

Emanuel, Kerry. 2012. *What We Know about Climate Change.* 2nd ed. MIT Press.

Engel, Stefanie, Stefano Pagiola, and Sven Wunder. 2008. "Designing Payments for Environmental Services in Theory and Practice: An Overview of the Issues." *Ecological Economics* 65 (4). Elsevier: 663–74. http://www.sciencedirect.com/science/article/pii/S0921800908001420.

Engelman, Robert. 2009. "Population and Sustainability: Can We Avoid Limiting the Number of People?" *Scientific American*, no. June 2009. http://www.scientificamerican.com/article.cfm?id=population-and-sustainability.

Enkvist, Per-Anders, Jens Dinkel, and Charles Lin. 2010. "Impact of the Financial Crisis on Carbon Economics: Version 2.1 of the Global Greenhouse Gas Abatement Cost Curve." McKinsey & Company. http://www.mckinsey.com/clientservice/sustainability/pdf/Impact_Financial_Crisis_Carbon_Economics_GHGcostcurveV2.1.pdf.

Environmental Protection Agency. 2003. "Drinking Water Advisory: Consumer Acceptability Advice and Health Effects Analysis." www.epa.gov/safewater/ccl/pdf/sodium.pdf.

EPA. 1998. "Puzzled About Recycling's Value? Look Beyond the Bin." Edited by United States Environmental Protection Agency. U.S. Government.

———. 2015. "Advancing Sustainable Materials Management: 2013 Fact Sheet." https://www.epa.gov/sites/production/files/2015-09/documents/2013_advncng_smm_fs.pdf.

EPRI. 2011. "Estimating the Costs and Benefits of the Smart Grid: Technical Report 1022519." www.rmi.org/Content/Files/EstimatingCostsSmartGRid.pdf.

Falkenmark, Malin. 1989. "The Massive Water Scarcity Now Threatening Africa: Why Isn't It Being Addressed?" *Ambio*. JSTOR, 112–18.

FAO. 2005. "AQUASTAT Information System on Water and Agriculture." Food and Agriculture Organization of the United Nations (FAO), Land and Water Development Division, Rome. www.fao.org/nr/water/aqustat/maps/index.stm.

Farley, Joshua. 2010. "Ecological Economics." In *The Post Carbon Reader: Managing the 21st Century's Sustainability Crises*, 259–78.

Faure, G. 1998. *Principles and Applications of Geochemistry: A Comprehensive Textbook for Geology Students*. 2nd ed. Prentice Hall.

Fearnside, Philip M, and Salvador Pueyo. 2012. "Greenhouse-Gas Emissions from Tropical Dams." *Nature Clim. Change* 2 (6). Nature Publishing Group. 382–84. http://dx.doi.org/10.1038/nclimate1540.

Feinberg, Matthew, and Robb Willer. 2011. "Apocalypse Soon? Dire Messages Reduce Belief in Global Warming by Contradicting Just-World Beliefs." *Psychological Science* 22 (1): 34–38. doi:10.1177/0956797610391911.

Feng, Shuaizhang, Alan B Krueger, and Michael Oppenheimer. 2010. "Linkages among Climate Change, Crop Yields and Mexico–US Cross-Border Migration." *Proceedings of the National Academy of Sciences* 107 (32): 14257–62. doi:10.1073/pnas.1002632107.

Ferreira, C, A Ribeiro, and L Ottosen. 2003. "Possible Applications for Municipal Solid Waste Fly Ash." *Journal of Hazardous Materials* 96 (2–3): 201–16. doi:http://dx.doi.org/10.1016/S0304-3894(02)00201-7.

Finer, Matt, and Clinton N Jenkins. 2012. "Proliferation of Hydroelectric Dams in the Andean Amazon and Implications for Andes-Amazon Connectivity." *PLoS ONE* 7 (4). Public Library of Science: e35126. http://dx.doi.org/10.1371/journal.pone.0035126.

Fischer, Carolyn, and Richard G Newell. 2008. "Environmental and Technology Policies for Climate Mitigation." *Journal of Environmental Economics and Management* 55 (2): 142–62. doi:http://dx.doi.org/10.1016/j.jeem.2007.11.001.

Fischer, Corinna. 2008. "Feedback on Household Electricity Consumption: A Tool for Saving Energy?" *Energy Efficiency* 1 (1): 79–104. doi:10.1007/s12053-008-9009-7.

Flannery, T. 2008. *Now or Never: A Sustainable Future for Australia?* Black Inc.

Foden, Wendy B, Stuart H M Butchart, Simon N Stuart, Jean-Christophe Vié, H Resit Akçakaya, Ariadne Angulo, Lyndon M DeVantier et al. 2013. "Identifying the World's Most Climate Change Vulnerable Species: A Systematic Trait-Based Assessment of All Birds, Amphibians and Corals." *PLoS ONE* 8 (6). Public Library of Science: e65427. http://dx.doi.org/10.1371%252Fjournal.pone.0065427.

Foley, Jonathan A, Navin Ramankutty, Kate A Brauman, Emily S Cassidy, James S Gerber, Matt Johnston, Nathaniel D Mueller et al. 2011. "Solutions for a Cultivated Planet." *Nature* 478 (7369). Nature Publishing Group, a division of Macmillan Publishers Limited. All Rights Reserved.: 337–42.

Food & Water Watch and Network for New Energy Choices. 2007. "The Rush to Ethanol: Not All Fuels Are Created Equal." http://www.newenergychoices.org/uploads/RushToEthanol-rep.pdf.

Food and Agriculture Organization of the United Nations. 2005. *Understanding Aquaculture*. Rome: Food and Agriculture Organization of the United Nations.

Frank, Andrew A. 2007. "Plug-in Hybrid Vehicles for a Sustainable Future." *American Scientist* 95 (2): 158–65.

Freeman, Benjamin G, and Alexandra M Class Freeman. 2014. "Rapid Upslope Shifts in New Guinean Birds Illustrate Strong Distributional Responses of Tropical Montane Species to Global Warming." *Proceedings of the National Academy of Sciences* 111 (12): 4490–94. doi:10.1073/pnas.1318190111.

Friedman, Thomas. 2008. *Hot, Flat, and Crowded: Why We Need a Green Revolution—And How It Can Renew America*. Farrar, Strauss and Giroux.

Friedman, Thomas L. 2006. *The World Is Flat: A Brief History of the Twenty-First Century*. New York: Farrar, Straus and Giroux.

Friedrichs, Jörg. 2010. "Global Energy Crunch: How Different Parts of the World Would React to a Peak Oil Scenario." *Energy Policy* 38 (8): 4562–69. doi:http://dx.doi.org/10.1016/j.enpol.2010.04.011.

Frohlich, Cliff. 2012. "Two-Year Survey Comparing Earthquake Activity and Injection-Well Locations in the Barnett Shale, Texas." *Proceedings of the National Academy of Sciences* 109 (35): 13934–38. doi:10.1073/pnas.1207728109.

Frumkin, Howard. 2005. *Environmental Health: From Global to Local*. Vol. 10. Jossey-Bass.

Fthenakis, Vasilis. 2009. "Sustainability of Photovoltaics: The Case for Thin-Film Solar Cells." *Renewable and Sustainable Energy Reviews* 13 (9): 2746–50. doi:http://dx.doi.org/10.1016/j.rser.2009.05.001.

Fubini, Bice, and Ivana Fenoglio. 2007. "Toxic Potential of Mineral Dusts." *Elements* 3 (6). GeoScienceWorld: 407–14.

Gallopín, Gilberto C. 2006. "Linkages between Vulnerability, Resilience, and Adaptive Capacity." *Global Environmental Change* 16 (3): 293–303. doi:10.1016/j.gloenvcha.2006.02.004.

Gardner, Gary. 2013. "Conserving Nonrenewable Resources." In *State of the World 2013: Is Sustainability Still Possible?*, 99–109. Island Press.

Gatehouse, Angharad M R, Natalie Ferry, and Romaan J M Raemaekers. 2002. "The Case of the Monarch Butterfly: A Verdict Is Returned." *Trends in Genetics* 18 (5): 249–51. doi:http://dx.doi.org/10.1016/S0168-9525(02)02664-1.

Gemenne, Francois. 2015. "Migration as a Climate Adaptation Strategy." In *State of the World 2015: Confronting Hidden Threats to Sustainability*, 117–26. Washington, DC: Island Press.

Gerland, Patrick, Adrian E Raftery, Hana Ševčíková, Nan Li, Danan Gu, Thomas Spoorenberg, Leontine Alkema et al. 2014. "World Population Stabilization Unlikely This Century." *Science* 346 (6206): 234–37. doi:10.1126/science.1257469.

GFN. 2009. "Global Footprint Network Response to The 'Commission on the Measurement of Economic Performance and Social Progress' ('Stiglitz Commission') Report." http://www.footprintnetwork.org/images/uploads/Global Footprint Network Stiglitz response.pdf.

Gibbs, H K, A S Ruesch, F Achard, M K Clayton, P Holmgren, N Ramankutty, and J A Foley. 2010. "Tropical Forests Were the Primary Sources of New Agricultural Land in the 1980s and 1990s." *Proceedings of the National Academy of Sciences* 107 (38): 16732–37. doi:10.1073/pnas.0910275107.

Gieré, Reto, and Xavier Querol. 2010. "Solid Particulate Matter in the Atmosphere." *Elements* 6 (4). Mineralogical Association of Canada: 215–22.

Gilbert, Natasha. 2009. "Environment: The Disappearing Nutrient." *Nature News* 461 (7265). Nature Publishing Group: 716–18.

Gilbert, Richard, and Anthony Perl. 2010. "Transportation in the Post-Carbon World." *The Post Carbon Reader*, 347–60.

Gleeson, Tom, Yoshihide Wada, Marc F P Bierkens, and Ludovicus P H van Beek. 2012. "Water Balance of Global Aquifers Revealed by Groundwater Footprint." *Nature* 488 (7410). Nature Publishing Group. 197–200. doi:http://www.nature.com/nature/journal/v488/n7410/abs/nature11295.html#supplementary-information.

Gleick, P H. 2009. *The World's Water 2008–2009: The Biennial Report on Freshwater Resources*. Island Press.

———. 2010. *Bottled and Sold: The Story behind Our Obsession with Bottled Water*. Shearwater.

Gleick, Peter H. 2003. "Global Freshwater Resources: Soft-Path Solutions for the 21st Century." *Science* 302 (5650): 1524–28. doi:10.1126/science.1089967.

Gleick, Peter H, and Newsha Ajami. 2011. *The World's Water Volume 7: The Biennial Report on Freshwater Resources*. Vol. 7. Island Press.

Gleick, Peter H, and Meena Palaniappan. 2010. "Peak Water Limits to Freshwater Withdrawal and Use." *Proceedings of the National Academy of Sciences* 107 (25): 11155–62. doi:10.1073/pnas.1004812107.

Global Footprint Network. 2016. "National Footprint Accounts, 2016 Edition." http://www.footprintnetwork.org/en/index.php/GFN/page/footprint_data_and_results/.

Godfray, H Charles J. 2011. "Food for Thought." *Proceedings of the National Academy of Sciences* 108 (50): 19845–46. doi:10.1073/pnas.1118568109.

Goodland, Robert, and Herman Daly. 1996. "Environmental Sustainability: Universal and Non-Negotiable." *Ecological Applications* 6 (4). Ecological Society of America: 1002–17. http://www.jstor.org/stable/2269583.

Gore, Al. 2009. *Our Choice: A Plan to Solve the Climate Crisis*. Rodale, Inc.

Graedel, T E, E M Harper, N T Nassar, and Barbara K Reck. 2015. "On the Materials Basis of Modern Society." *Proceedings of the National Academy of Sciences* 112 (20): 6295–6300. doi:10.1073/pnas.1312752110.

Granade, Hannah Choi, John Creyts, Anton Derkach, Philip Farese, Scott Nyquist, and Ken Ostrowski. 2009. "Unlocking Energy Efficiency in the U.S. Economy." McKinsey & Co. http://www.mckinsey.com/clientservice/electricpowernaturalgas/downloads/us_energy_efficiency_full_report.pdf.

Gray, Clark L, and Valerie Mueller. 2012. "Natural Disasters and Population Mobility in Bangladesh." *Proceedings of the National Academy of Sciences* 109 (16): 6000–6005. doi:10.1073/pnas.1115944109.

Greenblatt, Jeffery B, and Max Wei. 2016. "Assessment of the Climate Commitments and Additional Mitigation Policies of the United States." *Nature Climate Change* 6 (September). Nature Publishing Group: 1090–93. http://dx.doi.org/10.1038/nclimate3125.

Greening, Lorna A, David L Greene, and Carmen Difiglio. 2000. "Energy Efficiency and Consumption—The Rebound Effect—A Survey." *Energy Policy* 28 (6). Elsevier: 389–401.

Greer, J M. 2009. *The Ecotechnic Future: Envisioning a Post-Peak World.* New Society Pub.

Grobéty, Bernard, Reto Gieré, Volker Dietze, and Peter Stille. 2010. "Airborne Particles in the Urban Environment." *Elements* 6 (4). GeoScienceWorld: 229–34.

Grorud-Colvert, Kirsten, and Sarah E Lester. 2012. "Marine Protected Areas (MPAs)." In *Berkshire Encyclopedia of Sustainability*, edited by Robin Kundis Craig, John Copeland Nagle, Bruce Pardy, Oswald J Schmitz, and William K Smith, 5: Ecosyst: 252–55. Ecosystems Management and Sustainability. Great Barrington, MA: Berkshire.

Grubert, Emily A, Fred C Beach, and Michael E Webber. 2012. "Can Switching Fuels Save Water? A Life Cycle Quantification of Freshwater Consumption for Texas Coal- and Natural Gas-Fired Electricity." *Environmental Research Letters* 7 (4): 45801. doi:10.1088/1748-9326/7/4/045801.

Gunda, Thushara, Laura Benneyworth, and Emily Burchfield. 2014. "Exploring Water Indices and Associated Parameters: A Case Study Approach." *Water Policy* 17. IWA Publishing: 98–111. doi:10.2166/wp.2014.022.

Hails, Chris. 2008. "Living Planet Report." World Wildlife Federation, Zoological Society of London, and the Global Footprint Network. http://www.footprintnetwork.org/download.php?id=505.

Hall, Charles S A, and John W Jr. Day. 2009. "Revisiting the Limits to Growth After Peak Oil." *American Scientist* 97: 230–37.

Hall, Kevin D, Juen Guo, Michael Dore, and Carson C Chow. 2009. "The Progressive Increase of Food Waste in America and Its Environmental Impact." *PLoS ONE* 4 (11). Public Library of Science: e7940. http://dx.doi.org/10.1371/journal.pone.0007940.

Halweil, Brian. 2006. "Can Organic Farming Feed Us All?" *World Watch Magazine*, May/June: 18–24.

Halweil, Brian, and Danielle Nierenberg. 2011. "Charting a New Path to Eliminating Hunger." In *State of the World 2011: Innovations That Nourish the Planet*, edited by Linda Starke, 3–12. Washington, DC: Worldwatch Institute.

Hammarstrom, Jane M, Robert R Seal, Allen L Meier, and John C Jackson. 2003. "Weathering of Sulfidic Shale and Copper Mine Waste: Secondary Minerals and Metal Cycling in Great Smoky Mountains National Park, Tennessee, and North Carolina, USA." *Environmental Geology* 45 (1): 35–57. doi:10.1007/s00254-003-0856-4.

Hanna-Attisha, Mona, Jenny LaChance, Richard Casey Sadler, and Allison Champney Schnepp. 2015. "Elevated Blood Lead Levels in Children Associated With the Flint Drinking Water Crisis: A Spatial Analysis of Risk and Public Health Response." *American Journal of Public Health* 106 (2). American Public Health Association: 283–90. doi:10.2105/AJPH.2015.303003.

Hansen, James. 2006. "Swift Boating, Stealth Budgeting, and Unitary Executives." World Watch Magazine, November/December: 26,28-29,31.

Hansen, James, Makiko Sato, Gary Russell, and Pushker Kharecha. 2013. "Climate Sensitivity, Sea Level and Atmospheric Carbon Dioxide." *Philosophical Transactions of the Royal Society of London A: Mathematical, Physical and Engineering Sciences* 371 (2001). http://rsta.royalsocietypublishing.org/content/371/2001/20120294.abstract.

Hansen, J, M Sato, P Kharecha, D Beerling, R Berner, V Masson-Delmotte, M Pagani, M Raymo, D L Royer, and J C Zachos. 2008. "Target Atmospheric CO_2: Where Should Humanity Aim?" *Open Atmospheric Science Journal* 2: 217–31.

Hanski, Ilkka, Leena von Hertzen, Nanna Fyhrquist, Kaisa Koskinen, Kaisa Torppa, Tiina Laatikainen, Piia Karisola et al. 2012. "Environmental Biodiversity, Human Microbiota, and Allergy Are Interrelated." *Proceedings of the National Academy of Sciences* 109 (21): 8334–39. doi:10.1073/pnas.1205624109.

Hardin, Garrett. 1968. "The Tragedy of the Commons." Science 162 (3859): 1243–48. doi:10.1126/science.162.3859.1243.

———. 1998. "Extensions of 'The Tragedy of the Commons'." *Science* 280 (5364): 682–83. http://science.sciencemag.org/content/280/5364/682.abstract.

Havala, Suzanne. 2001. *Being Vegetarian for Dummies.* Wiley Publishing, Inc.

Hayes, Tyrone B, Vicky Khoury, Anne Narayan, Mariam Nazir, Andrew Park, Travis Brown, Lillian Adame et al. 2010. "Atrazine Induces Complete Feminization and Chemical Castration in Male African Clawed Frogs (Xenopus Laevis)." *Proceedings of the National Academy of Sciences* 107 (10): 4612–17. doi:10.1073/pnas.0909519107.

Heede, Richard. 2014. "Tracing Anthropogenic Carbon Dioxide and Methane Emissions to Fossil Fuel and Cement Producers, 1854–2010." *Climatic Change* 122 (1–2). Springer: 229–41.

Heinberg, R. 2004. *Powerdown: Options and Actions for a Post-Carbon World*. New Society Pub.

Heinberg, R, and D Lerch. 2010. *The Post Carbon Reader: Managing the 21st Century's Sustainability Crises*. University of California Press.

Heinberg, Richard. 2010. "Beyond the Limits to Growth." *Post Carbon Institute, Foundation Series Concepts*. Jul 28: 3–12.

Heinberg, Richard, and David Fridley. 2010. "The End of Cheap Coal." *Nature* 468 (7322). Nature Publishing Group. 367–69. http://dx.doi.org/10.1038/468367a.

Henry, Mickaël, Maxime Béguin, Fabrice Requier, Orianne Rollin, Jean-François Odoux, Pierrick Aupinel, Jean Aptel et al. 2012. "A Common Pesticide Decreases Foraging Success and Survival in Honey Bees." *Science* 336 (6079): 348–50. doi:10.1126/science.1215039.

Herring, Horace. 2008. "Rebound Effect." Edited by Cutler Cleveland. *Encyclopedia of Earth*. Washington, DC: Environmental Information Coalition, National Council for Science and the Environment. http://www.eoearth.org/article/Rebound_effect.

Hill, Graham, and Meaghan O'Neill. 2008. *Ready, Set, Green: Eight Weeks to Modern Eco-Living*. Villard.

Hirsch, R L, R Bezdek, and R Wendling. 2005. "Peaking of World Oil Production: Impacts, Mitigation, & Risk Management." doi:10.2172/939271.

Holland, Robert Alan, Kate A Scott, Martina Flörke, Gareth Brown, Robert M Ewers, Elizabeth Farmer, Valerie Kapos et al. 2015. "Global Impacts of Energy Demand on the Freshwater Resources of Nations." *Proceedings of the National Academy of Sciences* 112 (48): E6707–16. doi:10.1073/pnas.1507701112.

Homer-Dixon, T. 2006. *The Upside of down: Catastrophe, Creativity and the Renewal of Civilization*. Island Press.

Hopenhayn, Claudia. 2006. "Arsenic in Drinking Water: Impact on Human Health." *Elements* 2 (2): 103–7. doi:10.2113/gselements.2.2.103.

Hopkins, R. 2008. *The Transition Handbook: From Oil Dependency to Local Resilience*. Chelsea Green Publishing.

Hornberger, George M, Patricia L Wiberg, Paolo D'Odorico, and Jeffrey P Raffensperger. 2014. *Elements of Physical Hydrology*. JHU Press.

Howarth, R W, D M Anderson, T M Church, H Greening, C S Hopkinson, W C Huber, N Marcus et al. 2000. *Clean Coastal Waters: Understanding and Reducing the Effects of Nutrient Pollution*. Washington, DC: National Academy of Sciences.

Howarth, Robert W, Renee Santoro, and Anthony Ingraffea. 2011. "Methane and the Greenhouse-Gas Footprint of Natural Gas from Shale Formations." *Climatic Change* 106 (4): 679–90. doi:10.1007/s10584-011-0061-5.

Howe, Peter D, Matto Mildenberger, Jennifer R Marlon, and Anthony Leiserowitz. 2015. "Geographic Variation in Opinions on Climate Change at State and Local Scales in the USA." *Nature Clim. Change* 5 (6). Nature Publishing Group: 596–603. http://dx.doi.org/10.1038/nclimate2583.

Hubbert, M K. 1956. "Nuclear Energy and the Fossil Fuel." *Drilling and Production Practice*.

Hubbert, M King. 1974. "Statements on the Relations between Industrial Growth, the Monetary Interest Rate, and Price Inflation." *Hearing before the Sub-Committee on the Environment, Committee on Interior and Insular Affairs*, House of Representatives, Serial, no. 93–55.

———. 1987. "Exponential Growth as a Transient Phenomenon in Human History." In *Societal Issues, Scientific Viewpoints*, edited by Margaret A Strom, 75–84. New York: American Institute of Physics.

Hughes, J David. 2010. "Hydrocarbons in North America." In *The Post Carbon Reader: Managing the 21st Century's Sustainability Crises*, 211–28.

Hutchings, Jeffrey A. 2000. "Collapse and Recovery of Marine Fishes." *Nature* 406 (6798). Nature Publishing Group: 882–85.

IPCC. 2007. Climate Change 2007: The Physical Science Basis. Contribution of Working Group I to the Fourth Assessment Report of the Intergovernmental Panel on Climate Change. Edited by S Solomon, D Qin, M Manning, Z Chen, M Marquis, K B Averyt, M Tignor et al. Cambridge, U.K. and New York, NY: Cambridge University Press.

———. 2014. "Summary for Policymakers." In Climate Change 2014: Impacts, Adaptation, and Vulnerability. Part A: Global and Sectoral Aspects. Contribution of Working Group II to the Fifth Assessment Report of the Intergovernmental Panel on Climate Change, edited by C B Field, V R Barros, D J Dokken, K J Mach, M D Mastrandrea, T E Bilir, M Chatterjee et al., 1–32. Cambridge, United Kingdom, and New York : Cambridge University Press.

Ives, Anthony R, and Stephen R Carpenter. 2007. "Stability and Diversity of Ecosystems." *Science* 317 (5834). American Association for the Advancement of Science: 58–62.

Jack, B Kelsey, Carolyn Kousky, and Katharine R E Sims. 2008. "Designing Payments for Ecosystem Services: Lessons from Previous Experience with Incentive-Based Mechanisms." *Proceedings of the National Academy of Sciences* 105 (28). National Acad Sciences: 9465–70.

Jackson, Robert B, Avner Vengosh, Thomas H Darrah, Nathaniel R Warner, Adrian Down, Robert J Poreda, Stephen G Osborn et al. 2013. "Increased Stray Gas Abundance in a Subset of Drinking Water Wells near Marcellus Shale Gas Extraction." *Proceedings of the National Academy of Sciences of the United States of America* 110 (28): 11250–55. doi:10.1073/pnas.1221635110.

Jackson, Robert B, Avner Vengosh, J William Carey, Richard J Davies, Thomas H Darrah, Francis O'Sullivan, and Gabrielle Pétron. 2014. "The Environmental Costs and Benefits of Fracking." *Annual Review of Environment and Resources* 39. Annual Reviews: 327–62. doi:10.1146/annurev-environ-031113-144051.

Jackson, Tim. 2008. "The Challenge of Sustainable Lifestyles." In *State of the World 2008: Innovations for a Sustainable Economy*, edited by Linda Starke, 45–60. W. W. Norton & Company.

Jackson, Wes. 2010. "Tackling the Oldest Environmental Problem: Agriculture and Its Impact on Soil." In *The Post Carbon Reader*, edited by Richard Heinberg and Daniel Lerch, 128–39. Watershed Media.

Jacobson, Mark Z. 2009. "Review of Solutions to Global Warming, Air Pollution, and Energy Security." *Energy & Environmental Science* 2 (2). The Royal Society of Chemistry: 148–73. http://dx.doi.org/10.1039/B809990C.

Jacobson, Mark Z, and Mark A Delucchi. 2011. "Providing All Global Energy with Wind, Water, and Solar Power, Part I: Technologies, Energy Resources, Quantities and Areas of Infrastructure, and Materials." *Energy Policy* 39 (3): 1154–69. doi:10.1016/j.enpol.2010.11.040.

Jacobson, Mark Z, Mark A Delucchi, Mary A Cameron, and Bethany A Frew. 2015. "Low-Cost Solution to the Grid Reliability Problem with 100% Penetration of Intermittent Wind, Water, and Solar for All Purposes." *Proceedings of the National Academy of Sciences* 112 (49): 15060–65. doi:10.1073/pnas.1510028112.

Jaffe, Adam B, Richard G Newell, and Robert N Stavins. 2005. "A Tale of Two Market Failures: Technology and Environmental Policy." *Ecological Economics* 54 (2–3): 164–74. doi:http://dx.doi.org/10.1016/j.ecolecon.2004.12.027.

Jeffery, Yvonne, Liz Barclay, and Michael Grosvenor. 2008. *Green Living for Dummies*. For Dummies.

John, Sepkoski J. 1998. "Rates of Speciation in the Fossil Record." *Philosophical Transactions of the Royal Society of London B: Biological Sciences* 353 (1366): 315–26. http://rstb.royalsocietypublishing.org/content/353/1366/315.abstract.

Johnston, Richard B, Michael Berg, C Annette Johnson, Elizabeth Tilley, and Janet G Hering. 2011. "Water and Sanitation in Developing Countries: Geochemical Aspects of Quality and Treatment." *Elements* 7 (3): 163–68. doi:10.2113/gselements.7.3.163.

Joos, Fortunat, and Renato Spahni. 2008. "Rates of Change in Natural and Anthropogenic Radiative Forcing over the Past 20,000 Years." *Proceedings of the National Academy of Sciences* 105 (5): 1425–30. doi:10.1073/pnas.0707386105.

Jordan, T E et al. 2010. "Recommendations for Interdisciplinary Study of Tipping Points in Natural and Social Systems." *EOS* 91 (16): 143–44. http://www.agu.org/journals/eo/eo1016/2010EO160005.pdf#anchor.

Kahan, Dan M, Ellen Peters, Maggie Wittlin, Paul Slovic, Lisa Larrimore Ouellette, Donald Braman, and Gregory Mandel. 2012. "The Polarizing Impact of Science Literacy and Numeracy on Perceived Climate Change Risks." *Nature Clim. Change* 2 (10). Nature Publishing Group: 732–35. http://dx.doi.org/10.1038/nclimate1547.

Kahneman, Daniel. 2011. *Thinking, Fast and Slow*. Macmillan.

Karl, Thomas R, Jerry M Melillo, and Thomas C Peterson. 2009. *Global Climate Change Impacts in the United States*. Cambridge University Press.

Kates, R W, C E Colten, S Laska, and S P Leatherman. 2006. "Reconstruction of New Orleans after Hurricane Katrina: A Research Perspective." *Proceedings of the National Academy of Sciences* 103 (40): 14653–60. doi:10.1073/pnas.0605726103.

Kathage, Jonas, and Matin Qaim. 2012. "Economic Impacts and Impact Dynamics of Bt (Bacillus Thuringiensis) Cotton in India." *Proceedings of the National Academy of Sciences* 109 (29): 11652–56. doi:10.1073/pnas.1203647109.

Kaufman, David Scott, and Jay Noller. 1999. "The Effect of Pine Afforestation on Copper and Iron Movement Through the Recovering Soils of the Copper Basin Mining District, Ducktown, Tennessee." *Geology*. Nashville, TN: Vanderbilt University.

Kaya, Y, and K Yokoboi. 1993. *Environment, Energy, and Economy: Strategies for Sustainability*. Tokyo, Japan: Tokyo Conference on Global Environment.

Kelemen, Peter B, and J Matter. 2008. "In Situ Carbonation of Peridotite for CO_2 Storage." *Proceedings of the National Academy of Sciences* 105 (45): 17295–300. doi:10.1073/pnas.0805794105.

Keller, Edward A. 2005. *Introduction to Environmental Geology*. 3rd ed. Prentice-Hall, Inc.

———. 2011. *Environmental Geology*. 9th ed. Pearson Prentice Hall.

Kellogg, Scott, and Stacy Pettigrew. 2008. Toolbox for Sustainable City Living. Cambridge, MA: South End Press.

Kelly, Anne E, and Michael L Goulden. 2008. "Rapid Shifts in Plant Distribution with Recent Climate Change." *Proceedings of the National Academy of Sciences* 105 (33): 11823–26. doi:10.1073/pnas.0802891105.

Kelly, Erin N, David W Schindler, Peter V Hodson, Jeffrey W Short, Roseanna Radmanovich, and Charlene C Nielsen. 2010. "Oil Sands Development Contributes Elements Toxic at Low Concentrations to the Athabasca River and Its Tributaries." *Proceedings of the National Academy of Sciences* 107 (37): 16178–83. doi:10.1073/pnas.1008754107.

Kennedy, Christopher A, Iain Stewart, Angelo Facchini, Igor Cersosimo, Renata Mele, Bin Chen, Mariko Uda et al. 2015. "Energy and Material Flows of Megacities." *Proceedings of the National Academy of Sciences* 112 (19): 5985–90. doi:10.1073/pnas.1504315112.

Kennedy, Christopher, John Cuddihy, and Joshua Engel-Yan. 2007. "The Changing Metabolism of Cities." *Journal of Industrial Ecology* 11 (2). Wiley Online Library: 43–59.

Khan, A E, Andrew Ireson, and Sari Kovats. 2011. "Drinking Water Salinity and Maternal Health in Coastal Bangladesh: Implications of Climate Change." *Environmental Health Perspectives* 119 (9): 1328–32. http://www.ncbi.nlm.nih.gov/pmc/articles/PMC3230389/.

Kharaka, Y K, J J Thordsen, C H Conaway, and R B Thomas. 2013. "The Energy-Water Nexus: Potential Groundwater-Quality Degradation Associated with Production of Shale Gas." *Procedia Earth and Planetary Science* 7 (January): 417–22. doi:10.1016/j.proeps.2013.03.132.

Kharecha, Pushker A, and James E Hansen. 2013. "Prevented Mortality and Greenhouse Gas Emissions from Historical and Projected Nuclear Power." *Environmental Science & Technology* 47 (9). American Chemical Society: 4889–95. doi:10.1021/es3051197.

King, George E. 2012. "Hydraulic Fracturing 101: What Every Representative, Environmentalist, Regulator, Reporter, Investor, University Researcher, Neighbor and Engineer Should Know About Estimating Frac Risk and Improving Frac Performance in Unconventional Gas." In *SPE Hydraulic Fracturing Technology Conference*, 1–80. The Woodlands, Texas, USA: Society of Petroleum Engineers. doi:10.2118/152596-MS.

Klaver, Irene. 2009. "Water." Edited by J Callicott and Robert Frodeman. *Encyclopedia of Environmental Ethics and Philosophy*. Detriot, MI: Macmillan Reference USA.

Komanoff, Charles. 2006. "Whither Wind? A Journey through the Heated Debate over Wind Power." *Orion*, September/October.

Koomey, J G, S Berard, M Sanchez, and H Wong. 2011. "Implications of Historical Trends in the Electrical Efficiency of Computing." *Annals of the History of Computing, IEEE* 33 (3): 46–54. doi:10.1109/mahc.2010.28.

Kopp, Robert E, Andrew C Kemp, Klaus Bittermann, Benjamin P Horton, Jeffrey P Donnelly, W Roland Gehrels, Carling C Hay et al. 2016. "Temperature-Driven Global Sea-Level Variability in the Common Era." *Proceedings of the National Academy of Sciences* 113 (11): E1434–41. doi:10.1073/pnas.1517056113.

Krkošek, Martin, Brendan M Connors, Alexandra Morton, Mark A Lewis, Lawrence M Dill, and Ray Hilborn. 2011. "Effects of Parasites from Salmon Farms on Productivity of Wild Salmon." *Proceedings of the National Academy of Sciences* 108 (35): 14700–704. doi:10.1073/pnas.1101845108.

Kroeker, Kristy J, Fiorenza Micheli, Maria Cristina Gambi, and Todd R Martz. 2011. "Divergent Ecosystem Responses within a Benthic Marine Community to Ocean Acidification." *Proceedings of the National Academy of Sciences* 108 (35): 14515–20. doi:10.1073/pnas.1107789108.

Kuhn, Thomas S. 1970. *The Structure of Scientific Revolutions*. 2nd ed. The University of Chicago Press.

Kwong, Jo. 2005. "Globalization's Effects on the Environment." *Society* Jan/Feb: 21–28.

Lal, Rattan. 2004. "Soil Carbon Sequestration Impacts on Global Climate Change and Food Security." *Science* 304 (5677). American Association for the Advancement of Science: 1623–27.

Lambin, Eric F, and Patrick Meyfroidt. 2011. "Global Land Use Change, Economic Globalization, and the Looming Land Scarcity." *Proceedings of the National Academy of Sciences* 108 (9): 3465–72. doi:10.1073/pnas.1100480108.

Lamborg, Carl H, Chad R Hammerschmidt, Katlin L Bowman, Gretchen J Swarr, Kathleen M Munson, Daniel C Ohnemus, Phoebe J Lam et al. 2014. "A Global Ocean Inventory of Anthropogenic Mercury Based on Water Column Measurements." *Nature* 512 (7512). Nature Publishing Group. 65–68. http://dx.doi.org/10.1038/nature13563.

Langmuir, Donald. 1997. *Aqueous Environmental Geochemistry*. Upper Saddle River, NJ: Prentice Hall.

Larson, Christina. 2010. "Growing Shortages of Water Threaten China's Development." *Yale Environment 360*. http://e360.yale.edu/feature/growing_shortages_of_water_threaten_chinas_development/2298/.

Lattemann, Sabine, and Thomas Höpner. 2008. "Environmental Impact and Impact Assessment of Seawater Desalination." *Desalination* 220 (1). Elsevier: 1–15.

Lavelle, Marianne. 2012. "Good Gas Bad Gas." *National Geographic* 222: 90.

Lean, J L, and D H Rind. 2009. "How Will Earth's Surface Temperature Change in Future Decades?" *Geophysical Research Letters* 36 (15): L15708.

Leape, James P, and Sarah Humphrey. 2010. "Towards a Sustainable Future." In *Global Sustainability: A Nobel Cause*, edited by Hans Joachim Schellnhuber, Mario Molina, Nicholas Stern, Veronika Huber, and Susanne Kadner, 49–64. Cambridge, UK: Cambridge University Press.

Lelieveld, J, J S Evans, M Fnais, D Giannadaki, and A Pozzer. 2015. "The Contribution of Outdoor Air Pollution Sources to Premature Mortality on a Global Scale." *Nature* 525 (7569). Nature Publishing Group. 367–71. http://dx.doi.org/10.1038/nature15371.

Lewis, Simon L, and Mark A Maslin. 2015. "Defining the Anthropocene." *Nature* 519 (7542). Nature Publishing Group: 171–80.

Li, Duo. 2014. "Effect of the Vegetarian Diet on Non-Communicable Diseases." *Journal of the Science of Food and Agriculture* 94 (2). John Wiley & Sons, Ltd: 169–73. doi:10.1002/jsfa.6362.

Lindberg, T Ty, Emily S Bernhardt, Raven Bier, A M Helton, R Brittany Merola, Avner Vengosh, and Richard T Di Giulio. 2011. "Cumulative Impacts of Mountaintop Mining on an Appalachian Watershed." *Proceedings of the National Academy of Sciences* 108 (52): 20929–34. doi:10.1073/pnas.1112381108.

Little, Jane Braxton. 2009. "The Ogallala Aquifer: Saving a Vital U.S. Water Source." *Scientific American Earth 3.0*. http://www.sciam.com/article.cfm?id=the-ogallala-aquifer.

Llewellyn, Garth T, Frank Dorman, J L Westland, D Yoxtheimer, Paul Grieve, Todd Sowers, E Humston-Fulmer et al. 2015. "Evaluating a Groundwater Supply Contamination Incident Attributed to Marcellus Shale Gas Development." *Proceedings of the National Academy of Sciences* 112 (20): 6325–30. doi:10.1073/pnas.1420279112.

Long, Jane C S, and Rodney C Ewing. 2004. "YUCCA MOUNTAIN: Earth-Science Issues at a Geologic Repository for High-Level Nuclear Waste." *Annual Review of Earth and Planetary Sciences* 32 (1): 363–401. http://arjournals.annualreviews.org/loi/earth.

Long, Stephen P, Elizabeth A Ainsworth, Andrew D B Leakey, Josef Nösberger, and Donald R Ort. 2006. "Food for Thought: Lower-than-Expected Crop Yield Stimulation with Rising CO2 Concentrations." *Science* 312 (5782). American Association for the Advancement of Science: 1918–21.

Losey, John E, Linda S Rayor, and Maureen E Carter. 1999. "Transgenic Pollen Harms Monarch Larvae." *Nature* 399 (6733): 214. http://dx.doi.org/10.1038/20338.

Lottermoser, Bernd. 2010. *Mine Wastes: Characterization, Treatment and Environmental Impacts*. Springer Science & Business Media.

Lotze, Heike K, Hunter S Lenihan, Bruce J Bourque, Roger H Bradbury, Richard G Cooke, Matthew C Kay, Susan M Kidwell et al. 2006. "Depletion, Degradation, and Recovery Potential of Estuaries and Coastal Seas." *Science* 312 (5781): 1806–9. doi:10.1126/science.1128035.

Lovelock, James. 2006. *The Revenge of Gaia: Earth's Climate Crisis & The Fate of Humanity*. Basic Books.

Lowe, Sarah, Michael Browne, Souyad Boudjelas, and Maj De Poorter. 2000. "100 of the World's Worst Invasive Alien Species: A Selection from the Global Invasive Species Database." Invasive Species Specialist Group Auckland. www.issg.org/booklet.pdf.

Ludden, John, Denis Peach, and Dee Flight. 2015. "Geochemically Based Solutions for Urban Society: London, A Case Study." *Elements* 11 (4): 253–58. doi:10.2113/gselements.11.4.253.

Luoma, John R. 2011. "Why Does Energy Efficiency's Promise Remain Unfulfilled?" *Yale Environment 360*. http://e360.yale.edu/content/print.msp?id=2367.

Maavara, Taylor, Christopher T Parsons, Christine Ridenour, Severin Stojanovic, Hans H Dürr, Helen R Powley, and Philippe Van Cappellen. 2015. "Global Phosphorus Retention by River Damming." *Proceedings of the National Academy of Sciences* 112 (51): 15603–8. doi:10.1073/pnas.1511797112.

MacKay, David J C, Peter Cramton, Axel Ockenfels, and Steven Stoft. 2015. "Price Carbon—I Will If You Will." *Nature* 526: 315–16. http://www.nature.com/news/price-carbon-i-will-if-you-will-1.18538.

MacKay, David J C. 2009. *Sustainable Energy—Without the Hot Air*. Cambridge, England: UIT Cambridge Ltd. www.withouthotair.com.

Majer, Jonathan David. 2012. "Indicator Species." In *Berkshire Encyclopedia of Sustainability*, edited by Robin Kundis Craig, John Copeland Nagle, Bruce Pardy, Oswald J Schmitz, and William K I B Smith, 5: Ecosyst:200–203. Ecosystems Management and Sustainability. Great Barrington, MA: Berkshire.

Makhijani, Shakuntala, and Alexander Ochs. 2013. "Renewable Energy's Natural Resource Impacts." In *State of the World* 2013, 84–98. Springer.

Malaj, Egina, Peter C von der Ohe, Matthias Grote, Ralph Kühne, Cédric P Mondy, Philippe Usseglio-Polatera, Werner Brack et al. 2014. "Organic Chemicals Jeopardize the Health of Freshwater Ecosystems on the Continental Scale." *Proceedings of the National Academy of Sciences* 111 (26): 9549–54. doi:10.1073/pnas.1321082111.

Manahan, Stanley. 2009. *Fundamentals of Environmental Chemistry*. 3rd ed. CRC Press.

Manahan, Stanley E. 2006. *Green Chemistry and the Ten Commandments of Sustainability*. 2nd ed. Columbia, MO: ChemChar Research, Inc. http://www.asdlib.org/onlineArticles/ecourseware/Manahan/GreenChem-2.pdf.

———. 2013. *Fundamentals of Environmental and Toxicological Chemistry: Sustainable Science*. CRC Press.

Mann, Michael E, and Lee R Kump. 2009. *Dire Predictions: Understanding Global Warming*. New York: DK Publishing, Inc.

Mann, Michael E, Stefan Rahmstorf, Byron A Steinman, Martin Tingley, and Sonya K Miller. 2016. "The Likelihood of Recent Record Warmth." *Scientific Reports* 6 (January). The Author(s): 19831. http://dx.doi.org/10.1038/srep19831.

Manning, David A C. 2008. "Phosphate Minerals, Environmental Pollution and Sustainable Agriculture." *Elements* 4 (2): 105–8. doi:10.2113/GSELEMENTS.4.2.105.

Manning, Richard. 2004. "The Oil We Eat: Tracing the Food Chain Back to Iraq." *Harper's Magazine*.

Mareels, Iven, Erik Weyer, Su Ki Ooi, Michael Cantoni, Yuping Li, and Girish Nair. 2005. "Systems Engineering for Irrigation Systems: Successes and Challenges." *Annual Reviews in Control* 29 (2): 191–204. doi:DOI: 10.1016/j.arcontrol.2005.08.001.

Marshall, George. 2014. *Don't Even Think about It: Why Our Brains Are Wired to Ignore Climate Change*. Bloomsbury Publishing USA.

Mathez, E A. 2009. *Climate Change: The Science of Global Warming and Our Energy Future*. Columbia University Press.

Maupin, M A, J F Kenny, S S Hutson, J K Lovelace, N L Barber, and K S Linsey. 2014. "Estimated Use of Water in the United States in 2010, U.S. Geological Survey Circular 1405." http://pubs.usgs.gov/circ/1405/.

Mazur, Laurie. 2013. "Cultivating Resilience in a Dangerous World." In *State of the World 2013: Is Sustainability Still Possible?*, 353–62. Island Press.

McBride, J P, R E Moore, J P Witherspoon, and R E Blanco. 1978. "Radiological Impact of Airborne Effluents of Coal and Nuclear Plants." *Science* 202 (4372): 1045–50. doi:10.1126/science.202.4372.1045.

McCann, Kevin Shear. 2000. "The Diversity-Stability Debate." *Nature* 405 (6783). Macmillan Magazines Ltd.: 228–33. http://dx.doi.org/10.1038/35012234.

McCauley, Douglas J, Malin L Pinsky, Stephen R Palumbi, James A Estes, Francis H Joyce, and Robert R Warner. 2015. "Marine Defaunation: Animal Loss in the Global Ocean." *Science* 347 (6219). American Association for the Advancement of Science: 1255641. http://science.sciencemag.org/content/347/6219/1255641.

McDonough, W, and M Braungart. 2002. *Cradle to Cradle: Remaking the Way We Make Things*. North Point Press.

McGlade, Christophe, and Paul Ekins. 2015. "The Geographical Distribution of Fossil Fuels Unused When Limiting Global Warming to 2 [deg]C." *Nature* 517 (7533). Nature Publishing Group. 187–90. http://dx.doi.org/10.1038/nature14016.

McKibben, B. 2007. *Deep Economy: The Wealth of Communities and the Durable Future*. Macmillan.

McLellan, R, L Iyengar, B Jeffries, and N Oerlemans. 2014. "Living Planet Report 2014: Species and Spaces, People and Places." http://bit.ly/1ssxx5m.

McMichael, Anthony J. 2012. "Insights from Past Millennia into Climatic Impacts on Human Health and Survival." *Proceedings of the National Academy of Sciences* 109 (13): 4730–37. doi:10.1073/pnas.1120177109.

Meadows, Donella H. 2008. *Thinking in Systems: A Primer*. Chelsea Green Publishing.

Meadows, Donella H, Dennis L Meadows, Jorgen Randers, and William W I I I Behrens. 1972. *The Limits to Growth*. New York: Universe Books.

Meadows, Donella H, Jorgen Randers, and Dennis L Meadows. 2004. *Limits to Growth: The 30-Year Update*. Chelsea Green.

Messner, J J, Nate Haken, P Taft, H Blyth, K Lawrence, S P Graham, and F Umana. 2015. "Fragile States Index 2015." *The Fund for Peace*. http://library.fundforpeace.org/library/fragilestatesindex-2015.pdf.

Michalak, A M, E J Anderson, D Beletsky, S Boland, N S Bosch, T B Bridgeman, J D Chaffin et al. 2013. "Record-Setting Algal Bloom in Lake Erie Caused by Agricultural and Meteorological Trends Consistent with Expected Future Conditions." *Proceedings of the National Academy of Sciences* 110 (16): 6448–52. doi:10.1073/pnas.1216006110.

Millennium Ecosystem Assessment. 2005. Ecosystems and Human Well-Being: Synthesis. Washington, DC: Island Press. http://www.maweb.org/documents/document.356.aspx.pdf.

Miller, Jennie B, and Oswald J Schmitz. 2012. "Food Webs." In *Berkshire Encyclopedia of Sustainability*, edited by Robin Kundis Craig, John Copeland Nagle, Bruce Pardy, Oswald J Schmitz, and William K Smith, 5: Ecosyst:152–55. Ecosystems Management and Sustainability. Great Barrington, MA: Berkshire.

Milly, P C D, Julio Betancourt, Malin Falkenmark, Robert M Hirsch, Zbigniew W Kundzewicz, Dennis P Lettenmaier, and Ronald J Stouffer. 2008. "Stationarity Is Dead: Whither Water Management?" *Science* 319 (5863): 573–74. doi:10.1126/science.1151915.

MIT Energy Initiative. 2011. "The Future of the Nuclear Fuel Cycle." http://mitei.mit.edu/publications/reports-studies/future-nuclear-fuel-cycle.

Mitchell, Valerie L. 2014. "Health Risks Associated with Chronic Exposures to Arsenic in the Environment." *Reviews in Mineralogy and Geochemistry* 79 (1). Mineral Soc America: 435–49.

Molina, Mario J, and F S Rowland. 1974. "Stratospheric Sink for Chlorofluoromethanes: Chlorine Atom Catalysed Destruction of Ozone." *Nature* 249 (5460): 810–12. http://dx.doi.org/10.1038/249810a0.

Moore, Jennie, and William E Rees. 2013. "Getting to One-Planet Living." In *State of the World* 2013, 39–50. Springer.

Morgan, Faith. 2006. *The Power of Community: How Cuba Survived the Peak Oil Crisis*. The Community Solution.

Morrison, Jim. 2005. "How Much Is Clean Water Worth?" *National Wildlife* 24: 26–28.

Morse, Stephen S. 2001. "Factors in the Emergence of Infectious Diseases." In *Plagues and Politics: Infectious Disease and International Policy*, edited by Andrew T Price-Smith, 8–26. London: Palgrave Macmillan UK. doi:10.1057/9780230524248_2.

Moser, Susanne C, and Julia A Ekstrom. 2010. "A Framework to Diagnose Barriers to Climate Change Adaptation." *Proceedings of the National Academy of Sciences* 107 (51): 22026–31. doi:10.1073/pnas.1007887107.

Murphy, Pat, and Faith Morgan. 2013. "Cuba: Lessons from a Forced Decline." *In State of the World 2013: Is Sustainability Still Possible?*, 332–42. The Worldwatch Institute and Island Press.

Myrtveit, Magne. 2005. "The World Model Controversy." *Working Papers In System Dynamics*. https://bora.uib.no/bitstream/1956/1974/1/WPSD1.05WorldControversy.pdf.

Nagle, John Copeland. 2011. "Endangered Species Act." In *Berkshire Encyclopedia of Sustainability*, edited by Klaus Bosselmann, Daniel S Fogel, and J B Ruhl, 3: The Law:151–54. The Law and Politics of Sustainability. Great Barrington, MA: Berkshire.

Naill, R F. 1973. "The Discovery Life Cycle of a Finite Resource: A Case Study of US Natural Gas." In *Toward Global Equilibrium: Collected Papers*, edited by Donnella H Meadows and Dennis L Meadows, 213–56. Cambridge, MA: Wright-Allen Press.

Nakicenovic, Nebojsa, Joseph Alcamo, Gerald Davis, Bert de Vries, Joergen Fenhann, Stuart Gaffin, Kenneth Gregory et al. 2000. "Special Report on Emissions Scenarios: A Special Report of Working Group III of the Intergovernmental Panel on Climate Change." Pacific Northwest National Laboratory, Richland, WA (US), Environmental Molecular Sciences Laboratory (US).

National Academy of Sciences. 2012. *Ecosystem Services: Charting a Path to Sustainability*. The National Academies Press. http://www.nap.edu/openbook.php?record_id=13331.

———. 2016. *Genetically Engineered Crops: Experiences and Prospects*. Washington, DC: The National Academies Press. doi:10.17226/23395.

National Academy of Sciences (U.S.). 2015a. "Climate Intervention: Carbon Dioxide Removal and Reliable Sequestration." doi:10.17226/18805.

———. 2015b. "Climate Intervention: Reflecting Sunlight to Cool Earth." doi:10.17226/18988.

National Research Council. 2006. *Health Risks from Dioxin and Related Compounds: Evaluation of the EPA Reassessment*. National Academies Press.

———. 2009. *Real Prospects for Energy Efficiency in the United States*. Washington, DC: National Academies Press. http://www.nap.edu/catalog.php?record_id=12621.

National Research Council (U.S.). 2009. *America's Energy Future: Technology and Transformation*. The National Academies Press. http://www.nap.edu/openbook.php?record_id=12091.

National Science Board, and National Science Foundation. 2009. Building a Sustainable Energy Future: U.S. Actions for an Effective Energy Economy Transformation. Arlington, VA: National Science Foundation.

Naylor, Rosamond L, Rebecca J Goldburg, Jurgenne H Primavera, Nils Kautsky, Malcolm C M Beveridge, Jason Clay, Carl Folke et al. 2000. "Effect of Aquaculture on World Fish Supplies." *Nature* 405 (6790): 1017–24. http://dx.doi.org/10.1038/35016500.

Naylor, Rosamond L, Ronald W Hardy, Dominique P Bureau, Alice Chiu, Matthew Elliott, Anthony P Farrell, Ian Forster et al. 2009. "Feeding Aquaculture in an Era of Finite Resources." *Proceedings of the National Academy of Sciences* 106 (36): 15103–10. doi:10.1073/pnas.0905235106.

Nelson, Gerald C, Mark W Rosegrant, Jawoo Koo, Richard Robertson, Timothy Sulser, Tingju Zhu, Claudia Ringler et al. 2009. *Climate Change: Impact on Agriculture and Costs of Adaptation*. Vol. 21. Intl Food Policy Res Inst.

Nicholson, Simon. 2013. "The Promises and Perils of Geoengineering." In *State of the World 2013: Is Sustainability Still Possible?*, 317–31. The Worldwatch Institute and Island Press.

Nierenberg, Danielle. 2013. "Agriculture: Growing Food—And Solutions." In *State of the World 2013*, 190–200. Springer.

Nordhaus, William D. 2007. "A Review of the 'Stern Review on the Economics of Climate Change.'" *Journal of Economic Literature*. JSTOR, 686–702.

Normile, Dennis. 2016a. "Epidemic of Fear." *Science* 351 (6277): 1022–23. http://science.sciencemag.org/content/351/6277/1022.abstract.

———. 2016b. "Slow Burn." *Science* 351 (6277): 1018–20. http://science.sciencemag.org/content/351/6277/1018.abstract.

Northrup, Joseph M, and George Wittemyer. 2013. "Characterising the Impacts of Emerging Energy Development on Wildlife, with an Eye towards Mitigation." *Ecology Letters* 16 (1): 112–25. doi:10.1111/ele.12009.

NRC (National Research Council). 2013. *Induced Seismicity Potential in Energy Technologies*. Washington, DC: National Academies Press. http://dels.nas.edu/Report/Induced-.

Nuáez, Martin A, and Romina D Dimarco. 2012. "Keystone Species." In *Berkshire Encyclopedia of Sustainability*, edited by Robin Kundis Craig, John Copeland Nagle, Bruce Pardy, Oswald J Schmitz, and William K Smith, 5: Ecosyst: 226–30. Ecosystems Management and Sustainability. Great Barrington, MA: Berkshire.

OECD. 2010. *Uranium 2009: Resources, Production and Demand*. OECD Publishing.

OECD, and NEA. 2010. "Comparing Nuclear Accident Risks with Those from Other Energy Sources." Nuclear Energy Agency, Organisation for Economic Co-Operation and Development. http://www.nea.fr/ndd/reports/2010/nea6862-comparing-risks.pdf.

Oelkers, Eric H, and David R Cole. 2008. "Carbon Dioxide Sequestration: A Solution to a Global Problem." *Elements* 4 (5). Mineralogical Association of Canada: 305–10.

Oelkers, Eric H, Janet G Hering, and Chen Zhu. 2011. "Water: Is There a Global Crisis?" *Elements* 7 (3): 157–62. doi:10.2113/gselements.7.3.157.

Oelkers, Eric H, and Eugenia Valsami-Jones. 2008. "Phosphate Mineral Reactivity and Global Sustainability." *Elements* 4 (2): 83–87. doi:10.2113/GSELEMENTS.4.2.83.

Oreskes, N, and E M Conway. 2010. *Merchants of Doubt: How a Handful of Scientists Obscured the Truth on Issues from Tobacco Smoke to Global Warming*. Bloomsbury Pub Plc USA.

Ostrom, E. 1990. *Governing the Commons: The Evolution of Institutions for Collective Action*. Cambridge University Press.

Ott, K. 2003. "The Case for Strong Sustainability." Edited by K Ott and P P Thapa. *Greifswalds Environmental Ethics*. Greifswald: Steinbecker Verlag. http://umwethik.botanik.uni-greifswald.de/booklet/8_strong_sustainability.pdf.

Pacala, S, and R Socolow. 2004. "Stabilization Wedges: Solving the Climate Problem for the Next 50 Years with Current Technologies." *Science* 305 (5686): 968–72. doi:10.1126/science.1100103.

Paez-Osuna, Federico. 2001. "The Environmental Impact of Shrimp Aquaculture: Causes, Effects, and Mitigating Alternatives." *Environmental Management* 28. Springer New York: 131–40. doi:10.1007/s002670010212.

Palmer, Lisa. 2016. "How Rising CO2 Levels May Contribute to Die-Off of Bees." *Yale Environment 360*. http://e360.yale.edu/feature/bee_collapse_co2_climate_change_agriculture/2991/.

Palmer, M A, E S Bernhardt, W H Schlesinger, K N Eshleman, E Foufoula-Georgiou, M S Hendryx, A D Lemly et al. 2010. "Mountaintop Mining Consequences." *Science* 327 (5962): 148–49. doi:10.1126/science.1180543.

Panchuk, K, A Ridgwell, and L R Kump. 2008. "Sedimentary Response to Paleocene-Eocene Thermal Maximum Carbon Release: A Model-Data Comparison." *Geology* 36 (4): 315–18. doi:10.1130/g24474a.1.

Paolo, Fernando S, Helen A Fricker, and Laurie Padman. 2015. "Volume Loss from Antarctic Ice Shelves Is Accelerating." *Science* 348 (6232): 327–31. doi:10.1126/science.aaa0940.

Papanicolaou, A, N Thanos, Kenneth M Wacha, Benjamin K Abban, Christopher G Wilson, Jerry L Hatfield, Charles O Stanier et al. 2015. "From Soilscapes to Landscapes: A Landscape-Oriented Approach to Simulate Soil Organic Carbon Dynamics in Intensively Managed Landscapes." *Journal of Geophysical Research: Biogeosciences* 120 (11): 2375–2401. doi:10.1002/2015JG003078.

Parsons, Simon A, and Jennifer A Smith. 2008. "Phosphorus Removal and Recovery from Municipal Wastewaters." *Elements* 4 (2): 109–12. doi:10.2113/GSELEMENTS.4.2.109.

Pearce, Fred. 2009. "Consumption Dwarfs Population As Main Environmental Threat." *Yale Environment 360*. http://www.e360.yale.edu/content/feature.msp?id=2140.

———. 2016. "Can We Reduce CO2 Emissions and Grow the Global Economy?" *Yale e360*. http://e360.yale.edu/feature/can_we_reduce_co2_emissions_and_grow_global_economy/2983/.

Pelletier, Nathan, and Peter Tyedmers. 2010. "Forecasting Potential Global Environmental Costs of Livestock Production 2000–2050." *Proceedings of the National Academy of Sciences* 107 (43): 18371–74. doi:10.1073/pnas.1004659107.

Perrone, Debra, and George Hornberger. 2016. "Frontiers of the Food–energy–water Trilemma: Sri Lanka as a Microcosm of Tradeoffs." *Environmental Research Letters* 11 (1): 14005. http://stacks.iop.org/1748-9326/11/i=1/a=014005.

Pershing, Andrew J, Michael A Alexander, Christina M Hernandez, Lisa A Kerr, Arnault Le Bris, Katherine E Mills, Janet A Nye et al. 2015. "Slow Adaptation in the Face of Rapid Warming Leads to Collapse of the Gulf of Maine Cod Fishery." *Science* 350 (6262): 809–12. http://science.sciencemag.org/content/350/6262/809.abstract.

Petit, J R, J Jouzel, D Raynaud, N I Barkov, J M Barnola, I Basile, M Bender, J Chappellaz, M Davis, and G Delaygue. 1999. "Climate and Atmospheric History of the Past 420,000 Years from the Vostok Ice Core, Antarctica." *Nature* 399 (6735): 429–36. doi:10.1038/20859.

Phalan, Ben, Malvika Onial, Andrew Balmford, and Rhys E Green. 2011. "Reconciling Food Production and Biodiversity Conservation: Land Sharing and Land Sparing Compared." *Science* 333 (6047). American Association for the Advancement of Science: 1289–91.

Phalkey, Revati K, Clara Aranda-Jan, Sabrina Marx, Bernhard Höfle, and Rainer Sauerborn. 2015. "Systematic Review of Current Efforts to Quantify the Impacts of Climate Change on Undernutrition." *Proceedings of the National Academy of Sciences* 112 (33): E4522–29. doi:10.1073/pnas.1409769112.

Pimentel, David, Rodolfo Zuniga, and Doug Morrison. 2005. "Update on the Environmental and Economic Costs Associated with Alien-Invasive Species in the United States." *Ecological Economics* 52 (3): 273–88. doi:http://dx.doi.org/10.1016/j.ecolecon.2004.10.002.

Pinker, Steven. 2011. *The Better Angels of Our Nature: Why Violence Has Declined*. Viking New York.

Pollack, Henry N. 2005. *Uncertain Science...Uncertain World*. Cambridge.

Pollan, Michael. 2007. *The Omnivore's Dilemma: A Natural History of Four Meals*.

———. 2008. "The Food Issue: Farmer in Chief." *New York Times*. http://www.nytimes.com/2008/10/12/magazine/12policy-t.html.

Pollnac, Richard, Patrick Christie, Joshua E Cinner, Tracey Dalton, Tim M Daw, Graham E Forrester, Nicholas A J Graham et al. 2010. "Marine Reserves as Linked Social–ecological Systems." *Proceedings of the National Academy of Sciences* 107 (43). National Acad Sciences: 18262–65. http://www.pnas.org/content/107/43/18262.short.

Postel, Sandra. 1999. *Pillar of Sand: Can The Irrigation Miracle Last?* W. W. Norton & Co., Inc.

Prasad, Naren. 2006. "Privatisation Results: Private Sector Participation in Water Services After 15 Years." *Development Policy Review* 24 (6). Blackwell Publishing Ltd: 669–92. doi:10.1111/j.1467-7679.2006.00353.x.

Prescott-Allen, Robert. 2001. *The Wellbeing of Nations: A Country-by-Country Index of Quality of Life and the Environment*. Island Press.

Primack, Richard B, and Elizabeth R Ellwood. 2012. "Biodiversity." In *Berkshire Encyclopedia of Sustainability*, edited by Robin Kundis Craig, John Copeland Nagle, Bruce Pardy, Oswald J Schmitz, and William K Smith, 5: Ecosyst:21–25. Ecosystems Management and Sustainability. Great Barrington, MA: Berkshire.

Princen, Thomas, Jack P Manno, and Pamela Martin. 2013. "Keeping Them in the Ground: Ending the Fossil Fuel Era." In *State of the World 2013: Is Sustainability Still Possible?*, 161–71. Island Press.

Prosek, James. 2010. "A Steady, Steep Decline for The Lowly, Uncharismatic Eel." *Yale Environment 360*. http://e360.yale.edu/feature/a_steady_steep_decline_for_the_lowly_uncharismatic_eel/2316/.

Prugh, Tom. 2015. "Childhood's End." In *State of the World 2015: Confronting Hidden Threats to Sustainability*, 129–40. Washington, D.C.: Island Press.

PyŠek, Petr, and David M Richardson. 2012. "Invasive Species." In *Berkshire Encyclopedia of Sustainability*, edited by Robin Kundis Craig, John Copeland Nagle, Bruce Pardy, Oswald J Schmitz, and William K Smith, 5: Ecosyst:211–19. Ecosystems Management and Sustainability. Great Barrington, MA: Berkshire.

Ragnarsdottir, Kristin Vala. 2008. "Rare Metals Getting Rarer." *Nature Geoscience* 1 (11): 720–21. http://www.nature.com/ngeo/journal/v1/n11/pdf/ngeo302.pdf.

Randers, Jorgen. 2012. *2052: A Global Forecast for the next Forty Years*. Chelsea Green Publishing.

Raudsepp-Hearne, Ciara, Garry D Peterson, Maria Tengö, Elena M Bennett, Tim Holland, Karina Benessaiah, Graham K MacDonald et al. 2010. "Untangling the Environmentalist's Paradox: Why Is Human Well-Being Increasing as Ecosystem Services Degrade?" *BioScience* 60 (8). Oxford University Press: 576–89.

Raven, P H, L R Berg, and G B Johnson. 1995. *Environment*. Saunders College Publishing.

Raymond, Peter A, and Neung-Hwan Oh. 2009. "Long Term Changes of Chemical Weathering Products in Rivers Heavily Impacted from Acid Mine Drainage: Insights on the Impact of Coal Mining on Regional and Global Carbon and Sulfur Budgets." *Earth and Planetary Science Letters* 284 (1–2): 50–56. http://www.sciencedirect.com/science/article/B6V61-4W99NRS-3/2/0885ccdac36263c39cdc31a2f87fd0b8.

Reed, Andra J, Michael E Mann, Kerry A Emanuel, Ning Lin, Benjamin P Horton, Andrew C Kemp, and Jeffrey P Donnelly. 2015. "Increased Threat of Tropical Cyclones and Coastal Flooding to New York City during the Anthropogenic Era." *Proceedings of the National Academy of Sciences* 112 (41): 12610–15. doi:10.1073/pnas.1513127112.

Rees, William E. 2010. "Thinking 'Resilience.'" *The Post Carbon Reader: Managing the 21st Century's Sustainability Crises*. Richard Heinberg, Daniel Lerch, Eds. Healdsburg, CA: Watershed Media, 24–40.

Renner, Michael. 2013. "Climate Change and Displacements." In *State of the World 2013: Is Sustainability Still Possible?*, 342–52. Island Press.

Richter, B. 2010. *Beyond Smoke and Mirrors: Climate Change and Energy in the 21st Century*. Cambridge University Press.

Ridgwell, Andy, and Daniela N Schmidt. 2010. "Past Constraints on the Vulnerability of Marine Calcifiers to Massive Carbon Dioxide Release." *Nature Geosci* 3 (3). Nature Publishing Group: 196–200. doi:http://www.nature.com/ngeo/journal/v3/n3/suppinfo/ngeo755_S1.html.

Rockstrom, Johan, Will Steffen, Kevin Noone, Asa Persson, F Stuart Chapin, Eric F Lambin, Timothy M Lenton et al. 2009. "A Safe Operating Space for Humanity." *Nature* 461 (7263). Nature Publishing Group: 472–75. http://dx.doi.org/10.1038/461472a.

Rogers, John J W, and Geoffrey P Feiss. 1998. *People and the Earth: Basic Issues in the Sustainability of Resources and the Environment*. Cambridge, UK: Cambridge University Press.

Rosenthal, Elisabeth. 2009. "What Makes Europe Greener than the U.S.?" *Yale Environment 360*. Yale University. http://www.e360.yale.edu/content/feature.msp?id=2193.

Ruddiman, W F, J E Kutzbach, and S J Vavrus. 2011. "Can Natural or Anthropogenic Explanations of Late-Holocene CO2 and CH4 Increases Be Falsified?" *The Holocene* 21 (5): 865–8879. doi:10.1177/0959683610387172.

Ruhl, Laura, Avner Vengosh, Gary S Dwyer, Heileen Hsu-Kim, Amrika Deonarine, Mike Bergin, and Julia Kravchenko. 2009. "Survey of the Potential Environmental and Health Impacts in the Immediate Aftermath of the Coal Ash Spill in Kingston, Tennessee." *Environmental Science & Technology* 43 (16). American Chemical Society: 6326–33. doi:10.1021/es900714p.

Rulli, Maria Cristina, Antonio Saviori, and Paolo D'Odorico. 2013. "Global Land and Water Grabbing." *Proceedings of the National Academy of Sciences* 110 (3): 892–97. doi:10.1073/pnas.1213163110.

Sachs, Jeffrey D, and John W McArthur. 2005. "The Millennium Project: A Plan for Meeting the Millennium Development Goals." *The Lancet* 365 (9456). Elsevier: 347–53.

Sahagian, D L, F W Schwartz, and D K Jacobs. 1994. "Direct Anthropogenic Contributions to Sea Level Rise in the Twentieth Century." *Nature* 367. Nature Publishing Group: 54–57.

Sala, Osvaldo E, F Stuart Chapin, III, Juan J Armesto, Eric Berlow, Janine Bloomfield, Rodolfo Dirzo et al. 2000. "Global Biodiversity Scenarios for the Year 2100." *Science* 287 (5459): 1770–74. http://science.sciencemag.org/content/287/5459/1770.abstract.

Savage, Ian. 2013. "Comparing the Fatality Risks in United States Transportation across Modes and over Time." *Research in Transportation Economics* 43 (1): 9–22. doi:http://dx.doi.org/10.1016/j.retrec.2012.12.011.

Sawin, Janet L, and William R Moomaw. 2009. "An Enduring Energy Future." In *State of the World 2009: Into a Warming World*, edited by Linda Starke. Worldwatch Institute.

Schafft, Kai A, Yetkin Borlu, and Leland Glenna. 2013. "The Relationship between Marcellus Shale Gas Development in Pennsylvania and Local Perceptions of Risk and Opportunity." *Rural Sociology* 78 (2): 143–66. doi:10.1111/ruso.12004.

Schaltegger, Stefan, and Roger Burritt. 2000. *Contemporary Environmental Accounting: Issues, Concepts and Practice*. Sheffield, GB: Greenleaf Publishing. http://site.ebrary.com/lib/vanderbilt/docDetail.action?docID=10650139.

Schellnhuber, H J, M Molina, N Stern, V Huber, and S Kadner. 2010. *Global Sustainability: A Nobel Cause*. Cambridge, UK: Cambridge.

Schendel, Willem van. 2009. *A History of Bangladesh*. Cambridge University Press.

Schindler, David W, R E Hecky, D L Findlay, M P Stainton, B R Parker, M J Paterson, K G Beaty et al. 2008. "Eutrophication of Lakes Cannot Be Controlled by Reducing Nitrogen Input: Results of a 37-Year Whole-Ecosystem Experiment." *Proceedings of the National Academy of Sciences* 105 (32): 11254–58. doi:10.1073/pnas.0805108105.

Schlenker, Wolfram, and Michael J Roberts. 2009. "Nonlinear Temperature Effects Indicate Severe Damages to U.S. Crop Yields under Climate Change." *Proceedings of the National Academy of Sciences* 106 (37): 15594–98. doi:10.1073/pnas.0906865106.

Schloss, Carrie A, Tristan A Nuñez, and Joshua J Lawler. 2012. "Dispersal Will Limit Ability of Mammals to Track Climate Change in the Western Hemisphere." *Proceedings of the National Academy of Sciences* 109 (22): 8606–11. doi:10.1073/pnas.1116791109.

Schmidt-Bleek, F. 2007. *The Earth: Natural Resources and Human Intervention. The Sustainability Project*. Haus Publishing Limited, London.

Schrag, Daniel P. 2012. "Is Shale Gas Good for Climate Change?" *Daedalus* 141 (2): 72–80. doi:10.1162/DAED_a_00147.

Schwartz, Franklin W, and Motomu Ibaraki. 2011. "Groundwater: A Resource in Decline." *Elements* 7 (3): 175–79. doi:10.2113/gselements.7.3.175.

Sheehan, Bill, and Helen Spiegelman. 2010. "Climate Change, Peak Oil, and the End of Waste." In *The Post Carbon Reader*, edited by Richard Heinberg and Daniel Lerch, 363–84. Watershed Media and the Post Carbon Institute.

Shindell, Drew, Johan C I Kuylenstierna, Elisabetta Vignati, Rita van Dingenen, Markus Amann, Zbigniew Klimont, Susan C Anenberg et al. 2012. "Simultaneously Mitigating Near-Term Climate Change and Improving Human Health and Food Security." *Science* 335 (6065): 183–89. doi:10.1126/science.1210026.

Silliman, Brian R, Johan van de Koppel, Michael W McCoy, Jessica Diller, Gabriel N Kasozi, Kamala Earl, Peter N Adams et al. 2012. "Degradation and Resilience in Louisiana Salt Marshes after the BP–Deepwater Horizon Oil Spill." *Proceedings of the National Academy of Sciences* 109 (28): 11234–39. doi:10.1073/pnas.1204922109.

Simms, Andrew. 2008. "Trickle-down Myth: Arguing That Economic Growth Is the Only Way to Make Poor People Richer Is Misguided." *New Scientist* 200 (2678): 49.

Sipes, J L. 2010. *Sustainable Solutions for Water Resources: Policies, Planning, Design, and Implementation*. Wiley.

Siriri, D, J Wilson, R Coe, M M Tenywa, M A Bekunda, C K Ong, and C R Black. 2013. "Trees Improve Water Storage and Reduce Soil Evaporation in Agroforestry Systems on Bench Terraces in SW Uganda." *Agroforestry Systems* 87 (1): 45–58. doi:10.1007/s10457-012-9520-x.

Smil, V. 2004. Enriching the Earth: Fritz Haber, Carl Bosch, and the Transformation of World Food Production. MIT Press.

Smil, Vaclav. 2002. *The Earth's Biosphere: Evolution, Dynamics, and Change*. The MIT Press.

———. 2008. "Water News: Bad, Good and Virtual." *American Scientist* 96 (5): 399–407.

Smith-Spangler, Crystal, Margaret L Brandeau, Grace E Hunter, J Clay Bavinger, Maren Pearson, Paul J Eschbach, Vandana Sundaram et al. 2012. "Are Organic Foods Safer or Healthier Than Conventional Alternatives? A Systematic Review." *Annals of Internal Medicine* 157 (5): 348–66. http://dx.doi.org/10.7326/0003-4819-157-5-201209040-00007.

Sohns, A, and L. Crowder. 2013. "Sustainable Fisheries and Seas: Preventing Ecological Collapse." In *State of the World 2013: Is Sustainability Still Possible?*, 63–72. The Worldwatch Institute and Island Press.

Solomon, Susan, Diane J Ivy, Doug Kinnison, Michael J Mills, Ryan R Neely, and Anja Schmidt. 2016. "Emergence of Healing in the Antarctic Ozone Layer." *Science* 353 (6296): 269–74.

Solomon, Susan, Gian-Kasper Plattner, Reto Knutti, and Pierre Friedlingstein. 2009. "Irreversible Climate Change due to Carbon Dioxide Emissions." *Proceedings of the National Academy of Sciences* 106 (6): 1704–9. doi:10.1073/pnas.0812721106.

Speth, James Gustave. 2008. *The Bridge at the Edge of the World: Capitalism, the Environment, and Crossing from Crisis to Sustainability*. New Haven, CT: Yale University Press.

Starke, Linda. 2009. *State of the World 2009: Into a Warming World*. New York/London: W. W. Norton & Company.

Starr, Douglas. 2016. "The Carbon Accountant." *Science* 353 (6302): 858 LP-861. http://science.sciencemag.org/content/353/6302/858.abstract.

Steckel, Jan Christoph, Ottmar Edenhofer, and Michael Jakob. 2015. "Drivers for the Renaissance of Coal." *Proceedings of the National Academy of Sciences* 112 (29): E3775–81. doi:10.1073/pnas.1422722112.

Steffen, Alex. 2006. *World Changing: A User's Guide for the 21st Century*. New York: Abrams.

Steffen, Will, Katherine Richardson, Johan Rockström, Sarah E Cornell, Ingo Fetzer, Elena M Bennett, Reinette Biggs et al. 2015. "Planetary Boundaries: Guiding Human Development on a Changing Planet." *Science* 347 (6223). doi:10.1126/science.1259855.

Stern, N H, and G B Treasury. 2007. *The Economics of Climate Change: The Stern Review*. Cambridge University Press.

Steward, David R, Paul J Bruss, Xiaoying Yang, Scott A Staggenborg, Stephen M Welch, and Michael D Apley. 2013. "Tapping Unsustainable Groundwater Stores for Agricultural Production in the High Plains Aquifer of Kansas, Projections to 2110." *Proceedings of the National Academy of Sciences* 110 (37): E3477–86. doi:10.1073/pnas.1220351110.

Stiglitz, Joseph E, Amartya Sen, and Jean-Paul Fitoussi. 2009. "Report by the Commission on the Measurement of Economic Performance and Social Progress (Stiglitz Commission)." http://www.stiglitz-sen-fitoussi.fr/documents/rapport_anglais.pdf.

Stossel, John. 2004. *Give Me a Break*. HarperCollins.

Struzik, Ed. 2014. "Loss of Snowpack and Glaciers In Rockies Poses Water Threat." *Yale Environment 360*. http://e360.yale.edu/feature/loss_of_snowpack_and_glaciers_in_rockies_poses_water_threat/2785/.

Suweis, Samir, Joel A Carr, Amos Maritan, Andrea Rinaldo, and Paolo D'Odorico. 2015. "Resilience and Reactivity of Global Food Security." *Proceedings of the National Academy of Sciences* 112 (22): 6902–7. doi:10.1073/pnas.1507366112.

Suweis, Samir, Andrea Rinaldo, Amos Maritan, and Paolo D'Odorico. 2013. "Water-Controlled Wealth of Nations." *Proceedings of the National Academy of Sciences* 110 (11): 4230–33. doi:10.1073/pnas.1222452110.

Syed, Tajdarul H, James S Famiglietti, Don P Chambers, Josh K Willis, and Kyle Hilburn. 2010. "Satellite-Based Global-Ocean Mass Balance Estimates of Interannual Variability and Emerging Trends in Continental Freshwater Discharge." *Proceedings of the National Academy of Sciences* 107 (42): 17916–21. doi:10.1073/pnas.1003292107.

Syvitski, James P M, Albert J Kettner, Irina Overeem, Eric W H Hutton, Mark T Hannon, G Robert Brakenridge, John Day et al. 2009. "Sinking Deltas due to Human Activities." *Nature Geosci* 2. Nature Publishing Group: 681–86. http://dx.doi.org/10.1038/ngeo629.

Szulczewski, Michael L, Christopher W MacMinn, Howard J Herzog, and Ruben Juanes. 2012. "Lifetime of Carbon Capture and Storage as a Climate-Change Mitigation Technology." *Proceedings of the National Academy of Sciences* 109 (14): 5185–89. doi:10.1073/pnas.1115347109.

Tal, Alon. 2006. "Seeking Sustainability: Israel's Evolving Water Management Strategy." *Science* 313 (5790): 1081–84. http://science.sciencemag.org/content/313/5790/1081.abstract.

Thomas, Chris D, Alison Cameron, Rhys E Green, Michel Bakkenes, Linda J Beaumont, Yvonne C Collingham, Barend F N Erasmus et al. 2004. "Extinction Risk from Climate Change." *Nature* 427 (6970). Nature Publishing Group: 145–48.

Tilman, David, Christian Balzer, Jason Hill, and Belinda L Befort. 2011. "Global Food Demand and the Sustainable Intensification of Agriculture." *Proceedings of the National Academy of Sciences* 108 (50): 20260–64. doi:10.1073/pnas.1116437108.

Tilman, David, Kenneth G Cassman, Pamela A Matson, Rosamond Naylor, and Stephen Polasky. 2002. "Agricultural Sustainability and Intensive Production Practices." *Nature* 418 (6898). Nature Publishing Group: 671–77.

Trawick, Paul. 2001. "The Moral Economy of Water: Equity and Antiquity in the Andean Commons." *American Anthropologist* 103 (2). Wiley Online Library: 361–379. http://onlinelibrary.wiley.com/doi/10.1525/aa.2001.103.2.361/abstract.

Troell, Max, Rosamond L Naylor, Marc Metian, Malcolm Beveridge, Peter H Tyedmers, Carl Folke, Kenneth J Arrow et al. 2014. "Does Aquaculture Add Resilience to the Global Food System?" *Proceedings of the National Academy of Sciences* 111 (37): 13257–63. doi:10.1073/pnas.1404067111.

Tsao, J Y et al. 2010. "Solid-State Lighting: An Energy-Economics Perspective." *Journal of Physics D: Applied Physics* 43 (35): 354001. http://stacks.iop.org/0022-3727/43/i=35/a=354001.

Tukker, Arnold, and Bart Jansen. 2006. "Environmental Impacts of Products: A Detailed Review of Studies." *Journal of Industrial Ecology* 10 (3). MIT Press: 159–82. doi:10.1162/jiec.2006.10.3.159.

UN. 2000. *United Nations Millennium Declaration: Resolution.* http://www.un.org/ga/search/view_doc.asp?symbol=A/RES/55/2.

———. 2012. "UN World Population Prospects: The 2012 Revision." http://esa.un.org/wpp/Excel-Data/Interpolated.htm.

UN-Habitat. 2004. "The Challenge of Slums: Global Report on Human Settlements 2003." *Management of Environmental Quality: An International Journal* 15 (3). Emerald Group Publishing Limited: 337–38.

UN Water. 2013. "Water Security and the Global Water Agenda: A UN-Water Analytical Brief." Hamilton, ON: UN University. http://mdgs.un.org/unsd/mdg/data.aspx.

UNDP. 2015. "Human Development Report 2015: Work for Human Development." http://report.hdr.undp.org/.

UNEP. 2007. *Global Environment Outlook: Environment for Development (GEO-4).* United Nations Environment Programme. http://www.unep.org/geo/geo4/report/GEO-4_Report_Full_en.pdf.

U.S. EIA Annual Energy Outlook 2014. http://www.eia.gov/outlooks/aeo/0383(2014).pdf.

USDA. 2011. "Dietary Guidelines for America 2010." United States Department of Agriculture. www.dietaryguidelines.gov.

USGS. 2000. "Land Subsidence in the United States." http://water.usgs.gov/ogw/pubs/fs00165/.

USNRC. 2010. *Hidden Costs of Energy: Unpriced Consequences of Energy Production and Use.* National Academy of Science.

Valentine, David L, G Burch Fisher, Sarah C Bagby, Robert K Nelson, Christopher M Reddy, Sean P Sylva, and Mary A Woo. 2014. "Fallout Plume of Submerged Oil from Deepwater Horizon." *Proceedings of the National Academy of Sciences* 111 (45): 15906–11. doi:10.1073/pnas.1414873111.

Vallero, D A, and C Brasier. 2008. Sustainable Design: The Science of Sustainability and Green Engineering. Wiley.

Vance, Robert. 2006. "What Can 40 Years of Red Books Tell Us?" *World Nuclear Association Annual Symposium*. http://www.world-nuclear.org/sym/2006/vance.htm.

Vandenbergh, Michael P, and Jonathan M Gilligan. 2014. "Beyond Gridlock." *Social Science Research Network*. doi:10.2139/ssrn.2533643.

Vaux Jr., Henry. 2011. "Water Conservation, Efficiency, and Reuse." *Elements* 7 (3): 187–91. doi:10.2113/gselements.7.3.187.

Vermeer, Eduard B. 2012. "China." In *Berkshire Encyclopedia of Sustainability*, edited by Sam Geall, Jingjing Liu, and Sony Pellissery, 7:46–54. China, India, and East and Southeast Asia: Assessing Sustainability. Great Barrington, MA: Berkshire.

Vermeer, Martin, and Stefan Rahmstorf. 2009. "Global Sea Level Linked to Global Temperature." *Proceedings of the National Academy of Sciences* 106 (51): 21527–32. doi:10.1073/pnas.0907765106.

Vidic, R D, S L Brantley, J M Vandenbossche, D Yoxtheimer, and J D Abad. 2013. "Impact of Shale Gas Development on Regional Water Quality." *Science (New York, N.Y.)* 340 (6134): 1235009. doi:10.1126/science.1235009.

Viebahn, Peter, Daniel Vallentin, and Manfred Fischedick. 2009. "Carbon Capture and Storage." In *State of the World 2009: Into a Warming World*, edited by Linda Starke.

Vitousek, P M, H A Mooney, J Lubchenco, and J M Melillo. 1997. "Human Domination of Earth's Ecosystems." *Science* 277 (5325): 494.

Vorosmarty, C J, P B McIntyre, M O Gessner, D Dudgeon, A Prusevich, P Green, S Glidden et al. 2010. "Global Threats to Human Water Security and River Biodiversity." *Nature* 467 (7315). Nature Publishing Group, a division of Macmillan Publishers Limited. All Rights Reserved.: 555–61.

Waggoner, P E, and J H Ausubel. 2002. "A Framework for Sustainability Science: A Renovated IPAT Identity." *Proceedings of the National Academy of Sciences of the United States of America* 99 (12): 7860–65. doi:10.1073/pnas.122235999.

Waldron, Anthony, Arne O Mooers, Daniel C Miller, Nate Nibbelink, David Redding, Tyler S Kuhn, J Timmons Roberts et al. 2013. "Targeting Global Conservation Funding to Limit Immediate Biodiversity Declines." *Proceedings of the National Academy of Sciences* 110 (29): 12144–48. doi:10.1073/pnas.1221370110.

Walker, B, and D Salt. 2006. *Resilience Thinking: Sustaining Ecosystems and People in a Changing World.* Island Press.

Wallace, Michael J. 2005. "Testimony before the U.S. Senate Committee on Energy and Natural Resources, Hearing on the Department of Energy's Nuclear Power 2010 Program." http://www.energy.senate.gov/public/index.cfm/hearings-and-business-meetings?Id=873cf7b4-c9d8-45e1-bf40-dbeb9eff294f&Statement_id=bdd3a910-5252-4268-abca-cb8bf35d09fd.
Walljasper, Jay. 2010. *All That We Share: A Field Guide To The Commons*. New York: The New Press.
Walther, Gian-Reto, Eric Post, Peter Convey, Annette Menzel, Camille Parmesan, Trevor J C Beebee, Jean-Marc Fromentin et al. 2002. "Ecological Responses to Recent Climate Change." *Nature* 416 (6879): 389–95. http://dx.doi.org/10.1038/416389a.
Walther, John V. 2009. *Essentials of Geochemistry*. 2nd ed. Jones & Bartlett Learning.
———. 2014. *Earth's Natural Resources*. Jones & Bartlett Learning.
Wann, David. 2007. *Simple Prosperity: Finding Real Wealth in a Sustainable Lifestyle*. New York: St. Martins' Press.
Waters, Colin N, Jan Zalasiewicz, Colin Summerhayes, Anthony D Barnosky, Clément Poirier, Agnieszka Gałuszka, Alejandro Cearreta et al. 2016. "The Anthropocene Is Functionally and Stratigraphically Distinct from the Holocene." *Science* 351 (6269). doi:10.1126/science.aad2622.
WCED. 1987. *Our Common Future*. Oxford, UK: United Nations World Commission on Environment and Development, Oxford University Press.
Weart, Spencer R. 1987. "Nuclear Fear: A History and an Experiment." *Scientific Controversies: Case Studies in the Resolution and Closure of Disputes in Science and Technology*.
Weber, Christopher L, and H Scott Matthews. 2008. "Food-Miles and the Relative Climate Impacts of Food Choices in the United States." *Environmental Science & Technology* 42 (10). American Chemical Society: 3508–13. doi:10.1021/es702969f.
Weiner, Jonathan. 1994. *The Beak of the Finch: A Story of Evolution in Our Time*. Vintage.
Weisman, Alan. 1999. *Gaviotas: A Village to Reinvent the World*. Chelsea Green Publishing Company.
Westerhoff, Paul, Sungyun Lee, Yu Yang, Gwyneth W Gordon, Kiril Hristovski, Rolf U Halden, and Pierre Herckes. 2015. "Characterization, Recovery Opportunities, and Valuation of Metals in Municipal Sludges from U.S. Wastewater Treatment Plants Nationwide." *Environmental Science & Technology* 49 (16). American Chemical Society: 9479–88. doi:10.1021/es505329q.
Whitehorn, Penelope R, Stephanie O'Connor, Felix L Wackers, and Dave Goulson. 2012. "Neonicctinoid Pesticide Reduces Bumble Bee Colony Growth and Queen Production." *Science* 336 (6079): 351–52. doi:10.1126/science.1215025.
Whitty, Julia. 2007. "Animal Extinction—The Greatest Threat to Mankind." *The Independent*. http://www.independent.co.uk/environment/animal-extinction--the-greatest-threat-to-mankind-397939.html.
Willett, Walter. 2011. *Eat, Drink, and Be Healthy: The Harvard Medical School Guide to Healthy Eating*. Simon and Schuster.
Wilson, Edward O. 1998. *Consilience: The Unity of Knowledge*. New York: Vintage Books.
———. 2002. The Future of Life. Knopf.
Witte, Wolfgang. 1998. "Medical Consequences of Antibiotic Use in Agriculture." *Science* 279 (5353): 996 LP-997. http://science.sciencemag.org/content/279/5353/996.abstract.
Woolf, Aaron, Michael Pollan, Ian Cheney, and Curtis Ellis. 2007. King Corn.
World Bank. 2007. *Changing the Face of the Waters: The Promise and Challenge of Sustainable Aquaculture. Agriculture and Rural Development*. Washington, DC: World Bank. http://documents.worldbank.org/curated/en/322181468157766884/Changing-the-face-of-the-waters-the-promise-and-challenge-of-sustainable-aquaculture.
World Health Organization. 2002. "The World Health Report 2002: Reducing Risks, Promoting Healthy Life." Vol. 16. Geneva: Carfax Publishing.
———. 2011. "Guidelines for Drinking-Water Quality 4th Ed." http://www.who.int/water_sanitation_health/publications/2011/dwq_guidelines/en/.
———. 2014. "World Health Statistics 2014." apps.who.int/iris/bitstream/10665/112738/1/9789240692671_eng.pdf.
———. 2015. "World Health Statistics 2015." http://www.who.int/gho/publications/world_health_statistics/en/.
Wright, Richard T. 2005. *Environmental Science: Toward a Sustainable Future*. 9th ed. Pearson Prentice Hall.
Wuebbles, Donald J, and Katharine Hayhoe. 2000. "Atmospheric Methane: Trends and Impacts." In *Non-CO2 Greenhouse Gases: Scientific Understanding*, edited by J van Ham, A P M Baede, L A Meyer, and R Ybema, 1:425–32. Noordwijkerhout, The Netherlands: Springer Netherlands. doi:10.1007/978-94-015-9343-4.
WWF International. 2015. "Living Blue Planet Report 2015." www.worldwildlife.org/publications/living-blue-planet-report-2015.

Xie, Jian, Liangliang Hu, Jianjun Tang, Xue Wu, Nana Li, Yongge Yuan, Haishui Yang et al. 2011. "Ecological Mechanisms Underlying the Sustainability of the Agricultural Heritage Rice–Fish Coculture System." *Proceedings of the National Academy of Sciences* 108 (50). National Acad Sciences: E1381–87.

York, R, E A Rosa, and T Dietz. 2003. "Footprints on the Earth: The Environmental Consequences of Modernity." *American Sociological Review*, 279–300.

York, Richard. 2012. "Do Alternative Energy Sources Displace Fossil Fuels?" *Nature Climate Change* 2 (6). Nature Publishing Group: 441–43.

York, Richard. 2012. "I = P × A × T Equation." In *Berkshire Encyclopedia of Sustainability*, edited by Ian Spellerberg, Daniel S Fogel, Sarah E Fredericks, and Lisa M Butler Harrington, 6: Measure:194–97. Measurements, Indicators, and Research Methods for Sustainability. Great Barrington, MA: Berkshire.

Zalasiewicz, Jan, Mark Williams, Will Steffen, and Paul Crutzen. 2010. "The New World of the Anthropocene." *Environmental Science & Technology* 44 (7). American Chemical Society: 2228–31. doi:10.1021/es903118j.

Zeebe, Richard E, Andy Ridgwell, and James C Zachos. 2016. "Anthropogenic Carbon Release Rate Unprecedented during the Past 66 Million Years." *Nature Geoscience* 9 (March). Nature Publishing Group: 325–29. http://dx.doi.org/10.1038/ngeo2681.

Zheng, Hua, Brian E Robinson, Yi-Cheng Liang, Stephen Polasky, Dong-Chun Ma, Feng-Chun Wang, Mary Ruckelshaus et al. 2013. "Benefits, Costs, and Livelihood Implications of a Regional Payment for Ecosystem Service Program." *Proceedings of the National Academy of Sciences* 110 (41): 16681–86. doi:10.1073/pnas.1312324110.

Zhu, Chen, and Franklin W Schwartz. 2011. "Hydrogeochemical Processes and Controls on Water Quality and Water Management." *Elements* 7 (3): 169–74. doi:10.2113/gselements.7.3.169.

Zimmer, C. 2010. "A Looming Oxygen Crisis and Its Impact on World's Oceans." *Yale Environment 360*. http://e360.yale.edu/content/feature.msp?id=2301.

Zoback, Mark D, and Steven M Gorelick. 2012. "Earthquake Triggering and Large-Scale Geologic Storage of Carbon Dioxide." *Proceedings of the National Academy of Sciences* 109 (26): 10164–68. doi:10.1073/pnas.1202473109.

Index

A

B

C

D

E

F

G

H

I

J

K

L

M

N

S

T

U

Y

Z